Topology for Analysis

Topology for Analysis

Albert Wilansky

Lehigh University

Ginn A Xerox Company

The College Division

Waltham, Massachusetts · Toronto · London

CONSULTING EDITOR

George Springer, Indiana University

Library of Congress Catalog Card Number: 72–84650
Printed in the United States of America.

To my mother, *Esther (Sidel) Wilansky,*
although a widow with six children,
she saw to the completion of my education.

Preface

This book is intended to serve the needs both of the beginning student and of the mature mathematician. Also it is intended as a reference and handbook. As an elementary text it begins with first principles and develops without haste all that part of topology which may be described as generalized analysis; for the mathematician, the subject is carried further in examples and problems. The handbook function of this text is carried out in an experimental concept: the Tables of Theorems and Counterexamples in the Appendix. For a discussion of the use of these tables, the reader is referred to the introduction immediately preceding the tables.

In reading or lecturing from the book, it should be noted that all examples and problems marked with a star ★ are essential for later development. Unmarked examples and problems are optional, and those marked ▲ are harder and more special. Each problem section is broken into three parts —the second and third parts are numbered from 100 and 200, respectively. The "100" problems are special and somewhat challenging; the "200" problems may be extremely difficult.

Both nets and filters are explained and used throughout the text. All the standard counterexamples of the subject are presented; an almost successful effort was made to use the Stone–Cech compactification as a "universal" counterexample—thus the use of ordinal spaces is reduced to a minimum, almost zero.

NOTE. I have followed the custom, now quite respectable, of not using more separation axioms than needed. Arguments developed for T_2 spaces are adapted in several ways. One is by the device of using retractions extensively (see, for example, Sec. 6.7, Problem 102). Another consists of using closed graph instead of continuous functions (see, for example, Sec. 7.1, Problem 109).† Metrization theorems, usually proved by embedding in products, are replaced by semimetrization theorems, proved by means of weak topologies. Occasionally the hypothesis "compact" must be replaced by "locally compact," an unrelated condition in non-Hausdorff spaces (see, for example, Lemma 13.1.1 and Theorem 13.1.1). This procedure is set forth in detail in my article "Life without T_2," *Amer. Math. Monthly*, *77* (1970), pp. 157–161.

AXIOMATICS. One should beware of the disease called "Axiomatics," which consists of wasting time wondering whether a, b, and c imply d, where a, b, c, and d are properties selected at random. We can do no better than quote Edwin Hewitt's remark made at the 1955 Madison Summer Institute on Set Theoretic Topology and printed on p. 17 of its Proceedings: "I am happy to say that it (the disease of axiomatic topology) has been almost totally cured. Right now I don't care a bit whether every beta capsule of type delta is also a T-spot of the second kind." However, every young mathematician must have an injection of the live virus; and every mathematician should be aware of a few basic implications such as "regular, Lindelöf implies normal," and "regular, second countable implies semimetrizable"; results of this basic nature are included in the text. For those who absolutely *must* know whether every locally compact, completely regular, separable space is σ-compact, a search in the Tables will reveal a counterexample.

† For further evidence of the value of using closed graph functions see F. J. Murray, "Quasi-complements and closed projections in reflexive Banach spaces," *Trans. Amer. Math. Soc. 58* (1945), p. 77.

Acknowledgments

It has always been my custom to write to everybody about everything. The kindness of the mathematical community in replying to my letters has been so great that I could not even begin to list the names of those people who made significant contributions to this book. In addition to this correspondence, I have incorporated hundreds of results from the mathematical literature. In some cases the authors are cited at appropriate places in the text. The greatest source of help and encouragement has been the Lehigh University Mathematics Department. My familiar notes on our bulletin board with the large letters HELP on top were invariably answered by the elegant examples of Jerry Rayna, or the erudite references and proofs of Bill Ruckle, to mention only two of my very helpful colleagues. Murray Kirch used my typescript in a course at SUNY Buffalo and made helpful criticisms. John W. Taylor made an enormous contribution! Unfortunately some of his remarks came too late to be used.

The typing was done by Helen Farrell, Rosemarie Ehser, and Judy Arroyo; I thank these three ladies for their excellent work and their devotion to a tedious task.

The administration of Lehigh University was at all times sympathetic and helpful. My deepest thanks go to all.

In addition I want to express my great appreciation to the staff of the college division of Ginn and Company and most especially to Mrs. Claire Felts whose contributions to this book occur in every line.

A.W.

Contents

Topology for Analysis

Introduction

1.1 Explanatory Notes

There are certain standard and all-pervasive notations and terminologies used by mathematicians. In addition, we use a few special notations with less currency.

The following notations will be used:

$\mathbf{R}$	real numbers
$\mathbf{Q}$	rational numbers
$\mathbf{J}$	irrational numbers
ω	positive integers
$\mathbf{Z}$	integers
$\varnothing$	empty set
$\tilde{A}$	the complement of A
$\{x: \ldots\}$	the set of all x such that
$x \in A$	x is a member of A
$x \notin A$	x is not a member of A
$A \cup B$	$\{x: x \in A \text{ or } x \in B\}$
union of A and B	$A \cup B$
$A \cap B$	$\{x: x \in A \text{ and } x \in B\}$
intersection of A and B	$A \cap B$
$\bigcup \{S: S \in \Sigma\}$	$\{x: x \in S \text{ for some } S \in \Sigma\}$ (Σ is a collection of sets)
$\bigcup \{S_\alpha: \alpha \in A\}$	$\{x: x \in S_\alpha \text{ for some } \alpha \in A\}$ (A is some indexing set)

$\bigcap \{S: S \in \Sigma\}$	$\{x: x \in S \text{ for all } S \in \Sigma\}$
$\bigcap \{S_\alpha: \alpha \in A\}$	$\{x: x \in S_\alpha \text{ for all } \alpha \in A\}$
$A \subset B$, $B \supset A$ A is included in B, B includes A, A is a subset of B, A is a set in B, B is a superset of A	every member of A is a member of B
A is a proper subset of B	$A \subset B$, $B \supset A$ $A \subset B$, $\varnothing \neq A \neq B$
$A \setminus B$	$A \cap \tilde{B}$
$A \pitchfork B$	$A \cap B = \varnothing$
A does not meet B	$A \pitchfork B$
A meets B	$A \cap B \neq \varnothing$
A meets B in x	$x \in A \cap B$
singleton	set with one member
$\{x\}$	set whose only member is x
disjoint family	a family of sets, each pair of which has empty intersection
$[a, b]$	$\{x: a \leq x \leq b\}$
(a, b)	$\{x: a < x < b\}$
$[a, b)$	$[a, b] \setminus \{b\}$
$(a, b]$	$(a, b) \cup \{b\}$
$(-\infty, a)$	$\{x: x < a\}$
$[a, \infty)$	$\{x: x \geq a\}$
characteristic function of S	f, where $f(x) = 1$ if $x \in S$, $f(x) = 0$ if $x \notin S$
$f[S]$	$\{f(x): x \in S\}$
$f^{-1}[S]$	$\{x: f(x) \in S\}$
$(f < a)$	$\{x: f(x) < a\}$
$(f = a)$	$\{x: f(x) = a\}$
$f^{\perp}$	$(f = 0)$
f is *one-to-one*	$x_1 \neq x_2$ implies $f(x_1) \neq f(x_2)$
$f: X \to Y$ is *onto*	$f[X] = Y$
one-to-one correspondence	function which is one-to-one and onto

We draw the reader's attention to the following:

$A \pitchfork B$, read "A does not meet B" meaning $A \cap B = \varnothing$.

When the notation $\tilde{A}$ is used, it is assumed that a set X has been designated and $\tilde{A} = X \setminus A$. When the presence of X is not clear from the context, the notation $X \setminus A$ will be used.

The words "space," "set," "family," and "collection" are synonymous.

When a space X has been designated the members of X will be called *points*.

The words "map," "mapping," and "function" are synonymous. If $f: X \to Y$ we call X the *domain*, and Y the *range* of f; "f" will sometimes be written as "$x \to f(x)$."

ITALICS. A word in italics is being used for the first time and is defined by the sentence in which it appears.

PROOF BRACKETS. Part of a discussion enclosed in square brackets means the statement immediately preceding the brackets is being proved. As an example, suppose the text reads, "Since x is not zero ⟦if $x = 0$, it follows that $\cos x = 1$ contradicting the hypothesis⟧, we may cancel x from both sides." The reader should first absorb "Since x is not zero, we may cancel x from both sides." He may then proceed with the text, or, if desired, return to the proof in brackets.

STARRED PROBLEMS AND EXAMPLES. Problems marked ★ must be done as they form part of the development of the text and, in extreme cases, are used later without citation. These are simple or are supplied with hints. Examples marked ★ are part of the development; those unmarked may be omitted, and those marked ▲ are special and of limited interest.

100 PROBLEMS. In each section Problems 101, 102, . . . are devoted to extending the text, and exposing interesting results beyond the scope of the book.

200 PROBLEMS. Problems 201, 202, . . . may be extremely challenging.

END OF PROOF. The end of a proof is indicated by ▮.

BIBLIOGRAPHY. References to the bibliography at the back of the book are indicated by [ ].

THE EMPTY SET. We state our conventions concerning the empty set $\varnothing$. These cannot be proved since we do not set up our set theory formally. There is only one empty set $\varnothing$; $\varnothing \subset A$ for all sets A. (This makes the statement "$A \subset B$ implies $\tilde{B} \subset \tilde{A}$, for subsets of a space X," true if $B = X$ since $\tilde{X} = \varnothing$.) Wherever some property of sets is tested for a set A by examining an arbitrary point of A, then this property is true of the empty set. For example, "all positive integers a, b, c satisfying $a^n + b^n = c^n$ for some integer $n > 2$ are larger than 3" is a true statement (and very easily proved!) even though there may be no such integers. (Here $A = \bigcup_{n=3}^{\infty} \{(a, b, c): a^n + b^n = c^n, a, b, c \in \omega\}$.) The empty set is finite.

FOUNDATIONS. We do not state our set theoretical foundations. These are based on a naive belief that we know what sets are like. In any particular argument the context will always indicate some fixed space of which all sets mentioned are subsets or members. The only explicit statements are those on the empty set, just given, and on maximal chains in Section 7.3. Our use of cardinal and ordinal numbers is restricted to examples marked ▲.

Problems

★1. If $f: X \to Y$, $g: Y \to Z$, define $g \circ f: X \to Z$ by $(g \circ f)(x) = g[f(x)]$. If f, g are one-to-one and onto prove that $(g \circ f)^{-1} = f^{-1} \circ g^{-1}$. ($f^{-1}$ means the map from Y to X satisfying $f^{-1} \circ f = i_X$, $f \circ f^{-1} = i_Y$ where $i_X(x) = x$ for all $x \in X$.)

★2. Let $\{S_\alpha: \alpha \in A\}$ be a family of subsets of a set X, and let $f: X \to Y$. Prove that

(a) $f[\bigcup \{S_\alpha: \alpha \in A\}] = \bigcup \{f[S_\alpha]: \alpha \in A\}$;
(b) $f[\bigcap \{S_\alpha: \alpha \in A\}] \subset \bigcap \{f[S_\alpha]: \alpha \in A\}$;
(c) if f is one-to-one, inclusion may be replaced by equality in (b), but not in general, even if A has only two members;
(d) if f is one-to-one, $f[\tilde{S}] \subset \{f[S]\}^\sim$;
(e) if f is onto, $\{f[S]\}^\sim \subset f[\tilde{S}]$;
(f) "one-to-one" cannot be omitted in (d), and "onto" cannot be omitted in (e).

★3. Let $\{S_\alpha: \alpha \in A\}$ be a family of subsets of Y, and let $f: X \to Y$. Then (in contrast with Problem 2), $f^{-1}[\bigcup S_\alpha] = \bigcup f^{-1}[S_\alpha]$, $f^{-1}[\bigcap S_\alpha] = \bigcap f^{-1}[S_\alpha]$, $f^{-1}[\tilde{S}] = \{f^{-1}[S]\}^\sim$.

★4. Let A, B be two collections of subsets of a set X with $A \subset B$. Prove that $\bigcap \{S: S \in A\} \supset \bigcap \{S: S \in B\}$. What is the corresponding result for union?

★5. Let $f: X \to Y$ and let S be a subset of X or Y. Prove that $f[f^{-1}[S]] \subset S$, $f^{-1}[f[S]] \supset S$, $f[\{f^{-1}[S]\}^\sim] \not\subset S$.

6. What assumption about f would produce equality in the first two parts of Problem 5?

★7. Let X, Y be sets. Let $X \times Y$ be the set of all ordered pairs (x, y) with $x \in X$, $y \in Y$. Show that $R^2 = R \times R$.

★8. Show that $\tilde{A} \subset \tilde{B}$ if and only if $A \supset B$.

★9. Prove the following formulas (given by the 19th-century mathematician, A. de Morgan).

$$[\bigcup \{S_\alpha: \alpha \in A\}]^\sim = \bigcap \{\tilde{S}_\alpha: \alpha \in A\},$$
$$[\bigcap \{S_\alpha: \alpha \in A\}]^\sim = \bigcup \{\tilde{S}_\alpha: \alpha \in A\}.$$

★10. Let $f: X \to X$. Let A, B be disjoint subsets of X, and set $G = A \cap f^{-1}[B]$. Show that $f[G] \not\subset G$.

★11. Let $f: X \to Y$, $g: Y \to X$. We say that g is a *left inverse* of f, and f is a *right inverse* of g if $g \circ f$ is the identity on X. Show that f has a left inverse if and only if f is one-to-one, and a right inverse if and only if f is onto.

1.2 n-Space

The space $\mathbf{R}$ of real numbers will not be defined in this book. It will be assumed to allow the usual operations of arithmetic, to have its usual ordering (in short, it is a totally ordered field), and to have the property that every bounded set S has a least upper bound, written sup S. Such facts as $x^2 \geq 0$ for all $x \in \mathbf{R}$ will be used without scruple.

A set S is called *countably infinite* if it can be put in one-to-one correspondence with ω; that is, there exists $f: S \to \omega$ which is one-to-one and onto. A set is called *countable* if it is finite or countably infinite. A set which is not countable is called *uncountable*. We shall assume the existence of an uncountable set. (Some are shown in Problems 201, 202, etc., and $\mathbf{R}$ is also proved uncountable in Sec. 9.3, Problem 114.)

We denote by $\mathbf{R}^n$ the set of all ordered n-tuples of real numbers, where n is a positive integer; $\mathbf{R}^1$ is the same as $\mathbf{R}$. For $x, y \in \mathbf{R}^n$, say, $x = (x_1, x_2, \ldots, x_n)$, $y = (y_1, y_2, \ldots, y_n)$, we define $\|x\|$, pronounced norm x, by the formula

$$\|x\| = \left(\sum_{i=1}^{n} |x_i|^2\right)^{1/2},$$

and $x \cdot y$, pronounced x dot y, by the formula

$$x \cdot y = \sum_{i=1}^{n} x_i y_i,$$

so that, in particular, $x \cdot x = \|x\|^2$.

Note that $\|x + y\|^2 = (x + y)\cdot(x + y) = x \cdot x + 2x \cdot y + y \cdot y$. Thus, for all x, y,

$$2x \cdot y = \|x + y\|^2 - \|x\|^2 - \|y\|^2,$$

and, similarly

$$-2x \cdot y = \|x - y\|^2 - \|x\|^2 - \|y\|^2.$$

Since $\|x \pm y\|^2 \geq 0$ we get $\pm 2x \cdot y \leq \|x\|^2 + \|y\|^2$, hence,

$$|x \cdot y| \leq \frac{\|x\|^2 + \|y\|^2}{2}. \tag{1.2.1}$$

Now let x, y be different from 0. (By 0 is meant the n-tuple $(0, 0, \ldots, 0)$.) Let $x' = x/\|x\|$, $y' = y/\|y\|$. Then $\|x'\| = \|y'\| = 1$ and so, by (1.2.1), we have $|x' \cdot y'| \leq 1$, and so

$$|x \cdot y| \leq \|x\| \cdot \|y\|. \tag{1.2.2}$$

Formula (1.2.2), called Cauchy's inequality (named for the famous 19th-

for abs. value replace with negative

century mathematician, A. Cauchy), was proved for x, y different from 0, but obviously holds in this case also.

We now have

$$\begin{aligned}\|x + y\|^2 &= \|x\|^2 + \|y\|^2 + 2x\cdot y \le \|x\|^2 + \|y\|^2 + 2\|x\|\cdot\|y\| \\ &= (\|x\| + \|y\|)^2.\end{aligned}$$

Taking positive square roots we obtain

$$\|x + y\| \le \|x\| + \|y\|. \tag{1.2.3}$$

If now we define $d(x, y) = \|x - y\|$, the familiar distance (familiar at least for $n = 1, 2, 3$), we have, for any x, y, z,

$$\begin{aligned}d(x, y) &= \|x - y\| = \|x - z + z - y\| \\ &\le \|x - z\| + \|z - y\| = d(x, z) + d(z, y).\end{aligned}$$

This function d is called the *Euclidean distance for* $\mathbf{R}^n$, also the *Euclidean metric for* $\mathbf{R}^n$.

Problems on n-Space

★1. Show that $\|x - y\| \ge |\, \|x\| - \|y\| \,|$.

2. Fix m, n with $m > n$ and define $f\colon \mathbf{R}^n \to \mathbf{R}^m$ by $(x_1, x_2, \ldots, x_n) \to (x_1, x_2, \ldots, x_n, 0, 0, \ldots, 0)$. Is f one-to-one? Is it onto? Show that $\|f(x)\| = \|x\|$ for all $x \in \mathbf{R}^n$.

3. Let $a \vee b$ (read: a sup b) and $a \wedge b$ (read: a inf b) be the larger and smaller, respectively of $a, b \in \mathbf{R}$. Prove that

$$a \vee b = \tfrac{1}{2}(a + b + |a - b|), \qquad a \wedge b = \tfrac{1}{2}(a + b - |a - b|).$$

---

101. Show that $\mathbf{R}$ and $\mathbf{R}^2$ can be put in one-to-one correspondence. ⟦$(x, y) \to z$ where $z = .x_1\, y_1\, x_2\, y_2\, x_3\, y_3 \cdots$ in decimal notation. Make $x_i, y_i = 0$ if possible. This map is one-to-one but not onto.⟧

102. If there exists a one-to-one $f\colon S \to \omega$; then S must be countable.

103. If there exists an onto $f\colon \omega \to S$; then S must be countable.

104. Show that $\omega = \bigcup_{n=1}^{\infty} S_n$ where each S_n is an infinite set and $\{S_n\}$ is a disjoint family.

---

201. Prove that the set S of all sequences of 0's and 1's is not countable. (A sequence of 0's and 1's is a sequence $\{x_n\}$ with $x_n = 0$ or 1 for all n.) ⟦Let $f\colon \omega \to S$. Then the characteristic function of $\{n \in \omega\colon y_n = 0$ if $y = f(n)\}$ does not belong to $f[\omega]$.⟧ Deduce that $\mathbf{R}$ is uncountable.

202. The set of all subsets of a set S is denoted by 2^S. Prove that 2^ω is uncountable. ⟦Apply Problem 201 to characteristic functions of subsets.⟧

203. If there is a one-to-one map from A to B and no one-to-one map from B to A, we write $|B| > |A|$ (say: B has a *larger cardinality* than A). Show that $|2^S| > |S|$ for every set S.
204. Show that $S \times S$ can be put into one-to-one correspondence with S if and only if S is an infinite set or is empty.
205. Show that the set of all permutations of ω is uncountable. (Indeed there are exactly $2^{|\omega|}$ of them.)
206. A closed interval in $\mathbf{R}$ cannot be the union of a disjoint countable family, with more than one member, of closed intervals.
207. Show that there is room in $\mathbf{R}^2$ for uncountably many disjoint L's but only countably many disjoint T's. ⟦MR *21*(1960) #852.⟧
208. Given a countable set $S \subset \mathbf{R}$, show that there exists $t \in \mathbf{R}$ such that $x + t$ is irrational for every $x \in S$. ⟦$\mathbf{Q} - S$ is countable, hence there exists $t \in \mathbf{R} \setminus (\mathbf{Q} - S)$.⟧
209. Let s be the set of all sequences of real numbers, and J^ω the set of all sequences of irrational numbers. Let A be a countable subset of s. Show that there exists $t \in s$ such that $x + t \in J^\omega$ for all $x \in A$. ⟦For each n, let S_n be the set of all nth terms of members of A. Choose t_n as in Problem 208, and let $t = \{t_n\}$.⟧

1.3 Abstraction

In the course of his mathematical education, the student encounters various structures. By a structure we mean a set together with some rules concerning its members and subsets. For example, the set of real numbers together with the operation of addition, the set of positive real numbers with multiplication, and the set of integers with addition, are three structures. In each of these three examples, there is a set and one operation. When we say that each of them is a group we are performing an *abstraction*; forgetting the differences between these examples, we consider only their similarities. If a remark is made to the effect that a group must have a certain property, this means that every group, in particular the three just mentioned, must have this property. The three examples, and any others, are called *realizations*, or *examples*, of the abstraction: *group*. Even more briefly, they are called groups.

The study of the real numbers has led to many abstractions; for example, the concept of ring is suggested by the two operations $+$ and $\times$, while the concepts of ordered system and lattice are suggested by the fact that real numbers may be compared in size. In every case, the abstraction has turned out to have realizations, other than the real number system, the study of which has been fruitful. For example, there exist finite groups.

The abstraction *topological space* is suggested by those aspects of the real number system which are studied in "advanced calculus" courses, namely those which enter into discussions of such topics as continuity and convergence. When "topological space" is defined (Sec. 2.1, Definition 2), it will be pointed out that the set of real numbers is a topological space (Sec. 2.1, Example 7).

Topological Space

2.1 Topological Space

After reading Definition 1, below, the student may not recognize that, as promised in Section 1.3, we are speaking of things involved in discussions of continuity and convergence. The only remedy for such doubts is the actual pursuit, carried on in the text, of such topics in an arbitrary topological space, that is, using only ideas introduced in Definition 1.

DEFINITION 1. *Let X be a set. Let T be a collection of subsets of X satisfying*

(i) $\varnothing \in T$;
(ii) $X \in T$;
(iii) *If* $G_1 \in T$ *and* $G_2 \in T$, *then* $G_1 \cap G_2 \in T$; (finite intersection)
(iv) *If* $\Sigma \subset T, \Sigma \neq \varnothing$, *then* $\bigcup \{G: G \in \Sigma\} \in T$. (arbitrary union)

Then T is called a topology for X.

Thus a topology for X is a certain collection of subsets of X. By definition, the empty set and X itself are included. Condition (iii) says that the intersection of any two sets in T must be in T, hence the intersection of any finite collection of sets in T must be in T. Condition (iv) says that if any collection, no matter how numerous, of members of T be given, then the union of this collection is also a member of T. Briefly, a topology for X is a collection of subsets of X containing $\varnothing$, X, and closed under finite intersections and arbitrary unions.

To *topologize* a set X means to specify a topology for X.

★EXAMPLE 1. *The indiscrete topology.* Let X be a set. Let $T = \{\varnothing, X\}$. Then T is a topology for X. ⟦Conditions (i) and (ii) of Definition 1 clearly hold. Conditions (iii) and (iv) are also clear since $\varnothing \cap X = \varnothing$ and $\varnothing \cup X = X$.⟧ This particular topology is called the *indiscrete* topology. Any set may be given the indiscrete topology. For example, let $X = \{1, 2, 3, 4\}$. Then $T = \{\varnothing, \{1, 2, 3, 4\}\}$ is the indiscrete topology for X.

★EXAMPLE 2. *The discrete topology.* Let X be a set. Let T be the collection of all subsets of X. Then T is a topology for X, called the *discrete* topology.

★EXAMPLE 3. *The cofinite topology.* Let X be a set. Let

$$T = \{G \subset X: X \setminus G \text{ is a finite set}\} \cup \{\varnothing\}.$$

Thus T consists, except for $\varnothing$, of complements of finite sets. To check Condition (iii) of Definition 1, let $G_1, G_2 \in T$. If either G_1 or G_2 is empty, $G_1 \cap G_2 = \varnothing \in T$; otherwise, $X \setminus (G_1 \cap G_2) = (X \setminus G_1) \cup (X \setminus G_2)$ is finite, being the union of two finite sets. To check Condition (iv), let $\Sigma \subset T$. If $\Sigma = \{\varnothing\}$, there is nothing to prove; otherwise, let $\varnothing \neq G \in \Sigma$. Then $T \setminus \bigcup \{S: S \in \Sigma\} \subset T \setminus G$, hence is finite and so $\bigcup \{S: S \in \Sigma\} \in T$.

It should be noted that if X is a finite set, the cofinite topology is identical with the discrete topology.

EXAMPLE 4. Let $X = \omega$ (the positive integers) and let T be the cofinite topology for X. For $n = 1, 2, 3, \ldots$, let $G_n = \{1, n, n+1, n+2, \ldots\}$. Then each $G_n \in T$ but $\bigcap \{G_n: n = 1, 2, \ldots\} = \{1\} \notin T$. This illustrates the "finite intersection" part of the definition of topology.

EXAMPLE 5. Let X be an infinite set and T the collection of all finite subsets of X, together with X itself. Then T is not a topology since a union of finite sets need not be finite.

DEFINITION 2. *A topological space is a set together with a topology for the set. The members of the topology are called open sets.*

Thus we refer to a topological space (X, T), in which X is a set, and T is a topology for it. For $S \subset X$, the sentences "S is open" and "$S \in T$" are synonymous. When T is understood from the context, the space will be denoted by X, rather than (X, T). Our aim is to develop a usage which will not refer to T. Thus we shall speak of a topological space X together with its open subsets.

★EXAMPLE 6. If X is given the indiscrete topology, it has only two open subsets, $\varnothing$ and X. (But see Example 8.) With the discrete topology every subset of X is open. With the cofinite topology, a nonempty subset is open if and only if its complement is finite.

★EXAMPLE 7. *The Euclidean topology.* Let $X = \mathbf{R}$ (the real numbers) and let the empty set be called open; also any set G is called open if for every $x \in G$, there exists $\varepsilon > 0$ such that $(x - \varepsilon, x + \varepsilon) \subset G$. This defines a topology for X (namely, of course, the collection of open sets). ⟦To check Condition (iii) of Definition 1, let G_1, G_2 be open sets. If $G_1 \cap G_2 = \varnothing$ there is nothing to prove. If $G_1 \cap G_2 \neq \varnothing$, let $x \in G_1 \cap G_2$. There exist $\varepsilon_1 > 0, \varepsilon_2 > 0$ such that $(x - \varepsilon_i, x + \varepsilon_i) \subset G_i, i = 1, 2$. Let $\varepsilon = \min(\varepsilon_1, \varepsilon_2)$. Then $\varepsilon > 0$ and $(x - \varepsilon, x + \varepsilon) \subset G_1 \cap G_2$. Thus $G_1 \cap G_2$ is open. To check Condition (iv), let Σ be a collection of open sets, and we may assume that its union is nonempty. Let $x \in \bigcup \{G: G \in \Sigma\}$. Then $x \in S$ for some $S \in \Sigma$. There exists $\varepsilon > 0$ such that $(x - \varepsilon, x + \varepsilon) \subset S$. Then $(x - \varepsilon, x + \varepsilon) \subset \bigcup \{G: G \in \Sigma\}$.⟧

EXAMPLE 8. In Definition 1, it is possible that X is the empty set. Then the only topology for X is $\{\varnothing\}$. Needless to say, in any discussion we are always thinking of a nonempty set, and thus, on occasion, will make a statement which is not strictly true. For example, the indiscrete topology for $\varnothing$ has only one member, not two as stated in Example 6. It seems hardly worth while to be sufficiently careful at all times to cover this case, but the student is warned that such an occasion may arise.

If T, T' are topologies for a set X, it may happen that $T \supset T'$. In this case we shall say that T is *larger than* T', (including the possibility that $T = T'$), and T' is *smaller than* T. Very commonly, "*stronger*" and "*finer*" are used instead of "larger"; "*weaker*" and "*coarser*" instead of "smaller." Of course if T is both larger and smaller than T', then they are equal.

A set F in a topological space is called closed if its complement is open. (Use of the letters F—French *fermé*—and G—German *Gebiet*—for closed and open sets, respectively, has become traditional.)

★EXAMPLE 9. Let ω have the cofinite topology. Then each finite set is closed, indeed a proper subset is closed if and only if it is finite. The set of even integers is neither closed nor open; ω is both open and closed. (Two morals: A set can be both open and closed. Just because a set is not open, there is no reason to think that it is closed.)

Problems

★1. $[0, 1)$ is not open in $\mathbf{R}$, with the Euclidean topology. Neither is it closed.

★2. Let $S \subset X$. Let $T = \{\varnothing, S, X\}$. Show that T is a topology.

3. Find all sets for which the discrete and indiscrete topologies are the same.

4. Let $S_1 \subset X$, $S_2 \subset X$. Let $T = \{\varnothing, S_1, S_2, X\}$. Is T a topology?

5. Let
$$T = \{\varnothing\} \cup \{\mathbf{R}\} \cup \{(a, \infty): a \in \mathbf{R}\},$$
$$T' = \{\varnothing\} \cup \{\mathbf{R}\} \cup \{[a, \infty): a \in \mathbf{R}\}.$$
Show that T is a topology for $\mathbf{R}$, but T' is not.

★6. Let X be a topological space. Show that $\varnothing$, X, every finite union and every intersection of closed sets, are all closed.

---

101. Let (X, T) be a topological space and let T' be the collection of closed sets. Is T' a topology?
102. If, in Problem 101, X is finite, show that T' is a topology.
103. In Example 3 replace "finite" by "countable." Show that a topology is defined. It is called the *cocountable* topology.
104. The cocountable topology on a countable space is discrete.

---

201. Let $k(n)$ be the number of different topologies which can be placed on a set with n members. Show that $k(0) = k(1) = 1, k(2) = 4, k(3) = 29$. (*Note*: $k(4) = 355, k(5) = 6942, k(6) = 209527, k(7) = 9535241$. See [Evans, Harary, and Lynn].)
202. Let $d(n)$ be the number of topologies T which can be placed on a set with n members, satisfying $T = T'$ (Problem 102). Show that $d(0) = d(1), d(2) = 2, d(3) = 5$.

2.2 Semimetric and Metric Space

DEFINITION 1. *Semimetric and metric. Let X be a set and suppose given a real-valued function d of two variables, each selected from X. Thus d is a function from $X \times X$ to the real numbers. Any such function satisfying the following conditions is called a semimetric.*

(i) $d(x, y) = d(y, x) \geq 0$,

(ii) $d(x, x) = 0$,

(iii) $d(x, z) \leq d(x, y) + d(y, z)$,

for all $x, y, z \in X$. Condition (iii) *is called the triangle inequality. A metric is a semimetric, d, which satisfies the condition $d(x, y) > 0$ if $x \neq y$.*

DEFINITION 2. *A semimetric (or metric) space is a set together with a semimetric (or metric) for the set.*

Thus the pair (X, d), in which X is a set and d a semimetric, is a semimetric space. We shall usually denote it by X, when d is understood from the context.

★EXAMPLE 1. *The discrete metric.* Let X be a set. For $x, y \in X$, define $d(x, y) = 1$ if $x \neq y$, 0 if $x = y$. It is easily verified that this is a metric, for example, the triangle inequality may be checked for x, y, z by noting that it is trivial if $x \neq y$ and $y \neq z$, if $x = y$, or if $y = z$ ⟦Problem 1⟧.

★EXAMPLE 2. *The indiscrete semimetric.* Let X be a set. For $x, y \in X$, define $d(x, y) = 0$ ⟦Problem 1⟧.

★EXAMPLE 3. The Euclidean distance for $\mathbf{R}^n$ was proved, in Section 1.2, to obey the triangle inequality. Since it obviously has the other metric properties, it is a metric, called the *Euclidean metric*, and so $\mathbf{R}^n$ is, with this definition, a metric space.

EXAMPLE 4. For $w = (x, y, z) \in \mathbf{R}^3$ define $p(w) = |x| + |y|$. Then $d(w, w') = p(w - w')$ defines a semimetric for $\mathbf{R}^3$ which is not a metric. ⟦See Problem 4.⟧ For example, with $x = (1, 2, 3)$, $y = (1, 2, 4)$, $d(x, y) = 0$.

EXAMPLE 5. Let C be the set of continuous real functions on the closed interval $[0, 1]$. For $f \in C$, $g \in C$, define $\|f\| = \max\{|f(x)|: 0 \leq x \leq 1\}$ and $d(f, g) = \|f - g\|$. As in Section 1.2, the triangle inequality for d follows from Formula (1.2.3). ⟦To establish (1.2.3), we have, for any $x \in [0, 1]$, $|f(x) + g(x)| \leq |f(x)| + |g(x)| \leq \|f\| + \|g\|$. Choosing x so as to maximize $|f(x) + g(x)|$ gives the result.⟧

The following definitions apply to any semimetric space X. For $a \in X$ and $r \in \mathbf{R}$, the *cell of radius r and center a*, written $N(a, r)$, is $\{x: d(x, a) < r\}$; the *disc of radius r and center a*, written $D(a, r)$ is $\{x: d(x, a) \leq r\}$; and the *circumference of radius r and center a*, written $C(a, r)$, is $\{x: d(x, a) = r\}$. The word *sphere* is often used in place of circumference, and when $X = \mathbf{R}^{n+1}$ with the Euclidean metric, $C(0, 1)$ is called the *n-sphere*, written S_n. Thus S_1, the 1-sphere is the unit circumference in $\mathbf{R}^2$, $\{(x, y): x^2 + y^2 = 1\}$.

It is possible for a cell, disc, or circumference to be empty. For example, $N(a, 0)$ is certainly empty, while if X is a discrete space (that is, has the discrete metric), $C(a, \frac{1}{2})$ is empty for any a. Note also that a disc may have two different radii; for example, in a discrete space $D(a, \frac{1}{2}) = D(a, \frac{1}{3}) = \{a\}$. For $A \subset X$, $B \subset X$, $x \in X$, we define $d(x, A)$, *the distance from x to A*, to be

$$\inf\{d(x, a): a \in A\},$$

and $d(A, B)$, *the distance from A to B* to be

$$\inf\{d(a, b): a \in A, b \in B\}.$$

(See Problems 7, 8, and 9). The *diameter* of a set A is defined to be $\sup\{d(x, y): x \in A, y \in A\}$ (perhaps infinite). A set with finite diameter is said to be *metrically bounded.*

In the next result it is assumed that A is not empty.

LEMMA 2.2.1. $|d(x, A) - d(y, A)| \leq d(x, y)$.

We may assume that $d(x, A) \geq d(y, A)$ since $d(x, y) = d(y, x)$. Let $\varepsilon > 0$ be given and choose $a \in A$ with $d(y, a) < d(y, A) + \varepsilon$. Then

$$\begin{aligned} |d(x, A) - d(y, A)| &= d(x, A) - d(y, A) \\ &\leq d(x, a) - d(y, A) \\ &< d(x, a) - d(y, a) + \varepsilon \\ &\leq d(x, y) + \varepsilon \end{aligned}$$

⟦Problem 2⟧. ▮

Let X, Y be semimetric spaces and $f: X \to Y$. We call f an *isometry* if it is one-to-one and satisfies $d[f(x), f(x')] = d(x, x')$ for all $x, x' \in X$. (The assumption "one-to-one" is redundant if X is a metric space ⟦Problem 10⟧.) It is called an *isometry of X onto Y* if it is onto; if a particular isometry is not known to be onto, we shall sometimes write *isometry* (*into*), for emphasis. If there is an isometry of X onto Y we say that X, Y are *isometric*.

★EXAMPLE 6. Define $f: \mathbf{R} \to \mathbf{R}^2$ by $f(x) = (x, 0)$. Then f is an isometry of $\mathbf{R}$ into $\mathbf{R}^2$. ($\mathbf{R}$ and $\mathbf{R}^2$ are assumed to have the Euclidean metric, e.g., Example 3.)

★EXAMPLE 7. For $\omega = (x, y, z) \in \mathbf{R}^3$, define $p(\omega) = |z|$, $d(\omega, \omega') = p(\omega - \omega')$. Just as in Example 4, d is a semimetric for $\mathbf{R}^3$. Let $\mathbf{R}$ have the Euclidean metric. Define $f: \mathbf{R}^3 \to \mathbf{R}$ by $f(x, y, z) = z$. Then f is *distance preserving*, that is $d[f(\omega), f(\omega')] = d(\omega, \omega')$, but it is not an isometry since it is not one-to-one.

Problems

In this list, (X, d) is a semimetric space.

★1. Verify Examples 1 and 2.

J★2. Prove that $|d(x, a) - d(y, a)| \leq d(x, y)$.

★3. Prove that $|d(x, a) - d(y, b)| \leq d(x, y) + d(a, b)$.

4. Prove the triangle inequality for d in Example 4. ⟦Start with Formula (1.2.3) with p instead of $\|\cdot\|$.⟧
5. Let X be a discrete metric space. Show that $N(a, \frac{1}{2}) = \{a\}$ for all $a \in X$. What is $D(a, 1)$? $C(a, 1)$?
6. Prove that $d(x, S) = d(\{x\}, S)$.
7. Prove that $d(A, B) = \inf\{d(a, B): a \in A\}$.
8. Prove that $d(A, B) = d(B, A)$.
9. Prove that a set is metrically bounded if and only if it is included in some cell. (The center of the cell may be assigned arbitrarily.)

★10. Prove that a distance-preserving map from a metric space to a semimetric space is an isometry (into).

✓101. Let X be a set and $u: X \times X \to \mathbf{R}$ satisfy $u(x, y) = 0$ if and only if $x = y$; and $u(x, y) \leq u(x, z) + u(y, z)$. Show that u is a metric.

102. Give an example of a metric space which contains a nonempty disc whose diameter is less than its radius.

103. Give an example of a metric space in which every sphere has two centers. ⟦Easy examples are certain subspaces of $\mathbf{R}^2$, $\mathbf{R}^3$.⟧

104. $[0, \infty)$ and $\{-1\} \cup [0, \infty)$ are each isometric into the other, but the spaces are not isometric. (Thus isometry of metric spaces does not enjoy the *crisscross property*.)

2.3 Semimetric and Metric Topologies

There is a standard way of defining a topology for a semimetric space (X, d). The empty set is declared to be open; moreover, if $G \subset X$, G is called open if, for every $x \in G$, there exists $r > 0$ such that $N(x, r) \subset G$. Before proving that the collection of open sets is a topology, consider an example.

★EXAMPLE 1. Let $\mathbf{R}$ have the Euclidean metric, that is, $d(x, y) = |x - y|$. Let $S = (0, 1]$. Then S is *not* an open set since $N(1, r) \not\subset S$ for all $r > 0$. ⟦For example, $1 + (r/2) \in N(1, r) \setminus S$.⟧ However, $(0, 1)$ is open since if $0 < a < 1$, $N(a, r) \subset (0, 1)$ whenever $r = \min\{1 - a, a\}$.

We now prove that a topology has been defined. (It is called *the topology induced by the semimetric.*) To see that X is open, we merely note that for any $x \in X$, $N(x, 1) \subset X$. (Indeed, $N(x, r) \subset X$ for arbitrary r.) Next, let G_1, G_2 be open sets and let $a \in G_1 \cap G_2$. For $i = 1, 2$, choose $r_i > 0$ with $N(a, r_i) \subset G_i$. Then $N(a, r) \subset G_1 \cap G_2$ if $r = \min\{r_1, r_2\}$.

Finally let Σ be a collection of open sets and let $x \in \bigcup \{G: G \in \Sigma\}$. Then $x \in S$ for some $S \in \Sigma$, and so there exists $r > 0$ with $N(x, r) \subset S$. Then $N(x, r) \subset \bigcup \{G: G \in \Sigma\}$.

★EXAMPLE 2. The topology for $\mathbf{R}^n$ induced by the Euclidean metric (Sec. 2.2, Example 3) is called the *Euclidean topology for* $\mathbf{R}^n$ (see Problem 1).

THEOREM 2.3.1. *Cells in a semimetric space are open.*

Let $N(x, r)$ be a cell. It is empty, hence open, if $r \leq 0$, so we may assume $r > 0$. Let $y \in N(x, r)$. Then $N(y, s) \subset N(x, r)$ if $s < r - d(y, x)$. ⟦Let $a \in N(y, s)$. Then $d(a, x) \leq d(a, y) + d(y, x) \leq s + d(y, x) < r$. Thus $a \in N(x, r)$.⟧ ∎

It is natural to ask whether every topology can be obtained from a semimetric, and, if not, which ones can. (Those which can are called *semimetrizable.*) This question will be more fully treated in Chapter 10; however, it is possible to give a simple example (Example 3) of a topology which is not induced by any semimetric. The most instructive procedure is to see what properties a topology induced by a semimetric must have. The one we give here is only one of many possibilities.

THEOREM 2.3.2. *Let x, y be points in a semimetric space such that every open set containing x also contains y. Then every open set containing y also contains x.*

We first observe that $d(x, y) = 0$. ⟦For suppose on the contrary that $d(x, y) = \varepsilon > 0$. Then $N(x, \varepsilon/2)$ is an open set containing x but not y.⟧ Let G be an open set containing y. There exists $r > 0$ such that $N(y, r) \subset G$. Since $d(x, y) = 0 < r$, we have $x \in N(y, r)$, hence $x \in G$. ∎

Now it is an easy matter to construct a topology not obeying the conclusion of the preceding theorem.

EXAMPLE 3. Let X be a set, and x, y distinct points in X. Let $T = \{\varnothing, \{y\}, X\}$. This is a topology for X. ⟦Sec. 2.1, Problem 2.⟧ Every open set containing x contains y. ⟦There is only one.⟧ But $\{y\}$ contains y and not x. *Hence T is not semimetrizable* ⟦Theorem 2.3.2⟧.

REMARK ON USE OF ADJECTIVES. Adjectives are applied equally often to a topological space and to its topology. For example, "X is metrizable" can be written instead of "T (the topology of X) is metrizable." Later we shall speak of a connected space, a connected topology, a compact topology, and so on.

DEFINITION 1. *Let d, d' be semimetrics for a set X. Let T_d, $T_{d'}$ be the induced topologies. We say that d is stronger than d', and that d' is weaker than d, if $T_d \supset T_{d'}$. If $T_d = T_{d'}$ we say that d, d' are equivalent.*

Problems

1. Show that the Euclidean topology for $\mathbf{R}$ as given in Example 2 is the same as that given in Sec. 2.1, Example 7.
2. The discrete and indiscrete topologies are semimetrizable.
3. Let d, d' be semimetrics for X and suppose that there exists a number k such that $kd(x, y) \geq d'(x, y$ for all x, y. Prove that d is stronger than d'.
4. If d is a semimetric, and there exists a weaker metric, then d is a metric.
5. Problem 4 is false with "stronger" instead of "weaker."
6. Every disc is closed.

101. Give an example of a nonempty disc which is an open set.
102. Let d be a semimetric. Then $d/(1 + d)$ and $d \wedge 1$ are semimetrics equivalent to d. (By $d \wedge 1$ is meant the function whose value at (x, y) is $\min\{d(x, y), 1\}$.) ⟦For the triangle inequality use $d/(1 + d) = 1 - (1/(1 + d))$.⟧
103. A set may have different diameters with two equivalent metrics; it might be metrically bounded in one and not the other ⟦Problem 102⟧.
104. Every open set in **R**, with the Euclidean topology is the union of a disjoint family of open intervals.
105. Let $X = \{a, b, c, d, e\}$; $T = \{\varnothing, \{a\}, \{b, c\}, \{d, e\}, \{a, b, c\}, \{a, d, e\}, \{b, c, d, e\}, X\}$. Show that T is a semimetrizable topology for X.
106. Suppose that the condition $d(x, y) = d(y, x)$ is omitted from the definition of semimetric. Would Theorems 2.3.1 and 2.3.2 be true?
107. Give an example of a metric inducing the discrete topology such that there are points arbitrarily close together. ⟦Consider $\{1/n\}$ in **R**.⟧
108. Let X be a noncountable set and d a metric which induces the discrete topology. Then X has a noncountable subset on which $d \geq \varepsilon D$, where D is the discrete metric, and $\varepsilon > 0$. ⟦With $S_n = \{x : N(x, 1/n) = \{x\}\}$ for $n = 1, 2, \ldots, \bigcup S_n$ is noncountable and so some S_n is noncountable.⟧
109. With X, d, as in Problem 108, there exists $\varepsilon > 0$ such that X is not a countable union of sets of diameter $< \varepsilon$. ⟦This is surely true of the subspace of Problem 108.⟧

---

201. The converse of Problem 3 is false; that is, d might be stronger than d' and no such k exist.
202. Let X be the set of all closed convex curves in $\mathbf{R}^2$, symmetric with respect to the origin. Does $d(C_1, C_2) = \sup_{P \in C_1} \inf_{Q \in C_2} d(P, Q)$ define a metric on X?

2.4 Natural Topologies and Metrics

Certain spaces will be assigned topologies which will, throughout this book, be referred to as the *natural topologies* for these spaces. For example, *the natural topology for* **R** *is the Euclidean topology*; wherever **R** is mentioned, it will be assumed to have this topology. If for some reason we want to consider **R** with some other topology, the topology will be explicitly stated. The phrase "let G be an open subset of **R**" will mean "let $G \in T$, where T is the Euclidean topology for **R**."

The natural topology for $\mathbf{R}^n$ *is the Euclidean topology. The natural topology for* **Z** *is the discrete topology.* We hasten to add that this is also the Euclidean topology for **Z**. ⟦For $m, n \in \mathbf{Z}$ let $d(m, n) = |m - n|$. Then $N(n, \frac{1}{2}) = \{n\}$, that is, for each n, $\{n\}$ is an open set. Hence d induces the discrete topology.]] *The natural topology for* **Q**, *and* **J**, *will be that induced by the Euclidean metric.*

Similarly certain spaces will be assigned natural metrics. The natural metric for $\mathbf{R}^n$, **Z**, **Q**, **J** is the Euclidean metric. In each case, the natural metric induces the natural topology.

EXAMPLE 1. Let $X = \{1, \frac{1}{2}, \frac{1}{3}, \ldots\}$. Then X inherits from **R** the metric $d(1/m, 1/n) = |1/m - 1/n|$. This gives X the discrete topology [[$\{1/m\}$ is open for each m since it is $(1/m - \varepsilon, 1/m + \varepsilon) \cap X$ for suitably chosen $\varepsilon > 0$]]. Similarly **Z** has the discrete topology. However **Z** also has the discrete metric as its natural metric, while the metric given for X is not the discrete metric.

★EXAMPLE 2. We shall usually identify $\mathbf{R}^2$ and the complex plane, writing $|z|$ for $(x^2 + y^2)^{1/2}$ if $z = (x, y)$.

2.5 Notation and Terminology

Let x, N be, respectively, a point, and a set in a topological space. We say that N is a *neighborhood* of x if there exists an open set G with $x \in G \subset N$. We call N a *deleted* neighborhood of x if $x \notin N$, and $N \cup \{x\}$ is a neighborhood of x. (Note that a neighborhood need not be open.)

THEOREM 2.5.1. *A set is open if and only if it is a neighborhood of each of its points.*

Let G be open. Then for every $x \in G$, $x \in G \subset G$. Conversely, let G be a neighborhood of each of its points. If G is empty it is open. If G is not empty, for each $x \in G$, let G_x be open with $x \in G_x \subset G$. Then $G = \bigcup \{G_x : x \in X\}$, a union of open sets. Hence G is open. ∎

If S, N are sets and there exists an open set G with $S \subset G \subset N$, we say that N is a *neighborhood* of S. Thus N is a neighborhood of $\{x\}$ if and only if it is a neighborhood of x.

We call x an *accumulation point* of a set S if every deleted neighborhood of x meets S; in other words, if every neighborhood of x meets S in a point other than x.

THEOREM 2.5.2. *A set is closed if and only if it contains all of its accumulation points. In particular, a set with no accumulation points is closed.*

Let F be closed, and x an accumulation point. Then x must belong to F, for, if not, $\tilde{F} \setminus \{x\}$ is a deleted neighborhood of x which does not meet F, so that x is not an accumulation point. Conversely, let F contain all of its accumulation points. Let $x \in \tilde{F}$. Then x is not an accumulation point of F, hence x has a deleted neighborhood N not meeting F. Thus $\tilde{F} \supset N \cup \{x\}$, making $\tilde{F}$ a neighborhood of x. By Theorem 2.5.1, $\tilde{F}$ is open. ▮

The *interior*, S^i, of a set S, is the set of all points x such that S is a neighborhood of x. As a generalization of Theorem 2.5.1, and proved in the same way, we can see that S^i is open for all S. The statement: "*S has interior*" will mean "$S^i \neq \varnothing$."

The *closure*, $\bar{S}$, of a set S, is the intersection of all closed sets F such that $F \supset S$. (Compare Problem 2.) Thus $\bar{S} \supset S$ and $\bar{S}$ is closed ⟦Sec. 2.1, Problem 6⟧. When there is need to specify the topology T we shall write $\bar{S}$ as $\text{cl}_T\, S$. The most important fact about closure is Problem 3 which we shall use immediately.

★EXAMPLE 1. *Let S be a set in a semimetric space X and $x \in X$. Then $x \in \bar{S}$ if and only if $d(x, S) = 0$.* If $x \in \bar{S}$, let $\varepsilon > 0$. Then $N(x, \varepsilon)$ meets S so that $d(x, S) < \varepsilon$. Since ε is arbitrary, $d(x, S) = 0$. Conversely, suppose that $d(x, S) = 0$ and let N be a neighborhood of x. There exists $\varepsilon > 0$ such that $N \supset N(x, \varepsilon)$ and there exists $s \in S$ with $d(x, s) < \varepsilon$. Thus $s \in N$, that is N meets S. Hence $x \in \bar{S}$. ▮

It is often convenient to compare topologies by means of closures in the following way.

LEMMA 2.5.1. *Let T, T' be topologies for a set X. Then $T \supset T'$ if and only if for every $S \subset X$, it is true that $\text{cl}_T\, S \subset \text{cl}_{T'}\, S$.*

Let $T \supset T'$. Then $\text{cl}_{T'}\, S = \bigcap \{F: F \supset S, F \text{ is } T' \text{ closed}\} \supset \bigcap \{F: F \supset S, F \text{ is } T \text{ closed}\} = \text{cl}_T\, S$. ⟦The inclusion holds since every T' closed set is T closed, thus the second intersection is taken over a larger class.⟧ Conversely, suppose that $\text{cl}_T\, S \subset \text{cl}_{T'}\, S$ for all S. Let $G \in T'$. Then $\tilde{G}$ is T' closed, hence $\text{cl}_T\, \tilde{G} \subset \text{cl}_{T'}\, \tilde{G} = \tilde{G}$ so that $\tilde{G}$ is T closed ⟦Problem 4⟧. Hence $G \in T$. ▮

THEOREM 2.5.3. *Let T, T' be topologies for a set X. Then $T = T'$ if and only if $\text{cl}_T\, S = \text{cl}_{T'}\, S$ for all $S \subset X$.*

This is immediate from Lemma 2.5.1. ▮

A set S in a topological space X is called *dense* if $\bar{S} = X$. For example, **Q** and **J** are dense in **R**, while **Z** is not. A discrete space has no dense proper subsets, while every nonempty subset of an indiscrete space is dense.

A *cover* of a set S in a space X is a collection C of subsets of X such that $\bigcup \{A: A \in C\} \supset S$. If all members of C are open, we call it an *open cover* of S. A *subcover of C* is merely a subset of C which still *covers* S (is a cover of S).

★EXAMPLE 2. $\{(-1, x): 0 < x < 2\}$ is an open cover of $(0, 2)$, but is not a cover of $[0, 2]$. This cover of $(0, 2)$ has, as a subcover, the collection $\{(-1, x): 0 < x < 2, x \text{ rational}\}$.

Problems on Topological Space

In this list, (X, T) is a topological space, and N, S, G, A, B, D are subsets of X.

1. N is a neighborhood of S if and only if N is a neighborhood of each point of S. (Thus a set is open if and only if it is a neighborhood of itself.)

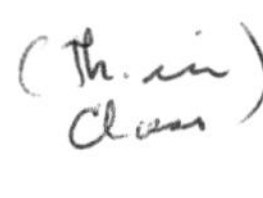

★2. S^i is the largest open set included in S. (This means that S^i is open; that $S^i \subset S$; and if G is open, and $G \subset S$, then $G \subset S^i$.) Also S^i is the union of all open sets which are included in S. Prove also that $(S^i)^i = S^i$.

★3. $x \in \bar{S}$ if and only if every neighborhood of x meets S.

★4. $\bar{S}$ is the smallest closed set which includes S (compare with Problem 2), and S is closed if and only if $\bar{S} \subset S$. Prove also that $\bar{\bar{S}} = \bar{S}$.

★5. If G is open, and $G \not\pitchfork S$, then $G \not\pitchfork \bar{S}$. ⟦$\tilde{G} \supset S$; apply Problem 4.⟧

6. $\overline{A \cup B} = \bar{A} \cup \bar{B}$, $\overline{A \cap B} \subset \bar{A} \cap \bar{B}$.

7. Let $\{S_\alpha : \alpha \in A\}$ be a collection of sets in X. Show that $\bigcup \{\bar{S}_\alpha : \alpha \in A\} \subset \overline{\bigcup \{S_\alpha : \alpha \in A\}}$. ⟦The right-hand side includes each S_α, and is closed; use Problem 4.⟧

8. Equality need not hold in the second part of Problem 6 and in Problem 7. (Give examples in $\mathbf{R}$.)

★9. $S^{i\sim} = S^{\sim -}$, and $S^{-\sim} = S^{\sim i}$ ⟦$S^{i\sim}$ is closed and includes $S^\sim$ since $S^i \subset S$. Thus $S^{i\sim} \supset S^{\sim -}$ by Problem 4. $S^{\sim - \sim}$ is open and is included in $S^{\sim\sim} = S$ since $S^{\sim -} \supset S^\sim$. Thus $S^{\sim - \sim} \subset S^i$ by Problem 2, and so $S^{\sim -} \supset S^{i\sim}$⟧.

10. Let T, T' be topologies for a set X. Show that $T' \supset T$ if and only if every T closed set is T' closed.

★11. A set is dense if and only if it meets every nonempty open set.

★12. If D is dense and G is open, then $\overline{G \cap D} = \bar{G}$. (Thus $\overline{G \cap D} = \bar{G} \cap \bar{D}$, compare Problem 8.)

13. Let S be a set in a semimetric space. Then S, $\bar{S}$ have the same diameter. ⟦Let $x, y \in \bar{S}$ with $d(x, y) > D(\bar{S}) - \varepsilon$; choose $a, b \in S$ with $d(a, x) < \varepsilon$, $d(b, y) < \varepsilon$. Then $D(S) \geq d(a, b) > D(\bar{S}) - 3\varepsilon$.⟧

14. S is dense if and only if $\tilde{S}$ has no interior.

15. The boundary of S, written bS, is $\bar{S} \cap \bar{\tilde{S}}$. Show that b$S = \bar{S} \setminus S^i$; b$S$ is closed; b$S = \varnothing$ if and only if S is both open and closed; G open $\Rightarrow$ bG has no interior.

16. Find the interior and all cluster points of $\{b\}$ if $X = \{a, b, c\}$ with topology $\{\varnothing, X, \{a\}, \{b\}, \{a, b\}\}$.

★17. A topological space is indiscrete if and only if every singleton is dense.

---

101. $\bar{S} = S \cup S'$, where S' is the set of accumulation points of S.

102. Give an example of a metric space in which $\overline{N(x, r)} \neq D(x, r)$ for some x and $r > 0$. Is there any necessary relationship between these sets?
103. The following are equivalent: X is the union of two closed proper subsets. X contains two disjoint open proper subsets. X has a closed proper subset with nonempty interior. X has a subset with nonempty interior which is not dense.
104. **R** has the property of Problem 103 while an infinite cofinite space does not.
105. Can Problem 12 be generalized to: "$\overline{G \cap D} = \bar{G} \cap \bar{D}$ whenever G is open?"
106. A space is called *extremally disconnected* if the closure of every open set is open. Show that exactly three of the following four spaces are extremally disconnected: any discrete space, any cofinite space, any cocountable space, **R**.
107. Consider the statement, denoted by (*): If $A \not\pitchfork \bar{B}$ then $B \not\pitchfork \bar{A}$. Show that (*) is false in **R**; that if A is finite (*) still may not be true, but is true in a semimetric space.
108. Give an example of an uncountable subset of **J** which is closed in **R**.
109. A set is called *nowhere dense* if its closure has empty interior. Show that the union of finitely many nowhere-dense sets is nowhere dense.
110. The intersection of two dense open sets is dense. ⟦Every open set must meet one of them in an open set, which then must meet the other.⟧

---

201. Call x a *condensation point* of S if every neighborhood of x meets S in an uncountable set. Call S *self-dense* if every $x \in S$ is an accumulation point of S, and *perfect* if it is closed and self-dense. Let $S \subset \mathbf{R}$. Show that the set of condensation points of S is perfect. Show that every closed set in **R** is the union of a perfect and a countable set.
202. There does not exist any topological space with the property that a set has nonempty interior if and only if it is infinite.
203. An extremally disconnected metric space is discrete.
204. Suppose that every nonempty T' open set has T interior. Does it follow that $T \supset T'$?
205. Suppose that every T-dense subset of X is T'-dense. Does it follow that $T \supset T'$?

2.6 Base and Subbase

It is convenient to have some method of defining a specific topology other than describing *all* the open sets. What we want is to be able to mention only *some* of the open sets, but enough to identify the topology uniquely. For example, if two sets are listed, there is no need to mention their union; that is automatically open. But we need some way to ensure that enough open sets are given, for example if we say "Let T be the topology for $\mathbf{R}$ in which $(0, 1)$ is open" we would certainly not have adequately described the Euclidean topology, and it is a little doubtful whether we have described anything. (A generous interpretation might be that $T = \{\varnothing, (0, 1), \mathbf{R}\}$.) However if we say "Let T be the topology for $\mathbf{R}$ in which (a, b) is open for all a, b" we have what appears to be the Euclidean topology. These remarks are clarified in the rest of this section.

DEFINITION 1. *A collection $\mathscr{B}$ of subsets of a topological space is called a local base at a point x, or simply, a base at x, if every member of $\mathscr{B}$ is a neighborhood of x, and if, for every neighborhood N of x, there exists a set $S \in \mathscr{B}$ with $S \subset N$.*

★EXAMPLE 1. Let X be a semimetric space, and fix $x \in X$. Let $\mathscr{B}_1$ be the collection of cells $N(x, 1/n)$, $n = 1, 2, \ldots$; and let $\mathscr{B}_2$ be the collection of discs $D(x, 1/n)$, $n = 1, 2, \ldots$. Then $\mathscr{B}_1$, $\mathscr{B}_2$, and $\mathscr{B}_1 \cup \mathscr{B}_2$ are all local bases at x.

DEFINITION 2. *A collection $\mathscr{B}$ of subsets of a topological space is called a base for the topology if every member of $\mathscr{B}$ is open and $\mathscr{B}$ includes a local base at each point.*

By a *basic set* will be meant a base. A *basic sequence* is a countable basic set. We shall also refer to a base for the topology as a *base for the space.*

★EXAMPLE 2. The set of all $N(x, 1/n)$, $x \in X$, $n = 1, 2, \ldots$, is a base for a semimetric space X ⟦Theorem 2.3.1⟧.

Notice that we ask a little more of the members of a base than of a local base, namely that they be open sets. Distinctions of this sort are usually for technical convenience, and other definitions could easily be set up leading to the same end result (see Problem 101). However, we must use precisely the definitions given, once we have committed ourselves to them.

THEOREM 2.6.1. *Let T, T' be topologies for a set X which have a common base $\mathscr{B}$. Then $T = T'$.*

Let $G \in T$ and $x \in G$. There exists $S \in \mathscr{B}$ with $x \in S \subset G$. Since $S \in \mathscr{B}$ it follows that $S \in T'$ and so G is a T' neighborhood of x. Since x is arbitrary, $G \in T'$ ⟦Theorem 2.5.1⟧. Thus $T \subset T'$ and, by symmetry $T' \subset T$. ▌

THEOREM 2.6.2. *Let X be a set and $\mathscr{B}$ a collection of subsets of X with the following properties:* (i) $\bigcup \{A : A \in \mathscr{B}\} = X$, (ii) *for each $U \in \mathscr{B}$, $V \in \mathscr{B}$, and each $x \in U \cap V$, $\mathscr{B}$ contains a member W with $x \in W \subset U \cap V$. Then there is a unique topology for X which has $\mathscr{B}$ as a base.*

Note that we are asserting both the existence and the uniqueness of the topology. Uniqueness follows from Theorem 2.6.1. The topology is defined explicitly: Let $\varnothing$ be open, and let a nonempty set G be called open if and only if there is a subset $\mathscr{B}_1$ of $\mathscr{B}$ with $G = \bigcup \{A\colon A \in \mathscr{B}_1\}$. Then $\varnothing$ and X are open ⟦for X take $\mathscr{B}_1 = \mathscr{B}$⟧. The union of open sets is open. ⟦Let C be a collection of open sets. For each $G \in C$, let $G = \bigcup \{A\colon A \in \mathscr{B}_G\}$. Then $\bigcup C = \bigcup \{A\colon A \in \bigcup \{\mathscr{B}_G\colon G \in C\}\}$.⟧ Finally, let G, H be open sets and $x \in G \cap H$. There exists $U \in \mathscr{B}$ with $x \in U \subset G$. ⟦Since $x \in G$, x must be in one of the sets whose union G is, G being open.⟧ Also there exists $V \in \mathscr{B}$ with $x \in V \subset H$, and so, by hypothesis, there exists $W \in \mathscr{B}$ with $x \in W \subset U \cap V$. Denoting this W by W_x, we have $W_x \in \mathscr{B}$ and $x \in W_x \subset G \cap H$. Clearly $G \cap H \supset \bigcup \{W_x\colon x \in G \cap H\}$, and

$$G \cap H = \bigcup \{\{x\}\colon x \in G \cap H\} \subset \bigcup \{W_x\colon x \in G \cap H\}.$$

Hence $G \cap H$ is equal to this union and is thus open. These remarks show that a topology has been defined. To see that $\mathscr{B}$ is a base, let N be a neighborhood of some x. Then N includes an open neighborhood of x, which is, by definition, a union of certain members of $\mathscr{B}$. Selecting one of these, V, which contains x we have $x \in V \subset N$. ▮

DEFINITION 3. *A collection $\mathscr{B}$ of subsets of a topological space is called a subbase for the topology if all its members are open and the collection of all finite intersections of members of $\mathscr{B}$ is a base.*

Here is the easiest way to recognize a subbase. The proof is trivial.

THEOREM 2.6.3. *A collection $\mathscr{B}$ of subsets of a topological space is a subbase for the topology if and only if its members are open, and for each x and each neighborhood N of x there exists a finite collection $\{A_1, A_2, \ldots, A_n\}$ of members of $\mathscr{B}$ with $x \in \bigcap A_i \subset N$.*

The analogue of Theorem 2.6.2 omits Condition (ii).

THEOREM 2.6.4. *Let X be a set and $\mathscr{B}$ a collection of subsets of X whose union is X. Then there is a unique topology for X which has $\mathscr{B}$ as subbase.*

The collection of all finite intersections of members of $\mathscr{B}$ satisfies the two conditions of Theorem 2.6.2 ⟦take $W = U \cap V$⟧, hence is a base for a unique topology. ▮

The collection $\mathscr{B}$ in the preceding theorem is said to *generate* the topology. (A slight stretching of these ideas is often used; namely, if $\mathscr{B}$ is an arbitrary collection of subsets of X, then $\mathscr{B} \cup \{X\}$ satisfies the condition of Theorem 2.6.4. The resulting topology is also said to be *generated by* $\mathscr{B}$. Points outside $\bigcup \{S\colon S \in \mathscr{B}\}$ would have only one neighborhood, X.)

EXAMPLE 3. *The* RHO *topology*. We shall describe a topology for **R** called the *right half-open interval topology*, or RHO topology, for short. This is the topology generated by the set of all intervals of the form $[a, b)$. By Theorem 2.6.2, this collection of intervals is a base for the topology. Each interval $[a, b)$ is open ⟦by definition of base⟧ and closed. ⟦Let $x < a$, then $[x, (x + a)/2)$ is a neighborhood of x not meeting $[a, b)$; if $x \geq b$, then $[x, x + 1)$ is a neighborhood of x not meeting $[a, b)$.⟧ Thus the RHO topology is zero-dimensional. (A *zero-dimensional space* is one which has a base of open and closed sets.) The RHO topology is strictly larger than the Euclidean topology for **R**. ⟦Let G be a Euclidean open set and $x \in G$. There exist a, b with $a < x < b$, $(a, b) \subset G$. Then $[x, b) \subset G$ so that x is RHO-interior to G. Thus G is RHO-open. RHO is strictly larger because $[0, 1)$ is not Euclidean open.⟧

Problems on Topological Space

1. Every base has the properties given in Theorem 2.6.2.
2. Let $\mathscr{B}$ be a collection of open subsets of X. Show that $\mathscr{B}$ is a base for the topology if and only if every open set is a union of members of $\mathscr{B}$.
3. The collection of open neighborhoods of a point is a local base at that point.
4. What topology is generated by $\{X\}$?
5. What topology is generated by $\{\{x\}: x \in X\}$?
6. Let $\mathscr{B}_1$, $\mathscr{B}_2$ be bases for topologies T_1, T_2 on a given set. Show that $T_1 \subset T_2$ if and only if for every point x and $V \in \mathscr{B}_1$ with $x \in V$, there exists $W \in \mathscr{B}_2$ with $x \in W \subset V$.

★7. If a set meets every member of a base, it is dense. It is not sufficient that it meet every member of a subbase. ⟦The line $(y = x)$ meets every horizontal and every vertical strip in $\mathbf{R}^2$.⟧

---

101. Show that a topology may be defined by specifying neighborhoods in the following way. A set X is given. For each $x \in X$, certain sets are specified and called vicinities of x. It is assumed that x belongs to each vicinity of x, that X is a vicinity of every $x \in X$, that the intersection of two vicinities of x is a vicinity of x, that any set which includes a vicinity of x is itself a vicinity of x, and finally that each vicinity of x includes a vicinity of x which is a vicinity of all its points. Show that there is a unique topology for X with the property that a set is a neighborhood of x if and only if it is a vicinity of x. ⟦Call G open if it is a vicinity of each of its points.⟧
102. The collection of all closed intervals $[a, b]$ is not a base for the Euclidean topology of **R**. However, it satisfies Theorem 2.6.2. For what topology is it a base?

103. What topology for $\mathbf{R}^2$ is generated by the set of all horizontal and vertical lines.
104. Show that the discrete and indiscrete topologies are zero-dimensional.
105. $\mathbf{Q}$ is zero-dimensional.
106. The cofinite topology for a set is zero-dimensional if and only if the set is finite.
107. Show that one of the following two sets is RHO-closed and one is not: $(1, \frac{1}{2}, \frac{1}{3}, \ldots), (-1, -\frac{1}{2}, -\frac{1}{3}, \ldots)$.
108. Find the RHO closure and interior of each of $(0, 1)$, $(0, 1]$.
109. A collection $\mathscr{B}$ of nonempty open sets is called a *pseudobase* if every nonempty open set includes a member of $\mathscr{B}$. Find a pseudobase for $\mathbf{R}$ which is not a base.
110. Show by Theorem 2.6.2 that there is a topology T for $\mathbf{Z}$ which has as a base the collection of all arithmetic progressions $\{a + kd: k = 0, \pm 1, \pm 2, \ldots\}$. Show that T is zero-dimensional. ⟦Each arithmetic progression is open and closed.⟧
111. For prime p, let A_p be the arithmetic progression $(\ldots, -2p, -p, 0, p, 2p, \ldots)$. Show that $\bigcup \{A_p : p \text{ prime}, p \geq 2\} = \mathbf{Z} \setminus \{-1, 1\}$. Thus this union is not closed in the topology T of Problem 110, and so cannot be a finite union. Deduce that $\mathbf{Z}$ contains infinitely many primes.
112. In Theorem 2.6.2 the assumption $x \in W$ cannot be omitted. ⟦Let $X = \{a, b, c, d\}$, $U = \{a, b, c\}$, $V = \{b, c, d\}$, $W = \{b\}$, $B = \{U, V, W\}$. If B were a base for a topology $U \cap V$ would be open and yet not a neighborhood of c.⟧

---

201. Show that $\{(x, y): x \in \mathbf{J}, y \in \mathbf{J}\}$ is a zero-dimensional space (with the Euclidean metric of $\mathbf{R}^2$); and the same is true of $\{(x, y): x \in \mathbf{R}, y \in \mathbf{R},$ either x or $y \in \mathbf{Q}\}$.
202. $\mathbf{R}$ is not zero-dimensional.
203. Find the RHO closure of $\mathbf{Q}$.
204. Is the union of a chain of topologies (on a fixed space) a topology?

Convergence

3

3.1 Sequences

Suppose that y is a sequence in a topological space X. This means that y is a function from ω (the positive integers) to X. For $n \in \omega$, we usually write y_n instead of $y(n)$; y is sometimes written as $\{y_n\}$, sometimes as y_n, where there is no danger of confusion. We say that y *converges to* x, in symbols, $y_n \to x$, if, for every neighborhood N of x, $y_n \in N$ *eventually*, that is, there exists $n_0 \in \mathbf{R}$ such that $n \geq n_0$ implies $y_n \in N$. Any point x such that $y_n \to x$ is called a *limit* of $\{y_n\}$. A sequence in $\mathbf{R}$ which converges to 0 is called a *null sequence*.

★EXAMPLE 1. If y is a sequence in $\mathbf{R}$, $y_n \to x$ if and only if for every $\varepsilon > 0$, there exists $n_0 \in \mathbf{R}$ such that $n > n_0$ implies $|y_n - x| < \varepsilon$. If y is a sequence in a semimetric space (X, d), $y_n \to x$ if and only if $d(y_n, x) \to 0$. Note that $\{d(y_n, x)\}$ is a real sequence.

EXAMPLE 2. Define a semimetric d for $\mathbf{R}^2$ by $d[(x, y), (u, v)] = |x - u|$. Let $z_n = (1/n, 0)$. Then $z = \{z_n\}$ is a sequence in $\mathbf{R}^2$ and it is clear that $z_n \to (0, 0)$. But it is equally clear that $z_n \to (0, 1)$. Thus a convergent sequence may converge to several points.

When a sequence y converges to a unique point x, we also write $\lim y = x$ or $\lim y_n = x$ or $\lim_n y_n = x$.

We have now seen that in a topological space, it is possible to discuss convergence of sequences. We now investigate to what extent knowledge of which sequences are convergent specifies the topology. In other words, we ask under what circumstances two different topologies on a set may, or may not, have exactly the same convergent sequences and corresponding limits. We begin with an example which shows that this may happen. Let us call a topology T *sequentially stronger* than a topology T' if $x_n \to x$ in T implies $x_n \to x$ in T'; T, T' are called *sequentially equivalent* if each is sequentially stronger than the other.

EXAMPLE 3. Let X be any uncountable set. Let T be the cocountable topology. We shall show that a sequence y is convergent if and only if it is eventually constant, that is $y_n \to x$ if and only if $y_n = x$ eventually. ⟦Half of this is easy (Problem 1). Now suppose that y is a sequence in X, $x \in X$, and it is false that $y_n = x$ eventually. Let $F = \{y_n : y_n \neq x\}$. Then F is countable (perhaps finite), hence $\tilde{F}$ is a neighborhood of x. But it is false that $y_n \in \tilde{F}$ eventually, hence y_n does not converge to x.⟧ Now T is not the discrete topology ⟦singletons are not open since the complement of a singleton is uncountable⟧, but, as we have just seen, T and the discrete topology are sequentially equivalent ⟦Problem 2⟧.

Another topology which is sequentially equivalent to the discrete topology is shown in Theorem 14.1.6. Further, using language to be explained later, this topology, in contrast with the cocountable topology, is compact, T_4.

DEFINITION 1. *A topological space is called first countable at x if there is a countable local base at x, and first countable if it is first countable at each of its points.*

★EXAMPLE 4. A finite space is first countable.

★EXAMPLE 5. Every semimetric space is first countable since, for each x, $\{N(x, 1/n)\}$ is a countable local base at x.

THEOREM 3.1.1. *Let X be a topological space, $x \in X$, $S \subset X$, and suppose that X is first countable at x. Then $x \in \bar{S}$ if and only if there is a sequence in S which converges to x.*

Suppose first that such a sequence exists. Every neighborhood of x contains a point of this sequence, hence a point of S. Conversely, let $x \in \bar{S}$. Let $\{V_n\}$ be a shrinking ⟦Problems 5 and 7⟧ basic sequence at x. Each V_n meets S so we may choose $s_n \in V_n \cap S$. Then $s_n \to x$ ⟦Problem 5⟧. ∎

THEOREM 3.1.2. *Let T and T' be topologies for a set X with T first countable. Then $T \supset T'$ if and only if T is sequentially stronger than T'.*

Suppose first that $T \supset T'$ and $x_n \to x$ in T. Let V be a T' neighborhood of x. Then V is a T neighborhood of x and so $x_n \in V$ eventually. Conversely, let T be sequentially stronger than T'. It follows from Theorem 3.1.1 and Problem 3, that $\text{cl}_T\, S \subset \text{cl}_{T'}\, S$ for all $S \subset X$. The result follows from Lemma 2.5.1. ▮

THEOREM 3.1.3. *Let T, T' be first countable topologies for a set X. Then $T = T'$ if and only if T, T' are sequentially equivalent.*

This follows immediately from Theorem 3.1.2. ▮

★EXAMPLE 6. Let d, d' be semimetrics. Then d is stronger than d' if and only if $x_n \to x$ in d implies $x_n \to x$ in d' ⟦Example 5, Theorem 3.1.2, and Sec. 2.3, Definition 1⟧. A sufficient (but not necessary ⟦Problem 12⟧) condition that d be stronger than d' is $d(x, y) \geq d'(x, y)$ for all x, y ⟦$d'(x_n, x) \leq d(x_n, x) \to 0$⟧.

★EXAMPLE 7. Let d be a semimetric and let $d' = d/(1 + d)$. Then d' is a semimetric ⟦for example,

$$d'(x, z) = \frac{d(x, z)}{1 + d(x, z)} = 1 - \frac{1}{1 + d(x, z)} \leq 1 - \frac{1}{1 + d(x, y) + d(y, z)}$$

$$= \frac{d(x, y)}{1 + d(x, y) + d(y, z)} + \frac{d(y, z)}{1 + d(x, y) + d(y, z)}$$

$$\leq d'(x, y) + d'(y, z).⟧$$

Also, d' is equivalent to d. ⟦Since $d \geq d'$, half is trivial by Example 6. Conversely, if $d'(x_n, x) \to 0$ we have $d(x_n, x) = d'(x_n, x)/(1 - d'(x_n, x)) \to 0$.⟧

Problems on Topological Space

★1. If $y_n = x$ eventually, then $y_n \to x$.

★2. $y_n \to x$ in the discrete topology if and only if $y_n = x$ eventually. ⟦$\{x\}$ is a neighborhood of x.⟧

★3. Half of Theorem 3.1.1 holds without assuming that X is first countable. ⟦The part proved first.⟧

★4. Theorem 3.1.1 fails if X is not assumed first countable. ⟦Let X be an uncountable space with the cocountable topology. Let $S \subset X$ with $\tilde{S}$ countable, and $x \notin S$. Then $x \in \bar{S}$ but no sequence in S converges to x by Example 3.⟧

★5. Suppose that $\{V_n\}$ is a countable *shrinking* (that is, $V_{n+1} \subset V_n$ for all n) local base at x, and that, for each n, $y_n \in V_n$. Prove that $y_n \to x$. ⟦Every neighborhood of x includes some V_n, hence all V_k for $k > n$.⟧

6. Show that "shrinking" cannot be omitted in Problem 5. (Give an example in which $X = \mathbf{R}$.)

★7. Let X be first countable at x. Show that X has a countable shrinking local base at x. ⟦Consider $\{\bigcap_{i=1}^{n} V_i\}$.⟧

8. If $y_n \to x$, every subsequence of y also converges to x.

9. Half of Theorem 3.1.2 holds without assuming that X is first countable. ⟦The part proved first.⟧

10. Theorem 3.1.2 fails if T is not assumed first countable. ⟦T = co-countable, T' = discrete. Here T' is first countable.⟧

11. Suppose given two sequentially equivalent topologies. If one of them is first countable, it must be larger.

12. For $x, y \in \mathbf{R}$ let $d'(x, y) = |x - y|/(1 + |x - y|)$. Show that d' is equivalent to the Euclidean metric d ⟦Example 7⟧, and that there exists no $k > 0$ such that $d(x, y) \le kd'(x, y)$ for all x, y.

---

101. Let d be a semimetric and let $d' = d \wedge 1$. (This means $d'(x, y) = \min\{1, d(x, y)\}$). Show that d' is a semimetric and is equivalent to d.

102. Let X be an infinite cofinite space, and let y be a one-to-one sequence in X. (This means that $y_i \ne y_j$ if $i \ne j$.) Show that $y_n \to x$ for all $x \in X$.

103. Does the result of Problem 102 hold if, instead of one-to-one, it is assumed that y has infinite range?

104. Let T, T' be topologies for a set X with T' first countable. Suppose that every sequence which T' converges to a point has a subsequence which T converges to that point. Show that $T' \supset T$. ⟦Let $x \in \mathrm{cl}_{T'} S$. Apply Theorem 3.1.1 and Problem 3 to show that $x \in \mathrm{cl}_T S$; then apply Lemma 2.5.1.⟧

105. One cannot define a topology, first countable or not, by specifying that certain sequences shall be convergent, even if the class of convergent sequences is very well behaved (see Problem 106). Illustrate this by showing that $\mathbf{R}$ has no topology T such that $x_n \xrightarrow{T} x$ if and only if $|x_n - x| < 1/n$ eventually. ⟦Apply Problem 104 with T' = Euclidean topology, T = supposed topology. Theorem 3.1.2, with Problem 9, is contradicted.⟧

106. Which of the four properties (a), (b), (c), (d) of Problem 206 does the convergence of Problem 105 have?

107. Give an example of a first countable space which is not semimetrizable.

108. Let $X = \{0, 1, 2, 3, \ldots\}$. Let every set which does not contain 0 be open and let a set G containing 0 be open if and only if G has the property $u(n)/n \to 1$ as $n \to \infty$, where $u(n)$ is the number of members of G in the interval $[1, n]$. (Sometimes phrased, G has *density* 1.) Check that this is a topology, and that 0 is the only point which is an accumulation point of its complement.

109. A space is called *closure-sequential* if whenever $x \in \bar{S}$ there is a sequence of points in S converging to x. (Theorem 3.1.1 shows that every first countable space is closure-sequential.) Show that X is closure-sequential if and only if every sequential neighborhood of any point is a neighborhood of that point. (N is a *sequential neighborhood* of x if whenever $x_n \to x$, $x_n \in N$ eventually.)
110. A set is called *sequentially closed* if it contains all its sequential limit points, and *sequentially open* if it is a sequential neighborhood (Problem 109) of all its points. A space is called *sequential* if every sequentially closed set is closed. Show that a space is sequential if and only if every sequentially open set is open.

---

201. In the space X of Problem 108, 0 is not a *sequential accumulation point.* (This means that no sequence of positive integers converges to 0.) Deduce that X is not first countable.
202. Describe the RHO convergent sequences.
203. A closure-sequential space (Problem 109) need not be first countable. ⟦$\omega \cup \{0\}$ with ω having the discrete topology and a set G containing 0 open if and only if there exists an increasing sequence $\{m_k\}$ of positive integers with $G \supset \{2^k(2n - 1): n > m_k\}$. An easier example is given in Sec. 8.1, Problem 131.⟧
204. A closure-sequential space is sequential but not conversely. ⟦Add to the Euclidean topology for **R** all sets $\{0\} \cup N$ where N is a Euclidean neighborhood of the sequence $\{1/n\}$.⟧
205. Generalize Problem 108 by replacing density by a countably additive measure.
206. Suppose that on a set X certain sequences are said to converge to certain points, written $x_n \xrightarrow{c} x$. Suppose that (a) if $x_n = x$ for all n, then $x_n \xrightarrow{c} x$; (b) if $x_n \xrightarrow{c} x$, and $\{y_n\}$ is a subsequence of $\{x_n\}$, then $y_n \xrightarrow{c} x$; (c) if $\{x^k\}$ is a sequence of sequences, $x^k = \{x_n^k\}$ for $k = 1, 2, \ldots, x_n^k \xrightarrow{c} a_k$ as $n \to \infty$ for each k, and $a_k \xrightarrow{c} a$, then there exist increasing sequences $\{k(n)\}$, $\{m(n)\}$ of positive integers such that $x_{m(n)}^{k(n)} \xrightarrow{c} a$; (d) if $x_n \not\xrightarrow{c} x$, $\{x_n\}$ has a subsequence, no subsequence of which c converges to x. Let S be called closed if $x_n \in S$, $x_n \xrightarrow{c} x$ implies $x \in S$. Show that this yields a topology such that $x_n \to x$ if and only if $x_n \xrightarrow{c} x$.
207. For measurable real functions on $[0, 1]$, pointwise convergence almost everywhere implies convergence in measure, and if $f_n \to f$ in measure, $\{f_n\}$ has a subsequence $\{g_n\}$ with $g_n \to f$ pointwise almost everywhere. Compare Problem 104. Show that the procedure of Problem 206 applied to either of these two convergences yields the same topology T, but that $f_n \to f$ in T if and only if $f_n \to f$ in measure.

208. Construct a countable but not first countable space as follows: $X = (0, 1, 2, \ldots)$, ω is a discrete subspace of X, and neighborhoods of 0 are all sets of the form $S \cup \{0\}$ such that $S_n \setminus S$ is finite for every n, where S_n is given in Sec. 1.2, Problem 104.

3.2 Filters

If one deals only with first countable spaces it is possible to describe all topological occurrences in terms of sequences, instead of in terms of open sets or neighborhoods. For example, $x \in \bar{S}$ if and only if every neighborhood of x meets S, and (in a first countable space) if and only if S contains a sequence converging to x. Other examples are given in Theorems 3.1.2, 3.1.3, and Sec. 4.2, Problem 10. In general, however, such descriptions are impossible; we saw in Sec. 3.1, Problem 4, a case in which $x \in \bar{S}$ but this fact could not be ascertained by examination of sequences. We concede then that sequences are inadequate to describe topologies in general. Because of the very natural form taken by sequential arguments in classical analysis, mathematicians were unwilling to replace them by the more cumbersome direct arguments with neighborhoods, and two convergence theories were devised during the period from 1915 to 1940; these were nets [E. H. Moore, H. L. Smith, J. L. Kelley, and others] and filters [H. Cartan, and others.].

There is a sense in which these theories are equivalent. Each is adequate for topology in the same sense that sequences are not, as just pointed out. Nets resemble sequences strongly, and are handier to use in discussions of continuity of functions, and algebraic operations; while filters are preferable in dealing with compactness and completeness. One must know both theories, and it is pointless to become emotionally attached to one or the other. We shall expose both theories, and make use of whichever is more natural in any given situation. It must be emphasized that both nets and filters could be dispensed with in topological arguments; see, for example, Theorem 5.4.5, and the alternate proof in Sec. 7.1, Problem 111; but their use is fully justified by the resulting brevity, fluency, and elegance.

DEFINITION 1. *A collection $\mathscr{F}$ of subsets of a set X is called a filter in X if*
(i) $X \in \mathscr{F}$;
(ii) $\varnothing \notin \mathscr{F}$;
(iii) $A \in \mathscr{F}$, $B \in \mathscr{F}$ *imply* $A \cap B \in \mathscr{F}$;
(iv) $A \in \mathscr{F}$, $B \supset A$ *imply* $B \in \mathscr{F}$.

Condition (i) ensures that $\mathscr{F}$ is nonempty. Some simple consequences of the definition are given in Problems 1 to 4.

★EXAMPLE 1. Fix a point x in a set X. Let D_x be the collection of all subsets of X which contain x. Then D_x is a filter. It is called the *discrete filter at x*, for reasons which will become clear in Example 2. Let $I_x = \{X\}$. This is called the *indiscrete filter at x*; it happens to be independent of x.

★EXAMPLE 2. Let X be a topological space and $x \in X$. Let N_x be the collection of all neighborhoods of x. If X has the discrete topology, $N_x = D_x$ (Example 1). If X has the indiscrete topology, $N_x = I_x$. In any case N_x is a filter, called the *neighborhood filter at x*.

Now let $\mathscr{F}$ be a filter in a topological space X. We say that $\mathscr{F}$ *converges* to x, in symbols, $\mathscr{F} \to x$, if $\mathscr{F} \supset N_x$, the neighborhood filter at x. This definition, perhaps, does not seem as reasonable as the definition of convergence of a sequence. We shall see, however, in Example 4, and in Theorem 4.2.2, that it has the right properties. If $\mathscr{F}$ converges to a unique point x, we also write $\lim \mathscr{F} = x$. Any point x such that $\mathscr{F} \to x$ is called a *limit* of $\mathscr{F}$. We shall often use the obvious fact that *if $\mathscr{F} \to x$ then any larger filter than $\mathscr{F}$ (that is, any filter which includes $\mathscr{F}$), also converges to x.*

★EXAMPLE 3. Let X be a topological space and $x \in X$. Then $N_x \to x$. ⟦This is trivial.⟧

EXAMPLE 4. Let $\{x_n\}$ be a sequence in X. Let $\mathscr{F} = \{A \subset X : x_n \in A$ eventually$\}$. Then $\mathscr{F}$ is a filter ⟦Problem 1⟧ and $x_n \to x$ if and only if $\mathscr{F} \to x$. ⟦If $x_n \to x$ and N is a neighborhood of x, $x_n \in N$ eventually; hence $N \in \mathscr{F}$. This shows that $N_x \subset \mathscr{F}$ and so $\mathscr{F} \to x$. Conversely, if $\mathscr{F} \to x$, let N be a neighborhood of x. Now $N \in \mathscr{F}$ by hypothesis, thus $x_n \in N$ eventually; hence $x_n \to x$.⟧

To specify a filter, it is not necessary to name all of its sets. Whenever it is known that a given set S belongs to a filter $\mathscr{F}$, then it is known that any superset of S belongs to $\mathscr{F}$. We now define the concept *filterbase* and show how a filterbase leads to a filter.

DEFINITION 2. *A collection $\mathscr{B}$ of subsets of a set X is called a filterbase in X if*

(i) $\mathscr{B} \neq \varnothing$;
(ii) $\varnothing \notin \mathscr{B}$;
(iii) *$A \in \mathscr{B}$, $B \in \mathscr{B}$ imply that $\mathscr{B}$ contains a set C with $C \subset A \cap B$.*

Thus a filter is a filterbase. A significant example of a filterbase is the set of discs in the plane containing the origin in their interiors. A filterbase $\mathscr{B}$ *generates* a filter $\mathscr{F}$ in the obvious way, $\mathscr{F} = \{A : A \supset B$ for some $B \in \mathscr{B}\}$.

We say of a filterbase $\mathscr{B}$, that $\mathscr{B} \to x$ if the filter generated by $\mathscr{B}$ converges to x. Thus $\mathscr{B} \to x$ if and only if, for each neighborhood N of x, $\mathscr{B}$ has a member A with $A \subset N$.

★**EXAMPLE 5.** Let S be a proper subset of a set X. Let $\mathscr{F}$ be a filter on S. Then $\mathscr{F}$ is not a filter on X $[\![X \notin \mathscr{F}]\!]$, but $\mathscr{F}$ is a filterbase on X. Indeed if $\mathscr{B}$ is a filterbase on S, then $\mathscr{B}$ is also a filterbase on X. This property, and that given in problem 5, make filterbases easier to use than filters.

EXAMPLE 6. Let X be a topological space, and $x \in X$. Then $D_x \to x$ (Example 1). $[\![D_x \supset N_x$, obviously.$]\!]$ An interesting filterbase which generates D_x is $\{\{x\}\}$; this is a filterbase containing just one set, and that set is a singleton.

We can now obtain an improved form of Theorem 3.1.1. (Recall Sec. 3.1, Problem 4, which shows that this result is false if filterbase is replaced by sequence.)

THEOREM 3.2.1. *Let X be a topological space, $x \in X$, $S \subset X$. Then $x \in \bar{S}$ if and only if there is a filterbase in S which converges to x.*

Suppose first that such a filterbase $\mathscr{B}$ exists, and let N be a neighborhood of x. There exists $A \in \mathscr{B}$ with $A \subset N$. Since $A \subset S$, this means that N must meet S. Conversely, let $x \in \bar{S}$ and let $\mathscr{B} = \{N \cap S: N \in N_x\}$. We check the definition of filterbase. $\mathscr{B} \neq \varnothing$ since $S \in F$ $[\![S = X \cap S]\!]$; $\varnothing \notin \mathscr{B}$ since, for every $N \in N_x$, $N \cap S \neq \varnothing$ $[\![x \in \bar{S}]\!]$; and, if $A = N_1 \cap S$, $B = N_2 \cap S$, then $A \cap B = N_1 \cap N_2 \cap S \in \mathscr{B}$. Finally $\mathscr{B} \to x$. $[\![$Let $N \in N_x$. Then $N \cap S \in \mathscr{B}$ and $N \cap S \subset N$.$]\!]$ ∎

The succeeding results of Section 3.1 follow in an entirely similar way.

THEOREM 3.2.2. *Let T, T' be topologies for a set X. Then $T \supset T'$ if and only if $\mathscr{F} \to x$ in T implies $\mathscr{F} \to x$ in T' for all filters $\mathscr{F}$ in X. Hence $T = T'$ if and only if T, T' have the same convergent filters and corresponding limits.*

Suppose first that $T \supset T'$ and $\mathscr{F} \to x$ in T. Then $N'_x \subset N_x \subset \mathscr{F}$, where N'_x, N_x are the sets of all T', T, neighborhoods of x, respectively. Hence $\mathscr{F} \to x$ in T'. Conversely, suppose that $\mathscr{F} \to x$ in T implies $\mathscr{F} \to x$ in T' and let $S \subset X$. If $x \in \text{cl}_T S$ there is a filterbase $\mathscr{B}$ in S with $\mathscr{B} \to x$ in T. Then $\mathscr{B} \to x$ in T' $[\![$in each case, $\mathscr{B} \to x$ if and only if the filter generated by $\mathscr{B} \to x]\!]$ and so $x \in \text{cl}_{T'} S$. We have proved that $\text{cl}_T S \subset \text{cl}_{T'} S$ for all S and so $T \supset T'$ $[\![$Lemma 2.5.1$]\!]$. ∎

There is a natural way to topologize a subset X of a topological space (Y, T); namely let $T_X = \{G \cap X; G \in T\}$. This is clearly a topology for X $[\![$Problem 7$]\!]$ and is called the *relative topology of Y for X*; when X has this topology it is called a *topological subspace of Y*. If S is an open subset of X in the relative topology of Y we say that S is open in X. (It may or may not be open in Y.)

THEOREM 3.2.3. *Let X be a topological subspace of Y, let $x \in X$, and let $\mathscr{B}$ be a filterbase on X. Then $\mathscr{B} \to x$ in X if and only if $\mathscr{B} \to x$ in Y.*

Suppose first that $\mathscr{B} \to x$ in Y. Every neighborhood of x in X is of the form $N \cap X$, where N is a neighborhood of x in Y. Now $N \supset A$ for some $A \in \mathscr{B}$; hence $N \cap X \supset A$. Thus $\mathscr{B} \to x$ in X. Conversely, let $\mathscr{B} \to x$ in X, and let N be a neighborhood of x in Y. Then $N \cap X$ is a neighborhood of x in X and so $N \cap X \supset A$ for some $A \in \mathscr{B}$. Thus $N \supset N \cap X \supset A$, and so $\mathscr{B} \to x$ in Y. ▮

COROLLARY. *Let X be a topological subspace of Y, and Y a topological subspace of Z. Then X is a topological subspace of Z.*

This is immediate from Theorems 3.2.2 and 3.2.3. ▮

Problems

In this list, $\mathscr{F}$ is a filter, and $\mathscr{B}$ is a filterbase.

★1. Check that $\mathscr{F}$ is a filter in Example 4.
★2. Check that $\{A: A \supset B \text{ for some } B \in \mathscr{B}\}$ is a filter.
★3. Any two members of $\mathscr{F}$ have nonempty intersection. The same is true of $\mathscr{B}$.
4. Any finite intersection of members of $\mathscr{F}$ belongs to $\mathscr{F}$.
★5. Let $f: X \to Y$, and let $\mathscr{B}$ be a filterbase in X. Show that $f[\mathscr{B}]$ is a filterbase in Y, and that if $\mathscr{F}$ is a filter in X, $f[\mathscr{F}]$ is a filter if and only if f is onto.
6. A filter is discrete if and only if it contains a singleton.
★7. The relative topology for X is a topology.
8. If X is an open set in Y, then a subset of X which is open in X is also open in Y. The assumption that X is open cannot be omitted.
9. For $S \subset X$, where X is a topological subspace of Y, if S is open in Y it is open in X.
★10. Let (Y, d) be a semimetric space and $X \subset Y$. Show that $d'(x, y) = d(x, y)$ for $x, y \in X$ makes (X, d') into a semimetric space with the relative topology of Y.
11. If X is a topological subspace of Y and $\mathscr{B}$ is a base for the topology of Y, then $\{A \cap X: A \in \mathscr{B}\}$ is a base for the relative topology.
12. Let X be a topological subspace of Y and $S \subset X$. Then S is closed in the relative topology if and only if there exists closed $F \subset Y$ with $S = F \cap X$. [[Take $F = \bar{S}$.]]
13. Modify the results of Problem 8 so as to refer to closed sets.
14. Let G be an open set in X and let D be a dense subset of X. Show that $G \cap D$ is dense in G.
15. In Problem 14, "open" cannot be omitted.

16. A base for a topology is usually not a filterbase. However, a local base is a filterbase.

---

101. If $\mathscr{B}$ is a filterbase, then $\{\bar{B}: B \in \mathscr{B}\}$ is also a filterbase, and generates a smaller filter.
102. The set of (circular) cells in the plane containing the origin is a filterbase but not a filter.
103. Let $f: X \to Y$ where X is a set and Y a topological space. Then $f[D_x] \to f(x)$.
104. Let $f: X \to Y$, and $x \in X$. Show that the filter generated by $f[D_x]$ is $D_{f(x)}$, and deduce the result of Problem 103.
105. Fix $x \in X$. Let $C_x = \{A \in D_x: \tilde{A}$ is finite.$\}$ Then C_x is a filter. It is called the *cofinite*, or *Fréchet filter at* x. Show that if X is finite $C_x = D_x$, while if X is a cofinite space, $C_x = N_x$.
106. Let X be an arbitrary infinite set. Let $C_\infty = \{A \subset X: \tilde{A}$ is finite$\}$. Then C_∞ is a filter, called the *cofinite*, or *Fréchet filter at* ∞.
107. Let X be a topological space, $x \in X$, and suppose that every filter on X converges to x. Show that x has only one neighborhood. ⟦Consider the indiscrete filter.⟧
108. Call a filterbase *discrete*, *indiscrete*, *cofinite*, if it generates a discrete, indiscrete, cofinite filter. Show that any singleton is a discrete filterbase, and given an example of a cofinite filterbase on **R** which is not a filter.
109. Let G be an open set in X and D a dense open set in G (with the relative topology.) Show that there exists a dense open subset V of X with $V \cap G = D$.
110. (*Duplication of a point.*) Given a topological space X, let x be a nonisolated point of X, and let $t \notin X$. Let $Y = X \cup \{t\}$. Let $T = \{G \subset Y$: either G or $(G \setminus \{t\}) \cup \{x\}$ is an open set in $X\}$. Show that T is a topology for Y and that x, t are not contained in disjoint open sets. (The end result is as if the point x has been duplicated.)
111. In Problem 110, Y is first countable at x if and only if it is first countable at t.

---

201. Any discrete subspace of **R** is countable.
202. Can the closure of a discrete subspace of **R** be uncountable?
203. Two different topologies may have the same convergent filters. (But see Theorem 3.2.2, and Sec. 4.1, Problem 121.)

3.3 Partially Ordered Sets

A sequence is a function on the positive integers. By considering functions on other sets, we obtain more general objects, called nets, which can be used to describe topological properties, as were filters, but which bear a greater resemblance to sequences. We begin by describing the sets on which the nets will be defined.

Let X be a set and assume that a relation, written $\geq$, is given such that for each pair x, y of members of X, the statement $x \geq y$ is either true or false. (Thus a *relation* is a set S of ordered pairs (x, y) with $x, y \in X$, and we write $x \geq y$ if and only if $(x, y) \in S$.) We assume that the relation is *reflexive* (that is, $x \geq x$, for all x), and *transitive* ($x \geq y \geq z$ implies $x \geq z$). We shall write $x \leq y$ to mean $y \geq x$. A set with a reflexive, transitive relation is called a *partially ordered set*, or *poset*, for short.

★EXAMPLE 1. $(\mathbf{R}, \geq)$ is a poset, where $\geq$ has its usual meaning. The same is true for $(\omega, \geq)$.

★EXAMPLE 2. *Ordering by containment.* Let S be the collection of all subsets of a set X. Then $(S, \supset)$ is a poset. ⟦$A \supset A$ for all $A \subset X$, and $A \supset B \supset C$ implies $A \supset C$.⟧

★EXAMPLE 3. *Ordering by inclusion.* Let S be the collection of all subsets of a set X. Then $(S, \subset)$ is a poset. ⟦$A \subset A$ for all $A \subset X$, and $A \subset B \subset C$ implies $A \subset C$.⟧

In each of Examples 2 and 3, there may be noncomparable elements; that is, x, y such that $x \geq y$ and $y \geq x$ are both false. ⟦Let S be the set of all subsets of $\mathbf{R}$, then $\mathbf{Q} \supset [0, 1]$, and $[0, 1] \supset \mathbf{Q}$ are both false.⟧

EXAMPLE 4. Let $X = (0, 2) \setminus \{1\}$. Let $x \geq y$ mean $|x - 1| \leq |y - 1|$, the second inequality being interpreted in the ordinary real number sense. Thus $x \geq y$ means that x is closer to 1 than y is (or, at least as close). This example shows that the ordering in a poset need not be *antisymmetric* (that is $x \geq y \geq x$ implies $y = x$); for example, $\frac{3}{2} \geq \frac{1}{2} \geq \frac{3}{2}$, but $\frac{3}{2} \neq \frac{1}{2}$.

A *directed set* is a poset X with the additional property that for each $x, y \in X$, there exists $z \in X$ with $z \geq x$, and $z \geq y$. Thus the posets of Examples 1, 2, 3, and 4 are directed sets. ⟦For example, in Example 3, given $A \subset X$, $B \subset X$, then $A \cap B \subset A$, $A \cap B \subset B$.⟧

★EXAMPLE 5. Let $\mathscr{B}$ be a filterbase. Then $(\mathscr{B}, \subset)$ is a directed set; that is *a filterbase is directed by inclusion.* ⟦Inclusion is automatically reflexive and transitive. Also, if $A \in \mathscr{B}$, $B \in \mathscr{B}$, then there exists $C \in \mathscr{B}$ with $C \subset A \cap B$; this implies $C \subset A$, $C \subset B$.⟧

Problems

1. Let X be a set and define $x \geq y$ to mean $x = y$. Show that this makes X a poset, but not a directed set (unless X has only one member.) This order is called the *discrete order*.
2. Let X be a set and define $x \geq y$ to be true for all x, y. Show that this makes X a directed set. This order is called the *indiscrete order*.
3. The discrete order is antisymmetric and the indiscrete order is not.
4. Let M be the set of all men. Let $x \rho y$ mean x is the brother of y. Show that, on M, the relation ρ is *symmetric* (that is, $x \rho y$ implies $y \rho x$), not reflexive, and not transitive.
5. Let N be the set of all nurses. Let $x \sigma y$ mean x is the sister of y. Show that the relation σ is not symmetric on N.
6. Let X be a directed set and S a nonempty finite subset. Show that there exists $x \in X$ with $x \geq s$ for $s \in S$.

★7. A reflexive, symmetric, transitive relation is called an *equivalence relation*. Show that if an equivalence relation $\sim$ is defined on a set S, then S can be written uniquely as a union of disjoint subsets (called *equivalence classes*) such that each subset is $\{x: x \sim y\}$ for some y.

8. Order $\mathbf{R}^2$ by: $(x, y) \geq (x', y')$ means $x > x'$, or $x = x'$ and $y \geq y'$. Show that this order is reflexive and transitive. This order is called the *lexicographic* order because it resembles alphabetical order for two-letter words.

★9. A poset is called *totally ordered* (or, *linearly ordered*) if it is antisymmetric and every two members are comparable. Show that the order of Problem 8 is total.

★10. A totally ordered subset of a poset is called a *chain*. Show that every subset of $\mathbf{R}$ is a chain.

★11. Prove that every countable poset contains a maximal chain. ⟦The poset is $\{x_n\}$. Proceed by induction, starting with x_1, and, at each step, adding that x_n with the smallest n such that addition of x_n yields a chain.⟧

---

101. Give an example of a relation on some nonempty set which is symmetric and transitive but not reflexive. (*Hint:* This *seems* impossible since $x \geq y$, $y \geq x$ imply $x \geq x$ by transitivity.)
102. If X, Y are posets, let $X \times Y = \{(x, y): x \in X, y \in Y\}$. Order $X \times Y$ by $(x, y) \geq (x', y')$ means $x \geq x'$ and $y \geq y'$. Show that this makes $X \times Y$ a poset, and, if X, Y are directed, a directed set.
103. In Problem 102, take $X = Y = \mathbf{R}$ with the usual order. In the Cartesian plane $\mathbf{R}^2 = \mathbf{R} \times \mathbf{R}$, sketch $\{(x, y): (x, y) \geq (0, 0)\}$.
104. Give an example of an uncountable family F of sets, no two equal, such that F is totally ordered by inclusion and the union of all the sets in F is countable. (*Hint:* This is easy.)

105. Find an infinite chain in $\mathbf{R}^2$ with the order given in Problem 8.
106. Find all chains in the discrete and indiscrete orders.
107. Let "word" mean "any finite set of letters listed in order." For example, *ATYYPV* and *APTYVY* are two different words. Suppose the set of all words arranged in alphabetical order. Show that between any two words, not both ending in *A*, there is another word.
108. It is easy to give an infinite ascending sequence of words in Problem 107; for example, *BC*, *BCC*, *BCCC*,.... Give an example of an infinite descending sequence.
109. Let X be the set of all sequences x of real numbers such that $x_n = 0$ eventually. Define $x \geq y$ to mean either $x = y$ or the last nonzero term of $x - y$ is positive. Show that X is a totally ordered set.
110. A set S in a poset X is called *cofinal* if for each $x \in X$, there exists $s \in S$ with $s \geq x$. In the space $\mathbf{R}^2$ of Problem 103, give an example of a maximal chain which is not cofinal.
111. Let A, B be directed sets and $u\colon A \to B$. We call u *finalizing* if for every $b \in B$, $u(a) \geq b$ *eventually*. (This means that there exists $a_0 \in A$ such that $u(a) \geq b$ whenever $a \geq a_0$.) Suppose that u is *isotone* ($a \geq a'$ implies $u(a) \geq u(a')$), show that u is finalizing if and only if $u[A]$ is cofinal.
112. No function $u\colon (0, 1] \to (0, 1)$ can be finalizing.

---

201. Let P be an antisymmetric poset with a largest member m. Fix two subsets A and B of P which contain m and which have the property that every subset of A (of B) has an inf (that is greatest lower bound) in A (in B). For $x \in P$ let $Ax = \inf\{y \in A : y \geq x\}$, similarly Bx. (For example if P is the set of subsets of a topological space, and A is the set of closed sets, Ax is the closure of x.) Assume that $Ba \in A$ for all $a \in A$, and set $C = A \cap B$. Show that $Cx = BAx$ for all $x \in P$ but that $Cx = ABx$ is not necessarily true.
202. Give an example of a directed set that has no cofinal chain. (One is given in Sec. 8.1, Problem 201.)
203. Suppose that a countable chain $C = \{a_n\}$ has no smallest or largest member and no two adjacent members. Show that C is order isomorphic to $\mathbf{Q}$. (An *order isomorphism* is a one-to-one isotone map.) ⟦Let $S = \{m/2^n : m, n \in \mathbf{Z}\}$. Let $u\colon C \to S$ have $u(a_1) = 0$, $u(b_1) = 1$, where b_1 is earliest $a_n > a_1$, $u(b_2) = 2$, where b_2 is earliest $a_n > b_1$, etc.; $u(c_1) = -1$, where c_1 is earliest $a_n < a_1$; $u(c_2) = -2$, where c_2 is earliest $a_n < c_1$, etc.; $u(d_1) = \frac{1}{2}$, where d_1 is earliest a_n between a_1 and b_1; $u(d_2) = \frac{3}{2}$, where d_2 is earliest a_n between b_1 and b_2, etc. Having shown C order isomorphic to S, it follows that $\mathbf{Q}$ is also, since $\mathbf{Q}$ is a chain of the cited type.⟧
204. For relations U, V on a set X, define $U \circ V$ by: $(x, y) \in U \circ V$ if and

only if $(a, y) \in U$ and $(x, a) \in V$ for some $a \in X$. Show that U is transitive if and only if $U \circ U \subset U$. Show also that a transitive relation U need not satisfy $U \circ U = U$.

205. Let U, V be equivalence relations. Then $U \circ V$ is an equivalence relation if and only if $U \circ V = V \circ U$.

3.4 Nets

A *net* is a function defined on some directed set. For example, a sequence is a special kind of net; for a sequence is a function defined on ω, the positive integers. Just as there are sequences of points, of real numbers, of functions, so are there nets of points, real numbers, functions; for example, a net of real numbers is a function from a directed set to $\mathbf{R}$. In general, if X is a set, a *net in X*, or, a *net of points of X*, is a function from some directed set D to X. We shall write such a net as $(x_\delta : D)$. When it is unnecessary to specify D we shall write x_δ for the net.

★EXAMPLE 1. Let $D = (0, 1)$ with its usual order. For $\delta \in D$, let $f_\delta(x) = \cos \delta x$ for all real x. Then $(f_\delta : D)$ is a net of real functions of a real variable. It is not a sequence.

If $(x_\delta : D)$ is a net in a set X, and $S \subset X$, we say that $x_\delta \in S$ *eventually* if there exists $\delta_0 \in D$ such that $x_\delta \in S$ for all $\delta \geq \delta_0$; and $x_\delta \in S$ *frequently* if, for any $\delta_0 \in D$, there exists $\delta \geq \delta_0$ with $x_\delta \in S$. For sequences, the definition of "eventually" just given is the same as that given in Section 3.1. A sequence x_n is frequently in S if and only if it is in S for arbitrarily large n (that is, for infinitely many n).

LEMMA 3.4.1 *Let $(x_\delta : D)$ be a net in a set X. Let $S_1, S_2, \ldots, S_n$ be (finitely many) subsets of X and suppose that for each $k = 1, 2, \ldots, n$, $x_\delta \in S_k$ eventually. Then $x_\delta \in \bigcap S_k$ eventually.*

For each k, choose δ_k such that $\delta \geq \delta_k$ implies $x_\delta \in S_k$. Since D is a directed set, there exists δ_o satisfying $\delta_o \geq \delta_k$ for each k. Then $\delta \geq \delta_o$ implies $\delta \geq \delta_k$ for each k, hence $x_\delta \in S_k$ for each k. ∎

Now let x_δ be a net in a topological space X. We say that *x_δ converges to x*, in symbols $x_\delta \to x$, if for every neighborhood N of x, $x_\delta \in N$ eventually. When x_δ converges to a unique point x, we also write $\lim x_\delta = x$. Any point x such that $x_\delta \to x$ is called a *limit* of x_δ.

EXAMPLE 2. Let $D = (0, 1)$ with the usual ordering, $\geq$. Any net $(x_\delta : D)$ in a topological space X is a function on $(0, 1)$ with values in X, and $x_\delta \to x$ means that for any neighborhood N of x, there exists δ_0, $0 < \delta_0 < 1$, such that $\delta_0 \leq \delta < 1$ implies $x_\delta \in N$; that is, $x_\delta \to x$ as $\delta \to 1-$ in the usual sense.

EXAMPLE 3. Let D be the punctured disc in the complex plane consisting of $\{z: 0 < |z| \leq 1\}$. Define $z_1 \geq z_2$ if $|z_1| \leq |z_2|$. Thus $z_1 \geq z_2$ means z_1 is closer to 0 than z_2. For a net $(x_\delta: D)$, $x_\delta \to x$ means $x_\delta \to x$ as $\delta \to 0$ in the usual sense.

★EXAMPLE 4. *In a semimetric space, $x_\delta \to x$ if and only if $d(x_\delta, x) \to 0$.* Suppose first that $x_\delta \to x$. Let $\varepsilon > 0$ be given. Then $x_\delta \in N(x, \varepsilon)$ eventually, that is, $d(x_\delta, x) < \varepsilon$ eventually, hence $d(x_\delta, x) \to 0$. Conversely, if $d(x_\delta, x) \to 0$, let N be a neighborhood of x. Then $N(x, \varepsilon) \subset N$ for some $\varepsilon > 0$. Since $d(x_\delta, x) < \varepsilon$ eventually, it follows that $x_\delta \in N(x, \varepsilon)$ eventually, and so $x_\delta \in N$ eventually. Thus $x_\delta \to x$. ▮

Suppose that $\mathscr{B}$ is a filterbase on a set X, and suppose that for each $\delta \in \mathscr{B}$, a certain point $x_\delta \in \delta$ is specified. (This makes sense since each δ is a subset of X.) Now $(\mathscr{B}, \subset)$ is a directed set ⟦Sec. 3.3, Example 5⟧, and so $(x_\delta: \mathscr{B})$ is a net in X. For a given filterbase $\mathscr{B}$, we shall refer to any net obtained in this way as *a net associated with $\mathscr{B}$*. All this illustrates precisely the difference between nets and filters. The construction of nets goes one step beyond that of filters, hence the net is a more complicated construct. The use of nets, when appropriate, is justified by their similarity with sequences.

We now describe an important sufficient condition for net convergence. This generalizes the "shrinking base" condition given in Sec. 3.1, Problem 5.

THEOREM 3.4.1. *Let $\mathscr{B}$ be a convergent filterbase in a topological space; $\mathscr{B} \to x$. Let $(x_\delta: \mathscr{B})$ be any net associated with $\mathscr{B}$. Then $x_\delta \to x$.*

Let N be any neighborhood of x. Then N includes some member A of $\mathscr{B}$. Let $\delta_0 = A$. Then $\delta \geq \delta_0$ implies $x_\delta \in N$; ⟦$\delta \geq \delta_0$ means $\delta \subset A$, thus $x_\delta \in \delta \subset A \subset N$⟧; in other words $x_\delta \in N$ eventually. Thus $x_\delta \to x$. ▮

The most important special case of Theorem 3.4.1 is that in which $\mathscr{B} = N_x$, the filter of all neighborhoods of x. This yields:

COROLLARY 3.4.1 *Let X be a topological space and $x \in X$. For each neighborhood N of x choose some $x_N \in N$. Then $(x_N: N_x)$ is a net which converges to x.*

The results of Section 3.2 can now be translated into the language of nets. We show one such translation explicitly, leaving the others as problems.

THEOREM 3.4.2. *Let X be a topological space, $x \in X$, $S \subset X$. Then $x \in \bar{S}$ if and only if there is a net in S which converges to x.*

Suppose first that such a net exists. Every neighborhood of x contains a point of this net, hence meets S. Thus $x \in \bar{S}$. Conversely, let $x \in \bar{S}$. By Theorem 3.2.1, there exists a filterbase $\mathscr{B}$ in S which converges to x. Any associated net converges to x. ▮

★EXAMPLE 5. The fact that nets and filters are equivalent tools is demonstrated by using filters to construct nets and vice versa. We have already seen how to associate a net with a filterbase. In Example 6, this will be done in such a way that convergence of the two will be equivalent. Here we shall associate a filter with a net. Let $(x_\delta\colon D)$ be a net in X. Let $\mathscr{F} = \{A \subset X\colon x_\delta \in A \text{ eventually}\}$. It is clear that $\mathscr{F}$ is a filter, for example, let $A \in \mathscr{F}$, $B \in \mathscr{F}$. Then $A \cap B \in \mathscr{F}$ ⟦Lemma 3.4.1⟧. It is called *the filter associated with* x_δ. We shall prove that *$\mathscr{F} \to x$ if and only if $x_\delta \to x$, where $\mathscr{F}$ is the filter associated with x_δ.* Let $x_\delta \to x$ and $N \in N_x$. Then $x_\delta \in N$ eventually, hence $N \in \mathscr{F}$. Thus $N_x \subset \mathscr{F}$ and so $\mathscr{F} \to x$. Conversely, let $\mathscr{F} \to x$ and let N be a neighborhood of x. Then $N \in \mathscr{F}$ and so, by definition of $\mathscr{F}$, $x_\delta \in N$ eventually. Hence $x_\delta \to x$. ∎

▲EXAMPLE 6. Readers of a certain sophistication will be dissatisfied with the concept of the net associated with a filterbase. There are three valid reasons for such dissatisfaction. In the first place, our esthetic sense is displeased by the lack of uniqueness of the net associated with a filterbase $\mathscr{B}$. For each $\delta \in \mathscr{B}$, a point $x_\delta \in \delta$ is chosen at random. In the second place, there is an unfortunate lack of symmetry in that $\mathscr{B} \to X$ implies $(x_\delta\colon \mathscr{B}) \to x$, but the converse is false ⟦Problem 105⟧. Finally, $\mathscr{B}$ usually contains infinitely many members, and choice of one x_δ from each $\delta \in \mathscr{B}$ requires an infinite set of arbitrary choices. The possibility of such a set of choices (called the *Axiom of Choice*) is an unwelcome intruder into an elementary situation. All three of these objections are met by constructing a different sort of net. Let $\mathscr{B}$ be a filterbase on a set X. Let D be the set of all pairs (x, S) such that $x \in S \in \mathscr{B}$. Define $(x_1, S_1) \geq (x_2, S_2)$ to mean $S_1 \subset S_2$. (Ignore x_1, x_2.) Then D becomes a directed set ⟦for example, if $\delta = (x, S_1 \cap S_2)$, then $\delta \geq (x_1, S_1)$, $\delta \geq (x_2, S_2)$⟧. Now for $\delta \in D$, that is $\delta = (x, S)$, define $x_\delta = x$. This net, $(x_\delta\colon D)$ is called the *canonical net of the filterbase* $\mathscr{B}$. We shall prove that $x_\delta \to x$ if and only if $\mathscr{B} \to x$, where *x_δ is the canonical net of the filterbase $\mathscr{B}$.* It is sufficient by Example 5, to show that the filter associated with x_δ is the filter generated by $\mathscr{B}$. Denote these two filters by $\mathscr{F}_1, \mathscr{F}_2$. Let $A \in \mathscr{F}_1$. Then $x_\delta \in A$ eventually, say for $\delta \geq \delta_0 = (x, S)$, $x \in S \in \mathscr{B}$. Then $S \subset A$. ⟦Let $s \in S$. Then $(s, S) \geq (x, S)$ hence $x_{(s,S)} \in A$; that is $s \in A$.⟧ Hence $A \in \mathscr{F}_2$. Conversely, let $A \in \mathscr{F}_2$. Then $S \subset A$ for some $S \in \mathscr{B}$. Choose $x \in S$ and set $\delta_0 = (x, S)$. Then $\delta \geq \delta_0$ implies $x_\delta \in A$. ⟦Say $\delta = (y, B)$. Then $\delta \geq \delta_0$ implies $B \subset S \subset A$, but $x_\delta = y \in \mathscr{B}$.⟧ Since $x_\delta \in A$ eventually, we have $A \in \mathscr{F}_1$. ∎ However, it must be emphasized that the three objections mentioned are matters of taste only. We shall continue to use the associated net; indeed there are cases in which the canonical net cannot be used; one of these is the proof, Theorem 3.1.1, where a sequence was desired.

Problems

★1. State and prove the analogue for nets of Theorem 3.2.2. ⟦Consult Theorems 3.4.2, 3.1.2, and 3.1.3.⟧

★2. State and prove the analogue for nets of Theorem 3.2.3.

---

101. More than one net may lead to the same associated filter. Show that $\{n\}$ and $\{1, 1, 2, 2, 3, 3, \ldots\}$ lead to the same filter in **R**.
102. Let x_δ be the canonical net of the filterbase $\mathscr{B}$, and let $S \in \mathscr{B}$. Show that $x_\delta \in S$ eventually.
103. The filter associated with the canonical net of a filter $\mathscr{F}$ is $\mathscr{F}$.
104. Let $\mathscr{F}$ be a filter, $(x_\delta\colon \mathscr{F})$ an associated net, and $\mathscr{F}'$ the filter associated with $(x_\delta\colon \mathscr{F})$. Show that $\mathscr{F}' \supset \mathscr{F}$ and that possibly $\mathscr{F}' \neq \mathscr{F}$.
105. In Problem 104, it is possible that $x_\delta \to x$ (and hence $\mathscr{F}' \to x$) without $\mathscr{F}$ being convergent. ⟦Take $\mathscr{F} \subset N_x$ and $x_\delta = x$ for all $\delta \in \mathscr{F}$.⟧
106. Do Problem 107 of Section 3.2 with net instead of filter. Show in the course of the proof, a net associated with the indiscrete filter. (The problem could be solved by considering a sequence, but it is pointless to bring in $\boldsymbol{\omega}$ or any more complicated directed set.)
107. Solve Problem 106 using a net associated with a discrete filterbase $\{\{y\}\}$ for each $y \neq x$.
108. Let $S = \{x\}$. What net is produced by the construction of Theorem 3.4.2?
109. Let x_δ be a real net, and define $\liminf x_\delta = L$ to mean that for each $\varepsilon > 0$, $x_\delta > L - \varepsilon$ eventually, and $x_\delta < L + \varepsilon$ frequently. Define lim sup by interchanging "eventually" and "frequently". Show that a net may have at most one lim sup and one lim inf. Show also that $\limsup x_\delta = -\liminf(-x_\delta)$.

---

201. Let $u\colon D \to X$ be a net. Then $u \to x$ if and only if $x \in \overline{(u \circ f)[D]}$ for all $f\colon D \to D$ such that $f[D]$ is cofinal in D.
202. Let f be a bounded real function on $[0, 1]$. Consider the directed set D of partitions of the interval in the sense used in Riemann integration, with refinement as the order. For $\delta \in D$, let $\Sigma(\delta)$ be the Riemann sum. Show that $\lim \Sigma(\delta)$ is the Riemann integral $\int_0^1 f$ if it exists.
203. Let D be the set of all finite subsets of $[0, 1]$ directed by containment. Let f_δ be the continuous function with saw-tooth graph, $f_\delta(x) = 0$ for each $x \in \delta$, $f_\delta(x) = 1$ for x the midpoint of two adjacent points of δ. Show that $(f_\delta\colon D) \to 0$ pointwise but not uniformly.
204. Let D be uncountable, totally ordered, and such that for each $\delta_0 \in D$, $\{\delta\colon \delta < \delta_0\}$ is countable. Show that no net $(x_\delta\colon D)$ of positive numbers can converge to 0.

3.5 Arithmetic of Nets

Let $(x_\delta: D)$, $(y_\delta: D)$ be nets of complex numbers, defined on the same directed set D. Then we may consider the three nets $(x_\delta + y_\delta: D)$, $(x_\delta y_\delta: D)$ and $(x_\delta/y_\delta: D)$. For example, sequences can always be combined in this way $[\![D = \omega]\!]$. The next result is an entirely expected extension of classical sequence theory. This result is actually a commentary on the topology of the complex plane; see Problems 103, 104, 105 for details. In Chapter 12 we shall see a very general setting for these ideas.

THEOREM 3.5.1. *Let $(x_\delta: D)$, $(y_\delta: D)$ be nets of complex numbers defined on same directed set D, and assume that $x_\delta \to x$, $y_\delta \to y$. Then,* (i) $x_\delta + y_\delta \to x + y$, (ii) $x_\delta y_\delta \to xy$, *and,* (iii) *if $y \neq 0$, $x_\delta/y_\delta \to x/y$ in the sense that for each $\varepsilon > 0$, there exists δ_0 such that $\delta \geq \delta_0$ implies $y_\delta \neq 0$ and $|x_\delta/y_\delta - x/y| < \varepsilon$.*

Let $\varepsilon > 0$. Choose δ_1, δ_2, so that $\delta \geq \delta_1, \delta \geq \delta_2$, imply, respectively, that $|x_\delta - x| < \varepsilon/2, |y_\delta - y| < \varepsilon/2$. Then choose δ_0 with $\delta_0 \geq \delta_1, \delta_0 \geq \delta_2$. For $\delta \geq \delta_0$ we have $|(x_\delta + y_\delta) - (x + y)| \leq |x_\delta - x| + |y_\delta - y| < \varepsilon$. This proves (i). To prove (ii) we first consider the case $x = y = 0$. It is immediate that $x_\delta y_\delta \to 0$. $[\![$Making $|x_\delta| < \varepsilon^{1/2}$, $|y_\delta| < \varepsilon^{1/2}$, forces $|x_\delta y_\delta| < \varepsilon$.$]\!]$ Next, consider the case $x_\delta = x$ for all δ. We may assume $x \neq 0$. $[\![$If $x = 0$, $x_\delta y_\delta = 0$ for all δ and the result is trivial.$]\!]$ Choose δ_0 so that $\delta \geq \delta_0$ implies $|y_\delta - y| < \varepsilon/|x|$. Then $\delta \geq \delta_0$ implies $|x_\delta y_\delta - xy| = |x|\cdot|y_\delta - y| < \varepsilon$. To prove (ii) in general, let $u_\delta = x_\delta - x$, $v_\delta = y_\delta - y$; then $x_\delta y_\delta - xy = u_\delta v_\delta + y u_\delta + x v_\delta \to 0$ by (i) and the earlier cases of (ii). Finally, $x_\delta/y_\delta - x/y = (u_\delta + x)/(v_\delta + y) - x/y = u_\delta/(v_\delta + y) - t v_\delta/(v_\delta + y)$, where $t = x/y$. It now suffices to prove that $u_\delta/(v_\delta + y) \to 0$ in the sense given in the statement of the theorem. $[\![$The same argument will apply to $v_\delta/(v_\delta + y)$ and application of (i), (ii) will yield the result.$]\!]$ Since $v_\delta \to 0$, it follows that $|v_\delta + y| \geq |y| - |v_\delta| \geq \frac{1}{2}|y|$ eventually. Also $|u_\delta| < \varepsilon|y|/2$ eventually, and the result follows. ∎

Infinite combinations of nets can also be formed. The one now given is of the first importance.

DEFINITION 1 (THE FRÉCHET COMBINATION). *Let D be a directed set, and for each $n = 1, 2, \ldots$, let $(x_\delta^n: D)$ be a net of nonnegative real numbers. For each $\delta \in D$, let*

$$y_\delta = \sum_{n=1}^{\infty} \frac{1}{2^n} \cdot \frac{x_\delta^n}{1 + x_\delta^n}.$$

The net $(y_\delta: D)$ is called the Fréchet combination of the sequence $\{x_\delta^n\}$ of nets.

The series must converge for each δ since its general term lies between 0 and $1/2^n$.

In reading the following theorem keep in mind that the various nets are nets of nonnegative numbers, hence absolute value signs do not appear.

THEOREM 3.5.2. *With the notation of Definition* 1, $y_\delta \to 0$ *if and only if* $x_\delta^n \to 0$ *for each* n.

Suppose first that $x_\delta^n \to 0$ for each n and let $\varepsilon > 0$. Let m be a positive integer such that $2^{-m} < \varepsilon/2$. Then

$$y_\delta = \sum_{n=1}^{m} \frac{1}{2^n} \cdot \frac{x_\delta^n}{1 + x_\delta^n} + \sum_{n=m+1}^{\infty} \frac{1}{2^n} \cdot \frac{x_\delta^n}{1 + x_\delta^n}.$$

The second term on the right is less than $\sum_{n=m+1}^{\infty} 1/2^n = 2^{-m} < \varepsilon/2$, while the first term, call it A_δ, tends to 0 by Theorem 3.5.1. Thus if $\delta \geq \delta_0$ implies $A_\delta < \varepsilon/2$, we have $\delta \geq \delta_0$ implies $y_\delta < \varepsilon$. Conversely suppose that $y_\delta \to 0$. For each n, δ, $y_\delta \geq (1/2^n) \cdot (x_\delta^n/(1 + x_\delta^n))$. Thus $2^n y_\delta + 2^n x_\delta^n y_\delta \geq x_\delta^n$ and so $x_\delta^n \leq 2^n y_\delta/(1 - 2^n y_\delta)$. The right-hand side tends to 0 by Theorem 3.5.1. ▌

Problems

101. Let $\{d_n)$ be a sequence of semimetrics on a set X. Let

$$d(x, y) = \sum_{n=1}^{\infty} \frac{1}{2^n} \frac{d_n(x, y)}{1 + d_n(x, y)}.$$

Then d is a semimetric, and for any net x_δ in X, $x_\delta \to x$ in (X, d) if and only if $x_\delta \to x$ in (X, d_n) for each n.

102. Express the first half of the proof of Theorem 3.5.2 in the language of uniform convergence and interchange of "lim" and "Σ."
103. Place on **R** the topology which is gotten from the Euclidean topology by adjoining all singletons outside of $[0, 1]$. Let $x_n = 1/n$, $y_n = 1$, $z_n = \frac{1}{2} + 1/n$ for $n = 1, 2, \ldots$. Show that $x_n \to 0$, $y_n \to 1$, $z_n \to \frac{1}{2}$, but $x_n + y_n \not\to 1$, $-x_n \not\to 0$ and $2z_n \not\to 1$.
104. In the RHO topology, $1/n \to 0$, $-1/n \not\to 0$.
105. Let G be a group and a topological space, and suppose that for each neighborhood U of 1 (the identity) there is a neighborhood V of 1 with $VV \subset U$. ($VV = \{x \cdot y : x \in V,\ y \in V\}$.) Show that if $x_\delta \to 1$ and $y_\delta \to 1$, then $x_\delta y_\delta \to 1$.
106. Deduce the first two parts of Theorem 3.5.1 for the case $x = y = 0$ from Problem 105.
107. Let L be a complex vector space and a topological space, and suppose that for each neighborhood U of 0, there exists a neighborhood V of 0 and $\varepsilon > 0$ such that $tV \subset U$ for all complex t with $|t| < \varepsilon$. Show that if t_δ, x_δ are nets of complex numbers, and members of L, respectively, and if $t_\delta \to 0$, $x_\delta \to 0$, then $t_\delta x_\delta \to 0$.

108. Let D be a dense subset of a semimetric space X, and x_δ a net in X such that $d(x_\delta, y) \to d(x, y)$ for all $y \in D$. Show that $x_\delta \to x$. ⟦$d(x_\delta, x) \leq d(x_\delta, y) + d(y, x) \to 2d(x, y)$ for each y.⟧

---

201. Prove the converse of Problem 105: if $x_\delta \to 1$, $y_\delta \to 1$ implies $x_\delta y_\delta \to 1$, then for every neighborhood U of 1, there exists a neighborhood V of 1 with $VV \subset U$. ⟦If not, let D be the poset of neighborhoods of 1. There exists $U \in D$ with $VV \not\subset U$. Choose $x_V \in VV \setminus U$.⟧
202. Prove the converse of Problem 107 (as in Problem 201). ⟦If not, there exists U such that for every $n \in \omega$ and neighborhood V of 0, $tV \not\subset U$ for some t with $|t| < 1/n$. Choose $x_{n, V} \in t \cdot V \setminus U$.⟧
203. In Problems 107 and 202 show that the results are correct if ε is put equal to 1 everywhere.

Separation Axioms

4

4.1 Separation by Open Sets

In this chapter we define nine kinds of topological space, each characterized by the possibility of "separating" certain kinds of sets. Six of these are designated by a symbol like T_n, and they form a hierarchy in the sense that each T_m space is also a T_n space for all $n < m$. As we define T_0, T_1, T_2, T_3, $T_{3\frac{1}{2}}$, and T_4 space, the reader should check the hierarchy (for example, every T_2 space is a T_1 space). As each definition is given, an example shows that the class of spaces is strictly larger than the preceding (for example, not every T_1 space is a T_2 space).

The letter T was chosen for "Trennung," the German word for "separation."

We call a topological space *a T_0 space* if, for each pair of points x, y with $x \neq y$, there is either a neighborhood of x not containing y, or a neighborhood of y not containing x.

EXAMPLE 1. Let X be a set with more than one member, and give X the indiscrete topology. Then X is not a T_0 space.

EXAMPLE 2. Let $X = \mathbf{R}$ and $T = \{\varnothing\} \cup \{\mathbf{R}\} \cup \{(a, \infty): a \in \mathbf{R}\}$. Then X is a T_0 space since if $x < y$, (x, ∞) is a neighborhood of y, but not of x.

★EXAMPLE 3. *A semimetric T_0 space must be a metric space.* If X is semimetric but not metric, there are distinct points x, y with $d(x, y) = 0$.

Then $y \in N(x, r)$ for all $r > 0$, and so y belongs to every neighborhood of x. Also x belongs to every neighborhood of y, and so X is not a T_0 space.

We call a topological space *a T_1 space* if, for each pair of points x, y with $x \neq y$, there is a neighborhood of x not containing y.

The space of Example 2 is not a T_1 space since every neighborhood of 1 contains 2. The cofinite topology is always T_1, for if $x \neq y$, $\{y\}^\sim$ is open ⟦surely $\{y\}^{\sim\sim}$ is finite⟧, contains x but not y.

THEOREM 4.1.1. *A topological space is a T_1 space if and only if every finite set is closed.*

Let X be a T_1 space and $x \in X$. For any $y \neq x$, there is a neighborhood of y not containing x, thus $y \notin \overline{\{x\}}$. ⟦Sec. 2.5, Problem 11.⟧ Hence $\{x\}$ is closed. But any finite set is, then, a finite union of closed sets, namely, singletons. Conversely, let every finite set be closed and let $x \neq y$. Then $\{y\}^\sim$ is a neighborhood of x not containing y. ∎

We now introduce the abbreviation: "sets A, B can be *separated by neighborhoods*" to mean "there exist disjoint sets N_1, N_2 which are neighborhoods of A, B, respectively." (Of course, separation by neighborhoods is equivalent to separation by open sets.)

We call a topological space *a T_2 space*, or, very commonly, *a Hausdorff space*, if each pair of points x, y with $x \neq y$, can be separated by neighborhoods.

✓ The cofinite topology on an infinite set X is not T_2 since any two nonempty open sets must meet. ⟦Each has a finite complement which, thus, cannot contain the other.⟧ The Euclidean topology for $\mathbf{R}$ is T_2; indeed most of the historical examples of topological spaces are T_2 spaces, and there are many mathematicians who disdain to work at any lower level. There are good reasons for this, for example, the following theorem, but in many situations the extra generality is worth having. (See also the remarks following the definition of "regular," below.)

THEOREM 4.1.2. *The following conditions on a topological space X are equivalent.*

(i) *X is a Hausdorff space.*
(ii) *Every convergent net in X has exactly one limit.*
(iii) *Every convergent filter in X has exactly one limit.*

Proof. (i) *implies* (ii). Let X be T_2 and $x_\delta \to x$. Let $y \neq x$. Let N_1, N_2 be → X is T2
disjoint neighborhoods of x, y respectively. Now $x_\delta \in N_1$ eventually, hence it is false that $x_\delta \in N_2$ eventually. ⟦Otherwise, Lemma 3.4.1 would yield the ridiculous conclusion that $x_\delta \in N_1 \cap N_2$, the empty set, eventually.⟧ Thus $x_\delta \not\to y$.

(ii) *implies* (iii). If (iii) were false we would have a filter with two limits, and any net associated with it would have two limits ⟦Theorem 3.4.1⟧, so that (ii) would be false.

(iii) *implies* (i). Suppose that X is not a T_2 space. Then there are points x, y, with $x \neq y$, such that every neighborhood of x meets every neighborhood of y. Let $\mathscr{F} = \{A \cap B: A, B$ neighborhoods of x, y respectively$\}$. Then $\mathscr{F}$ is a filter ⟦$X \in \mathscr{F}$ since $X = X \cap X$; $\varnothing \notin \mathscr{F}$ by the condition on x, y; if $A_1 \cap B_1, A_2 \cap B_2$ are members of $\mathscr{F}$, then $(A_1 \cap B_1) \cap (A_2 \cap B_2) = (A_1 \cap A_2) \cap (B_1 \cap B_2) \in \mathscr{F}$; $S \supset A \cap B \in \mathscr{F}$ implies $S = (S \cup A) \cap (S \cup B) \in \mathscr{F}$⟧, $\mathscr{F} \to x$. ⟦Let N be a neighborhood of x. Then $N = N \cap X \in \mathscr{F}$.⟧ Also $\mathscr{F} \to y$ so condition (iii) is false. ▮

We call a topological space *regular* if each point x and each closed set F with $x \notin F$ can be separated by neighborhoods.

The "good" topological spaces are those which are regular, in the sense that many useful results hold for regular spaces. As examples we cite Theorem 5.4.7; Theorem 5.4.11 (with Sec. 8.1, Problem 6); Sec. 5.4, Problem 210 (with Sec. 8.1, Problem 114); Theorem 8.1.2; and Lemma 12.2.5 (with Sec. 4.2, Problem 203). Often, when a space is assumed to be Hausdorff, this is simply to ensure (together with other assumptions) that it will be regular, as, for example, Theorem 5.4.7.

Any indiscrete topological space is regular. ⟦If F is closed and $x \notin F$ we must have $F = \varnothing$. Then X and $\varnothing$ are disjoint neighborhoods of x, F.⟧ The Euclidean topology for **R** is regular. ⟦If F is closed and $x \notin F$, $\tilde{F}$ is a neighborhood of x, thus there exists $\varepsilon > 0$ with $(x - \varepsilon, x + \varepsilon) \subset \tilde{F}$. Then $(x - \frac{1}{2}\varepsilon, x + \frac{1}{2}\varepsilon)$, and $(-\infty, x - \frac{1}{2}\varepsilon) \cup (x + \frac{1}{2}\varepsilon, \infty)$ are disjoint neighborhoods of x, F.⟧ The cofinite topology on an infinite set is not regular since any two nonempty open sets must meet and the space has a proper closed subset.

THEOREM 4.1.3. *A topological space is regular if and only if the set of closed neighborhoods of any point is a local base at that point.*

Let X be regular and N a neighborhood of $x \in X$. There exists an open set G with $x \in G \subset N$. Since $\tilde{G}$ is closed, there are disjoint open sets G_1, G_2 with $x \in G_1$, $\tilde{G} \subset G_2$. Then $\tilde{G}_2$ is a closed set, a neighborhood of x, ⟦$G_1 \subset \tilde{G}_2$⟧, and is included in G.

Conversely, assume that the closed neighborhoods form a local base. Let x, F be, respectively, a point and a closed set not containing it. Then $\tilde{F}$ is a neighborhood of x, thus there exists a closed neighborhood, N, of x with $N \subset \tilde{F}$. Then N includes an open neighborhood, G_1, of x, and G_1, $\tilde{N}$ are disjoint open sets separating x, F. ▮

We call a topological space *a T_3 space* if it is a regular T_1 space. The Euclidean topology for **R** is T_3; the indiscrete and cofinite topologies on an infinite set are not T_3 because, respectively, they are not T_1, regular.

We call a topological space *normal* if each disjoint pair of closed sets can be separated by neighborhoods. A normal T_1 space is called *a T_4 space*.

Any indiscrete topological space is normal. ⟦The only two disjoint closed sets are $\varnothing$, X and they are also open.⟧ Numerous examples of normal spaces will be seen later. For example, in Section 4.3 it will be seen that every semimetric space is normal, hence every metric space is T_4.

It is easy to see, using Theorem 4.1.1, that $T_4 \Rightarrow T_3 \Rightarrow T_2 \Rightarrow T_1 \Rightarrow T_0$. It is not true that every normal space is regular ⟦Problem 8⟧, nor that every regular space is T_2 ⟦indiscrete⟧; the counterexamples work because a point need not be closed. To round out this discussion we mention that a T_2 space need not be T_3, and a T_3 space need not be T_4. Counterexamples will be presented when they fall easily out of wider contexts. They are Sec. 6.2, Example 2 and Sec. 6.7, Example 3.

Problems on Topological Space

1. X is a T_0 space if and only if for each pair of points x, y, if $\{x\}^{\sim}$ is not a neighborhood of y, then $\{y\}^{\sim}$ is not a neighborhood of x.
★2. $T_4 \Rightarrow T_3 \Rightarrow T_2 \Rightarrow T_1 \Rightarrow T_0$.
3. There is only one T_1 topology for a finite space ⟦discrete⟧.
★4. If (X, T) is a T_0 space, so is (X, T') where T' is any topology larger than T. The same is true for T_1, T_2.
★5. The cofinite topology is minimum among T_1 topologies. (This means that every T_1 topology for a set includes the cofinite topology for that set.)
★6. A property P is called *hereditary* if every subspace of a space with property P also has property P. It is called *F-hereditary* if every closed subspace of a space with property P also has property P. *G-hereditary* is defined by replacing "closed" with "open." Show that T_0, T_1, T_2, T_3, regular, are hereditary properties, and that normal is F-hereditary. ⟦To show a subspace S of a normal space is normal, it is sufficient to show that if F_1, F_2 are disjoint closed sets in S, their closures in X are disjoint.⟧
★7. Prove an analogue of Theorem 4.1.3 for normal spaces: a space is normal if and only if for every closed set F and every neighborhood N of F, N includes a closed neighborhood of F.
8. Let X be a set and fix $x \in X$. Let a nonempty set be called closed if and only if it contains x. Show that this defines a topology for X which is normal but not regular.
9. Call X *symmetric* if $x \in \overline{\{y\}}$ implies $y \in \overline{\{x\}}$. Show that every regular space is symmetric.
10. A symmetric T_0 space is T_1.

11. A regular T_0 space is T_3 ⟦Problems 9 and 10⟧ but a normal T_0 space need not be T_4 ⟦Example 2⟧.
12. A symmetric normal space must be regular.
13. A space is symmetric if and only if every open set includes the closure of each of its singleton subsets. ⟦If this holds and $x \notin \overline{\{y\}}$, then $\{x\} \subset \widetilde{\overline{\{y\}}}$. Conversely, if X is symmetric and $x \in G$, $y \in \overline{\{x\}}$ implies $x \in \overline{\{y\}}$, hence $y \in G$ since it is a neighborhood of x.⟧

---

101. In a T_2 space, must three points have disjoint neighborhoods?
102. If $\{x\}$ is open, we call x an *isolated point*. Show that a finite T_0 space must have an isolated point.
103. A zero-dimensional space must be regular but need not be T_3 ⟦indiscrete⟧.
104. A finite regular space is zero-dimensional. ⟦An open neighborhood of x is either closed or properly includes a closed neighborhood of x. The latter is either open or properly includes an open neighborhood of x. Continue.⟧
105. Let X be finite. Each point x is included in a smallest open set $O(x)$: $y \in O(x)$ if and only if $x \in \overline{\{y\}}$. Say $x \leq y$ if and only if $x \in O(y)$; prove that X is T_0 if and only if $\leq$ is antisymmetric.
106. Let $k_0(n)$ be the number of different T_0 topologies which can be placed on a set with n members. Show that $k_0(1) = 1$, $k_0(2) = 3$. (*Note:* $k_0(3) = 19$, $k_0(4) = 219$, $k_0(5) = 4231$, $k_0(6) = 130023$, $k_0(7) = 6129859$. See [Evans, Harary and Lynn].)
107. A set is called *regular open* if it is the interior of its closure. Show that the interior of any closed set and the intersection of two regular open sets are regular open.
108. A cell in a metric space need not be regular open.
109. In a noncountable cocountable space, there are only two regular open sets.
110. A space is called *semiregular* if it has a basis of regular open sets. Show that every regular space is semiregular.
111. Show that the following condition is equivalent to regularity of X: if $\mathscr{F} \to x$ then $\overline{\mathscr{F}} \to x$. ($\mathscr{F}$ is a filter and $\overline{\mathscr{F}} = \{\bar{A} : A \in \mathscr{F}\}$.)
112. In a symmetric space $\overline{\{x\}} = \bigcap N_x$.
113. For a finite space, the properties symmetric, regular, and semimetrizable are equivalent. ⟦Use Problems 105 and 112; $d(x, y) = 1$ if $y \notin \overline{\{x\}}$.⟧
114. For $A \subset X$, let $n(A)$ be the intersection of all neighborhoods of A. Show that if X is T_1, $n(A) = A$; and if X is regular, $n(A) \subset \bar{A}$.
115. In a normal space, disjoint closed sets can be separated by closed neighborhoods of those sets.

116. Suppose that every set which has nonempty interior is open. Show that such a space may be T_1 and not T_2; and, if T_2, must be discrete.
117. X is called a *US* space if a sequence may converge to at most one limit. Show that $T_2 \Rightarrow US \Rightarrow T_1$ but no converse holds ⟦cofinite; cocountable⟧. However, a first countable *US* space is T_2.
118. In duplicating a point (Sec. 3.2, Problem 110), Y is T_1 if X is, but Y cannot be T_2.
119. An *almost discrete* space (exactly one nonisolated point) must be normal.
120. First countability is a hereditary property, but non-first-countability is not.
121. Let (X, T) be a T_1 space and let T' be a topology for X such that a net is T-convergent if and only if it is T'-convergent. Show that $T = T'$. (See Sec. 3.2, Problem 203.)
122. Let $\{G_1, G_2, \ldots, G_n\}$ be an open cover of a normal space X. Then there exists an open cover $\{H_1, H_2, \ldots, H_n\}$ with $\bar{H}_i \subset G_i$ for $i = 1, 2, \ldots, n$. ⟦Separate $\tilde{G}_1$ and $\bigcap \{\tilde{G}_i: i = 2, 3, \ldots, n\}$ by open sets H, H_1. Then $H_1 \subset \tilde{H} \subset G_1$ so that $\bar{H}_1 \subset G_1$. Also $H_1 \cup G_2 \cup G_3 \cup \cdots \cup G_n = X$. Continue.⟧
123. A space is normal if it is the union of subsets, each of which is open, closed, and normal. ⟦For disjoint closed A, B, separate $A \cap S$ from $B \cap S$ in each subspace S.⟧
124. Two different Hausdorff topologies may agree on each of two complementary subsets. ⟦Let A be a nonopen proper subset of (X, T). Say $G \in T'$ if $G \cap A$ and $G \cap B$ are relatively T open in A, B, respectively. Then $T' \supset T$; $T' \neq T$ since A is T' open. *Note:* A may be a singleton!⟧

---

201. Let X be a T_2 space and $\{x_n\}$ a convergent sequence in X. Suppose that a certain point x has the property that x_n is eventually in each closed neighborhood of x. Show that $x_n \to x$.
202. The word "convergent" cannot be omitted from Problem 201, although it can be if X is regular.
203. Let (X, T) have the property that distinct points have disjoint closed neighborhoods. Then X can be given a smaller semiregular topology T' such that T, T' have the same isolated points.
204. Let D be dense in a T_2-space X. Show that $|X| \leq 2^{2^{|D|}}$.
205. Problem 4 becomes false if T_0 is replaced by T_3 or T_4.

4.2 Continuity

The standard real variables definition of continuity of a function f contains the phrase "$|x - a| < \delta$ implies $|f(x) - f(a)| < \varepsilon$"; in other words (replacing $f(a)$ by b for convenience), "$f^{-1}[(b - \varepsilon, b + \varepsilon)] \supset (a - \delta, a + \delta)$"; thus, for each neighborhood N of b, $f^{-1}[N]$ is a neighborhood of a. We have now expressed continuity in terms sufficiently general for topology. Let X, Y be topological spaces and $f: X \to Y$. We say that *f is continuous at* $x \in X$ if, for each neighborhood N of $f(x)$, $f^{-1}[N]$ is a neighborhood of x. We say that *f is continuous on X* if it is continuous at each point of X.

THEOREM 4.2.1. *A function $f: X \to Y$ is continuous on X if and only if whenever G is an open set in Y, $f^{-1}[G]$ is an open set in X.*

Let f be continuous and G an open set in Y. Let $x \in f^{-1}[G]$. Since G is a neighborhood of $f(x)$, $f^{-1}[G]$ is a neighborhood of x. Thus $f^{-1}[G]$ is open, by Theorem 2.5.1. Conversely, suppose that $f^{-1}[G]$ is open whenever G is open; let $x \in X$, and let N be a neighborhood of $f(x)$. Then there exists an open set G with $f(x) \in G \subset N$, by definition of neighborhood. Now $f^{-1}[G]$ is open by hypothesis, contains x, hence is a neighborhood of x. Thus f is continuous at x. ∎

★EXAMPLE 1. Let X be indiscrete and Y a T_0 space. *If $f: X \to Y$ is continuous it must be constant.* If f takes on two values a, b, let N be a neighborhood of a not containing b (or vice versa). Then $f^{-1}[N]$ is a proper subset of X, hence is not a neighborhood of anything. Thus f is not continuous.

LEMMA 4.2.1. *If a function $f: X \to Y$ has the property that $f[\bar{S}] \subset \overline{f[S]}$ for all $S \subset X$, then f is continuous on X. More precisely, if f is not continuous at $x \in X$, there exists $S \subset X$ with $x \in \bar{S}$, $f(x) \notin \overline{f[S]}$.*

If f is not continuous at $x \in X$, there is a neighborhood N of $f(x)$ such that $f^{-1}[N]$ is not a neighborhood of x. Let $S = X \setminus f^{-1}[N]$. Then $x \in \bar{S}$ ⟦Sec. 2.5, Problem 3⟧. Hence $f(x) \in f[\bar{S}]$. But $f(x) \notin \overline{f[S]}$ since N is a neighborhood of $f(x)$ not meeting $f[S]$ ⟦Sec. 1.1, Problem 5⟧. ∎

THEOREM 4.2.2. *For a function $f: X \to Y$ the following conditions are equivalent:*

(i) *f is continuous at x;*
(ii) *for every filter $\mathscr{F}$ converging to x, $f[\mathscr{F}] \to f(x)$;*
(iii) *for every net x_δ converging to x, $f(x_\delta) \to f(x)$.*

(See also Problems 6 and 20.) In this statement we are using the obvious facts that if $(x_\delta: D)$ is a net and $\mathscr{F}$ a filter, then $(f(x_\delta): D)$ is a net and $f[\mathscr{F}] = \{f[S]: S \in \mathscr{F}\}$ is a filterbase ⟦Sec. 3.2, Problem 5⟧.

Proof. (i) *implies* (ii). Let $\mathscr{F}$ be a filter converging to x, and N a neighborhood of $f(x)$. Then $f^{-1}[N]$ is a neighborhood of x, hence belongs to $\mathscr{F}$.

Since $N \supset ff^{-1}[N]$ and the latter set belongs to $f[\mathscr{F}]$, it follows that N includes a member of $f[\mathscr{F}]$.

(ii) *implies* (iii). Let x_δ be a net converging to x. Let $\mathscr{F}$ be the associated filter. Then $\mathscr{F} \to x$. ⟦Sec. 3.4, Example 5.⟧ By hypothesis, $f[\mathscr{F}] \to f(x)$. The result follows by Sec. 3.4, Example 5, when we observe that $f[\mathscr{F}]$ is a filterbase associated with $f(x_\delta)$. (That is, it generates the associated filter.) ⟦Let $B \in f[\mathscr{F}]$; say $B = f[A]$, $A \in \mathscr{F}$. Then $x_\delta \in A$ eventually, hence $f(x_\delta) \in B$ eventually. Conversely, let $B \subset Y$ and suppose that $f(x_\delta) \in B$ eventually. Then $x_\delta \in f^{-1}[B]$ eventually, hence $f^{-1}[B] \in \mathscr{F}$ and so $B \supset ff^{-1}[B]$, that is B includes a member of $f[\mathscr{F}]$.⟧

(iii) *implies* (i). Assume that f is not continuous at x. Choose $S \subset X$ with $x \in \bar{S}$, $f(x) \notin \overline{f[S]}$ ⟦Lemma 4.2.1⟧. Then Theorem 3.4.2 yields both of the following facts: there exists a net x_δ in S with $x_\delta \to x$, and $f(x_\delta) \not\to f(x)$. ▌

▲ REMARK ON NOTATION. Suppose that $f: X \to Y$, and $x \in X$. Let us write "$f(t) \to L$ as $t \to x$" or, if Y is a Hausdorff space, "$\lim_{t \to x} f(t) = L$" to mean that $f(t_\delta) \to L$ whenever t_δ is a net and $t_\delta \to x$; equivalently ⟦Problem 101⟧ that $f[\mathscr{F}] \to L$ whenever $\mathscr{F}$ is a filter and $\mathscr{F} \to x$. Then Theorem 4.2.2 takes the intuitively appealing form: f is continuous at x if and only if $f(t) \to f(x)$ as $t \to x$.

Use of Theorem 4.2.2 simplifies some tedious, though necessary computations. Suppose $f: X \to Y$ is continuous, and $S \subset X$. Then $f \mid S$ (read: f restricted to S) is that function from S to Y whose value at $s \in S$ is $f(s)$. Giving S the relative topology we prove that $f \mid S$ is continuous. Let $\mathscr{B}$ be a filterbase on S with $\mathscr{B} \to s$. Then $\mathscr{B} \to s$ in X ⟦Theorem 3.2.3⟧, hence $f[\mathscr{B}] \to f[S]$ ⟦Theorem 4.2.2⟧ and so $f \mid S$ is continuous ⟦Theorem 4.2.2⟧. This, together with Theorem 3.2.3, proves the following result.

THEOREM 4.2.3. *Let $f: X \to Y$ be continuous. Let $A \subset X$, and $f[A] \subset B \subset Y$. Then $f \mid A: A \to B$ is continuous.*

THEOREM 4.2.4. *A function $f: X \to Y$ is continuous if and only if $f[\bar{S}] \subset \overline{f[S]}$ for all $S \subset X$.*

Half of this is Lemma 4.2.1. Conversely, suppose that f is continuous and $y \in f[\bar{S}]$, say $y = f(x)$, $x \in \bar{S}$. There is a net s_δ in S with $s_\delta \to x$ ⟦Theorem 3.4.2⟧. By Theorem 4.2.2, $f(s_\delta) \to y$ and so $y \in \overline{f[S]}$. ▌

THEOREM 4.2.5. *Let T, T' be topologies for a set X. Then $T \supset T'$ if and only if the identity map $i: (X, T) \to (X, T')$ is continuous.*

The following statements are equivalent: $T \supset T'$, every T' open set is T open, $i^{-1}[G] \in T$ for all $G \in T'$, i is continuous. ▌

We now consider the possibility that various combinations of continuous functions are continuous. Under certain circumstances it is possible to consider the sum of two functions, for example, when they are real-valued and have the same domain. The sum of two continuous functions need not be continuous, and the same is true for a scalar multiple of a continuous function ⟦Problem 19⟧. However, for sufficiently well-behaved range spaces, we obtain expected results. These will be shown here for $\mathbf{R}$; very general ranges are considered in Chapter 12.

THEOREM 4.2.6. *Let f, g be continuous functions from X to $\mathbf{R}$, and let $t \in \mathbf{R}$. Then $tf + g$ and $f \cdot g$ are continuous. Moreover f/g is continuous at any point where g is not zero.*

It is clear that $tf + g$ is, by definition, the function whose value at x is $tf(x) + g(x)$. Similarly $(f \cdot g)(x) = f(x) \cdot g(x)$. The result follows from Theorem 4.2.2 and Theorem 3.5.1. ∎

EXAMPLE 2. The collection of all continuous real-valued functions on X is written $C(X)$. It is a commutative ring with identity ⟦Theorem 4.2.6⟧.

THEOREM 4.2.7. *Let $f\colon X \to Y$, $g\colon Y \to Z$ be continuous. Then $g \circ f\colon X \to Z$ is continuous.*

Let $\mathscr{F}$ be a filter in X with $\mathscr{F} \to x$. Then $(g \circ f)(\mathscr{F}) = g[f[\mathscr{F}]] \to g(f(x)) = (g \circ f)(x)$ using successively the facts that f and g are continuous. ∎

THEOREM 4.2.8. *Let $f\colon X \to \mathbf{R}$ be continuous. Then $|f|$ is continuous, where $|f|(x) = |f(x)|$.*

In Theorem 4.2.7, take $Y = Z = \mathbf{R}$, and let g be the function $x \to |x|$. ∎

THEOREM 4.2.9. *Let f, g be continuous real-valued functions. Then $f \vee g$ and $f \wedge g$ are continuous.*

For any two real numbers a, b we have

$$a \vee b = \max(a, b) = \tfrac{1}{2}(a + b + |a - b|),$$
$$a \wedge b = \tfrac{1}{2}(a + b - |a - b|).$$

Thus $f \vee g$ and $f \wedge g$ are expressed in terms of functions known, by Theorems 4.2.8 and 4.2.6, to be continuous. ∎

An important sufficient condition for continuity is uniform convergence. Let S be a set and Y a semimetric space. Let f_δ be a net of functions, each

$f_\delta: S \to Y$, and let $f: S \to Y$. We say that $f_\delta \to f$ *uniformly* if

$$\sup\{d(f_\delta x, fx): x \in S\} \to 0.$$

This means that for each $\varepsilon > 0$, there exists δ_0 such that $\delta \geq \delta_0$ implies $d(f_\delta x, fx) < \varepsilon$ for all $x \in S$.

THEOREM 4.2.10. *Let f_δ be a net of continuous functions from a topological space X to a semimetric space Y such that $f_\delta \to f$ uniformly. Then f is continuous.*

Let $t \in X$ and $\varepsilon > 0$. Choose δ such that $\sup_X d(f_\delta x, fx) < \varepsilon/3$. Choose a neighborhood N of t such that $x \in N$ implies $d(f_\delta x, f_\delta t) < \varepsilon/3$. ⟦Possible because f_δ is continuous.⟧ Then for $x \in N$ we have $d(fx, ft) \leq d(fx, f_\delta x) + d(f_\delta x, f_\delta t) + d(f_\delta t, ft) < \varepsilon$. ∎

An important method of studying a space X is by investigation of $C(X)$ (Example 2). This involves restricting our attention to those X for which genuine information is available; for example, if X is indiscrete, $C(X)$ is precisely the set of constant functions ⟦Example 1⟧, in other words $C(X) = C(Y)$ if X, Y are any indiscrete spaces; thus even the cardinality of X cannot be determined from knowledge of $C(X)$. This situation persists even for some T_3 spaces (see [Herrlich]), but, as we now see, T_4 spaces allow a rich supply of continuous real-valued functions. (Use of Theorems 4.2.6 to 4.2.10 increases the supply.) This result was given by P. Urysohn in 1924.

THEOREM 4.2.11 (URYSOHN'S LEMMA). *Let F, F_1 be disjoint closed sets in a normal space X. Then there exists a continuous real-valued function f on X with $0 \leq f(x) \leq 1$ for all $x \in X$, $f(x) = 0$ for all $x \in F$, $f(x) = 1$ for all $x \in F_1$.*

Let $G = \tilde{F}_1$. Then G is a neighborhood of F and the third assertion of the theorem is that the resulting function satisfies $f(x) = 1$ for all $x \notin G$. We introduce the following abbreviation: "S is *between* A and B" shall mean "S is an open set, $A \subset S$, and $\bar{S} \subset B$." We shall also use the result of Sec. 4.1, Problem 7 which says that a set can be found between a set and any neighborhood of its closure.

Let D be the set of all numbers in the interval $[0, 1]$ which are of the form $k/2^n$, $n = 0, 1, 2, \ldots$; $k = 0, 1, 3, 5, 7, \ldots$. For each $d \in D$ we shall define an open set $G(d)$ which includes F, such that for any two members, d, d' of D with $d < d'$, we have $\overline{G(d)} \subset G(d')$. We begin by taking $G(1) = G$. Next let $G(0)$ be between F, G. Let $G(\frac{1}{2})$ be between $G(0)$, $G(1)$; let $G(\frac{1}{4})$ be between $G(0)$, $G(\frac{1}{2})$; $G(\frac{3}{4})$ between $G(\frac{1}{2})$, $G(1)$; and, in general, $G(k/2^n)$ is chosen between $G[(k-1)/2^n]$ and $G[(k+1)/2^n]$. The latter two sets are of the form $G(d)$ with $d = r/2^{n-1}$ hence, have already been defined. (The induction is on n.) We have thus arrived at an "expanding" sequence of "concentric rings" "going around" F, running from an innermost one $G(0)$ to an outermost one $G(1) = G$ and arranged in an order like that of D. With each $x \in G$ we can now associate a number $f(x) = \inf\{d: x \in G(d)\}$; thus if $d > f(x)$, $G(d)$ contains x, while if $d < f(x)$, $G(d)$ does not contain x. We complete the definition of f by setting $f(x) = 1$ for $x \notin G$. Then surely $0 \leq f(x) \leq 1$ for all $x \in X$, and $f(x) = 0$ for $x \in F$. ⟦Since $x \in F$ implies $x \in G(0)$.⟧ Thus we have only to show that f is continuous. Let $x \in X$, let x_δ be a net with $x_\delta \to x$, and let $\varepsilon > 0$. We shall prove that

$$f(x_\delta) < f(x) + \varepsilon \quad \text{eventually.} \tag{4.2.1}$$

Since (4.2.1) is trivial if $f(x) = 1$, we may assume $f(x) < 1$. Choose $d \in D$ with $f(x) < d < f(x) + \varepsilon$. Then $x \in G(d)$, and so $x_\delta \in G(d)$ eventually. Thus $f(x_\delta) \leq d$ eventually, proving (4.2.1). We next prove

$$f(x_\delta) > f(x) - \varepsilon \quad \text{eventually.} \tag{4.2.2}$$

Since (4.2.2) is trivial if $f(x) = 0$, we may assume $f(x) > 0$. Choose d, d' in D with $f(x) - \varepsilon < d < d' < f(x)$. Then $x \notin G(d')$ and so $x \notin \overline{G(d)}$. Hence $x_\delta \notin \overline{G(d)}$ eventually. ⟦Since $\overline{G(d)}$ is a neighborhood of x.⟧ Thus $f(x_\delta) \geq d$ eventually. ⟦$a \notin \overline{G(d)}$ implies $a \notin G(d)$, which implies $f(a) \geq d$.⟧ This proves (4.2.2). Inequalities (4.2.1) and (4.2.2) yield that $|f(x_\delta) - f(x)| < \varepsilon$ eventually. ∎

If $f: X \to Y$ is continuous and one-to-one, it may happen that $f^{-1}: f[X] \to X$ is also continuous; if so, we call f a *homeomorphism*. If also, f is onto, it is called a *homeomorphism onto*. For emphasis, we shall often write homeomorphism (into) to remind the reader that a homeomorphism need not be onto.

★EXAMPLE 3. If $X \subset Y$, $i: X \to Y$, the inclusion map, is a homeomorphism (into).

★EXAMPLE 4. Let D be the discrete topology for $\mathbf{R}$; then $i: (\mathbf{R}, D) \to \mathbf{R}$ is continuous, one-to-one, onto, but not a homeomorphism. ⟦Let S be a nonopen set in $\mathbf{R}$; then $(i^{-1})^{-1}[S] = S$ is not open while S is an open set in $(\mathbf{R}, D)$. Hence i^{-1} is not continuous.⟧

★EXAMPLE 5. $\mathbf{R}$ *is homeomorphic with* $(-1, 1)$, that is, there is a homeomorphism from $\mathbf{R}$ onto $(-1, 1)$. For $x \in \mathbf{R}$, let $f(x) = x/(1 + |x|)$. Then f is continuous and, clearly, $-1 < f(x) < 1$ for all x. For $-1 < x < 1$, let $g(x) = x/(1 - |x|)$. Then g is continuous on $(-1, 1)$. Finally $f[g(x)] = x$ for $x \in (-1, 1)$, and $g[f(x)] = x$ for $x \in \mathbf{R}$, hence $f = g^{-1}$, and so f is a homeomorphism onto.

★EXAMPLE 6. A very important type of continuous map is the *retraction*. For a subspace S of X we say that S is a *retract* of X if there exists a continuous map $r: X \to S$ such that $r(s) = s$ for all $s \in S$. Such a map r is called a *retraction* of X onto S. ⟦It must be onto since for each $s \in S$, $r(s) = s$.⟧ Some properties of retractions are given in Problems 24 to 32.

Problems

In this list, X, Y are topological spaces, and $f: X \to Y$.

★1. Show that if f is continuous, it remains continuous if X is given a larger, and Y a smaller, topology.

★2. Show that f is always continuous if either X is discrete or Y is indiscrete.

★3. A constant function must be continuous.

★4. f is continuous if and only if $f^{-1}[F]$ is closed for every closed set $F \subset Y$.

★5. Let $f: X \to \mathbf{R}$ be continuous, and $a \in \mathbf{R}$. Show that $(f < a)$, $(f > a)$ are open, and that $(f = a)$, $(f \leq a)$, $(f \geq a)$ are closed.

★6. Show that Theorem 4.2.2 remains correct if filter is replaced by filterbase.

★7. Let Y be a Hausdorff space and f, g continuous functions from X to Y. Show that $\{x: f(x) = g(x)\}$ is a closed subset of X. Thus if $f(x) = g(x)$ for all x in a dense subset of X, it follows that $f = g$.

8. The assumption "Hausdorff" cannot be omitted in Problem 7 ⟦Y indiscrete, g constant⟧.

9. If S is dense in X, and f is continuous and onto, then $f[S]$ is dense in Y ⟦Theorem 4.2.4⟧.

10. f is called *sequentially continuous* at x if $f(x_n) \to f(x)$ whenever $\{x_n\}$ is a sequence converging to x. Show that if X is first countable, f is continuous if and only if it is sequentially continuous. ⟦Half is a special case of Theorem 4.2.2. Conversely, imitate the proof that (iii) implies (i) in Theorem 4.2.2 using Theorem 3.1.1.⟧

11. Give an example of a sequentially continuous noncontinuous function. ⟦Theorem 4.2.5; Sec. 3.1, Example 3.⟧

★12. Let X be a semimetric space and fix $a \in X$. Let $g(x) = d(x, a)$. Show that g is continuous ⟦Sec. 2.2, Problem 2⟧. Do the same for $d(x, A)$ where $A \subset X$ ⟦Lemma 2.2.1⟧.

★13. If $f: X \to Y$ is continuous, so is $f: X \to W$ for all W satisfying $W \supset Y$, Y having the relative topology ⟦Theorems 4.2.2 and 3.2.3⟧.

★14. The *characteristic function* of a set $S \subset X$ is defined by $u: X \to \mathbf{R}$, $u(x) = 1$ if $x \in S$, 0 if $x \notin S$. Show that the characteristic function of a closed and open set is continuous. ⟦$x_\delta \to x \Rightarrow u(x_\delta) = u(x)$ eventually; $\mathscr{F} \to x \Rightarrow u[\mathscr{F}] = \{u(x)\}$ eventually.⟧

★15. (*The Weierstrass M test.*) Let $\{f_n\}$ be a sequence of real-valued function on a set S. Suppose that there exists a sequence $\{M_n\}$ of real numbers such that $\sum M_n < \infty$ and $|f_n(x)| < M_n$ for all n, and all $x \in S$. Show that the series $\sum f_n(x)$ is uniformly convergent; this means that the sequence $\{\sum_{k=1}^{n} f_k\}$ is uniformly convergent.

★16. If $f | A$ is continuous, it does not follow that f is continuous at any point of A. ⟦f = characteristic function of $\mathbf{Q} \subset \mathbf{R}$.⟧

17. If $f | A$ is continuous, $x \in A$, and N is a neighborhood of $f(x)$, there exists a neighborhood V of x in X such that $f[V \cap A] \subset N$.

18. True or false? "The function defined in the proof of Urysohn's lemma (Theorem 4.2.11) takes on only rational values."

19. Show how to give $\mathbf{R}$ a topology such that the sum of two continuous real-valued functions need not be continuous. Do the same for a scalar multiple ⟦Sec. 3.5, Problems 103, 104, 105⟧.

20. Prove that (ii) implies (i) in Theorem 4.2.2 thus: Let N be a neighborhood of $f(x)$. Since $N_x \to x$, it follows that $f[N_x] \to f(x)$, hence $f^{-1}[N]$ is a neighborhood of x.

21. Let $f: X \to Y$ be called an *open map* if $f[G]$ is open whenever G is open. Supposing that f is one-to-one and onto, show that f is open if and only if f^{-1} is continuous.

22. A continuous open map from X onto Y carries a base for X onto a base for Y.

23. Give an example of a continuous open map from a Hausdorff space onto an indiscrete space with more than one point. ⟦Characteristic function of $\mathbf{Q}$.⟧

24. Every singleton in a space X is a retract of X, and X is a retract of itself.

25. A retract of a retract is a retract ⟦Theorem 4.2.7⟧.

26. $[0, \infty)$ is a retract of $\mathbf{R}$.

★27. A retract of a Hausdorff space must be a closed subspace. ⟦In Example 6, $S = \{x: r(x) = x\}$ is closed, by Problem 7.⟧

28. Every nonempty open subset of an infinite cofinite space X is a retract of X. (Thus "Hausdorff" cannot be replaced by "T_1" in Problem 27.)

29. A retraction need not be an open map. ⟦Let $f(x) = x$ for $0 \leq x \leq 1$, $f(2) = \frac{1}{2}$.⟧

30. Let S be a retract of X and $f: S \to Y$ a continuous map. Show that f can be *extended* to a continuous map $F: X \to Y$. ⟦Take $F = f \circ r$.⟧

31. If such an extension (Problem 30) exists for every Y, then S must be a retract of X. ⟦Take $Y = S, f =$ inclusion.⟧

★32. Let X, Y be topological spaces. Show that Y is homeomorphic with a retract of X if and only if there exist continuous maps $f: X \to Y$, $g: Y \to X$ with $f \circ g$ equal to the identity; that is g is a right inverse for f. ⟦If f, g exist, let $S = g[Y]$, $r = g \circ f$. Conversely if $h: S \to Y$, let $g = h^{-1}, f = h \circ r$.⟧

★33. Condition (iii) of Theorem 4.2.2 may be replaced by: "for every net x_δ converging to x with $x_\delta \neq x$ for all δ, $f(x_\delta) \to f(x)$." ⟦The net constructed for Theorem 4.2.2 has this property.⟧

---

101. Show that f satisfies (a) $f(t_\delta) \to L$ whenever $t_\delta \to x$, if and only if it satisfies (b) $f(\mathscr{F}) \to L$ whenever $\mathscr{F} \to x$. Here t_δ is a net, and $\mathscr{F}$ a filter, both in X.
102. The converse of Problem 9 is false, that is $f[S]$ may be dense, and S not.
103. Let $C^*(X)$ be the set of all bounded continuous real functions on X. Call X *pseudocompact* if $C(X) = C^*(X)$. Show that every finite space, and every closed interval in **R**, are pseudocompact, and that **R** is not pseudocompact.
104. The continuous image of a pseudocompact space is pseudocompact. ⟦$g \in C[Y] \Rightarrow g \circ f \in C(X) = C^*(X)$.⟧
105. X is pseudocompact if and only if every $f \in C[X]$ assumes a maximum on X. ⟦Consider $1/(f - m)$, where $m = \sup f$. Consider $-1/(1 + |f|^2)$.⟧
106. Let X have the cofinite topology. Show that every one-to-one map from X to itself is a homeomorphism (into).
107. S is called a zero-set of X if there exists $g \in C(X)$ with $S = g^\perp$. Show that for each $g \in C(X)$, $(g \leq a)$ is a zero-set ⟦use Theorem 4.2.9⟧.
108. A T_1 space with more than one point must be the union of two open proper subsets. "T_1" may not be replaced by "normal and non-indiscrete" ⟦Sec. 4.1, Problem 8⟧.
109. Let X be normal; then $\dim C(X) > 1$ (that is, X allows a nonconstant continuous real function) if and only if X is the union of two open proper subsets.
110. Give an example of a continuous g: $[0, 1] \to \mathbf{R}$ such that g is one-to-one on a dense subset of $[0, 1]$, but g is not one-to-one.
111. Let X, Y be semimetric spaces. We call f: $X \to Y$ *uniformly continuous* if for each $\varepsilon > 0$, there exists $\delta > 0$ such that $d(a, b) < \delta$ implies $d(fa, fb) < \varepsilon$. Show that a uniformly continuous function is continuous but not conversely.
112. A function f: $X \to Y$ is uniformly continuous if and only if whenever A, B are subsets of X satisfying $d(A, B) = 0$, it follows that $d(fA, fB) = 0$.
113. A semimetric d for X is called a *u-semimetric* if whenever A, B are disjoint closed subsets of X, it follows that $d(A, B) > 0$. Show that an infinite discrete topological space can be metrized with a u-metric and with a non-u-metric. ⟦Consider the discrete metric. Next let $\{x_n\}$ be a sequence, make it isometric with $\{1/n\}$ and consider $\{x_{2n}\}$, $\{x_{2n+1}\}$.⟧

114. The Euclidean topology for (0, 1] cannot be given by a u-metric. ⟦Take $A = \{1/n\}$, $B = \{1/n + \varepsilon_n\}$, where $\varepsilon_n > 0$ is very small.⟧
115. Let d be a u-semimetric for X, and let Y be a semimetric space. Show that every continuous $f: X \to Y$ is uniformly continuous ⟦Problem 112⟧. (These ideas are continued in Sec. 4.3, Problem 203; Sec. 7.1, Problem 117; Sec. 8.5, Problem 116; Sec. 9.1, Problem 114; Sec. 11.2, Problem 102; as well as in [Mrowka]; [Waterhouse].)
116. Call $f: X \to Y$ *almost open* if for every $x \in X$ and neighborhood U of x, $\overline{f[U]}$ is a neighborhood of $f(x)$. Show that the inclusion map from a dense proper subspace is almost open and may or may not be open.
117. Suppose that $f: X \to Y$ has the property that $f[N_x] \to f(x)$ for each x, where N_x is the neighborhood filter of x. Show that f is continuous.

---

201. Let g be defined on X and with values in $\mathbf{R} \cup \{+\infty\}$. We call g *lower semicontinuous* if $(g \leq t)$ is closed for each $t \in \mathbf{R}$. (The convention is that for all t $(g \leq t)$ contains no x for which $g(x) = +\infty$.) Show that g is lower semicontinuous if and only if $x_\delta \to x$ implies $\liminf g(x_\delta) \geq g(x)$. Show that the characteristic function of an open set is lower semicontinuous. (This generalizes Problem 14.)
202. Let Φ be a family of real continuous functions and $u(x) = \sup\{g(x): g \in \Phi\}$. Show that u is lower semicontinuous.
203. Suppose that Y is regular and f is not continuous. Show that for every dense subspace D of X, there exists $x \notin D$ such that $f|(D \cup \{x\})$ is not continuous. ⟦Otherwise, with N a closed neighborhood of $f(x)$, $f^{-1}[N]$ is a neighborhood of x.⟧ ("Regular" cannot be dropped, [Bourbaki(b), Vol. 1, p. 137, Example 19].)
204. Let X, Y be finite. Then f is continuous if and only if it is *isotone*, that is $x \leq y \Rightarrow f(x) \leq f(y)$. (See Sec. 4.1, Problem 105, for definitions.)
205. If there exists a sequence $\{f_n\}$ of continuous functions such that $f_n \to f$ *pointwise* (that is, $f_n(x) \to f(x)$ for each x), then f is said to be of *Baire class* 1. Let $f: X \to \mathbf{R}$. Suppose that f is of Baire class 1. Show that $f^{-1}[F]$ is a G_δ for all closed $F \subset \mathbf{R}$. (A G_δ is a set which is the intersection of a sequence of open sets.)
206. Let a be an infinite cardinal number. Let a topological space be called co-a if it has the property that a proper subset is closed if and only if it has less than a members. Show that the discrete, cofinite, and co-countable topologies are co-a topologies. Show that a space X has the property that every permutation of X is continuous if and only if X has a co-a topology.
207. Every nonempty closed subset of $\mathbf{J}$ is a retract of $\mathbf{J}$. ⟦See [Frolik(a), p. 171].⟧
208. Every function from an infinite cofinite space to $\mathbf{R}$ which is of Baire class 1 is constant. (Hence the converse to Problem 205 is false.)

4.3 Separation by Continuous Functions

We shall use the abbreviation: "sets A, B in X *can be completely separated*" to mean "there exists a continuous real-valued function f defined on X with $0 \leq f(x) \leq 1$ for all x, $f(x) = 0$ for all $x \in A$, $f(x) = 1$ for all $x \in B$." We shall say that the function f *separates* A, B. Thus, Urysohn's lemma says that in a normal space, any two disjoint closed sets can be completely separated. A topological space is called *completely regular* if every point and closed set not containing the point can be completely separated. (More precisely, two disjoint sets, one a singleton and the other closed, can be completely separated.) A completely regular T_1 space is called *a $T_{3\frac{1}{2}}$ space*, or, a *Tychonoff space*. This completes our list of nine separation axioms. Devising other separation axioms is a popular sport, and (perhaps) hundreds of them have appeared in print. They are of varying degrees of practical and historic importance; a number of these axioms will be found in the problems of this text, and a list may be found in the index under "separation axioms."

The reader will have noticed that Urysohn's lemma precludes introducing a complete separation axiom for normal spaces similar to complete regularity for regular spaces. In contrast to the situation for regular spaces, *every* normal space has the complete separation property.

Some remarks on the relative roles of complete regularity and normality are given in Section 14.7.

THEOREM 4.3.1. *Every T_4 space is a $T_{3\frac{1}{2}}$ space, and so on, according to the following scheme:*

$$T_4 \Rightarrow T_{3\frac{1}{2}} \Rightarrow T_3 \Rightarrow T_2 \Rightarrow T_1 \Rightarrow T_0.$$

None of the implications can be reversed. The Tables may be consulted for the appropriate references.

That $T_3 \Rightarrow T_2 \Rightarrow T_1 \Rightarrow T_0$ is trivial. ⟦In a T_1 space each point is closed.⟧ Next, let X be a completely regular space, $a \in X$, F a closed subspace of X, $a \notin F$. By hypothesis, there exists f separating $\{a\}$, F. Let $N_1 = \{x : f(x) < \frac{1}{2}\}$, $N_2 = \{x : f(x) > \frac{1}{2}\}$. Then N_1, N_2 are disjoint, open ⟦Sec. 4.2, Problem 5⟧ neighborhoods of a, F respectively. Thus X is regular. That a T_4 space is $T_{3\frac{1}{2}}$ is a special case of Urysohn's lemma. ∎

Completely regular spaces are distinguished in their very definition by the richness of the supply of continuous real functions defined on them; allowing the hope that topological properties can be phrased in terms of these functions, leading to a so-called *duality theory*; that is, study of a space by means of functions defined on the space. An important example of such a result is the following characterization of convergence.

THEOREM 4.3.2. *Let X be completely regular, and $\mathscr{F}$ a filter in X. Then $\mathscr{F} \to x$ if and only if $f[\mathscr{F}] \to f(x)$ for every continuous real function f on X.*

A similar result holds for nets. The statement of the Theorem remains true if "continuous real function f on X" is replaced by "continuous function $f\colon X \to [0, 1]$."

Half of the main statement follows trivially from Theorem 4.2.2. Conversely, suppose that $\mathscr{F} \nrightarrow x$. Then x has an open neighborhood N which does not belong to $\mathscr{F}$. There exists a continuous function $f\colon X \to [0, 1]$ with $f(x) = 0, f(y) = 1$ for $y \notin N$ ⟦since x and $\tilde{N}$ are completely separated⟧. Then $f[\mathscr{F}] \nrightarrow f(x)$. ⟦$f(x) = 0$, and $(-\frac{1}{2}, \frac{1}{2})$ is a neighborhood of 0 which does not include any member of $f[\mathscr{F}]$ since $S \in \mathscr{F}$ implies $S \not\subset N$ and so $1 \in f[S]$.⟧ Finally if x_δ is a net and $f(x_\delta) \to f(x)$, then $f[\mathscr{F}] \to f(x)$ where $\mathscr{F}$ is the filter associated with x_δ. If this is true for all f, then $\mathscr{F} \to x$, hence $x_\delta \to x$. ⟦Sec. 3.4, Example 5⟧. ∎

THEOREM 4.3.3. *Every semimetric space is normal and completely regular. Every metric space is a T_4 space.*

Let A, B be disjoint closed sets in a semimetric space X. For $x \in X$, set

$$f(x) = \frac{d(x, A)}{d(x, A) + d(x, B)}.$$

By Theorem 4.2.6 and Problem 12 of Section 4.2, f is continuous. ⟦The denominator never vanishes since $d(x, A)$, $d(x, B)$ are nonnegative and could be simultaneously zero only in the impossible circumstance that $x \in A \cap B$; this follows from Sec. 2.5, Example 1.⟧ It is clear that f separates A, B. Then, just as in the proof of Theorem 4.3.1, $(f < \frac{1}{2})$ and $(f > \frac{1}{2})$ separate A, B by neighborhoods. This proves that X is normal. To prove complete regularity, repeat this argument taking A to be a singleton instead of a closed set. Then f separates A, B. Hence X is completely regular.

The second half of the theorem follows from the fact that a metric space must be a T_1 space. ⟦If $x \neq y$, let $r = d(x, y)$. Then $r > 0$ and $y \notin N(x, r)$.⟧ ∎

We thus have the remarkable fact that for semimetric spaces the weak separation axiom T_0 implies T_4 ⟦Sec. 4.1, Example 3⟧. We shall see a similar implication in a wider class of spaces in Chapter 11, and another instance in Theorem 6.3.3.

The following lemma, which will be used in Section 8.4, illustrates, along with Problem 105, that intuitively appealing results hold in the presence of sufficient separation (that is, with sufficiently many separation axioms).

LEMMA 4.3.1. *Let D be a dense subspace of a Hausdorff space X, and $f\colon X \to Y$ a continuous function such that $f \mid D$ is a homeomorphism. Then $f[\tilde{D}] \pitchfork f[D]$, that is $f(\tilde{D}) \subset [f(D)]^\sim$.*

Let $S = f^{-1}[f[D]]$. Then D is a dense subset of S. But D is also a retract of S. ⟦Let $u\colon S \to D$ be $u = f^{-1} \circ f$, where $f^{-1}\colon f[D] \to D$. For $d \in D$,

$u(d) = d$.⟧ Hence D is closed in S ⟦Sec. 4.2, Problem 27⟧ and so $D = S$ since it is dense. ▮

Problems

In this list X is a topological space.

1. X is completely regular if and only if for each z and each neighborhood N of z, there exists $f \in C(X)$, satisfying $0 \leq f(x) \leq 1$ for all x, $f(z) = 1$, $f(x) = 0$ for $x \notin N$.

★2. "Completely regular" and "$T_{3\frac{1}{2}}$" are hereditary. ⟦The crucial point is that if F is a closed set in $S \subset X$ and $x \in S \setminus F$ then $F = F_1 \cap S$ where F_1 is a closed set in X *not containing* x. Thus if f separates x, F_1, $f \mid S$ separates x, F.⟧

★3. A normal space need not be completely regular ⟦Sec. 4.1, Problem 8⟧. (Of course a normal T_1 space $= T_4$ space is completely regular.) However, a regular normal space is completely regular. ⟦To separate a point x from a closed set F, first enclose x in a closed neighborhood N not meeting F. Apply Urysohn's lemma to N, F.⟧

4. Every zero set is a G_δ. ⟦$f^\perp = \bigcap \{(|f| < 1/n) : n = 1, 2, \ldots\}$.⟧

5. A point in a Tychonoff space is a zero set if and only if it is a G_δ. ⟦Problem 4. Also, if $\{x\} = \bigcap G_n$, let f_n separate x from $\tilde{G}_n$, and set $g = \sum (|f_n| \wedge 2^{-n})$; $g^\perp = \{x\}$ and $g \in C(X)$ by Theorems 4.2.8, 4.2.9, and 4.2.10.⟧

★6. A *cozero* set is a set whose complement is a zero set. Show that in a completely regular space the cozero sets form a base for the topology. ⟦For $x \in N$, let $f(x) = 0$, $f = 1$ on $\tilde{N}$. Then $x \in (f > 0) \subset N$, and $(f > 0)$ is the complement of $f^\perp$.⟧

7. A retract of a normal space is normal. ⟦For disjoint closed A, B, separate $r^{-1}[A]$, $r^{-1}[B]$ by open G, H. Consider $G \cap S$, $H \cap S$.⟧

---

101. If A, B can be completely separated, so can $\bar{A}$, $\bar{B}$.

102. The following condition is sufficient that A, B can be completely separated: there exist real numbers u, v, with $u < v$, and $f \in C(X)$ with $f(x) \leq u$ for $x \in A$, $f(x) \geq v$ for $x \in B$ ⟦Theorem 4.2.9⟧.

103. A zero-dimensional space is completely regular ⟦Sec. 4.2, Problem 14⟧.

104. Lemma 4.3.1 becomes false if "a homeomorphism" is replaced by "one-to-one." ⟦$D = \{1/n\}$, $X = D \cup \{0\}$, $Y = D \setminus \{1\}$, $f(0) = f(1) = 0$, $f(1/n) = 1/n$ for $n > 1$.⟧

105. Lemma 4.3.1 becomes false if "Hausdorff" is replaced by "T_1." ⟦$f \mid D$ a permutation, where D is the complement of a singleton in a cofinite space.⟧

106. The space of Sec. 3.1, Problem 108 is normal.

107. Every closed set in a semimetric space is a zero set. ⟦Use the continuous function f defined in Theorem 4.3.3.⟧
108. A countable completely regular space is zero-dimensional. ⟦Let N be a neighborhood of x, $f(x) = 0$, $f = 1$ outside of N. Since $f[X]$ is countable, there exists t with $f(y) \neq t$ for all y. Now consider $(f < t) = (f \leq t)$.⟧
109. Let $(G_1, G_2, \ldots, G_n)$ be an open cover of a normal space X. Show that there exist continuous real functions $f_1, f_2, \ldots, f_n$ on X with $f_i(x) \geq 0$, $\sum_{k=1}^{n} f_k(x) = 1$ for all x, and $f_i(x) = 0$ for $x \notin G_i$, $i = 1, 2, \ldots, n$. Such a set of functions is called a *partition of unity subordinate to the cover*. ⟦Choose $H_1, H_2, \ldots, H_n$ as in Sec. 4.1, Problem 122. Let $g_i: X \to [0, 1]$ with $g_i = 1$ on H_i, 0 on $\tilde{G}_i$. Let $F_i = g_i / \sum_{k=1}^{n} g_k$.⟧

---

201. Let Γ be a family of topological spaces. Say that a space Y has the Γ property if whenever $S \subset X \in \Gamma$ and S is not dense in X, there exist two unequal continuous maps from X to Y which are equal on S. Thus **R** has the $T_{3\frac{1}{2}}$ property; no discrete Y has the connected property (see Section 5.2). Find a space Y which (a) does not have the $T_{3\frac{1}{2}}$ property, (b) does not have the discrete property. In case (a), make Y have more than one point; in case (b) find all such Y.
202. Disjoint closed sets which have disjoint neighborhoods may not be completely separated, even in a $T_{3\frac{1}{2}}$ space ⟦[Gillman and Jerison, 3.13]⟧.
203. Let (X, d) be a semimetric space. Then every $f \in C(X)$ is uniformly continuous if and only if d is a u-semimetric ⟦Sec. 4.2, Problems 115, 112, and Urysohn's lemma⟧. This result is generalized in Sec. 8.5, Problem 116.

Topological Concepts

5.1 Topological Properties

If there is a homeomorphism from X onto Y we say that X, Y are *homeomorphic*. This is an equivalence relation ⟦Theorems 4.2.5 and 4.2.7⟧. As we shall see in a moment, once two topological spaces are known to be homeomorphic, the topologist considers them to be identical, for the very good reason that one of these two spaces has any given (topological) property if and only if the other does; thus there is no (topological) test which will distinguish between them. To justify this remark, let $f: X \to Y$ be a homeomorphism onto. For each $S \subset X$, S is open if and only if $f[S]$ is open. ⟦S open implies $f[S] = (f^{-1})^{-1}[S]$ is open since f^{-1} is continuous. The same argument applied to f shows that $f[S]$ open implies S open.⟧ Thus f induces a one-to-one correspondence between the topologies of X, Y, namely $G \leftrightarrow f[G]$ for G an open set in X. We could define a topological property to be a property, which a given space may or may not have, which can be defined entirely in terms of open sets and the language of set theory. For example "T_2" is a topological property; so is "discrete". ⟦A space is discrete if and only if every subset is open.⟧ With this definition it is clear that a homeomorphism onto preserves topological properties since any argument used to check that one space had a certain property would simply be transferred to the other space by means of the homeomorphism. This definition of topological property is a little difficult to make precise, partly because of the vagueness of the phrase "language of set theory." Instead we shall simply define a *topological property* to be a property which is preserved by every homeomorphism onto. For example, "having 5 points" is a topological property, since it is clear that if X, Y are homeomorphic, and X has 5 points, then also Y has 5 points. Thus a topological property is also called an *invariant*. "*Topology*" is the study of topological properties or invariants (under homeomorphism).

To give some nontopological properties we consider first a metric space. "Being of finite diameter" is not a topological property; for example **R** and (0, 1), each with the Euclidean metric, are homeomorphic ⟦Sec. 4.2, Example 5⟧. (Compare Problem 103.) A very interesting example of a nontopological property is the property "being knotted" which certain simple closed curves in $\mathbf{R}^3$ have. A circular loop is not knotted, while we can easily envisage a simple closed curve in space which is knotted in the sense that if it were made of string, it could not be manipulated into a circular loop without breaking the string. Yet these two curves are homeomorphic under the map which matches points traversing each curve in the same time. The difference between the knotted and unknotted curves lies in the way in which they are embedded in $\mathbf{R}^3$, a subject which we shall not pursue further.

Because homeomorphic spaces are (topologically) indistinguishable, topologists often refrain from distinguishing them by name. For example, one might ask: if the points 0, 1 are removed from [0, 1], what space remains? Answer: **R**. (The answer is "really" (0, 1), which is homeomorphic with **R** ⟦Sec. 4.2, Example 5⟧.) As another example, one says: if a point is removed from **R**, the remaining set is the union of two disjoint copies of **R** ⟦Problem 1⟧.

A central problem of topology, called the *classification problem*, is to decide whether or not two spaces are homeomorphic. For example, a space, S, might arise in some natural way, and it might be suspected that S is homeomorphic with $\mathbf{R}^2$. The attempt to decide whether or not this is true, and if it is not, to find some other standard space homeomorphic with S, is called the classification of S. It is embarrassing to confess that, at this stage, we have not even "classified" $\mathbf{R}^n$, in the sense that we do not know for example that **R** and $\mathbf{R}^2$ are different (that is, not homeomorphic). This will be shown in Sec. 5.2, Problem 8. For purposes of classification it is important to develop methods of distinguishing between different spaces. What is done is to look for an invariant which will distinguish the spaces. For example, **R** is not homeomorphic with any cofinite space since **R** is an infinite Hausdorff space, while an infinite cofinite space must be non-Hausdorff. In succeeding portions of this book we shall introduce many such invariants, starting with connectedness in the next section. These invariants will play a dual role. As invariants, they help us distinguish between different spaces; as topological properties, they serve a more affirmative purpose, namely helping in the study and understanding of objects and spaces arising in classical analysis and geometry.

Remarks very similar to the preceding may be made about semimetric spaces and isometries.

Finally we turn to properties of subspaces. A subset S of a topological space X is said to have a certain property, P, say, if, considering S as a topological space (with the relative topology) S has property P. For example, we shall be speaking in the next section of connected topological spaces. The definition just given will then allow us to speak of connected subsets of topological spaces. The reader should realize the difference between this idea and the idea of an open subset of a topological space. For example $(0, 1)$ is an open subset of $\mathbf{R}$, and is homeomorphic with the open interval $(0 < x < 1, y = 0)$ in $\mathbf{R}^2$, which is not open. In contrast it is clear that if a P subset (P is any topological property) of a space is homeomorphic with a subset of some other space, then the second subset also has property P (see also Problem 5).

Problems

★1. Any two open intervals, (a, b), in $\mathbf{R}$ are homeomorphic with each other, hence ⟦Sec. 4.2, Example 5⟧ with $\mathbf{R}$. (Here $a \in \mathbf{R}$ or $a = -\infty$, $b \in \mathbf{R}$ or $b = +\infty$.)

2. An isometry between semimetric spaces is a homeomorphism.

3. $\mathbf{Q} \cap (0, 1)$ is homeomorphic with $\mathbf{Q}$, each having the relative topology of $\mathbf{R}$ ⟦Sec. 4.2, Example 5⟧.

★4. All convergent sequences in Hausdorff spaces are homeomorphic. More precisely, let X, Y be T_2 spaces. Let $A = (x_0, x_1, x_2, \ldots) \subset X$ with $x_i \neq x_j$ for $i \neq j$, $x_n \to x_0$. Let $B = (y_0, y_1, y_2, \ldots)$ be a sequence in Y satisfying similar conditions. Define $f: A \to B$ by $y_n = f(x_n)$ for all n. Show that f is a homeomorphism.

5. Like "open," "dense" is a property of the way a set is embedded in a space, and is not a topological property. Illustrate this by showing a dense subset D of a space X such that D is homeomorphic with a nondense subset of a space Y.

---

101. $\mathbf{R}$ is not homeomorphic with $(0, 1]$. ⟦Any continuous one-to-one map on $\mathbf{R}$ would have to be monotone.⟧

102. Semimetrizability is topological. ⟦If (X, d) is homeomorphic with Y, set $d(y_1, y_2) = d(x_1, x_2)$ where $y_i = f(x_i)$.⟧

103. Call a space P if its topology can be given by a semimetric in which the space has finite diameter. Show that P is a topological property.

104. $\mathbf{R}^2$ is homeomorphic with its subspace $\{(x, y): x^2 + y^2 < 1\}$.

105. (a) $\mathbf{Q}$ is homeomorphic into $\mathbf{J}$. (b) Are $\mathbf{Q}$, $\mathbf{J}$ homeomorphic?

106. Homeomorphism of topological spaces does not enjoy the crisscross property ⟦Sec. 2.2, Problem 104⟧.

107. In Problem 4, T_2 cannot be replaced by T_1. ⟦Let one space be cofinite.⟧

---

201. $\mathbf{R}^2$ is not homeomorphic with its subspace $\{(x, y): x^2 + y^2 < 1\} \cup \{(1, 0)\}$.
202. Any two convex open (nonempty) subsets of $\mathbf{R}^n$ are homeomorphic.
203. Find a space, which is neither discrete nor indiscrete, such that any two homeomorphic subsets must have their closures also homeomorphic.
204. Does $\mathbf{R}$ have a closed subset which is homeomorphic with $\mathbf{Q}$?
205. Call X *homogeneous* if for all $x, y \in X$, there is a homeomorphism f of X onto itself with $y = f(x)$. Show that $(0, 1)$ is and $(0, 1]$ is not homogeneous. Give an example of a homogeneous space X which has two points x, y such that no homeomorphism f of X onto itself has $y = f(x)$, $x = f(y)$.
206. Give an example of $X \subset \mathbf{R}$, $Y \subset \mathbf{R}$, with X, Y not homeomorphic, such that there exists a continuous one-to-one map from each of X, Y onto the other.
207. Call X *reversible* if every continuous one-to-one map from X onto itself must be a homeomorphism. Show that (X, T) is reversible if and only if it has no strictly larger topology T' such that (X, T), (X, T') are homeomorphic. Show that the following space is not reversible: let $\mathbf{R}$ have the Euclidean topology on $(x > 0)$ and the discrete topology on $(x \le 0)$. ⟦Consider a translation.⟧ Show that $\mathbf{R}$ is reversible. ⟦Continuous one-to-one functions must be monotone.⟧ For more on reversibility see [Rajagopalan and Wilansky].
208. Let X be an infinite cofinite space, Y a T_1 space, and $f: X \to Y$ a continuous map. Then either f is constant or X and Y are homeomorphic. ⟦Say f is onto and not constant. For any infinite closed $F \subset Y$, $f^{-1}[F] = X$, so $F = Y$. Thus Y is cofinite. Now $X = \bigcup \{f^{-1}[\{y\}]\}$, a union of finite sets, thus $|X| \le \aleph_0 |Y|$ so $|X| = |Y|$.⟧ (This result is due to V. V. Proizvolov.)
209. The only continuous real functions on an infinite cofinite space are the constants ⟦Problem 208⟧.

5.2 Connectedness

A topological space is said to be *disconnected* if it contains a proper subset which is both open and closed. The reason for the name is that if A is proper and both open and closed, then $\tilde{A}$ is also proper, open, closed, and so the space has "fallen into two pieces." A space which is not disconnected is called *connected.* The empty set is connected.

★EXAMPLE 1. Let $X = (0, 1) \cup (2, 3)$ with the Euclidean topology.

Then $(0, 1)$ is open and closed $[\![(0, 1) = [0, 1] \cap X]\!]$. Hence X is disconnected.

★EXAMPLE 2. **R** *is connected.* Let F be a closed proper subset of **R**. We shall show that F is not open. Let $x \notin F, y \in F$. $[\![\varnothing \neq F \neq \mathbf{R}.]\!]$ Suppose that $y < x$. (Without loss of generality; a similar argument, using inf instead of sup, deals with $y > x$.) Let $z = \sup\{t \in F: t < x\}$. Then $y \leq z \leq x$. Now $z \in F$. [[F is closed and every neighborhood of z meets F, by definition of z.]] Thus $z < x$, and since $(z, x) \not\!\!\pitchfork F$ it follows that z is not an interior point of F and so F is not open.

A subset of a topological space is called *connected* if it is connected as a topological space (with the relative topology). See the last part of Section 5.1. The phrase "removal of x *disconnects* X" means "$X \setminus \{x\}$ is disconnected."

★EXAMPLE 3. *A subset of* **R** *is connected if and only if it is an interval.* If S is not an interval, there exists $a \in \mathbf{R}$ such that $a \notin S$, and $A = (a, \infty) \cap S$ is a proper subset of S. Since also $A = [a, \infty) \cap S$, A is open and closed in S. The proof that each interval is connected is identical with the proof in Example 2. (Just add the remark that z lies in the interval since $y \leq z \leq x$.)

The following criterion for connectness is usually the easiest to apply. It is customary to denote by **2** any discrete space with exactly two points. (All such spaces are homeomorphic.)

LEMMA 5.2.1. *A space X is connected if and only if every continuous $f: X \to$* **2** *is constant.*

Suppose first that X is connected. For each $y \in f[X], f^{-1}[\{y\}]$ is closed, open and not empty. Hence it is all of X. Conversely, if X is not connected, let S be an open and closed proper subset of X. The characteristic function of S is a nonconstant continuous map of X into $\{0, 1\}$ with the discrete topology. [[Sec. 4.2, Problems 14, 13.]] ∎

The union of two connected sets need not be connected [[Example 1]], but unions with some extra properties are connected; Problem 106 and the following theorem are examples of this.

THEOREM 5.2.1. *Let C be a family of connected sets which has either one of the two following properties:*

(i) *Every pair of members of C has nonempty intersection.*
(ii) *C contains a set which meets every other member of C.*

Then $U = \bigcup \{A: A \in C\}$ is connected.

Let $f: U \to$ **2** be continuous. Then $f \mid A$ is continuous, hence constant, for each $A \in C$. This constant is obviously the same for all $A \in C$, since the contrary assumption leads immediately to two members of C which intersect and on which f has different values. The result follows from Lemma 5.2.1. ∎

EXAMPLE 4. $\mathbf{R}^n$ *is connected.* For it is the union of a family of lines passing through the origin. Each line is homeomorphic with $\mathbf{R}$, hence connected, and the result follows from Theorem 5.2.1.

By a *continuous image* of a topological space X is meant a topological space Y such that there exists a continuous map of X onto Y.

THEOREM 5.2.2. *A continuous image of a connected set is connected.*

(Note that in the preceding three lines we have used the words "set" and "space." A connected set is a connected topological space (with the relative topology) and, to remove all ambiguity, we note that the continuous map mentioned in the statement of Theorem 5.2.2 needs to be defined only on the set.) Let X be connected and $f: X \to Y$ a continuous function onto.
Let $g: Y \to \mathbf{2}$ be continuous. Then $g \circ f$ is continuous, hence constant. Thus g is constant. ⟦If $y_1, y_2 \in Y$, $y_1 = f(x_1)$, $y_2 = f(x_2)$, then $g(y_1) = g(fx_1) = g(fx_2) = g(y_2)$.⟧ By Lemma 5.2.1, Y is connected. ∎

COROLLARY 5.2.1. *A connected space remains connected if its topology is weakened (replaced by a smaller one).*

EXAMPLE 5. (*The intermediate-value theorem.*) *Let f be a continuous real-valued function on the interval $[a, b]$ such that $f(a) > 0, f(b) < 0$. Then there exists $x \in [a, b]$ such that $f(x) = 0$.* This follows immediately from the fact that $f[[a, b]]$ is connected ⟦Theorem 5.2.2 and Example 3⟧, hence is an interval ⟦Example 3⟧ and, by hypothesis contains both a positive and a negative number, hence contains 0. An immediate generalization is: *Let f be a continuous real-valued function on $[a, b]$ and assume that y is a number lying between $f(a), f(b)$. Then there exists $x \in [a, b]$ such that $f(x) = y$.* ⟦Apply the earlier result to $y - f$ or $f - y$.⟧ Theorem 5.2.2 is often cited as a generalization of the intermediate-value theorem; this is not quite fair, as the most difficult part of the proof of the latter theorem is not Theorem 5.2.2, but rather Example 3.

The next result is one of a very large collection, each of which asserts that some property is preserved under the operation of taking closures. The crucial application of such a fact is that any set which is maximal with respect to having this property must be closed since its closure has the property also. Here the application is Theorem 5.2.4.

THEOREM 5.2.3. *The closure of a connected set is connected.*

Let A be a connected subset of the space X, and let $f: \bar{A} \to \mathbf{2}$ be continuous.

Then $f[\bar{A}] \subset \overline{f[A]}$, ⟦Theorem 4.2.4⟧, and $f[A]$ contains only one point ⟦Lemma 5.2.1⟧. Thus $f[\bar{A}]$ contains only one point. ⟦Singletons in **2** are closed.⟧ The result follows from Lemma 5.2.1. ∎

A topological space may be subdivided into connected subsets, and when these are as large as possible they are called components. This decomposition is achieved in the following way. For $x \in X$, let

$$C_x = \bigcup \{S \subset X: x \in S \text{ and } S \text{ is connected}\}.$$

Then C_x is called the *component of x in X*, or, sometimes, the component of X containing x. Each component is connected ⟦Theorem 5.2.1⟧, includes every connected set which contains x, and, for $x, y \in X$, either $C_x \not\pitchfork C_y$ or $C_x = C_y$. ⟦If C_x meets C_y, $C_x \cup C_y$ is connected, by Theorem 5.2.1, hence is included in C_x.⟧ Since C_x includes every connected set which contains x, it follows that C_x is a maximal connected set containing x.

THEOREM 5.2.4. *Every component is closed.*

For its closure is connected ⟦Theorem 5.2.3⟧ and thus is included in it. ∎
Components need not be open however.

EXAMPLE 6. Let $X = \{0, 1, \frac{1}{2}, \frac{1}{3}, \ldots\}$ with the relative topology of **R**. Then $\{0\}$ is not open. ⟦Every neighborhood of 0 contains $1/n$ for sufficiently large n, hence is not equal to $\{0\}$.⟧ However, $\{0\}$ is a component of X. If C_0 contained some point $x = 1/n$, it would not be connected for the following reason: $\{x\} = (x - \varepsilon, x + \varepsilon) \cap X = [x - \varepsilon, x + \varepsilon] \cap X$ for some (sufficiently small) $\varepsilon > 0$. Thus $\{x\}$ is open and closed in X, hence in C_0.⟧

We now introduce the important concept of local property. We begin with the definition of the phrase "arbitrarily small." Let P be some property (such as connectedness) which certain subsets of a topological space X may have. We say that *arbitrarily small sets* containing a point x have property P if for every neighborhood N of x, there is a set S with property P such that $x \in S \subset N$. Then we say that X has property P *locally*, if for every $x \in X$ there are arbitrarily small neighborhoods of x with property P. Thus X is *locally connected* if each point has arbitrarily small connected neighborhoods; in other words, each neighborhood of a point includes a connected neighborhood of that point. Notice that a discrete space is locally connected. On the other hand, a connected space need not be locally connected ⟦Problem 109⟧.

REMARK. Since any discrete space is locally connected, it is trivial that continuous one-to-one maps do not preserve local connectedness.

Locally connected spaces do not allow the pathology of Example 6. The next result shows this, and the reader may note that it holds under the weaker hypothesis that each point has a connected neighborhood.

THEOREM 5.2.5. *Every component of a locally connected space is open and closed.*

Let C be a component and $x \in C$. Let N be a connected neighborhood of x. Then $C \supset N$. ⟦C is the component of x, hence includes every connected set which contains x.⟧ Thus C is a neighborhood of x, and so C is open ⟦Theorem 2.5.1⟧. C is closed by Theorem 5.2.4. ▮

This result is given wider applicability by means of the observation that *every open subset G of a locally connected space X is locally connected.* For let N be a neighborhood of x in G. Then N includes a connected X-neighborhood N_1 of x. ⟦Since G is open, N is a neighborhood of x in X.⟧ Then N_1 is also a G-neighborhood of x. As an application of this observation we obtain the amusing result that *in a locally connected space each neighborhood N of a point x includes a connected open neighborhood of x.* ⟦For N includes an open neighborhood G of x, and the component of G containing x is the required open neighborhood.⟧

▲ EXAMPLE 7. *Countable spaces.* Can a countable space (with more than one point) be connected? Such a countable connected space X would have to be pathological in several respects. It could not be $T_{3\frac{1}{2}}$ ⟦Problem 111, and Sec. 4.3, Problem 108⟧. It could not even be T_3 since, as we shall see a countable T_3 space must be T_4 ⟦Sec. 5.3, Example 4⟧. There are, however, (infinite) countable connected T_2 spaces ⟦Example 8⟧. (A finite connected space with more than one point could not even be T_1 ⟦Sec. 4.1, Problem 3⟧.) Another pathological fact about a countable connected space X is that the only continuous real-valued functions on X are the constants; that is, $C[X]$ is one-dimensional. ⟦$f[X]$ is a connected subset of $\mathbf{R}$ by Theorem 5.2.2, and is countable; hence by Example 3, $f[X]$ contains only one point.⟧ A further remark is that we may prove that a connected Tychonoff space has at least c points if it has more than one without using the continuum hypothesis. ⟦This follows from Theorem 5.2.2, Example 3, and the fact that there is a non-constant continuous real-valued function.⟧

▲ EXAMPLE 8. *A countably infinite connected T_2 space.* Many examples are known, the first was given by P. Urysohn in 1925. For some recent developments see [Roy]. We give an example due to S. W. Golomb and to M. Brown.

Let X be the positive integers. For a, b relatively prime positive integers, let $G(a, b) = \{a, a + b, a + 2b, \ldots\} = \{a + (n - 1)b\}$. Let T be the topology which has the set of all $G(a, b)$ as base. ⟦Theorem 2.6.2; for example, $7 \in G(1, 3) \cap G(2, 5)$, and $7 \in G(7, 15) \subset G(1, 3) \cap G(2, 5)$.⟧ T is a T_2 topology. ⟦Let $x \neq y$. Then $G(x, xy + 1)$, $G(y, xy + 1)$ are disjoint neighborhoods of x, y, since if $x + m(xy + 1) = y + n(xy + 1)$ and $y > x$, we would have $y - x = (m - n)(xy + 1) \geq xy + 1$.⟧ Now we observe that if b, d are relatively prime, $G(a, b)$ and $G(c, d)$ must meet. ⟦Find α, β with $\alpha b + \beta d = 1$. Then for arbitrary k, $[(c - a)\alpha + kd]b + [(c - a)\beta - kb]d = c - a$. Sufficiently large choice of k yields $mb - nd = c - a$ with $m > 0$, $n > 0$; that is, $a + mb = c + nd$.⟧ Finally, T is connected. ⟦Let A, B be disjoint, nonempty, open sets; A contains some $G(a, b)$; B contains some $G(c, d)$. Now if $bd \in A$ we get a contradiction thus: $bd \in G(h, k) \subset A$, thus $bd = h + nk$ and so d, k are relatively prime since otherwise h, k would not be. As just proved, this implies that $G(c, d)$ meets $G(h, k)$, hence B meets A. The assumption that $bd \in B$ leads in the same way to the same contradiction. Thus $bd \notin A \cup B$ so that $A \cup B \neq X$.⟧

▲ EXAMPLE 9. Let X be a closed interval in $\mathbf{R}$ and let G be a dense open subset. Then $F = \tilde{G}$ is *totally disconnected*, that is, none of its components has more than one point. ⟦Let $x \neq y$, $x, y \in F$. Say $y > x$. The interval (x, y) meets G; say $a \in (x, y) \cap G$. Then $(-\infty, a) \cap F = (-\infty, a] \cap F$ is open and closed in F, and contains x, but not y. Thus x, y do not belong to the same component of F.⟧ This shows that a totally disconnected nondense set may be quite large; it may be uncountable, indeed it may have Lebesgue measure arbitrarily near 1. ⟦Since the measure of $\mathbf{Q}$ is zero, we can find an open set containing it and of arbitrarily small measure.⟧

EXAMPLE 10. Like all other topological invariants, connectedness can be used to establish nonhomeomorphism. For example, $[0, 1]$ is not homeomorphic with $[0, 1] \cup [2, 3]$, and, a slightly more subtle example, $(0, 1) \cup (2, 3)$ is not homeomorphic with $(0, 1) \cup [2, 3)$ since the latter space possesses a point, 2, whose removal leaves a space with two components; the former space has no such point.

Problems

In this list, X is a topological space.

1. Suppose that $X = A \cup B$, with A, B nonempty. Then if $\bar{A} \not\pitchfork \bar{B}$, X is disconnected, but this does not follow if we know only that $\bar{A} \not\pitchfork B$.
2. $\mathbf{Q}$ is not connected. ⟦$(-\infty, \sqrt{2}) \cap \mathbf{Q} = (-\infty, \sqrt{2}] \cap \mathbf{Q}$.⟧
3. A space is locally connected if and only if the set of connected neighborhoods of each point is a local base at that point.
4. ★ No two of these three spaces are homeomorphic: $[0, 1)$, $[0, 1]$, the 1-sphere. ⟦Removal of two points may fail to disconnect $[0, 1]$. Addition of one point may fail to disconnect $[0, 1)$.⟧
5. Call sets A, B *separated* if $A \not\pitchfork \bar{B}$ and $\bar{A} \not\pitchfork B$. Let $A \cup B \subset S \subset X$. Show that A, B are separated in S if and only if they are separated in X.

Show also that A, B are separated if and only if they are disjoint and A is closed and open in $A \cup B$.

6. X is disconnected if and only if it is the union of two nonempty separated subsets.
7. Give an example of two connected sets in $\mathbf{R}^2$ whose intersection is not connected. Can this be done in $\mathbf{R}$?
8. $\mathbf{R}$ is not homeomorphic with $\mathbf{R}^2$. ⟦A point may be removed from $\mathbf{R}^2$ without disconnecting it.⟧
9. A connected T_1 space with more than one point is self-dense. "T_1" may not be omitted ⟦Sec. 4.1, Problem 8⟧.

---

101. Are connected, disconnected, and locally connected hereditary properties?
102. $A \subset B \subset \bar{A}$, A connected, implies B connected. ⟦Apply Theorem 5.2.3 with $X = B$.⟧ Hence every dense subset of a disconnected space is disconnected.
103. If a space has finitely many components, each component is open. If it has countably many components, each component is a G_δ ⟦Theorem 5.2.4⟧.
104. In the space of Example 8, the set of primes is a dense set with empty interior. ⟦Take as known Dirichlet's famous theorem: Every arithmetic progression contains a prime [Rademacher, Chapter 14].⟧
105. The 2-sphere is locally $\mathbf{R}^2$.
106. The union of a family of connected sets, no two of which are separated, is connected.
107. For $x, y \in X$, let $x \sim y$ mean that there is some connected set containing both x, y. Show that $\sim$ is an equivalence relation and that the equivalence classes are the components of X.
108. The following subsets of $\mathbf{R}^2$ are connected:

   (a) $\{(x, y): x = 0, -1 < y < 1\} \cup \{(x, y): y = \sin(1/x)\}$.
   (b) $\{(0, 0)\} \cup \{(x, y): y = 1\} \cup \{(x, y): x = 1/n, n = 1, 2, \ldots\}$.

   (These are examples of connected but not *arcwise connected* spaces; each space has a point which cannot be joined to $(0, 0)$ by an arc.)
109. The spaces of Problem 108 are not locally connected (at $(0, 0)$).
110. The space of Example 6 is the union of n disjoint nonempty open sets for any positive integer n, but not for $n = \infty$. The space has infinitely many components however.
111. A zero-dimensional T_0 space and an extremally disconnected T_2 space must be totally disconnected. (The indiscrete and cofinite topologies show that T_0 cannot be dropped, or T_2 replaced by T_1.)
112. $C(X)$ has more than two *idempotents* ($f(x) \cdot f(x) = f(x)$ for all x) if

and only if X is disconnected. If X is $T_{3\frac{1}{2}}$, $C(X)$ is an integral domain if and only if X has exactly one point.

113. The space of Example 8 is *second countable*. (That is, it has a countable base.)
114. $\mathbf{R}^2$ is not homeomorphic with $\mathbf{R}^3$. ⟦$\mathbf{R}^2$ is disconnected by removal of a subset which is homeomorphic with $\mathbf{R}$.⟧
115. The converse of Theorem 5.2.5 is false; that is the components of a space which is not locally connected may be all open and closed. ⟦Consider a connected space. See Problem 109.⟧

---

201. Every convex set in $\mathbf{R}^2$ is connected.
202. There exists a connected T_2 space of every infinite cardinality ⟦see [Anderson]⟧.
203. Show a shrinking sequence of connected sets in $\mathbf{R}^2$ whose intersection is not connected.
204. The complement of any countable subset of $\mathbf{R}^2$ is connected.
205. Let $c(n)$ (or $c_0(n)$) be the number of connected (or connected T_0) topologies which can be placed on a set with n members. Show that $c(1) = c_0(1) = 1$, $c_0(2) = 2$, $c(2) = 3$. (*Note:* $c_0(3) = 12$, $c(3) = 19$, $c_0(4) = 146$, $c(4) = 233$. See [Rankin].) Show that $c(n)$ is odd and $c_0(n)$ is even if $n \geq 2$.
206. Find two disjoint connected subsets of the unit disc in $\mathbf{R}^2$, $\{(x, y): x^2 + y^2 \leq 1\}$, one of which contains both $(\pm 1, 0)$, the other of which contains both $(0, \pm 1)$.
207. Every metric space is a (metric) subspace of a connected metric space ⟦[Sierpinski (a), p. 121]⟧.

5.3 Separability

A topological space is called *separable* if it has a dense countable subset. Thus $\mathbf{R}$ is separable since $\mathbf{Q}$ is dense. An uncountable discrete space is not separable since it has no dense proper subset. There is a topological property similar to first countability which, for semimetric spaces is equivalent to separability. A topological space is called *second countable* if it has a countable base. Thus $\mathbf{R}$ is second countable since the set of all intervals (a, b) with a, b both rational is a base for its topology. An uncountable discrete space is not second countable since it has an uncountable collection of disjoint open sets ⟦the singletons⟧. A second countable space is first countable, but not conversely ⟦discrete space⟧. A second countable space is separable ⟦Sec. 2.5, Problem 11; choose one point in each member of a base⟧, but the converse is false ⟦Example 1 or Problem 3⟧.

THEOREM 5.3.1. *A semimetric space is separable if and only if it is second countable.*

Any second countable space is separable. Conversely, let X be separable, and $\{x_n\}$ a dense sequence. The collection of cells $\{N(x_n, r): n \in \omega, r \in \mathbf{Q}\}$ is a countable base. ⟦Let $x \in G$, G open. Then $N(x, \varepsilon) \subset G$ for some $\varepsilon > 0$. For some n, $x_n \in N(x, \varepsilon/3)$. Let $r \in \mathbf{Q}$ with $\varepsilon/3 < r < \varepsilon/2$. Then $x \in N(x_n, r) \subset G$ because $d(x, x_n) < \varepsilon/3 < r$ and if $y \in N(x_n, r)$ it follows that $d(y, x) \leq d(y, x_n) + d(x_n, x) < r + \varepsilon/3 < \varepsilon$.⟧ ▮

▲ EXAMPLE 1. *The right half open interval topology is first countable, separable, not second countable, and not metrizable.* (See also Sec. 10.1, Problem 102.) For each x, the sequence $\{[x, x + 1/n)\}$ is a local base at x, so the topology is first countable. It is separable, since every nonempty open set includes an interval $[a, b)$ which contains a rational, hence $\mathbf{Q}$ is dense. Nonmetrizability will follow from Theorem 5.3.1 when it is shown that the topology is not second countable. To this end, let $\mathscr{B}$ be a base. For each x, $[x, x + 1)$ is a neighborhood of x, hence there exists $G_x \in \mathscr{B}$ with $x \in G_x$ and $[x, x + 1) \supset G_x$. Now $G_x \neq G_y$ if $x \neq y$ since their smallest members are x, y respectively; hence we have named an uncountable subset of $\mathscr{B}$, one for each $x \in \mathbf{R}$.

We now encounter the first example of one of the central techniques of topology, that of reducing and refining covers, an activity carried on, as we shall see, in discussions and applications of compactness, paracompactness, and separability, for example.

Theorem 5.3.2 was (essentially) given by E. Lindelöf in 1903.

THEOREM 5.3.2. *Let X be a second countable topological space, and C an open cover of X. Then C has a countable subcover.*

Let $\mathscr{B}$ be a countable base. For each $x \in X$ choose $A \in C$ with $x \in A$ and $S \in \mathscr{B}$ with $x \in S \subset A$. Let $\mathscr{B}_1$ be the collection of all S chosen in this way; $\mathscr{B}_1 \subset \mathscr{B}$, hence $\mathscr{B}_1$ is countable. For each $S \in \mathscr{B}_1$ choose one $A \in C$ with $A \supset S$. The collection of all A chosen in this way is countable, is a subset of C, and covers X since each x belongs to some S which is a subset of some $A \in C$. ▮

Being able to reduce every open cover of a space to a countable subcover is a useful technique, as we shall see. (For example, Theorem 5.3.5.) Hence spaces allowing this reduction are singled out for attention. A topological space X such that every open cover of X has a countable subcover is called a *Lindelöf space*. Theorem 5.3.2 says that *every second countable space is a Lindelöf space*; the converse is not true ⟦Example 4; Problem 1⟧. A slight extension of Theorem 5.3.2 says that every second countable space is *hereditarily Lindelöf*; that is, has the property that every subspace is a Lindelöf space ⟦Problem 1⟧. Of course we have assumed the obligation of showing that there are Lindelöf spaces that are not hereditarily Lindelöf. This will be pointed out in Section 8.1.

However, the Lindelöf property is F-hereditary.

THEOREM 5.3.3. *A closed subspace F of a Lindelöf space X is also a Lindelöf space.*

Let C be an open cover of F. Each member of C has the form $G \cap F$, where G is an open subset of X. Let C_1 be the set of all such G, together with the open set $\tilde{F}$. Then C_1 is an open cover of X. Reduce C_1 to a countable subcover of X, and, discarding $\tilde{F}$, if it is still there, we see that C has been reduced to a countable cover of F. ∎

THEOREM 5.3.4. *Let X be a semimetric space. Then X is a Lindelöf space if and only if it is second countable, and if and only if it is separable.*

Suppose that X is a Lindelöf space. For each $n = 1, 2, \ldots$, there is a countable set S_n of cells of radius $1/n$ which covers X. ⟦X is covered by the set of all cells of radius $1/n$; this cover may be reduced to a countable cover, by definition of Lindelöf space.⟧ Let C_n be the set of centers of the cells of S_n, and let $C = \bigcup C_n$. Then C is countable, and dense. ⟦For any open set G, $G \supset N(x, \varepsilon)$ for some $x \in X$, $\varepsilon > 0$. Let $n > 1/\varepsilon$. By definition of S_n, there exists $c \in C_n$ such that $x \in N(c, 1/n)$. Then $d(x, c) < 1/n < \varepsilon$ so that $c \in N(x, \varepsilon) \subset G$.⟧ This proves that X is separable, and the rest is contained in Theorems 5.3.1 and 5.3.2. ∎

Theorem 5.3.5 is an interesting and very important sufficient condition for normality. We begin with a computational lemma.

LEMMA 5.3.1. *Let A, B be sets in a topological space X. Suppose that A has a cover consisting of a sequence $\{V_n\}$ of open sets with $\overline{V}_n \not\pitchfork B$ for each n, and that B has a cover consisting of a sequence $\{W_n\}$ of open sets with $\overline{W}_n \not\pitchfork A$ for each n. Then A, B are separated by open sets.*

Case I. Assume that $\{V_n\}$, $\{W_n\}$ are expanding sequences; that is, $V_n \subset V_{n+1}$ and $W_n \subset W_{n+1}$ for all n. Let

$$V_n' = V_n \setminus \overline{W}_n, \qquad W_n' = W_n \setminus \overline{V}_n, \qquad G = \bigcup V_n', \qquad H = \bigcup W_n'.$$

Then G, H are open ⟦each $V_n' = V_n \cap \tilde{\overline{W}}_n$ is open⟧ and $A \subset G$. ⟦Let $x \in A$. Then $x \in V_n$ for some n. Hence $x \in V_n'$ since $\overline{W}_n \not\pitchfork A$.⟧ Similarly $B \subset H$, and it remains to show that G, H are disjoint. ⟦If $x \in G \cap H$, we would have $x \in V_n' \cap W_k'$ for some n, k. Suppose $n \geq k$. Now $x \notin W_n$ since $x \in V_n'$, hence $x \notin W_k$ since $W_k \subset W_n$. This contradicts $x \in W_k'$. A similar contradiction results if $n < k$.⟧

Case II. If $\{V_n\}$ is an open cover of A with $\bar{V}_n \not\cap B$ for each n, let $A_n = V_1 \cup V_2 \ldots \cup V_n$. Then $\{A_n\}$ is an expanding sequence of open sets covering A and with $\bar{A}_n = \bigcup \bar{V}_k$ not meeting B. Similarly the cover of B is replaced by an expanding sequence, and Case I applies. ▮

THEOREM 5.3.5. *A regular Lindelöf space is normal and completely regular.* It is sufficient to prove normality ⟦Sec. 4.3, Problem 3⟧. Let A, B be disjoint closed sets. For each $x \in A$ choose an open set V with $x \in V$ and $\bar{V} \not\cap B$. The set of all such V, one for each $x \in A$, is an open cover of A, which, by Theorem 5.3.3, may be reduced to a countable one. Repeating this process with B leads to the situation covered by Lemma 5.3.1. Hence A, B are separated by open sets. ▮

▲EXAMPLE 2. *With the right half open interval topology,* **R** *is a* T_4 *Lindelöf space.*

This topology is T_3 ⟦Sec. 2.6, Example 3, and Theorem 4.1.3⟧, so normality will follow from Theorem 5.3.5 when the topology is shown to be Lindelöf. To this end, let C be an open cover. Let $I = \{x$: there exist a, b with $a < x < b$, $[a, b) \subset G$ for some $G \in C\}$; thus I is the set of points Euclidean-interior to some member of C. Let $E = \mathbf{R} \setminus I$; the letter E is chosen since every member x of E can occur only as an end-point (on the left) of an interval $[a, b)$ which is a subset of some $G \in C$. The set E is countable. ⟦For each $x \in E$, choose one $b_x > x$ such that $[x, b_x)$ is a subset of some $G \in C$. For $x \neq y$, $x \in E$, $y \in E$, $[x, b_x)$ and $[y, b_y)$ must be disjoint by definition of E. Thus E can be matched with a disjoint family of intervals each of which contains a rational number. Since **Q** is countable, E must be.⟧ It remains to prove that C can be reduced to a countable subcover of I ⟦since I is all but a countable subset of **R**⟧. But $\{(a, b): (a, b) \subset G$ for some $G \in C\}$ is a cover of I, moreover, it is a Euclidean open cover! Since I is second countable in the Euclidean topology ⟦Problem 1⟧, Theorem 5.3.2 shows that this cover, and hence C, can be reduced to a countable subcover of I.

▲EXAMPLE 3. A property resembling separability is D-separability. We call a space *D-separable* if every discrete closed set is countable. We have the result, *every separable normal space is D-separable.* To prove this, let F be a discrete closed set in a separable normal space X, and let D be a dense set. For every subset S of F, both S and $F \setminus S$ are closed in X. Thus to each S we may associate an open set $G(S) \supset S$ such that $G(S) \not\cap G(F \setminus S)$. We shall now show that the map $S \to D \cap G(S)$ is one-to-one (from $\mathbf{2}^F \to \mathbf{2}^D$). ⟦Let $S_1 \neq S_2$. Say, for definiteness, that $S_1 \not\subset S_2$. There exists

$$d \in D \cap G(S_1 \setminus S_2) \cap G(S_1).$$

Then $d \in D \cap G(S_1)$ and $d \notin D \cap G(S_2)$.⟧ Since D is countable, it follows that F is. ▮ A different proof of this result is given in Sec. 8.5, Problem 203.

▲ EXAMPLE 4. A countable space is separable ⟦it, itself, is a dense subset⟧, also Lindelöf. ⟦For each point choose one set from the cover containing it.⟧ Hence a regular countable space is normal ⟦Theorem 5.3.5⟧. It is also true that a first countable countable space is second countable. ⟦A base can be made up of the union of all the (countably many) countable local bases.⟧ We have seen that a countable space need not be first countable ⟦Sec. 3.1, Problems 201, 208. A nicer example is given in Sec. 8.3, Problem 103⟧. Hence Theorem 5.3.1 fails for countable spaces.

Problems

1. Second countability is a hereditary property ⟦Sec. 3.2, Problem 11⟧. Hence ⟦Theorem 5.3.2⟧, a second countable space is hereditarily Lindelöf. But a hereditarily Lindelöf space need not be second countable ⟦Sec. 3.1, Problems 201, 208; Sec. 8.3, Problem 103.⟧
2. A second countable space is hereditarily separable, but a hereditarily separable space need not be second countable ⟦Sec. 3.1, Problems 201, 208; Sec. 8.3, Problem 103⟧.
3. An uncountable cofinite space is separable but not second countable. A countable cofinite space is second countable. ⟦Count its closed sets.⟧
4. Let X be an arbitrary space and $t \notin X$. Let $Y = X \cup \{t\}$ with the topology in which a set is open precisely if it is $G \cup \{t\}$, G an open set in X. Then $\{t\}$ is dense, hence Y is separable. Deduce that separability is not hereditary.
5. Give an example of a non-Lindelöf space. ⟦A suitable discrete space would do.⟧
6. A space is second countable if and only if it has a countable subbase.
7. A continuous image of a separable space is separable.
8. If a space is separable (or Lindelöf) it remains separable (or Lindelöf) if the topology is weakened. This is not true for second countability. ⟦Sec. 3.1, Problem 201, or Sec. 8.3, Problem 103. Compare with the discrete topology.⟧
9. A family of subsets of a set X is said to have the *countable intersection property* if the intersection of every countable subfamily is nonempty. Show that a space is Lindelöf if and only if every family of closed sets with the countable intersection property has nonempty intersection. ⟦Consider the family of complements.⟧ Compare Theorem 5.4.3.

---

101. RHO is hereditarily Lindelöf.

102. A countable regular space is zero-dimensional ⟦Sec. 4.3, Problem 108; Theorem 5.3.5⟧.
103. A pseudocompact T_2 space need not be T_3 ⟦Example 4; Sec. 5.2, Examples 7, 8⟧. (Compare Theorem 5.4.7.)
104. A separable T_2 space cannot have more than 2^c points ⟦Sec. 4.1, Problem 204⟧.
105. Let X be a regular space and G an open Lindelöf subspace. Then $\tilde{G}$ is a G_δ. ⟦For each $x \in G$, let $U(x)$, $V(x)$ be disjoint open neighborhoods of $\tilde{G}$, x. Reduce $\{V(x)\}$ to a countable cover $\{V(x_n)\}$ of G. Then $\tilde{G} = \bigcap U(x_n)$.⟧ *Note:* We cannot omit "open" here, see Sec. 9.3, Problem 11.
106. A *perfectly normal space* is a normal space in which every closed set is a G_δ. Show that every semimetric space is perfectly normal ⟦Sec. 4.3, Problem 107, and Theorem 4.3.3⟧.
107. A hereditarily Lindelöf regular space is perfectly normal ⟦Theorem 5.3.5; Problem 105⟧.

---

201. Let U be the upper half-plane ($y > 0$) of $\mathbf{R}^2$, X the X-axis ($y = 0$), and let $Z = U \cup X$. Let U have the Euclidean topology and let the neighborhoods of each point $P \in X$ be sets $\{P\} \cup N$, where N is the interior of an ordinary circle in U tangent to X at P. Show that Z is T_3 and separable, and that X is a discrete subspace. Deduce from Example 3 that Z is not normal, and from Theorem 5.3.5 that Z is not Lindelöf.
202. Let $X = C([0, 1])$ with metric $d(x, y) = \max\{|x(t) - y(t)| : 0 \le t \le 1\}$. Show that X is separable ⟦either polynomials or polygons with rational corners⟧.
203. If X is separable, $C(X)$ can have cardinality at most c. ⟦There are only c distinct real functions on a countable set.⟧

5.4 Compactness

A Lindelöf space is one in which every open cover has a countable subcover. This concept leads naturally to the consideration of spaces in which every open cover has a finite subcover, or in which every countable open cover has a finite subcover. Such properties are certainly first cousins to the Lindelöf property, and are related to separability, at least by the results of Theorems 5.3.1 and 5.3.2. We give formal definitions: A topological space X is called *compact* if every open cover of X has a finite subcover, *countably compact* if every countable open cover of X has a finite subcover. Thus a compact space is countably compact. The converse is false ⟦Sec. 14.1, Example 2⟧. The following result is a triviality (so is its converse).

THEOREM 5.4.1. *A countably compact Lindelöf space is compact.*

Reduce any open cover to a countable, thence to a finite, subcover. ▮

Of course a Lindelöf space need not be compact. ⟦**R** is Lindelöf, by Theorem 5.3.2, but the open cover $\{(-n, n): n = 1, 2, \ldots\}$ cannot be reduced to a finite one.⟧

★EXAMPLE 1. A standard result of elementary analysis is the Heine–Borel theorem which says that *a closed interval* $[a, b]$ *of* **R** *is compact*. (The theorem was given independently by E. Heine in 1872, and E. Borel in 1895.) Let C be an open cover of $[a, b]$. Let $S = \{x: a \leq x \leq b, [a, x]$ is covered by a finite subset of $C\}$. Then $S \neq \varnothing$ since $a \in S$.

S is open, (in $[a, b]$). ⟦Let $x \in S$; then x is in some $G \in C$, and for some $\varepsilon > 0$, $(x - \varepsilon, x + \varepsilon) \subset G$. Every $y \in (x - \varepsilon, x + \varepsilon)$ also belongs to S, since $[a, y]$ has the finite subcover $C' \cup \{G\}$, where C' is the finite subcover of $[a, x]$.⟧ Also, S is closed. ⟦Let $x \in \bar{S}$. Then x is in some $G \in C$, and, for some $\varepsilon > 0$, $(x - \varepsilon, x + \varepsilon) \subset G$. Now there exists $y \in (x - \varepsilon, x + \varepsilon) \cap S$, and, since $[a, y]$ has a finite subcover C', $[a, x]$ must have the finite subcover $C' \cup \{G\}$. Thus $x \in S$.⟧ Hence $S = [a, b]$ ⟦Sec. 5.2, Example 3⟧. In particular $b \in S$. ▮

THEOREM 5.4.2 *Compact, countably compact, and Lindelöf are F-hereditary.*

One of these is Theorem 5.3.3. The others are proved in the same way. ▮

It is useful to spell out the compactness definitions in terms of closed sets. These are obtained simply by taking complements as we now show. A collection of sets is said to have the *finite intersection property* if every finite subset has nonempty intersection. For example, $\{(0, 1/n): n = 1, 2, \ldots\}$ has the finite intersection property. It is of the utmost importance to recognize that every filter has the finite intersection property.

A collection of sets is called *fixed* if it has nonempty intersection, and *free* if its intersection is empty. This terminology is reasonable, for example $\{(n, \infty): n = 1, 2, \ldots\}$ disappears from the scene (is free), while $\{[1, 1 + 1/n]: n = 1, 2, \ldots\}$ is pinned down at 1 (is fixed).

THEOREM 5.4.3. *A topological space is compact if and only if every collection of closed sets with the finite intersection property is fixed. This statement remains true with "compact" replaced by "countably compact" and "collection" replaced by "countable collection."*

Let X be compact and C a free collection of closed sets. Then $\{G: \tilde{G} \in C\}$ is an open cover of X, hence can be reduced to a finite subcover $\{G_1, G_2, \ldots, G_n\}$. Since $X \subset \bigcup G_i$ we have $\bigcap \tilde{G}_i = \varnothing$, thus C does not have the finite intersection property. Conversely, let X be not compact; then

X has an open cover C which has no finite subcover. Then $\{F: \tilde{F} \in C\}$ is a free collection of closed sets with the finite intersection property.

$$[\![\bigcap F_i = X \setminus \bigcup \tilde{F}_i \neq \varnothing \text{ since } X \not\subset \bigcup \tilde{F}_i.]\!]$$

The other part of the theorem is proved in the same way with countable collections. ▮

THEOREM 5.4.4. *A continuous image of a compact space is compact. The same is true with compact replaced by countably compact or Lindelöf.*

We shall prove this only for compactness, by far the most important of the three cases. The other two are exactly similar. Let C be an open cover of $F[S]$, S compact. Then $\{f^{-1}[G]: G \in C\}$ is an open cover of S. It can be reduced to a finite cover, say $\{f^{-1}[G_1], \ldots, f^{-1}[G_n]\}$. Then $(G_1, G_2, \ldots, G_n)$ is a cover of $f[S]$. ▮

It is not hard to see (and will shortly be proved) that a compact subset of **R** must be closed. This is a universal property of compact sets, given enough separation [[Theorem 5.4.5]], but a compact set in a T_1 space need not be closed [[Problem 14]].

THEOREM 5.4.5. *A compact set K in a Hausdorff space is closed.*

Let $x \notin K$. For each $y \in K$ choose disjoint open neighborhoods of x, y, which we shall call $U(y)$, $V(y)$, respectively. Now $\{V(y): y \in K\}$ is an open cover of K, hence can be reduced to a finite open cover, $(V(y_1), V(y_2), \ldots, V(y_n))$.

Let

$$U = \bigcap_{i=1}^{n} U(y_i), \qquad V = \bigcup_{i=1}^{n} V(y_i).$$

Then U is a neighborhood of x and does not meet V [[$z \in V$ implies $z \in V(y_i)$ for some i. This implies $z \notin U(y_i) \supset U$]], hence U does not meet K. Hence $x \notin \bar{K}$. This shows that $\bar{K} \subset K$ and so K is closed. ▮

The reader will not have overlooked the stronger result which was obtained in the course of this proof. We state it as the first half of the following Theorem.

THEOREM 5.4.6. *In a Hausdorff space, a point and a compact set not containing it can be separated by open sets. In a regular space, every neighborhood of a compact set K includes a closed neighborhood of K.*

To prove the second half, suppose $K \subset G$ with G open. For each $x \in K$ choose a closed neighborhood F_x of x with $F_x \subset G$. The open cover $\{F_x^i: x \in K\}$ of K can be reduced to a finite cover $\{F_{x_k}^i: k = 1, 2, \ldots, n\}$. Then $\bigcup \{F_{x_k}: k = 1, 2, \ldots, n\}$ is the required neighborhood. ▮

THEOREM 5.4.7. *Every compact regular space, and every compact Hausdorff space, is normal and completely regular.*

That a compact Hausdorff space is regular follows from Theorem 5.4.6, since every closed set is compact ⟦Theorem 5.4.2⟧. A compact regular space must be normal ⟦Theorem 5.4.6; Sec. 4.1, Problem 7. Another proof consists of citing Theorem 5.3.5⟧, hence completely regular ⟦Sec. 4.3, Problem 3⟧. ∎

★EXAMPLE 2. *A set in* **R** *is compact if and only if it is closed and bounded.* Half of this follows from the Heine–Borel theorem ⟦Example 1⟧ and Theorem 5.4.2. Half of the other half follows from Theorem 5.4.5. Finally, let S be unbounded. The open cover $\{(-n, n): n = 1, 2, \ldots\}$ of S cannot be reduced to a finite cover. ∎

We can now deduce a standard result of analysis, namely that *a continuous real function on a compact set, in particular, any closed finite interval, is bounded and assumes its maximum and minimum.* This is a special case of Theorem 5.4.4, (take $Y = \mathbf{R}$) together with the facts just mentioned about **R**. (Some of this is contained in the statement: a compact space is pseudocompact. Sec. 7.1, Problem 114 generalizes this.)

EXAMPLE 3. *A compact set in a semimetric space is metrically bounded.* Let $f(x) = d(x, y)$ for some fixed y. Then f is continuous ⟦Sec. 4.2, Problem 12⟧ hence, as in Example 2, bounded on each compact set. Then $d(x_1, x_2) \leq f(x_1) + f(x_2)$ is also bounded and the set has finite diameter. ∎

The converse is false, indeed *a closed bounded set in a metric space need not be compact.* This is true because any metric space X can be given an equivalent metric which makes X have finite diameter ⟦Sec. 2.3, Problem 102⟧, and there are noncompact metric spaces ⟦Example 2⟧. ∎

We are now going to see how the assumption of compactness precludes various kinds of pathology. It does this by forcing certain functions to be continuous, and certain topologies to be equal (not to mention Theorem 5.4.7 and the second result of Example 2). A function f is called *closed* if $f[F]$ is closed whenever F is a closed set.

THEOREM 5.4.8. *A continuous map f from a compact space X to a Hausdorff space Y must be closed.*

If F is a closed subset of X it is compact ⟦Theorem 5.4.2⟧, hence $f[F]$ is compact ⟦Theorem 5.4.4⟧, hence closed ⟦Theorem 5.4.5⟧. ∎

THEOREM 5.4.9. *A one-to-one and continuous function f from a compact space X to a Hausdorff space Y is a homeomorphism (into).*

This result is generalized in Lemma 8.1.1.

Let $Z = f[X]$. If G is an open set in X, $f[G] = Z \setminus f[Z \setminus G]$ is open in Z by Theorem 5.4.8. Thus $f^{-1}: Z \to X$ is continuous. ▮

THEOREM 5.4.10. *Two comparable compact Hausdorff topologies are equal.*

Let T, T' be compact Hausdorff topologies for a set, with $T \supset T'$. Applying Theorem 5.4.9 to the identity map yields the result ⟦Theorem 4.2.5⟧. ▮

REMARK. In the proof of Theorem 5.4.10, it is sufficient to assume T compact and T' Hausdorff. But nothing is gained since this implies that T is Hausdorff, and T' compact ⟦Problem 4⟧.

▲EXAMPLE 4. (a) Call a space *KC* if every compact set is closed. Theorem 5.4.5 says that every Hausdorff space is *KC*. (b) Theorem 5.4.10 yields the results that a compact T_2 topology is *maximal compact* (it has no strictly larger compact topology), and minimal T_2. *A maximal compact topology need not be* T_2, *indeed a compact topology, is maximal compact if and only if it is KC*. The proofs of Theorems 5.4.8, 5.4.9, and 5.4.10 go through unchanged with *KC* for T_2. Conversely if (X, T) is compact and not *KC*, let S be a nonclosed compact subset and $T' = \{\varnothing, S, X\}$. Then $T \vee T'$ is compact ⟦[Levine, Theorem 6]; compare Sec. 6.2, Example 2⟧. A minimal T_2 topology need not be compact. ⟦Urysohn; see [Herrlich(a)].⟧

A *locally compact* space is a space in which each neighborhood of a point includes a compact neighborhood of that point. (See the discussion of local property in Section 5.2.)

THEOREM 5.4.11. *A compact regular space* (*hence, also, a compact Hausdorff space*) *is locally compact.*

The parenthesized remark follows from Theorem 5.4.7. Let N be a neighborhood of x in a compact regular space X. Let F be a closed neighborhood of x with $F \subset N$ ⟦Theorem 4.1.3⟧. Then F, as a closed subset of the compact space X, is compact ⟦Theorem 5.4.2⟧. ▮

REMARK. There are several definitions of local compactness current in the literature, all of which are equivalent for Hausdorff and regular spaces, but not in general. Our definition is a very popular one, and has the logic of language in its favor (local compactness is a property tested by examining arbitrarily small neighborhoods). It should be emphasized that a compact space need not be locally compact; that is, "Hausdorff" cannot be omitted in Theorem 5.4.11. For details see Sec. 8.1, Problems 6 and 126.

Local compactness is not hereditary. ⟦**Q** is a subspace of **R** but is not locally compact, as is obvious directly, or from Theorem 5.4.13.⟧ However, it is G-hereditary.

THEOREM 5.4.12. *An open subset G of a locally compact space X is locally compact.*

Let $x \in G$, and let N be a neighborhood of x in G. Then N is a neighborhood of x in X. ⟦By definition of the relative topology, $N = N_1 \cap G$, where N_1 is a neighborhood of x in X. But G is also a neighborhood of x in X.⟧ Thus N includes a set K, which in X is a compact neighborhood of x. Since $K = K \cap G$, the same is true of K in G. ▮

COROLLARY 5.4.1. *An open subset of a compact regular space is locally compact.*

This follows from Theorems 5.4.11 and 5.4.12. ▮

Problem 6 of Sec. 8.1, shows that "regular" cannot be replaced by "T_1" in Corollary 5.4.1. An important partial converse of Theorem 5.4.12 may be obtained.

THEOREM 5.4.13. *Let X be a Hausdorff space and D a dense locally compact subspace. Then D is open.*

Let $x \in D$. Let K be a compact neighborhood of x in D, then K includes an open neighborhood of x in D and so $K \supset G \cap D$ where G is an open neighborhood of x in X. Then, $D \supset K = \mathrm{cl}_x\, K$ ⟦K is compact in D, hence in X, hence closed, by Theorem 5.4.5⟧ $\supset \mathrm{cl}_X\,(G \cap D) = \mathrm{cl}_X\, G$ ⟦Sec. 2.5, Problem 12⟧ $\supset G$. Thus D is an X neighborhood of x, and so finally D is open. ▮

Problems on Topological Space

1. Let $\{x_n\}$ be a convergent sequence with $x_n \to x$. Show that $\{x, x_1, x_2, \ldots\}$ is compact. (x need not be the only limit.)
2. A countably compact countable space is compact ⟦Theorem 5.4.1; Sec. 5.3, Example 4⟧.
3. Locally compact is F-hereditary.
4. ★ A compact space remains compact when the topology is weakened (that is, replaced by a smaller topology) ⟦Theorem 5.4.4⟧.
5. In Theorems 5.4.8, 5.4.9, and 5.4.10, the assumption of compactness cannot be dropped. ⟦Consider discrete spaces.⟧
6. Let X be compact and $f \in C(X)$ satisfy $f(x) > 0$ for all x. Show that there exists $\varepsilon > 0$ with $f(x) \geq \varepsilon$ for all x.
7. Let X be compact and connected, and $f \in C(X)$. Show that $f(X)$ is a closed interval.
8. ★ Let C be a nonempty collection of sets with the finite-intersection property. Let $\mathscr{B}$ be the set of all finite intersections of members of C. Show that $\mathscr{B}$ is a filterbase.

★9. In a regular space, the closure of a compact set is compact. The same is (trivially) true in a Hausdorff space. ⟦Let C be an open cover of $\bar{K}$. Let $C_1 = \{G: G$ is open and $\bar{G}$ is included in some member of $C\}$. Reduce C_1 to a finite cover C_2 of K. For each $G \in C_2$ choose $H \in C$ with $\bar{G} \subset H$. The set of H is a finite cover of K which by its choice is also a cover of $\bar{K}$.⟧

10. The union of two compact sets is compact.

11. The intersection of a family of closed sets is compact if at least one of them is compact ⟦Theorem 5.4.2⟧. The same is true for countably compact and Lindelöf (instead of compact).

12. In Theorem 5.4.13, "Hausdorff" cannot be replaced by "T_1". ⟦Every infinite subspace of a cofinite space is dense and locally compact.⟧ It also cannot be replaced by "regular" ⟦indiscrete⟧.

13. An infinite discrete space is not countably compact. ⟦Let $S = \{x_n\}$ be a sequence of members of X. Then $(\tilde{S}, \{x_1\}, \{x_2\}, \ldots)$ is a countable open cover.⟧

14. Every subset of a cofinite space is compact. ⟦Any open set includes all but finitely many points. Thus an open cover can easily be reduced. It is also easy to apply Theorem 5.4.3.⟧

15. A continuous open image of a locally compact space is locally compact ⟦Sec. 4.2, Problem 22⟧.

16. A retract of a locally compact space is locally compact. ⟦For T_2 spaces use Problem 3. In general if N is a neighborhood of x in S, $r^{-1}[N]$ includes a compact neighborhood K of x. Then $r(K)$ is compact by Theorem 5.4.4, and $r(K) \supset K \cap S$, is a neighborhood of x in S.⟧

---

101. A locally compact Hausdorff space is regular. ⟦It has a base at each point of compact, hence closed sets. See Theorem 4.1.3. A different sort of proof is given near the beginning of Section 8.1.⟧

102. Every locally compact subset of a Hausdorff space is of the form $F \cap G$, F closed, G open.

103. Let X be locally compact T_2, and F, G closed, open subsets respectively. Show that $F \cap G$ is locally compact.

104. Find a compact set which has a cover by sets with nonempty interior not reducible to a finite cover. (An example may be given in **R**.)

105. "Hausdorff" cannot be replaced by "T_1" or "regular" in Theorem 5.4.9. ⟦Identity from [0, 1] to cofinite, or indiscrete.⟧ However, it can be replaced by "KC."

106. Every KC space is US, hence T_1. A T_1 pseudofinite space is KC. (A *pseudofinite* space is one in which every compact set is finite.) **Q**, **J** are not pseudofinite. A cocountable space is pseudofinite, hence KC. (Thus KC lies strictly between T_1, T_2.)

107. A lower semicontinuous function assumes a minimum value on every compact set.
108. A continuous function from a compact semimetric space to a semimetric space is uniformly continuous.
109. A countably compact subset of a Hausdorff space is sequentially closed; compare Theorem 5.4.5. ⟦If S is not sequentially closed, a sequence of points in S converging to a point outside S constitutes a discrete closed subset. With Problem 13 and Theorem 5.4.2 we see that S is not countably compact.⟧
110. Deduce from Problem 109, versions of Theorems 5.4.8, 5.4.9, and 5.4.10, in which "compact" is replaced by "countably compact" and appropriate spaces are assumed first countable.
111. The proof in Example 1 is deceptively simple. Replace, in it, $[a, b]$ by X, with $X = [0, 1)$. Then the same argument yields $S = X$. Yet X is not compact. Explain the apparent contradiction.
112. Let $K \subset G \subset X$ with X a semimetric space, G open, K compact. Prove that there exists $\varepsilon > 0$ such that $N(K, \varepsilon) \subset G$.

$$(N(K, \varepsilon) = \bigcup \{N(x, \varepsilon): x \in K\}.)$$

113. Let K be a compact, and G an open set in a regular space with $K \subset G$. Then there is a closed neighborhood F of K with $F \subset G$. Hence $\bar{K} \subset G$. In particular in a regular space, a compact open set K is closed, and $n(K) = \bar{K}$ (Sec. 4.1, Problem 114). ⟦For each $x \in K$, cover x by a closed neighborhood lying in G. Reduce to finite cover; its union must be in G.⟧ This result is false for the cofinite topology.
114. "Regular" cannot be replaced by "T_1" in Problem 9. ⟦Let $G \subset \omega$ be called open if it contains all but a finite number of even integers. With this topology ω is not compact since it has the odd numbers as a discrete closed subset; also the even numbers form a dense compact set. (This space is T_1.)⟧
115. In a T_2 space (indeed in any KC space), the intersection of two compact sets must be compact ⟦Theorems 5.4.2 and 5.4.5⟧, but not in a T_1 space. ⟦Let X be a compact T_2 space, x a nonisolated point, and $Y = X \setminus \{x\}$. Make x into two points x, x' as in Sec. 3.2, Problem 110. Then X and $Y \cup \{x'\}$ are compact but their intersection is not.⟧
116. In a hereditarily Lindelöf T_2 space, every compact set is a G_δ. ⟦Same as Sec. 5.3, Problem 105 with F compact. Use Theorem 5.4.6.⟧
117. Let X be a noncompact space. Then $\mathscr{B} = \{S: \tilde{S}$ is compact$\}$ is a filterbase. If X is T_2, the filter $\mathscr{F}$ generated by $\mathscr{B}$ is $\{S: \tilde{S}$ is relatively compact$\}$. (A *relatively compact* set is one whose closure is compact.) If X is pseudofinite, $\mathscr{B}$ is the cofinite filter.

---

201. Let X be regular, and let $x \in X$. If $\{x\}$ is a G_δ and x has a compact

neighborhood, then X is first countable at x. Compare Sec. 4.3, Problem 5. ⟦$\{x\} = \bigcap G_n$, each G_n a closed neighborhood of x. Let K be a compact neighborhood of x. Then with $N_n = K \cap G_1 \cap \cdots \cap G_n$, $\{N_n\}$ is a local base, for if N is an open neighborhood of x, $K \setminus N$ is compact and fails to meet $\bigcap N_n$.⟧

202. A countable, locally compact, T_3 space is second countable ⟦by Problem 201⟧.
203. Give an example of a subset of **R** which is not locally compact and which has exactly one accumulation point.
204. The countable space of Sec. 3.1, Problem 201 is pseudofinite.
205. (**R**, RHO) is not locally compact. Indeed every compact set has empty interior. ⟦$[a, b)$ is closed but not compact since not Euclidean compact.⟧
206. Let (X, T) be a Hausdorff space, and T' the *cocompact* topology. (A proper subset is T'-closed if and only if it is T-compact.) Show that $T' \subset T$, and that (X, T') is compact. Show also that the following three conditions are equivalent: T' is T_2, T is compact, $T' = T$.
207. Let X be compact and such that every point is the intersection of all the open-and-closed sets which contain it. Show that X is zero-dimensional and $T_{3\frac{1}{2}}$ ⟦see Problem 201⟧.
208. The *character* $\mathrm{ch}(x)$ of X at x is the smallest cardinal of a local base at x; for a T_1 space X, the *weight* $w(x)$ of X at x is the smallest cardinal of a collection of open sets whose intersection is x. Show that $\mathrm{ch}\, x \geq w(x)$, and, in a regular space, if x has a compact neighborhood, $\mathrm{ch}\, x = w(x)$. (Problem 201 is a special case.)

Sup, Weak, Product, and Quotient Topologies

6.1 Introduction

The studies of the first five chapters have been carried out in a rather rarefied atmosphere due to the shortage of examples to illustrate the various ideas. The abstractions of topology arose from (perhaps) a century and a half of experience with dozens of special spaces arising in classical analysis. Whereas in Section 5.4, we introduced compactness with the feeble excuse that it seemed a natural adjunct to the Lindelöf property, the historical reason for its study is the overwhelmingly good behavior of certain sets of real numbers and functions (examples: normal families, Dirichlet principle) when these sets have the property to which, as it developed, compactness specialized in their particular situations.

In this chapter we show techniques which, apart from their general utility and pervasiveness, also yield methods of constructing wide classes of examples.

6.2 Sup Topologies

Suppose that a set X has a nonempty collection Φ of topologies specified for it. Let $\mathscr{B} = \bigcup \{T: T \in \Phi\}$. Thus $\mathscr{B}$ is the collection of all sets, each of which is open in at least one $T \in \Phi$. The topology generated by $\mathscr{B}$ is called the *supremum* (for short, sup) of the collection Φ, written $\bigvee \Phi$ or $\bigvee \{T: T \in \Phi\}$. As usual, if Φ is finite, $\Phi = \{T_1, T_2, \ldots, T_n\}$, we write $\bigvee \Phi$

as $T_1 \vee T_2 \vee \cdots \vee T_n$ or $\bigvee_{i=1}^{n} T_i$; and if Φ is countably infinite, $\Phi = \{T_n\}$, we write $T_1 \vee T_2 \vee \cdots$ or $\bigvee_{i=1}^{\infty} T_i$.

The topology $\bigvee \Phi$ has the important property: *if T is a topology and $T \supset T'$ for all $T' \in \Phi$, then $T \supset \bigvee \Phi$.* ⟦Let $G \in \bigvee \Phi$ and $x \in G$. Then there exist sets $G_1, G_2, \ldots, G_n$, each G_i being open in at least one $T' \in \Phi$, such that $x \in \bigcap_{i=1}^{n} G_i \subset G$. Each $G_i \in T$, hence G is a T neighborhood of x. Since this is true for all $x \in G$, it follows that $G \in T$.⟧ This property is described by the sentence: "$\bigvee \Phi$ is the smallest topology which is larger than each member of Φ."

★EXAMPLE 1. On $\mathbf{R}^2$, let H be the topology generated by the set of horizontal open strips, $\{(x, y): a < y < b\}$, and V the topology generated by the set of vertical open strips. Then $H \vee V$ is the Euclidean topology for $\mathbf{R}^2$. To see this, it is sufficient to notice that for each point P and each Euclidean neighborhood N of P, N includes an open square centered at P, and this square is the intersection of a vertical and a horizontal open strip. This, along with the facts that the strips are (Euclidean) open and have $\mathbf{R}^2$ as their union, shows that the set of strips is a subbase for the Euclidean topology.

THEOREM 6.2.1. *Let $T' = \bigvee \Phi$, where Φ is a family of topologies on a set. Then a net $x_\delta \to x$ in T' if and only if $x_\delta \to x$ in every $T \in \Phi$. Also a filter $\mathscr{F} \to x$ in T' if and only if $\mathscr{F} \to x$ in every $T \in \Phi$.*

T' is stronger than each $T \in \Phi$; hence half of each result is trivial ⟦Theorem 3.2.2; Sec. 3.4, Problem 1⟧. Conversely, let $\mathscr{F}$ be a filter with $\mathscr{F} \to x$ in every $T \in \Phi$. Let N be a T' neighborhood of x. There exist $N_1, N_2, \ldots, N_k$, each N_i being a neighborhood of x in some $T \in \Phi$ with $x \in \bigcap N_i \subset N$. By hypothesis, each $N_i \in \mathscr{F}$, thus $N \in \mathscr{F}$, and so $\mathscr{F} \to x$. The same argument works for a net x_δ, the conclusion being that $x_\delta \in N$ eventually. ▌

REMARK 1. The student should carefully reread the part of this proof which says "There exist $N_1, N_2, \ldots, N_k$ with $x \in \bigcap N_i \subset N$." It is this property of the sup topology which is called on repeatedly. It follows from the definition of the phrase (used in the definition of the sup topology), "the topology generated by $\mathscr{B}$," that is, the topology of which $\mathscr{B}$ is a subbase.

One of the central problems of topology has been to identify those topologies which are semimetrizable (or, in case of T_0 spaces, metrizable). Several results of this nature will be given, of which Theorem 6.2.2 is the first. It is typical in that it yields a metrization theorem from a cardinality restriction. (For a finite collection, see Problem 102.)

THEOREM 6.2.2. *Let $\{T_n\}$ be a sequence of semimetrizable topologies for a set X. Then $\bigvee T_n$ is semimetrizable.*

Let T_n be induced by the semimetric d_n for $n = 1, 2, \ldots$. Let

$$d(x, y) = \sum_{n=1}^{\infty} \frac{1}{2^n} \cdot \frac{d_n(x, y)}{1 + d_n(x, y)}.$$

Then d is a semimetric ⟦for example, the triangle inequality is given in Sec. 3.1, Example 7, for each term of the sum⟧.

Let T be the topology induced by d; then $T = \bigvee T_n$ by Theorem 6.2.1, and Theorem 3.5.2. ∎

Theorem 6.2.2 fails for uncountable families ⟦Sec. 6.4, Problem 6⟧.

▲ EXAMPLE 2. *Simple extensions.* Let X be a set, S a proper subset, and $T_S = \{\varnothing, S, X\}$. Now if T is a topology for X, and $S \notin T$, the topology $T \vee T_S$ is called the *simple extension of T by S.* (If $S \in T$, then $T \vee T_S = T$; otherwise we are enlarging T by "declaring S open.") Simple extensions are useful for constructing counterexamples, for example, *a simple extension of a topology by a dense set S cannot be regular.* ⟦Let T, T' be the topology and its extension. Since $S \notin T$, we know that S contains a point x which is not T-interior to S. We shall show that, in (X, T'), x cannot be separated by neighborhoods from the closed set $\tilde{S}$. Let N_x be a T' neighborhood of x. Then there exists a T-open, T-neighborhood N_1 of x with $N_1 \cap S \subset N_x$. (Notice that every T_s neighborhood of x includes S.) Since x is not T-interior to S, N_1 meets $\tilde{S}$; say $y \in N_1 \cap \tilde{S}$. Let N_y be a T' neighborhood of y. Since $y \notin S$, N_y is a T neighborhood of y. Thus $N_x \cap N_y \supset N_1 \cap S \cap N_y \neq \varnothing$ since the T-dense set S must meet the set $N_1 \cap N_y$, the latter being a T neighborhood of y.⟧ Together with Sec. 4.1, Problem 4, this yields an example of a T_2 space which is not T_3. ⟦$X = \mathbf{R}$, $S = \mathbf{Q}$.⟧

▲ EXAMPLE 3. We show a generic method for constructing certain counterexamples. *Suppose that a property P is not hereditary but is either F-hereditary or G-hereditary. Then P is not preserved under strengthening of the topology.* This means that a space (X, T) may be found with property P such that (X, T') fails to have property P for certain $T' \supset T$. The proof consists of finding a space (X, T) with property P and a subspace S which does not have property P. ⟦This is the assumption.⟧ Let T' be the simple extension of T by S if P is G-hereditary, by $\tilde{S}$ if P is F-hereditary. Then $X, T')$ does not have the property P because it has the open (or closed) subspace S which does not have property P. An example is the remark that the simple extension by $\mathbf{Q}$ of the Euclidean topology for $\mathbf{R}$ is not locally compact.

Problems

In this list Φ is a nonempty collection of topologies for a set X.

1. In Example 1, a, b may be restricted to be rational.

2. If Φ has a maximum member, T, (that is $T \supset T'$ for all $T' \in \Phi$) then $\bigvee \Phi = T$.
3. If Φ contains two members T, T' with $T \subset T'$, then $\bigvee \Phi = \bigvee (\Phi \setminus \{T\})$.
4. Let $\bigwedge \Phi$ (read: inf Φ) be $\bigvee \{T: T \subset T' \text{ for all } T' \in \Phi\}$. (*Note:* The latter is the sup of a nonempty set. ⟦Consider the indiscrete topology.⟧) Show that $\bigwedge \Phi$ is the largest topology which is smaller than each member of Φ, and that $\bigwedge \Phi = \bigcap \Phi$.
5. If at least one T_i in Theorem 6.2.2 is metrizable, then $\bigvee T_n$ is metrizable.
6. Let $\{T_n\}$ be a sequence of first countable topologies for a set X. Show that $\bigvee T_n$ is first countable.
7. Extend Example 3 to a property P which is not hereditary but is such that if G is an open subset of a space with property P, $\bar{G}$ has property P.

---

101. A set may be dense in each of two topologies without being dense in their sup. ⟦Consider the line $(y = x)$ in Example 1.⟧ What is the situation with respect to inf?
102. Let d_1, d_2 be semimetrics for a set X. Let $d_3 = d_1 + d_2$, $d_4 = (d_1^2 + d_2^2)^{1/2}$. Show that d_3, d_4 are both equivalent with $D_1 \vee D_2$, where D_1, D_2 are the topologies generated by d_1, d_2.
103. Identify RHO $\vee$ LHO, and RHO $\cap$ LHO. (LHO is, of course, the left half-open interval topology.)
104. Let Φ be a chain of topologies on a set. (Of any two members, one includes the other.) Then $\bigvee \Phi \supset \bigcup \Phi$. Inequality may hold.
105. The inf of two T_1 topologies is T_1. ⟦It is larger than the cofinite.⟧
106. If $T \cap T'$ is Hausdorff and $\mathcal{F}$ is a filter with $\mathcal{F} \to x$ in T and $\mathcal{F} \to y$ in T', then $x = y$. ⟦$\mathcal{F} \to x$ and y in $T \cap T'$.⟧ This property can be used to show two Hausdorff topologies whose inf is not Hausdorff. Let T be the topology on $[0, 1]$ gotten by adding all singletons of $[0, 1)$ to the cofinite topology; for T', do the same with $(0, 1]$ instead of $[0, 1)$. Then any sequence of distinct points in $(0, 1)$ converges to 1, 0 in T, T' respectively.
107. Let T, T' be different compact Hausdorff topologies for a set X. Show that $T \vee T'$ cannot be compact, and $T \cap T'$ cannot be Hausdorff. ⟦Theorem 5.4.10. The second part also follows from Sec. 6.7, Problem 113, and Sec. 7.1, Problem 108.⟧
108. Fix a positive integer n. Let $T(n)$ be the topology for ω which is discrete on $\omega \setminus \{n\}$ and cofinite at n. (This means that G, containing n, is open if and only if $\omega \setminus G$ is finite.) Show that $T(n)$ is compact Hausdorff, and describe $T(1) \cap T(2)$. (In particular Problem 107 shows that it is not Hausdorff.)
109. Give an example of two topologies T, T' for a set X and a subset of X which is a T neighborhood of a certain point x, and a T' neighborhood of x, but not a $T \cap T'$ neighborhood of x. (On the other hand a

set which is both T-open and T'-open is $T \cap T'$-open.) ⟦In Example 1, let $S = \{(x, y): |x| < 1 \text{ or } |y| < 1.\}$⟧

110. Let T, T' be topologies for a set X such that (x, T) is pseudofinite, and in (X, T') every singleton is a G_δ. Show that $(X, T \vee T')$ has both properties. Hence show that **R** with Euclidean $\vee$ cocountable is not anywhere first countable and each singleton is a G_δ.
111. Let S be a set in (X, T) which is nowhere dense and not closed. Show that the simple extension of T by S is not extremally disconnected. ⟦$\text{cl}_{T'} S = \text{cl}_T S = (G_1 \cap S) \cup G_2$ with $G_1, G_2 \in T$, would imply $G_2 = \varnothing$ and so $\text{cl}_T S \subset S$.⟧
112. Let (X, T) be extremally disconnected and S a dense set. Show that the simple extension of T by S is extremally disconnected.
113. Two different topologies may have the same family of subsets with nonempty interior. ⟦Consider a simple extension by a set which is contained in the closure of its interior.⟧ Hence a noncontinuous function ⟦identity⟧ may have the property that the inverse image of every set with nonempty interior has nonempty interior.
114. Two different topologies may have the same dense sets ⟦Problem 113⟧.

---

201. The filter condition mentioned in Problem 106 is not sufficient that $T \cap T'$ be Hausdorff.
202. The inf of the family of all T_2 topologies for a set X is the cofinite topology. ⟦Let S be an infinite set and $x \notin S$. Let T be gotten by adding to the cofinite topology on X, the discrete topology on $X \setminus \{x\}$. Then S is not T closed.⟧
203. The *dispersion character* of a space is the smallest cardinal which a nonempty open set may have. Let (X, T) be a T_0 space. Show that T and $T \vee$ cofinite have the same dispersion character.
204. Let X be a set, $S \subset X$, T and T' Hausdorff topologies for X whose relative topology on S is discrete. Must the relative topology of $T \cap T'$ on S be discrete?
205. Is the space of Problem 110 regular? normal? Lindelöf?

6.3 Weak Topologies

If X, Y are topological spaces, and $f: X \to Y$, it is natural to inquire which topologies on X make f continuous with the given topology of Y, and the same question with X, Y interchanged. Both questions are of great importance; the first leading to weak, projective, and product topologies, and vast areas of functional analysis involving the study of function spaces; the second (see Section 6.5) leading to quotient or identification spaces, and some of the tools of geometric topology.

In this section we shall be interested in the smallest topology on X for which $f: X \to Y$ is continuous.

DEFINITION 1. *Let $f: X \to Y$, where X is a set and (Y, T) is a topological space. The weak topology by f for X, written $w(X, f)$, or $w(f)$, is $f^{-1}[T]$. Thus a set $S \subset X$ is open if and only if $S = f^{-1}[G]$, G open in Y.*

We omit the easy check that $w(f)$ is a topology ⟦Problem 1⟧. If X is given $w(f)$, $f: X \to Y$ is continuous ⟦$f^{-1}[G]$ is open if G is open⟧, moreover, if T' is a topology for X making f continuous, we have $T' \supset w(f)$. ⟦Let $S \in w(f)$, then $S = f^{-1}[G]$ with G open in Y. Then $S \in T'$ since f is T' continuous.⟧ Thus $w(f)$ fulfills the requirements which motivated it.

LEMMA 6.3.1. *With the notation of Definition 1, a net $x_\delta \to x$ in $w(f)$ if and only if $f(x_\delta) \to f(x)$ in Y, and a filter $\mathscr{F} \to x$ in $w(f)$ if and only if $f[\mathscr{F}] \to f(x)$ in Y.*

Since f is continuous, half of each result is trivial ⟦Theorem 4.2.2⟧. Conversely, suppose that $f(x_\delta) \to f(x)$, and let N be an open neighborhood of x, (in $w(f)$, of course). Then $N = f^{-1}[G]$, where G is an open neighborhood of $f(x)$; $f(x_\delta) \in G$ eventually, hence $x_\delta \in N$ eventually. If $\mathscr{F}$ is a filter and $f(\mathscr{F}) \to f(x)$, the G just mentioned includes a member of $f[\mathscr{F}]$, say $f[S]$, $S \in \mathscr{F}$. Then $N \supset S$, hence $N \in \mathscr{F}$. ∎

DEFINITION 2. *Let X be a set and Φ a family of functions f, each f defined on X and with range in some topological space Y_f. The weak topology by Φ for X, written $w(X, \Phi)$, or $w(\Phi)$, is $\bigvee \{w(X, f): f \in \Phi\}$.*

Thus $w(X, f) = w(X, \{f\})$. The comments following Definition 1 apply here also: $w(\Phi)$ is the smallest topology for which every $f \in \Phi$ is continuous ⟦Problem 2⟧.

THEOREM 6.3.1. *With the notation of Definition 2, a net $x_\delta \to x$ in $w(\Phi)$ if and only if $f(x_\delta) \to f(x)$ for every $f \in \Phi$, and a filter $\mathscr{F} \to x$ in $w(\Phi)$ if and only if $f[\mathscr{F}] \to f(x)$ for every $f \in \Phi$.*

This is an immediate consequence of Lemma 6.3.1 and Theorem 6.2.1. ∎

We now investigate a little of the separation character of a weak topology. With the notation of Definition 2, we call Φ *separating over* X if, whenever $x_1 \neq x_2$, there exists $f \in \Phi$ with $f(x_1) \neq f(x_2)$.

THEOREM 6.3.2. *Let X be a set and Φ a family of maps f, each f from X to some Hausdorff space Y_f. Then $w(X, \Phi)$ is a Hausdorff topology if and only if Φ is separating over X.*

Suppose that $w(\Phi)$ is not Hausdorff. Then there exists a net x_δ in X with $x_\delta \to x$, $x_\delta \to y$ in $w(\Phi)$, $x \neq y$ ⟦Theorem 4.1.2⟧. For every $f \in \Phi$, $f(x_\delta) \to f(x)$

and $f(x_\delta) \to f(y)$ ⟦Theorem 6.3.1⟧, hence $f(x) = f(y)$ ⟦Theorem 4.1.2⟧. Thus Φ is not separating. Conversely suppose that Φ is not separating; say, $f(x) = f(y)$ for all $f \in \Phi$, $x \neq y$. Let N be a $w(\Phi)$ neighborhood of x. There are $f_1, f_2, \ldots, f_n$ in Φ, and open sets $G_1, G_2, \ldots, G_n$, each $G_i \subset Y_{f_i}$ with $x \in \bigcap_{i=1}^{n} f^{-1}[G_i] \subset N$. ⟦See Sec. 6.2, Remark 1.⟧ Since $f_i(x) = f_i(y)$ for each i, it follows that $y \in \bigcap f_i^{-1}[G_i]$, hence $y \in N$. Thus $w(\Phi)$ is not even a T_0 space. ∎

The second part of the proof yielded a bonus separation result which we now state.

THEOREM 6.3.3. *If the weak topology by maps to Hausdorff spaces is* T_0, *then it is Hausdorff.*

Compare the similar bonus separation for semimetric spaces in Sec. 4.1, Example 3.

THEOREM 6.3.4. *The weak topology by a sequence of maps from a set* X *to semimetrizable spaces is semimetrizable. The weak topology by a sequence of maps to metrizable spaces is metrizable if and only if the sequence of maps is separating.*

The second part follows from the first part and Theorem 6.3.2. The first part will follow from Theorem 6.2.2, when we check that $w(X, f)$ is semimetrizable, where $f: X \to Y$ with (Y, d) a semimetric space. Set $d_1(x, y) = d[f(x), f(y)]$ for $x, y \in X$. Then $w(X, f)$ is induced by d_1 because $x_\delta \to x$ in $w(f)$ if and only if $f(x_\delta) \to f(x)$ ⟦Lemma 6.3.1⟧, this is true if and only if $d[f(x_\delta), f(x)] \to 0$ ⟦Sec. 3.4, Example 4⟧, that is $d_1(x_\delta, x) \to 0$, and this is true if and only if $x_\delta \to x$ in the topology induced by d_1. ∎

Problems

★1. With the notation of Definition 1, check that $w(f)$ is a topology.

★2. With the notation of Definition 2, $w(\Phi)$ is the smallest topology making every $f \in \Phi$ continuous. ⟦Use of Theorem 6.3.1 will simplify the calculations.⟧

★3. Let $f: X \to Y$ and $F \subset X$. Then F is $w(f)$ closed if and only if $F = f^{-1}[F']$ where F' is a closed subset of Y.

4. Let $X \subset Y$. Then $w(X, i)$ is the relative topology. (i is the inclusion map.)

5. What is the largest topology for X which makes $f: X \to Y$ continuous?

★6. Let X be completely regular. Then the topology of X is equal to the weak topology by $C(X)$ and also to the weak topology by $C^*(X)$ ⟦Theorems 6.3.1 and 4.3.2⟧.

101. Describe $w(f)$ if $f: X \to Y$ is a constant.
102. Let $f: X \to Y$ be onto. Then $f: [X, w(f)] \to Y$ is open and closed. In contrast show that if each f in a family Φ is onto, it does not follow that every $f: [X, w(\Phi)] \to Y_f$ is open. ⟦For example, Y_f may be indiscrete.⟧
103. Give an example of a sequence $\{T_n\}$ of topologies with $T_n \subset T_{n+1}$, $T_n \neq T_{n+1}$ for all n and such that T_1 has a simple extension T with $T \supset T_n$ for all n. ⟦T_n = simple extension of $\mathbf{R}$ by $\mathbf{Q} \cap (0, n)$.⟧

---

201. Can you find $f: \mathbf{R} \to \mathbf{R}$ such that $w(f)$ is discrete?
202. Let X be a topological space, $\Phi = C(X)$, $\Psi = C^*(X)$. Show that $w(\Phi)$ and $w(\Psi)$ are completely regular, and are equal. (With Problem 6, this characterizes completely regular spaces.)

6.4 Products

The concept of product, or Cartesian product, has its source in the view that $\mathbf{R}^2$ is formed by a certain way of combining $\mathbf{R}$ with itself, namely, $\mathbf{R}^2 = \{(x, y): x \in \mathbf{R}, y \in \mathbf{R}\}$ which is written $\mathbf{R} \times \mathbf{R}$. An obvious generalization is to write

$$X \times Y = \{(x, y): x \in X, y \in Y\}$$

or

$$X_1 \times X_2 \times \cdots \times X_n = \{(x_1, x_2, \ldots, x_n): x_i \in X_i, i = 1, 2, \ldots, n\}.$$

Here (x, y) is an ordered pair $(x_1, x_2, \ldots, x_n)$, an ordered n-tuple, and $X, Y, X_1, \ldots$ are sets. For a sequence $\{X_n\}$ of sets, $X_1 \times X_2 \times \cdots = \{\{x_n\}: x_i \in X_i \text{ for } i = 1, 2, \ldots\}$, that is, the collection of all sequences with ith term selected from X_i for each i. Alternative notations for the finite and countable products are $\prod_{i=1}^n X_i$, $\prod_{i=1}^\infty X_i$, respectively, or $\prod \{X_i: i = 1, 2, \ldots, n\}$, $\prod \{X_i: i = 1, 2, \ldots\}$, respectively. We now present a formal definition of product of an arbitrary family of sets, which specializes to the definitions just given if the family is finite or countably infinite.

We suppose given a certain nonempty index set A, and for each $\alpha \in A$, a nonempty set X_α. The *product*, or *Cartesian product*, of the X_α, written $\prod \{X_\alpha: \alpha \in A\}$ or $\prod X_\alpha$, is the collection of all functions $f: A \to \bigcup \{X_\alpha: \alpha \in A\}$ such that $f(\alpha) \in X_\alpha$ for each $\alpha \in A$.

For example, if A is countably infinite, we may take $A = \omega$. Then a typical member of $\prod \{X_n: n \in \omega\}$ is a function on ω, that is, a sequence, whose nth term is selected from X_n. If A has two members, we may take $A = \{1, 2\}$, then $\prod \{X_i: i \in A\} = X_1 \times X_2$ has, as a typical member, a function on A, that is, an ordered pair, whose first term belongs to X_1, and whose second term belongs to X_2.

A member x of $\prod \{X_\alpha: \alpha \in A\}$ is a function on A (a sort of generalized sequence) whose "αth coordinate" is $x(\alpha)$ and is selected from the αth factor X_α of the product, for each $\alpha \in A$.

If X_α is the same set for all $\alpha \in A$, say $X_\alpha = X$, the product $\prod\{X_\alpha : \alpha \in A\}$ is written X^A. Thus X^A is the set of all functions from A to X. For example $\mathbf{R} \times \mathbf{R} = \mathbf{R}^2$, where 2 stands for a set with two members. (The notation X^A is nicely consistent with the fact expressed in Problem 1. Exponentiation of infinite cardinals is *defined* this way.)

There is a natural map from a product $\prod\{X_\alpha : \alpha \in A\}$ onto each factor X_β, namely $P_\beta : \prod X_\alpha \to X_\beta$ defined by $P_\beta x = x(\beta)$ for each $x \in \prod X_\alpha$. This map, P_β, is called the *projection* on the βth factor. Suppose, for example, that $A = \{1, 2\}$, that $X_1 = X_2 = \mathbf{R}$. Then $\prod\{X_\alpha : \alpha \in A\} = \mathbf{R} \times \mathbf{R} = \mathbf{R}^2$, and if $z \in \mathbf{R}^2$, say $z = (x, y)$ (that is $z(1) = x$, $z(2) = y$), then $P_1 z = P_1(x, y) = x$, $P_2 z = P_2(x, y) = y$. Thus, in this case, P_1, P_2 are the usual projections on the X and Y axes. In case $A = \omega$, so that $\prod X_\alpha$ is a space of sequences, $P_n x = x(n)$, the nth term of the sequence x.

Finally we topologize a product of topological spaces. There are, of course, many ways to do this; we select, as the one deserving of most study, the weak topology by the set of all projections. This topology, namely $w(\prod\{X_\alpha : \alpha \in A\}, \{P_\alpha : \alpha \in A\})$, is called the *product topology*, and will be taken as the natural topology for the product. A central tool for operating with the product topology, and a strong reason for selecting it as the product topology are contained in the following result.

THEOREM 6.4.1. *Let x_δ be a net in a product $\prod\{X_\alpha : \alpha \in A\}$ of topological spaces. Then $x_\delta \to x$ if and only if $P_\alpha x_\delta \to P_\alpha x$ for each $\alpha \in A$. The same result holds for filters.*

This is precisely the specialization of Theorem 6.3.1, to the case $\Phi = \{P_\alpha : \alpha \in A\}$. ∎

★EXAMPLE 1. Let A be a set and X a topological space. Then each member of X^A is a function from A to X. Let g_δ be a net of such functions. *Then $g_\delta \to g$ in the product topology if and only if $g_\delta \to g$ pointwise, that is, $g_\delta(\alpha) \to g(\alpha)$ for every $\alpha \in A$.* This is true because for each α in A the projection P_α is defined by $P_\alpha(g) = g(\alpha)$ and Theorem 6.4.1 applies.

Henceforth we shall use without comment the fact that the projections P_α are continuous, and that the product topology is the smallest topology with this property.

THEOREM 6.4.2. *A countable product of semimetric (or metric) spaces is semimetrizable (or metrizable).*

This follows from Theorem 6.3.4. If the product spaces are metric, so is the product, by the same result, since the set of all projections is separating. ⟦Let $x \neq y$. Then $x_\alpha \neq y_\alpha$ for some α. That is $P_\alpha x \neq P_\alpha y$.⟧ ∎

There are some important computations which need to be carried out for the purpose of familiarity with and development of the product topology.

★EXAMPLE 2. *Boxes.* Let $S_\alpha \subset X_\alpha$ for each $\alpha \in A$, and suppose that each S_α is nonempty. Then $\prod \{S_\alpha : \alpha \in A\}$ is a nonempty subset of $\prod X_\alpha$. A subset of this kind is called a *box*. For example, consider $\mathbf{R}^2 = \mathbf{R} \times \mathbf{R}$, and let $S_1 = [1, 2]$, $S_2 = [5, 7]$. Then $S_1 \times S_2$ is a closed 1 by 2 rectangle; while if, instead $S_2 = \{5, 7\}$, $S_1 \times S_2$ is a pair of horizontal line segments of unit length. The various S_α are called *sides* of the box. We make two important observations, (i) for the box $B = \prod \{S_\alpha : \alpha \in A\}$ we have $P_\alpha[B] = S_\alpha$ for each α, and (ii) B can also be written $\bigcap \{P_\alpha^{-1}[S_\alpha] : \alpha \in A\}$.

★EXAMPLE 3. *Finite products.* We shall show that the set of boxes with open sides (Example 2) is a base for the product topology of a finite product. First, each such box is open. ⟦Suppose G_i is a nonempty open subset of X_i for $i = 1, 2, \ldots, n$. Then the box $G_1 \times G_2 \times \cdots \times G_n$ is precisely the set of all $x = (x_1, x_2, \ldots, x_n)$ such that $x_i \in G_i$ for each i, that is, such that $P_i x \in G_i$ for each i. Thus $G_1 \times G_2 \times \cdots \times G_n = \bigcap_{i=1}^n P_i^{-1}[G_i]$, the intersection of a finite number of sets, each of which is open because each P_i is continuous.⟧ Next, let x be a point in the product and N a neighborhood of x. There exist open sets $G', G'', G''', \ldots$ in some of the X_i such that $x \in \bigcap P^{-1}[G] \subset N$, where the intersection is taken over $G', G'', G''', \ldots$ and P runs through the projections on the various spaces to which $G', G'', G''', \ldots$ belong ⟦Sec. 6.2, Remark 1⟧. We can write $\bigcap P^{-1}[G]$ as $\bigcap_{i=1}^n P_i^{-1}[G_i]$ by choosing $G_i = X_i$ whenever X_i is not already represented ⟦$P_i^{-1}[X_i] = X$⟧ Thus we have placed a box with open sides between x and N.

In Example 3 we noted that in a finite product a box with open sides is open, using in the proof that the box is a finite intersection of open sets. This fails in an infinite product and we shall see that such a box is not necessarily open ⟦Theorem 6.4.4⟧.

▲EXAMPLE 4. The topology generated by the boxes with open sides is called the *box topology*. It is larger than the product topology. ⟦Since the product topology is the smallest making the projections continuous, it suffices to prove that they are continuous with the box topology. Let G_β be an open set in X_β. Then $P_\beta^{-1}[G_\beta] = \bigcap P_\alpha^{-1}[G_\alpha]$ where $G_\alpha = X_\alpha$ for $\alpha \neq \beta$. Thus $P_\beta^{-1}[G_\beta]$ is a box with open sides, hence is open in the box topology.⟧

It is usually strictly larger ⟦Theorem 6.4.4⟧. The box topology fails to have many useful properties, for example, Sec. 7.4, Problem 104, which are enjoyed by the product topology. On the other hand, it is in some ways more natural ⟦Problems 101 and 102⟧.

Let G be a nonempty open set in $\prod \{X_\alpha : \alpha \in A\}$ and let $x \in G$. As usual, there exist $\alpha(1), \alpha(2), \ldots, \alpha(n) \in A$ and open sets $G_1, G_2, \ldots, G_n$ with $G_i \in X_{\alpha(i)}$ for each i, such that, setting $S = \bigcap_{i=1}^{n} P_{\alpha(i)}^{-1}[G_i]$, we have $x \in S \subset G$. Now we can also write $S = \bigcap P_\alpha^{-1}[G_\alpha]$ where $G_\alpha = X_\alpha$ wherever α is not one of the $\alpha(i)$, and $G_\alpha = G_i$ when $\alpha = \alpha(i)$. Thus $S = \prod \{G_\alpha : \alpha \in A\}$ and so S is a box; but it is a box of a very special kind, namely with open sides, all but a finite number of which are the whole factor space X_α.

THEOREM 6.4.3. *A base for the product topology of $\prod \{X_\alpha : \alpha \in A\}$ is the collection of sets $\prod \{G_\alpha : \alpha \in A\}$ where all the G_α are nonempty open subsets of X_α with $G_\alpha = X_\alpha$ for all but a finite number of $\alpha \in A$.*

This has just been proved.

THEOREM 6.4.4. *A nonempty open set in a product projects onto almost all the factors; that is, if G is a nonempty open subset of $\prod \{X_\alpha : \alpha \in A\}$, $P_\beta[G] = X_\beta$ for all but finitely many $\beta \in A$.*

Consider the set S defined in the proof of Theorem 6.4.3. If β is not one of the numbers $\alpha(1), \alpha(2), \ldots, \alpha(n)$, then $P_\beta[S] = X_\beta$. ⟦Let $t \in X_\beta$. Define $x \in \prod X_\alpha$ thus: $x_\beta = t$, $x_{\alpha(i)}$ is an arbitrary member of G_i for $i = 1, 2, \ldots, n$, and for $\alpha \neq \beta$, $\alpha \neq \alpha(i)$, x_α is an arbitrary member of X_α. Then $P_{\alpha(i)} x = x_{\alpha(i)} \in G_i$ for each i, so that $x \in S$, while $P_\beta x = x_\beta = t$.⟧ Every nonempty open set includes a set like S ⟦Theorem 6.4.3⟧ and the result follows. ▌

The result of the preceding theorem should be emphasized thus; in $\mathbf{R}^2$, the line ($y = x$) projects onto both axes but has empty interior. A set G with interior in an infinite product $\prod \{X_\alpha, \alpha \in A\}$ is "large" in a much more definite way than this, namely if B is the finite exceptional set mentioned in Theorem 6.4.4, and $x \in G$, then G contains all y such that $y_\beta = x_\beta$ for $\beta \in B$. There is no restriction on y_α if $\alpha \notin B$; thus G contains a copy of each such X_α, namely $\{y : y_\gamma = x_\gamma \text{ for all } \gamma \neq \alpha\}$. This copy resembles a "line parallel to X_α passing through x."

Concerning the exceptional factors in Theorem 6.4.4 we can say something ⟦Theorem 6.4.5⟧ about the way open sets project onto them. Recall that a function f is called *open* if $f[G]$ is open for every open set G.

★EXAMPLE 5. Let C be the complex plane with the Euclidean topology (of $\mathbf{R}^2$), and let $f(z) = |z|$ for all $z \in C$. Then $f: C \to C$ is not an open map since, for example, $f[C]$ has no interior. But if we let $Y = [0, \infty) \subset \mathbf{R}$, then the same $f: C \to Y$ is an open map.

▲EXAMPLE 6. Let $f: C \to C$ be an open map, where C is as in Example 5. Then f obeys the *maximum modulus principle*, namely, for any set S, and $b \in S$, if $|f(b)| \geq |f(s)|$ for all $s \in S$, then b is on the boundary of S. Thus $|f|$

takes its maximum on S on the boundary of S. ⟦If s is an interior point of S, there is a cell $N(s, \varepsilon) \subset S$. Then $f[N(s, \varepsilon)]$ is an open set, hence contains a cell $N[f(s), \delta]$, thus S contains a point mapping to any given point of the latter cell; there are such points of absolute value greater than $|f(s)|$.⟧

THEOREM 6.4.5. *Each projection of a product onto one of its factors is a continuous open map.*

Continuity is part of Theorem 6.4.1. Next let G be a nonempty open set in $\prod \{X_\alpha : \alpha \in A\}$, fix $\beta \in A$ and $t \in P_\beta[G]$, then $t = P_\beta x$ for some $x \in G$. Let S be chosen as in the proof of Theorem 6.4.3. If β is not one of the $\alpha(i)$, $P_\beta[S] = X_\beta$. ⟦See the proof of Theorem 6.4.4.⟧ Thus $P_\beta[G] = X_\beta$ which is open. If $\beta = \alpha(i)$, $P_\beta[G] \supset P_\beta[S] = G_i$ ⟦Example 1⟧ which is a neighborhood of t. ⟦It is sufficient to show that $t \in G_i$. But $t = P_\beta x$ and $x \in S$.⟧ Thus $P_\beta[G]$, being a neighborhood of each of its points, is open. ▮

Note that Theorem 6.4.5 does not generalize to weak topologies by onto functions ⟦Sec. 6.3, Problem 102⟧.

A product of open sets need not be open; that is, if G_α is an open set in X_α for each α, it does not follow that $\prod G_\alpha$ is an open set in $\prod X_\alpha$ ⟦Theorem 6.4.4⟧. However, the corresponding result for closed sets does hold.

THEOREM 6.4.6. *Any product of closed sets is closed; that is, if F_α is closed in X_α for each α, $\prod F_\alpha$ is closed in $\prod \{X_\alpha : \alpha \in A\}$.*

Note that $\prod F_\alpha = \bigcap \{P_\alpha^{-1}[F_\alpha] : \alpha \in A\}$. ⟦$x \in \prod F_\alpha$ if and only if $x_\alpha \in F_\alpha$ for all α.⟧ Each $P_\alpha^{-1}[F_\alpha]$ is closed by Theorem 6.4.5 ⟦Sec. 4.2, Problem 4⟧ and the result follows since any intersection of closed sets is closed. ▮

The next result is useful in situations where it is desired to show that some property is preserved under closure. See, for example, Sec. 11.3, Problem 13.

THEOREM 6.4.7. *Let X be a topological space, and $A, B \subset X$. Then $\bar{A} \times \bar{B} = \overline{A \times B}$, the latter closure being taken in the product topology.*

Since $\bar{A} \times \bar{B}$ is closed ⟦Theorem 6.4.6⟧ and includes $A \times B$, it includes $\overline{A \times B}$. Conversely, let $x \in \bar{A} \times \bar{B}$, say $x = (u, v)$ with $u \in \bar{A}$, $v \in \bar{B}$. Let N be an arbitrary neighborhood of x; then $N \supset P_1^{-1}[U] \cap P_2^{-1}[V]$, where U, V are neighborhoods of u, v. Choose $a \in U \cap A$, $b \in V \cap B$. ⟦Possible since, for example, $u \in \bar{A}$.⟧ Then $(a, b) \in (A \times B) \cap N$ ⟦$P_1(a, b) = a \in U$, $P_2(a, b) = b \in V$.⟧ Since this latter intersection is thus proved nonempty, it follows that $x \in \overline{A \times B}$. ▮

Problems

1. Let X, A be finite sets with, respectively, x, a, members. Show that X^A has x^a members.

★2. If X and Y are given larger topologies, the topology on $X \times Y$ also becomes larger.

3. Let the set $\{0, 1\}$ be denoted by the symbol $\mathbf{2}$. For any set S, let $p(S)$ be the collection of all subsets of S. Define $f: \mathbf{2}^S \to p(S)$ by $f(x) = \{s \in S: x_s = 1\}$. Show that f is one-to-one and onto. (For this reason, the symbol $\mathbf{2}^S$ is often used to denote the set of all subsets of S.)

4. A projection need not be closed. ⟦Consider the curve $y = 1/x$ in $\mathbf{R} \times \mathbf{R}$.⟧ See Problem 112.

5. Both projections of a nonopen subset of $X \times Y$ may be open. (Give an example in $\mathbf{R}^2$.)

★6. An uncountable product of nonindiscrete spaces cannot be first countable. Hence ⟦Theorem 6.4.2⟧ a product of semimetric spaces is semimetrizable if it is first countable. ⟦If $\{G_n\}$ is a countable base at $x \in \prod\{X_\alpha: \alpha \in A\}$, let $B_n = \{\alpha: P_\alpha G_n \neq X_\alpha\}$. Each B_n is finite by Theorem 6.4.4. Choose $\alpha \in A \setminus \bigcup B_n$. Every neighborhood of $P_\alpha x$ includes $P_\alpha G_n$ for some n, hence is all of X_α. But for some x, this is true for no α.⟧

---

101. An infinite product of discrete spaces, each having more than one point, is not discrete ⟦Theorem 6.4.4⟧. Describe net convergence. (Compare the convergence of numbers expressed in decimal form: for for each n, the nth term in the decimal expansion of a sequence of numbers is eventually equal to the nth term in the expansion of the limit.)

102. An infinite product of discrete spaces is discrete in the box topology.

103. Let $\mathbf{2}$ have the discrete topology. (See Problem 3.) Define $f: \mathbf{2}^\omega \to [0, 1]$ by taking $f(x)$ to be that real number whose ternary expansion ("decimal" in the scale of 3) is $.a_1\, a_2\, a_3 \cdots$ with $a_n = 2x_n$. (Note that each point of $\mathbf{2}^\omega$ is a sequence of zeros and ones.) Show that f is a homeomorphism. (The range of f is called the *Cantor discontinuum*, or *middle-third set*.)

104. Let f be a nonconstant analytic function in the sense of the theory of functions of a complex variable. Then f is open and the maximum modulus theorem follows ⟦Example 6⟧.

105. An open map from the complex plane to itself can have its minimum modulus in a region only on the boundary of the region or at a point where the function is zero.

106. A *free union* of two spaces X, Y is any space which is the union of two disjoint open subspaces homeomorphic with X, Y respectively. Show that any two free unions of X, Y are homeomorphic.

107. In Problem 106, the word "open" may not be omitted. ⟦Let X, Y each have one point.⟧

108. Show that any two topological spaces have a free union.

109. Let x, G be a point and a nonempty open set in $\prod X_\alpha$. Show that G contains a point y such that $y_\alpha = x_\alpha$ for all but finitely many α ⟦Theorem 6.4.4⟧.
110. Let **s** be the set of all sequences of complex numbers. Show how to metrize **s** in such a way that $x^n \to x$ if and only if $x_k^n \to x_k$ for each k, where $\{x^n\}$ is a sequence of points in **s** ⟦Theorems 6.4.1 and 6.4.2⟧.
111. In a countable product, every box with open sides is a G_δ. ⟦See (ii) in Example 2.⟧
112. Let F be a closed set in $X_1 \times X_2$ such that $P_1[F]$ is compact. Show that $P_2[F]$ is closed.

---

201. Show that the Cantor discontinuum is nowhere dense and perfect. (It also has $2^{\aleph_0} = c$ members.)
202. Every two nowhere dense perfect bounded sets A, B in $\mathbf{R}$ are homeomorphic. ⟦Assume both sets lie in $[0, 1]$ and contain 0, 1. Let their complements be G, H; $G = \bigcup G_i$, $H = \bigcup H_i$. By Sec. 3.3, Problem 203, there exists an order preserving map from the chain $\{G_i\}$ onto the chain $\{H_i\}$. This leads to a map from part of A to B, namely the endpoints of the G_i.⟧
203. $\mathbf{J}$ is homeomorphic with $\mathbf{Z}^\omega$. ⟦Use continued fraction expansions of irrational numbers.⟧
204. $[0, 1]^\omega$ is homogeneous.
205. The Cantor discontinuum is homogeneous.
206. Call X *composite* if there exist spaces Y, Z, each with more than one point such that X is homeomorphic with $Y \times Z$. Is $\mathbf{R}$ composite?

6.5 Quotients

A natural complement to the discussion of Section 6.3 is the study of those topologies on Y which make $f: X \to Y$ continuous, in particular, the largest one. This will be called the *quotient* topology. Since only $f[X]$ would enter such a discussion, we shall assume that f is onto Y. As our discussion proceeds it will appear that the weak and quotient discussions are dual, in the sense that where one mentions domain, the other mentions range; where one says smallest, the other will say largest, and so on. We have already seen one instance of this in that a weak topology is a smallest topology for the domain of a function while the quotient topology is a largest topology for the range of a function. Other examples will appear.

DEFINITION 1. *Let X be a topological space, Y a set, and $f: X \to Y$ onto. The quotient topology by f for Y is $\{S: f^{-1}[S]$ is open in $X\}$.*

This topology makes f continuous and is the largest which does. ⟦Let $f: X \to (Y, T)$ be continuous and $G \in T$. Then $f^{-1}[G]$ is open, hence $G \in Q$, the quotient topology.⟧

Before giving any example, it will be useful to have some techniques for recognizing when a space has the quotient topology. We transfer our attention to the map in this way, a function $f\colon X \to Y$ is called a *quotient map* if Y has the quotient topology. Thus if X, Y are topological spaces and $f\colon X \to Y$, Y might not have the quotient topology, that is, f might not be a quotient map; but then we could alter the topology of Y (give it the quotient topology) so as to make f a quotient map. The conditions in the next two results are sufficient but not necessary ⟦Example 1 and Problem 7⟧.

THEOREM 6.5.1. *Every continuous open onto map is a quotient map.*

Let $f\colon X \to (Y, T)$ be continuous, open, onto, and let Q be the quotient topology. Then $Q \supset T$ since f is continuous. Conversely, if $G \in Q, f^{-1}[G]$ is open, by definition of Q, hence $G = ff^{-1}[G] \in T$ since f is open. ∎

It follows that if $X = \prod X_\alpha$, each X_α has the quotient topology by the projection P_α ⟦Theorem 6.4.5⟧. This does not extend to weak topologies ⟦Problem 114⟧.

THEOREM 6.5.2. *Every continuous closed onto map is a quotient map.*

Let $f\colon X \to (Y, T)$ be continuous, closed, onto, and let Q be the quotient topology. Then $Q \supset T$ since f is continuous. Conversely if $G \in Q, f^{-1}[G]$ is open, $X \setminus f^{-1}[G]$ is closed and so its image, call it I, is T-closed. Then $G = \tilde{I} \in T$. ∎

THEOREM 6.5.3. *Every continuous map from a compact space onto a Hausdorff space is a quotient map.*

This follows from Theorem 6.5.2, with Theorem 5.4.8. ∎

A quotient map is (intuitively) the nearest thing possible to a homeomorphism.

★EXAMPLE 1. Let $X = [0, 1]$, $Y =$ 1-sphere $=$ unit circumference in $\mathbf{R}^2 = \{(x, y)\colon x^2 + y^2 = 1\}$. Then X and Y are not homeomorphic ⟦Sec. 5.2, Problem 4⟧. Define $f\colon X \to Y$ by $f(r) = e^{2\pi ri}$, (take $\mathbf{R}^2 =$ complex plane); then f is a homeomorphism on $(0, 1)$, and is continuous on $[0, 1]$. It is a quotient map ⟦Theorem 6.5.3⟧. Note that f is not an open map. ⟦Consider $[0, \frac{1}{2})$.⟧

★EXAMPLE 2. *Let X, Y be semimetric spaces and let T be a function from X onto Y which preserves distances, $d(Tx, Tx') = d(x, x')$. Then T is a quotient map.* It is obvious that T is continuous. Moreover T is both open and closed. We shall omit the proof of openness ⟦Problem 1⟧ and show that T is closed. (The result then follows from Theorem 6.5.2.) Let F be a closed subset of X, and let $\{y_n\}$ be a sequence in $T[F]$ with $y_n \to y$. Say $y = Tx$

⟦T is onto⟧ and $y_n = Tx_n$ with $x_n \in F$. Then $x_n \to x$ ⟦$d(x_n, x) = d(Tx_n, Tx) = d(y_n, y) \to 0$⟧, hence $x \in F$ and so $y \in T[F]$. ∎

Next is given a method of recognizing continuity for a function defined on a quotient space.

THEOREM 6.5.4. *Let $f: X \to Y$ be a quotient map and $g: Y \to Z$. Then g is continuous if and only if $g \circ f$ is continuous.*

Half of this is trivial ⟦Theorem 4.2.7⟧. Conversely, suppose that $g \circ f$ is continuous and let G be an open set in Z. Then $g^{-1}[G]$ is open. ⟦$f^{-1}[g^{-1}[G]] = (g \circ f)^{-1}[G]$ is open in X and Definition 1 yields the result.⟧ Hence g is continuous. ∎

Example 1 describes a sense in which the 1-sphere Y is constructed out of the line segment X. If we ignore the topology of Y, take the given map of X onto Y, and give Y the quotient topology, we get the 1-sphere. Since f is one-to-one except that $f(0) = f(1)$, there is a sense in which Y is constructed out of X by identifying the end-points and disturbing the topology as little as possible. We now formalize this process.

Let ρ be an equivalence relation on a set X. This relation partitions X into a collection of disjoint subsets, namely the subset, $q(x)$, which contains x is $\{y: y \,\rho\, x\}$, or, in words, the set of all y which lie in the relation ρ to x.

EXAMPLE 3. Let X be a group, S a subgroup, and let $x \,\rho\, y$ mean $xy^{-1} \in S$. Then $q(x)$ is the right coset Sx. ⟦If $y \in q(x)$, then $yx^{-1} \in S$ and $y = yx^{-1}x \in Sx$. Conversely, if $y = sx$, $s \in S$, then $yx^{-1} = s \in S$, so $y \,\rho\, x$.⟧

Clearly $y \in q(x)$ if and only if $q(y) = q(x)$. We now give a name and symbol to the collection of all subsets $q(x)$, $x \in X$. It is called the *quotient space of X by ρ*, and written X/ρ. The map $q: X \to X/\rho$ is called the quotient map. Note that each point of X/ρ is a subset of X. In Example 3 this yields the set X/S (in the usual group theoretic notation) and, as usual, X/S becomes a group in a natural way when S is an invariant subgroup.

★EXAMPLE 4. Let $f: X \to Y$ onto, and define $x \,\rho\, x'$ to mean $f(x) = f(x')$. Then $q(x) = \{x': f(x') = f(x)\}$. Here we have the interesting fact that f is constant on each $q(x)$ and we may define a map $F: X/\rho \to Y$ by $F[q(x)] = f(x)$, confident that if $q(x) = q(x')$, then $f(x) = f(x')$, so that the definition of F makes sense. The relation $f = F \circ q$ is illustrated by the diagram

$$\begin{array}{ccc} X & \xrightarrow{\;f\;} & Y \\ & {\scriptstyle q}\searrow \quad \nearrow{\scriptstyle F} & \\ & X/\rho & \end{array}$$

The function F is onto since f is, but, more importantly, F is one-to-one. ⟦Let $F(a) = F(b)$ where $a, b \in X/\rho$, say $a = q(x)$, $b = q(x')$. Then $f(x) = F(a) = F(b) = f(x')$ hence $x \rho x'$ so that $q(x) = q(x')$, that is, $a = b$.⟧

★EXAMPLE 5. Instead of starting with a relation ρ we might assume that a set X is partitioned into a disjoint collection C of subsets. Then an equivalence relation ρ is defined by letting $x \rho y$ mean that x, y are in the same member of C. In this case $C = X/\rho$ by a trivial computation, and so this process is equivalent to starting with ρ.

We now topologize the quotient space. Let ρ be an equivalence relation on a topological space X, and $q: X \to X/\rho$ the quotient map. The quotient topology by the map q (which is clearly onto) is taken as the natural topology for X/ρ and is called the *quotient topology by the relation* ρ. We shall now see (Theorem 6.5.5) that this process fulfills our initial program in that, in Example 4, X/ρ is homeomorphic with Y; in particular, in Example 1, X/ρ is the 1-sphere, where ρ is formed from f as in Example 4.

THEOREM 6.5.5. *Let X be a topological space, Y a set, and $f: X \to Y$ onto. Defining ρ as in Example 4, X/ρ is homeomorphic with Y, each having its quotient topology.*

In fact, the homeomorphism is F, as given in Example 4. We have seen in Example 4 that F is one-to-one and onto. Next F is continuous ⟦by Theorem 6.5.4, since $F \circ q = f$ is continuous⟧ and F^{-1} is continuous ⟦by Theorem 6.5.4, since $F^{-1} \circ f = q$ is continuous⟧. ∎

Example 4 and Theorem 6.5.5 show that every quotient by a function is homeomorphic to a quotient by a relation, and so the latter is general enough to cover both cases.

★EXAMPLE 6. Returning to Example 1, let C be the pair $\{0, 1\}$, and the collection of all singletons other than $\{0\}$, $\{1\}$. As in Example 5 this defines a relation ρ. ⟦In fact, $x \rho x'$ if and only if $x = x'$, or $x = 0$ and $x' = 1$, or $x = 1$ and $x' = 0$.⟧ By Theorem 6.5.5, X/ρ is (homeomorphic with) the 1-sphere.

Intuitively, it seems that the space X/ρ is obtained from X by squeezing the various equivalence classes to points, or by identifying the various points of any given equivalence class, for example if we identify 0 and 1 in the space $[0, 1]$ we obtain the 1-sphere. For this reason the quotient topology is also called the *identification topology*, and q the *identification map*.

▲EXAMPLE 7. Consider the region in $\mathbf{R}^2$ given by

$$\{(x, y): -3 \le x \le 3, \quad -1 \le y \le 1\}.$$

Identifying each point $(3, y)$ with $(-3, y)$ yields a cylinder; in constructing a model we would do exactly this, paste the ends AD and BC together.

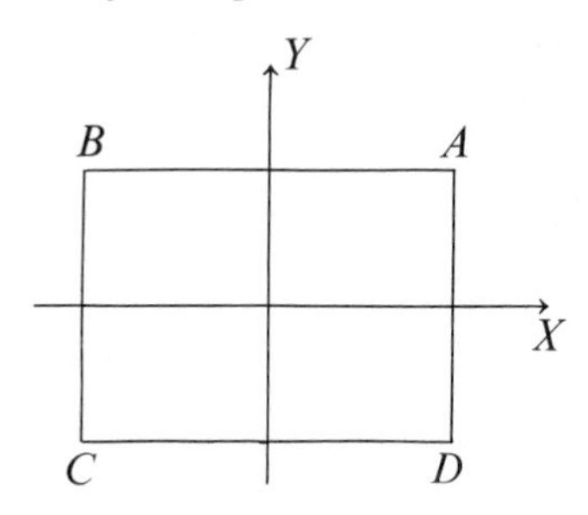

This could be taken as the definition of a cylinder; then it would be trivial that the cylinder is compact ⟦Theorem 5.4.4⟧, but difficult to see that the cylinder can be embedded in $\mathbf{R}^3$ (that is, is homeomorphic with a subset of $\mathbf{R}^3$).

▲ EXAMPLE 8. Using the region and diagram of Example 7, identify each point $(3, y)$ with $-3, -y)$. The resulting space is called the *Möbius strip*. The model made by twisting a rectangular piece of paper once and pasting BC and AD together, A on C and B on D, is one of the famous toys of mathematics.

▲ EXAMPLE 9. Using the region and diagram of Example 7, identify each point $(3, y)$ with $(-3, -y)$, and each point $(x, 1)$ with $(-x, -1)$. The resulting space is called the *projective plane*. It can also be considered, and more usually is, as a disc in $\mathbf{R}^2$ with diametrically opposite points identified.

The reader may have noticed that we did not characterize net and filter convergence in quotients. This is because there is no simple necessary and sufficient condition comparable with those of sup, weak, and product topologies, for example, Theorem 6.4.1. Of course, if q is the identification map: $X \to X/\rho$ and if $x_\delta \to x$ in X then $qx_\delta \to qx$ in X/ρ, but no sort of converse holds without restriction. Some details are given in Problems 107, 202, and 203. (There is a converse result for locally compact groups. See [Varoupolos].)

Problems

1. The map in Example 2 is open.
2. If the map in Example 2 had not been assumed to be onto, it need be neither open nor closed.
3. "Hausdorff" cannot be omitted in Theorem 6.5.3. ⟦$X = [0, 1]$, $Y = [0, 1]$ with the cofinite topology, $f =$ identity.⟧

★4. Let $f: X \to Y$ be a quotient map, and $S \subset Y$. Show that S is closed if and only if $f^{-1}[S]$ is closed.

5. A one-to-one quotient map must be a homeomorphism.

6. Let D be the unit disc ($|z| \leq 1$), C the circumference ($|z| = 1$), and E the circumference ($|z - 3| = 1$) in the complex plane. Let $X = D \cup E$. Show that C is homeomorphic with a retract of X. ⟦Namely E. For example, let $f(z) = 2$ for $z \in D, f(z) = z$ for $z \in E$.⟧ Compare Problem 210 which shows that C is not a retract of X. This point will be discussed in Sec. 14.3, Example 5.

7. Give an example of an open quotient map which is not closed ⟦the projection of $\mathbf{R}^2$ on $\mathbf{R}$⟧, and a closed quotient map which is not open ⟦Example 1⟧.

8. Let T be the simple extension of the Euclidean topology of $\mathbf{R}^2$ by the singleton $\{(0, 0)\}$. Let $f: (\mathbf{R}^2, T) \to \mathbf{R}$ be given by $f(x, y) = x$. Show that f is a quotient map which is neither closed nor open.

9. The restriction of a quotient map to an open subset need not be a quotient map. ⟦In Example 1, consider $[0, 1)$. See Problem 5.⟧

10. A continuous map with a continuous right inverse must be a quotient map. ⟦Let $f: X \to Y$, $g: Y \to X$ with $f \circ g =$ identity. If G is open, $f^{-1}[G]$ is open since f is continuous. Conversely, if $f^{-1}[G]$ is open, $G = (f \circ g)^{-1} = g^{-1}[f^{-1}[G]]$ is open.⟧

11. A retraction is a quotient map. ⟦Problem 10. Its right inverse is the inclusion map.⟧

---

101. For $x, y \in [0, 1]$, define $x \sim y$ by $x - y$ is rational. Show that the resulting quotient space is indiscrete.

102. Give an example of a continuous open map of $[0, 1]$ onto an infinite indiscrete space ⟦Problem 101⟧.

103. There exists an open map from $[0, 1]$ onto any topological space of cardinality no greater than c ⟦Problem 102⟧.

104. The *torus* is the space obtained if, in Example 7, each point of AD is identified with the point opposite it on BC (A is identified with B, D with C) and each point of BA is identified with the point on CD directly below it. Sketch a torus. Show that the torus is homeomorphic with the product of the one sphere with itself.

105. Let $x \,\rho\, y$ mean that x, y belong to the same component of X. Show that X/ρ is totally disconnected but need not be discrete. ⟦For example, let ρ be equality.⟧

106. Theorem 6.5.4 characterizes quotient maps in the sense that if it holds for all g, f must be a quotient map. ⟦Let Y_1 be Y with the quotient topology, and $g: Y \to Y_1$ the identity.⟧

107. Fix $a \in X$, $S \subset X$ with $a \in \bar{S} \setminus S$ and suppose that no sequence of members of S converges to a ⟦for example, Sec. 3.1, Problem 4⟧.

Define $x \rho y$ if either $x = y$ or both x and y belong to S. (The equivalence classes are S and singletons.) In X/ρ, we consider the constant sequence $u_n = S$ for all n. Show that $u_n \to \{a\}$ but no sequence of members of u_n converges to a.

108. Let T, T' be topologies for a set S. Let $X = (S, T)$, $Y = (S, T')$ and U the free union of X, Y. Define $f: U \to S$ by $f(s) = s$. (f is the identity on each copy of S.) Show that the quotient topology by f is $T \cap T'$. (Compare Sec. 6.7, Problem 114.)
109. Let $f: X \to Y$ with X locally connected and f onto. Show that Y need not be locally connected if f is continuous, but must be if f is a quotient map.
110. Let the equivalence relation ρ partition X into connected sets. Show that a closed set S in X/ρ is connected if and only if $q^{-1}S$ is connected.
111. In Problem 110, replace closed by open.
112. Let $f: X \to Y$ be onto, where Y is a topological space. If X is given the weak topology by f, show that f becomes a quotient map.
113. Let $f: X \to Y$ be onto. Show that X may have more than one topology which makes f a quotient map. ⟦f is open if X has $w(f)$; in Example 1, f is not open.⟧
114. Let f, g be maps from a set X onto two topological spaces Y_f, Y_g, and let X have $w(f, g)$. Show that g need not be a quotient map. ⟦Take Y_f discrete.⟧
115. Construct a continuous map from the unit disc $D = \{z: |z| \le 1\}$ onto its circumference. ⟦Map D onto a real segment which can be wrapped around S_1.⟧

---

201. Let S be a dense subgroup of a topological group X. Show that X/S is indiscrete.
202. In Problem 201, let $x \in X$. Define the (constant) sequence $\{u_n\}$ of points in X/S by $u_n = xS$ for all n. (A left coset.) Then $u_n \to S$ (the identity). However, if x is not a sequential accumulation point of S, no sequence $\{x_n\}$ with $x_n \in u_n$ exists such that x_n converges to the identity element of X or to any other member of S.
▲203. Give an example to illustrate Problem 202. ⟦See [Wilansky (a), Section 13.3, Problems 16, 28].⟧
204. Is there a continuous open map from **R** onto an infinite cofinite space? (Compare Sec. 5.1, Problem 208.)
205. Must a continuous almost open map onto be a quotient map?
206. Let Y be an infinite topological space and $f: X \to Y$ onto. Is it possible for X to have more than one Hausdorff topology making f a quotient map? (Compare Problem 113.)
207. Let D be the unit disc ($|z| \le 1$) in the complex plane, and S_1 its boundary (the 1-sphere). Let $f: D \to S_1$ be continuous. Show that

there exists continuous $g: D \to \mathbf{R}$ such that $f(z) = \exp[2\pi i g(z)]$. ⟦$g = (2\pi i)^{-1} \log f$.⟧

208. With f as in Problem 207 there must exist antipodal (diametrically opposite) points u, v of S_1 with $f(u) = f(v)$. ⟦Let $h(\theta) = g(-e^{i\theta}) - g(e^{i\theta})$. Then $h(0)$, $h(\pi)$ have opposite sign.⟧
209. (*The Borsuk–Ulam theorem.*) A continuous map F of S_2 into $\mathbf{R}^2$ must carry some two antipodal points to the same point. (Hence there are always antipodal points on Earth with the same temperature and pressure readings.) ⟦Otherwise, with $f = h/|h|$, where $h(x, y) = F(x, y, z) - F(-x, -y, -z)$, $z = (1 - x^2 - y^2)^{1/2}$, Problem 208 is contradicted.⟧
210. (*The Brouwer fixed-point theorem.*) S_1 is not a retract of D, the disc $(x^2 + y^2 \leq 1)$ ⟦Problem 208; compare Problem 115⟧, hence a continuous map f of D to itself must have a *fixed point* (a point u with $f(u) = u$). ⟦If not, map each u to the place where the line from $f(u)$ to u meets S_1.⟧
211. Let f map the complex plane continuously into itself, and satisfy $f(z)/z \to 1$ as $z \to \infty$. Show that $f(z) = 0$ for some z. ⟦With $g(z) = z - f(z)$, g maps some disc $\{z: |z| \leq n\}$ into itself. Apply Problem 210.⟧
212. Call $f: X \to Y$ *pseudo-open* if for all $y \in f[X]$ and neighborhood N of $f^{-1}(y)$, fN must be a neighborhood of y. Show that every open and every closed map is pseudo-open.
213. Every continuous pseudo-open onto map is a quotient map.

6.6 Continuity

One of the central tools of topology is the ability to recognize when a function is continuous. A particular aspect of this is the recognition of a homeomorphism, when a function and its inverse are both continuous. (Examples are Theorems 4.2.2., 6.5.4., and 5.4.9.) In part (i) of the following theorem, X is a topological space, ρ is an equivalence relation on X, and $q: X \to X/\rho$ is the quotient map; in parts (ii), (iii), and (iv), A is an indexing set, X, X_α are topological spaces for each $\alpha \in A$, Y is a set, T_α is a topology for Y for each $\alpha \in A$, f_α is a map from Y to some topological space Z_α for each $\alpha \in A$, and P_α is the projection of $\prod X_\alpha$ onto X_α.

THEOREM 6.6.1

(i) *A function $f: X/\rho \to Y$ is continuous if and only if $f \circ q$ is continuous.*

(ii) *A function $f: X \to (Y, \bigvee T_\alpha)$ is continuous if and only if $f: X \to (Y, T_\beta)$ is continuous for each $\beta \in A$.*

(iii) *A function $f: X \to (Y, w\{f_\alpha: \alpha \in A\})$ is continuous if and only if $f_\beta \circ f$ is continuous for each $\beta \in A$.*

(iv) *A function* $f: X \to \prod \{X_\alpha: \alpha \in A\}$ *is continuous if and only if* $P_\beta \circ f$ *is continuous for each* $\beta \in A$.

Part (i) is Theorem 6.5.4. Half of part (ii) is trivial since $T_\beta \subset \bigvee T_\alpha$. Conversely, suppose that $f: X \to (Y, T_\beta)$ is continuous for each β. Let $x_\delta \to x$ in X. Then $f(x_\delta) \to f(x)$ in (Y, T_β) for each β ⟦Theorem 4.2.2⟧. Hence $f(x_\delta) \to f(x)$ in $(Y, \bigvee T_\alpha)$ ⟦Theorem 6.2.1⟧. Thus f is continuous when Y has $\bigvee T_\alpha$.

Next we consider part (iii) in case A has one member. Thus we are asserting that $f: X \to (Y, w(g))$ is continuous, where $g: Y \to Z$, if and only if $g \circ f$ is continuous. Half of this is trivial ⟦Theorem 4.2.7⟧. Conversely, let $g \circ f$ be continuous. Let $x_\delta \to x$ in X. Then $g[f(x_\delta)] \to g[f(x)]$ and so $f(x_\delta) \to f(x)$ in $w(g)$ ⟦Lemma 6.3.1⟧. Thus f is continuous. Part (iii) now follows from part (ii) and the definition of $w\{f_\alpha\}$. Part (iv) follows from part (iii) and the definition of the product topology. ∎

EXAMPLE 1. *Joint continuity*. If $f: X \times Y \to Z$, f is a function of two variables. We call f *separately continuous* at (x', y') if $f(x, y)$ is continuous in x for $y = y'$, and continuous in y for $x = x'$. (More formally, the functions $g_{y'}: X \to Z$ and $h_{x'}: Y \to Z$ are continuous, where $g_{y'}(x) = f(x, y')$ and $h_{x'}(y) = f(x', y)$.) We call f *jointly continuous* (or just *continuous*, of course) if it is continuous. A standard example of calculus shows that $f: \mathbf{R}^2 \to \mathbf{R}$ can be separately continuous without being jointly continuous. ⟦$f(x, y) = xy/(x^2 + y^2)$ for $(x, y) \neq (0, 0)$; $f(0, 0) = 0$, defines a function f which is everywhere separately continuous, but is not jointly continuous at $(0, 0)$.⟧ *A function f is jointly continuous if and only if $x_\delta \to x$ in X, $y_\delta \to y$ in Y implies $f(x_\delta, y_\delta) \to f(x, y)$ in Z* ⟦Theorem 6.4.1⟧. Notice that all these nets are defined on the same directed set.

★EXAMPLE 2. *A semimetric is jointly continuous.* This means that if (X, d) is a semimetric space, $d: X \times X \to \mathbf{R}$ is continuous. This follows from the inequality $|d(x, y) - d(x', y')| \leq d(x, x') + d(y, y')$ ⟦Sec. 2.2, Problem 3⟧.

EXAMPLE 3. *Pointed spaces and injections.* A *pointed set* is merely a nonempty set and a distinguished point in it. For example we may distinguish the identity element in each group. The distinguished point is called the *base point*. Suppose given a family $\{X_\alpha: \alpha \in A\}$ of pointed topological spaces. In each X_α, let 0_α be the base point. The point $0 \in \prod X_\alpha$ defined by $P_\alpha 0 = 0_\alpha$ is called the *base point of the product*. For example, if each X_α is also a linear space we could take $0_\alpha = 0$. For each $\beta \in A$, define the *injection* $I_\beta: X_\beta \to \prod X_\alpha$ by $I_\beta(t) = x$, where $x_\alpha = 0_\alpha$ for $\alpha \neq \beta$, and $x_\beta = t$. Then $P_\beta[I_\beta(t)] = t$ so that P_β is a left inverse for I_β. It follows immediately from Theorem 6.6.1

that each I_β is continuous. More, since P_β is continuous, I_β is a homeomorphism (into).

Although the product is a special case of the weak topology by functions, there is a sense in which they are equivalent; namely, the weak topology by a family of maps can usually be looked upon as the relative topology of a subspace of a product ⟦Theorem 6.6.2⟧. This remark is important because the standard metrization theorems (Sections 10.1 and 10.3) will be proved by identifying topologies as weak topologies by countable families of maps to metric spaces. Theorem 6.6.2 then shows how these spaces can be embedded in countable products of metric spaces. See Corollary 10.1.2 and its following Remark for details and for a comparison of the two methods of metrization.

THEOREM 6.6.2. *Let X be a set and $\{f_\alpha : \alpha \in A\}$ a separating family of functions, each f_α from X to some topological space Y_α. Let X be given the topology $w\{f_\alpha : \alpha \in A\}$. Then X is homeomorphic with a subspace of $\prod Y_\alpha$.*

For each $x \in X$, define $y \in \prod \{Y_\alpha : \alpha \in A\}$ by $y_\alpha = f_\alpha(x)$. This defines a map $h: X \to \prod Y_\alpha$; h is continuous. ⟦For each $\beta \in A$, $P_\beta \circ h = f_\beta$ is continuous, hence h is continuous, by Theorem 6.6.1⟧ h is one-to-one ⟦if $h(x) = h(x')$, $f_\alpha(x) = f_\alpha(x')$ for all $\alpha \in A$, hence $x = x'$⟧, h is a homeomorphism. ⟦Let x_δ be a net in X and $h(x_\delta) \to h(x)$. Then $P_\alpha[h(x_\delta)] \to P_\alpha[h(x)]$ for each α since P_α is continuous; that is, $f_\alpha(x_\delta) \to f_\alpha(x)$ for each α, and so $x_\delta \to x$ by Theorem 6.6.1. Hence h^{-1} is continuous.⟧ ∎

WARNING. Theorem 6.6.2 does not show that X can be embedded except when X has $w\{f_\alpha\}$. If X has some other topology, there may be no such embedding even if all the f_α are continuous; for example, if X is not completely regular, there could be no embedding of X in $[0, 1]^A$ even though X might allow a separating family of continuous functions to $[0, 1]$. (An example can easily be constructed from Sec. 6.2, Example 2, starting with the Euclidean topology for **R**.)

It is interesting to ask which properties are preserved by the product operation. Separation, compactness, completeness, and so on are studied elsewhere in this book. Here we shall consider connectedness. The relevant results are given in Theorem 6.6.3 and in Sec. 6.7, Problem 102. A special case of Theorem 6.6.3 is that any finite product of connected spaces is connected. This also follows very easily from Theorem 5.2.1 (ii). ⟦Fix $z \in \prod X_i$. Let $A = \{x : x_i = z_i \text{ for } i = 2, 3, \ldots, n\}$, $B_t = \{x : x_1 = t\}$ for $t \in X_1$. Then A meets every B_t (in $(t, z_2, z_3, \ldots, z_n)$.) We may assume that each B_t is connected, and apply induction and Example 3.⟧ The proof of Theorem 6.6.3 given here was suggested to me by D. H. Taylor.

THEOREM 6.6.3. *A product of connected spaces is connected.*

Let $f: \prod \{X_\alpha : \alpha \in A\} \to \mathbf{2}$ be continuous, where each X_α is connected.

Say $\mathbf{2} = \{0, 1\}$ and $f(z) = 0$ for some z. It will be sufficient to show $f(x) = 0$ for all x ⟦Lemma 5.2.1⟧. Let $U = f^{-1}[\{0\}]$. Then U is a neighborhood of z, since $\{0\}$ is open. Thus U includes some set

$$V = \bigcap \{P_{\alpha_i}^{-1} [\{z_{\alpha_i}\}]: i = 1, 2, \ldots, n\}.$$

⟦Indeed this inclusion holds with each $\{z_{\alpha_i}\}$ replaced by some neighborhood of z_{α_i}.⟧ It follows that

$$f(z_{\alpha_1}, z_{\alpha_2}, \ldots, z_{\alpha_n}; x_\alpha) = 0 \quad \text{for all } x, \tag{6.6.1}$$

where $(a_1, a_2, \ldots, a_n; x_\alpha)$ is that member w of $\prod X_\alpha$ such that $w_{\alpha_i} = a_i$ for $i = 1, 2, \ldots, n$ and $w_\alpha = x_\alpha$ for all other α. ⟦Since such $w \in V$, $V \subset U$, and $f[U] = 0$.⟧ We shall now see that

$$f(x_{\alpha_1}, z_{\alpha_2}, z_{\alpha_3}, \ldots, z_{\alpha_n}; x_\alpha) = 0 \quad \text{for all } x. \tag{6.6.2}$$

⟦Let $H_1 = I_{\alpha_1}[X_{\alpha_1}]$ in the notation of Example 3, taking the base point of the product to be the argument of f in (6.6.2). Then H_1 is homeomorphic with X_{α_1}, hence is connected, and so f is constant on H_1, by Lemma 5.2.1. This constant must be zero by (6.6.1) since $(z_{\alpha_1}, z_{\alpha_2}, \ldots, z_{\alpha_n}; x_\alpha) \in H_1$. Hence (6.6.2) follows since the argument of f belongs to H_1.⟧

The same proof shows that

$$f(x_{\alpha_1}, x_{\alpha_2}, z_{\alpha_3}, z_{\alpha_4}, \ldots, z_{\alpha_n}; x_\alpha) = 0 \quad \text{for all } x \tag{6.6.3}$$

⟦now consider $I_{\alpha_2}[X_{\alpha_2}]$ with base point the argument of f in (6.6.3)⟧; and, continuing, that $f(x_{\alpha_1}, x_{\alpha_2}, \ldots, x_{\alpha_n}; x_\alpha) = 0$ for all x. But this says, simply, $f(x) = 0$ for all x. ∎

Problems

1. In Example 3, let each X_α be a T_1 space. Then the injection makes X_α homeomorphic with a closed subset of the product. ⟦It is $\bigcap \{P_\beta^\perp: \beta \neq \alpha\}$ where $P_\beta^\perp = \{x: x_\beta = 0_\beta\}$.⟧
2. T_1 cannot be omitted in Problem 1. ⟦Make the X axis not closed in $\mathbf{R}^2$.⟧
3. Each factor of a product is homeomorphic with a retract of the product ⟦Example 3 and Sec. 4.2, Problem 32⟧. (Problem 1 for T_2 spaces is a consequence ⟦Sec. 4.2, Problem 27⟧.)

★4. If $f: X \times Y \to Z$ is (jointly) continuous, it is separately continuous. ⟦$X \times \{y'\}$ is a subspace of $X \times Y$. Now use Theorem 4.2.3. Another proof is obtained by taking one of the nets in Example 1 to be constant.⟧

---

101. Let a group be given the cofinite topology. The map $x \to x^{-1}$ is

continuous, and the group operation is separately but not necessarily jointly continuous. ⟦In **R**, $1/n \to 2$, $n \to 3$ but $n \cdot (1/n) \not\to 6$.⟧

---

201. Is the product topology characterized by Theorem 6.6.1 in the same way as the quotient? (See Sec. 6.5, Problem 106.)
202. Every separable metric space is homeomorphic into **s**. ⟦With $\{r_n\}$ dense let $f(x) = \{d(x, r_n)\}$. Theorem 6.6.1 shows f continuous, and Sec. 3.5, Problem 108 shows f^{-1} continuous.⟧
203. Every separable metric space is homeomorphic into I^ω where I is the interval $[0, 1]$. ⟦Apply Problem 202 and the fact that $\mathbf{s} = \mathbf{R}^\omega$ is homeomorphic with $(0, 1)^\omega$.⟧
204. $\mathbf{J}^\omega$ is homeomorphic with **J**. ⟦Replace **J** by $\mathbf{J}_1 = \mathbf{J} \cap (0, 1)$. Write each x_k, for $x \in \mathbf{J}_1^\omega$, as a continued fraction and arrange the double sequence of coefficients as a single sequence. See Sec. 6.4, Problem 203, and [Sierpinski (b), p. 140].⟧
205. Every countable metric space is homeomorphic into **Q**. ⟦It is homeomorphic with a (countable) subset S of **J** by Problems 202, 204, and Sec. 1.2, Problem 209. Then $S \cup \mathbf{Q}$ is homeomorphic with **Q** by Sec. 3.3, Problem 203.⟧
206. There exists only one countable metric space without isolated points ⟦Problem 205 and its hint⟧.
207. **Q** and **J** are composite. ⟦$\mathbf{Q} = \mathbf{Q} \times \mathbf{Q}$, $\mathbf{J} = \mathbf{J} \times \mathbf{J}$. See Problems 206, 204.⟧
208. **J** is homeomorphic into $\mathbf{2}^\omega$. ⟦Map **Z** into $\mathbf{2}^\omega$, hence **J** into $(\mathbf{2}^\omega)^\omega$ by Sec. 6.4, Problem 203.⟧

6.7 Separation

We next study the separation axioms with a view to finding when they are enjoyed by various combinations (sup, weak, product, quotient), of spaces with various amounts of separation. Only the more important results will be given. A few more are in the problems, and still others in the Tables.

We shall follow the earlier plan of this chapter, starting with sups, and specializing to weak and product topologies.

THEOREM 6.7.1. *Let Φ be a family of topologies on a set X. If each $T \in \Phi$ is, respectively, T_0, T_1, T_2, regular, T_3, completely regular, $T_{3\frac{1}{2}}$; then $T' \equiv \bigvee \Phi$ is the same.*

For T_0, T_1, T_2, this is trivial since $T' \supset T$ for each $T \in \Phi$ ⟦Sec. 4.1, Problem 4⟧. Next suppose that each T is regular. Let $x \in G \in T'$. Then there exists $G_1, G_2, \ldots, G_n$ with each G_i open in some $T_i \in \Phi$, and $x \in \bigcap G_i \subset G$. For each i there is a T_i-closed, T_i-neighborhood N_i of x with $x \in N_i \subset G_i$

⟦Theorem 4.1.3⟧. Each N_i is a closed neighborhood of x in T', since T' is a larger topology than each T_i, hence $N = \bigcap N_i$ is a closed neighborhood of x in T', and $x \in N \subset G$. Thus T' is regular ⟦Theorem 4.1.3⟧.

Finally suppose that each T is completely regular. ⟦The cases T_3, $T_{3\frac{1}{2}}$ are trivial by collecting what has been done.⟧ Let F be a T' closed set, and $x \notin F$. There exist $G_1, G_2, \ldots, G_n$ with each G_i open in some $T_i \in \Phi$, and $x \in \bigcap G_i \subset \tilde{F}$. For each i, there is a T_i-continuous real-valued function f_i defined on X which separates x from $\tilde{G}_i$. Each f_i is T'-continuous since $T' \supset T_i$ ⟦Sec. 4.2, Problem 1⟧ and so f is T'-continuous, where $f = f_1 \wedge f_2 \wedge \cdots \wedge f_n$ ⟦Theorem 4.2.9⟧. Moreover, $0 \leq f \leq 1$, $f(x) = 1$, and $f(y) = 0$ for all y outside of any one G_i, hence for all $y \in F$. Thus T' is completely regular. ▌

For weak topologies in general we cannot expect T_0, T_1, T_2 separation; for example, with $f: \mathbf{R}^2 \to \mathbf{R}$ given by $f(x, y) = x$, $w(\mathbf{R}^2, f)$ is generated by vertical open strips and is not even T_0. We have seen what type of restriction is needed in Theorem 6.3.2, and this will be sufficient for our purposes.

THEOREM 6.7.2. *Let X be a set and Φ a family of functions f, each f defined on X and with range in some regular space Y_f. Then $w(X, \Phi)$ is regular. If each Y_f is completely regular, $w(X, \Phi)$ is completely regular.*

In view of Theorem 6.7.1, it is sufficient to prove these results for the case in which Φ contains only one function. So let $f: X \to Y$, with Y regular. Let $x \in G \in w(f)$. Then $G = f^{-1}[U]$, where U is some open neighborhood of $f(x)$. There exists a closed neighborhood N of $f(x)$ with $N \subset U$ ⟦Theorem 4.1.3⟧ and, since f is continuous on $[X, w(f)]$, $f^{-1}[N]$ is a closed neighborhood of x with $f^{-1}[N] \subset G$. Thus $w(f)$ is regular. Next, let Y be completely regular. Let F be a $w(f)$ closed set in X and $x \in X \setminus F$. Then $F = f^{-1}[V]$ where V is a closed set in Y ⟦Sec. 6.3, Problem 3⟧. Moreover $f(x) \notin V$, so there is a real-valued continuous function g defined on Y which separates $f(x)$ from V. Then $g \circ f$ is continuous on X ⟦Theorem 4.2.7⟧, and separates x from F. ▌

THEOREM 6.7.3. *A product of Hausdorff, regular, T_3, completely regular, or Tychonoff spaces is the same, that is, Hausdorff, regular, T_3, completely regular, or Tychonoff.*

For regular and completely regular spaces this follows from Theorems 6.7.1 and 6.7.2. For T_3 and $T_{3\frac{1}{2}}$ spaces the result is trivial from the other parts. For T_2 spaces it is sufficient to show that the set of projections is separating ⟦Theorem 6.3.2⟧. If $x, y \in \prod \{X_\alpha: \alpha \in A\}$ with $x \neq y$, it must be that $x(\alpha) \neq y(\alpha)$ for some $\alpha \in A$ ⟦otherwise x, y are identical functions, hence equal⟧. For this α, $P_\alpha(x) = x(\alpha) \neq y(\alpha) = P_\alpha(y)$. ▌

An important special case of Theorem 6.7.2 is that in which X is a topological space, and $\Phi = C(X)$; or $\Phi = C^*(X)$, the bounded members of $C(X)$.

THEOREM 6.7.4. *For a topological space X, the weak topologies by $C(X)$ and $C^*(X)$ are the same. This topology is always completely regular. It coincides with the original topology of X if and only if the latter is completely regular.*

Let w, w^* be the weak topologies by C, C^* respectively. Since $C(X) \supset C^*(X)$ we have $w \supset w^*$. Conversely, let $f \in C(X)$. Then every set of the form $(f \geq t)$, where t is a real number, is w^*-closed. ⟦Let

$$g(x) = |[f(x) - t] \wedge 0| \wedge 1.$$

Then $g \in C^*(X)$ since $0 \leq g(x) \leq 1$ and by Theorems 4.2.8 and 4.2.9. Moreover $(f \geq t) = g^{\perp}$.⟧ Similarly, every set $(f \leq t)$ is w^*-closed, and so f is w^*-continuous. ⟦For any open interval (a, b) in $\mathbf{R}$, $f^{-1}[(a, b)]$ is the complement of $(f \geq b) \cup (f \leq a)$, hence is w^*-open. But every open set in $\mathbf{R}$ is a union of such intervals.⟧ By definition of w, it follows that $w \subset w^*$. ⟦w is the smallest topology making each $f \in C(X)$ continuous.⟧ That w is completely regular is immediate from Theorem 6.7.2. ⟦$Y_f = \mathbf{R}$ for all f.⟧ Finally, suppose that (X, T) is completely regular. Then $T \supset w$. ⟦w is the smallest topology making each $f \in C(X)$ continuous.⟧ Conversely, let $\mathcal{F}$ be a filter, and $a \in X$, and suppose $\mathcal{F} \to a$ in w. Then $\mathcal{F} \to a$ in T ⟦Theorems 6.3.1 and 4.3.2⟧. Thus $w \supset T$ ⟦Theorems 4.2.2 and 4.2.5⟧. ∎

EXAMPLE 1. *The diagonal.* The product X^A, that is, $\prod \{X_\alpha : \alpha \in A\}$ in which all X_α are the same space X, is the set of all functions on A to X. It has an interesting subset known as the *diagonal*, namely the set of all constant functions. For example, the diagonal in $\mathbf{R} \times \mathbf{R}$ is the line $(y = x)$. *The diagonal D in $X \times X$ is closed if and only if X is a Hausdorff space.* ⟦Let x_δ be a net in X. Then D is closed if and only if $(x_\delta, x_\delta) \to (x, y)$ implies $y = x$ (that is, $(x, y) \in D$), and this is true if and only if $x_\delta \to x$, $x_\delta \to y$ implies $y = x$, by Theorem 6.4.1, and this is true if and only if X is T_2, by Theorem 4.1.2.⟧

▲EXAMPLE 2. The *graph* of a function $f: X \to Y$ is $\{(x, fx): x \in X\}$; it is a subset of $X \times Y$; if it is a closed subset we say that f is a *function with closed graph.* The result of Example 1 says that *X is a Hausdorff space if and only if the identity map on X has closed graph.* A useful equivalence is the result: *A function $f: X \to Y$ has closed graph if and only if whenever $x_\delta \to x$ in X and $f(x_\delta) \to y$ in Y, it follows that $y = f(x)$.* ⟦Suppose that f has closed graph, that $x_\delta \to x$ and $f(x_\delta) \to y$. Then $(x_\delta, fx_\delta) \in G$, the graph of f, and $(x_\delta, fx_\delta) \to (x, y)$ by Theorem 6.4.1. Hence $(x, y) \in G$; that is $y = f(x)$. Conversely, if the stated implication holds, suppose that $(x, y) \in \bar{G}$. There is a net $(x_\delta, y_\delta) \in G$ with $(x_\delta, y_\delta) \to (x, y)$, $y_\delta = f(x_\delta)$. An application of Theorem 6.4.1 reveals that $x_\delta \to x$, $y_\delta \to y$. Thus $y = f(x)$.⟧

▲EXAMPLE 3. *The product of two T_4 spaces need not be normal.* Indeed our example is the product of a T_4 Lindelöf space with itself. In view of Theorems 6.7.3 and 5.3.5 this yields the result that *the product of two Lindelöf spaces need not be a Lindelöf space* (even when they are both the same space and it is T_4). Let X be $\mathbf{R}$ with the right half open interval topology. Then $X \times X$ is separable ⟦a nonempty open set contains a rectangular region with two sides included and two not; such a region meets $\mathbf{Q} \times \mathbf{Q}$; hence $\mathbf{Q} \times \mathbf{Q}$ is dense⟧. Let D be the set $(y = -x)$, a straight line with slope -1. Then D is closed ⟦it is closed in the (smaller) Euclidean topology, see Sec. 6.4, Problem 2⟧, also D is discrete. ⟦Given $P \in D$, let S be a rectangular region with horizontal and vertical sides, with its lower and left sides included and the other two not, which has P as its lower left-hand corner. Then S is the product of two right half open intervals, hence is open; also $S \cap D = \{P\}$, thus $\{P\}$ is open.⟧ The result follows by Sec. 5.3, Example 3. ▮

Finally we turn to separation in X/ρ where X is a topological space, and ρ an equivalence relation. We shall also use q, the quotient map (Section 6.5).

THEOREM 6.7.5. *X/ρ is a T_1 space if and only if the members of X/ρ are closed subsets of X. X/ρ is indiscrete if all members of X/ρ are dense in X.*

If X/ρ is a T_1 space, let $S \in X/\rho$, then $\{S\}$ is closed ⟦Theorem 4.1.1⟧. Thus $q^{-1}[\{S\}]$ is a closed subset of X since q is continuous. This is S. Conversely, suppose that the members of X/ρ are closed, let $S \in X/\rho$, then $q^{-1}[\{S\}] = S$ is a closed subset of X, by hypothesis, hence $\{S\}$ is closed ⟦Sec. 6.5, Problem 4⟧, and so X/ρ is T_1 ⟦Theorem 4.1.1⟧. Finally, suppose that the equivalence classes are dense; let $S \in X/\rho$. Then $\overline{\{S\}} = \overline{q[S]} \supset q[\bar{S}]$ ⟦Theorem 4.2.4⟧ $= q[X] = X/\rho$. The result follows by Sec. 2.5, Problem 17. ▮

Note that in the first part of Theorem 6.7.5, X need not be a T_1 space.

EXAMPLE 4. Given a symmetric space X, we may define $x \,\rho\, y$ to mean $x \in \overline{\{y\}}$. ⟦For example, $x \in \overline{\{y\}}$, $y \in \overline{\{z\}}$ implies $\overline{\{y\}} \subset \overline{\{z\}}$ so that $x \in \overline{\{z\}}$.⟧ Then, by Theorem 6.7.5, X/ρ is a T_1 space. It may be expected to resemble X in many respects. (See Example 5, Problem 109, and Sec. 9.1, Example 5.)

★EXAMPLE 5. Let X be a semimetric space and say that $x \,\rho\, y$ if $d(x, y) = 0$. Thus $q(x) = \{y: d(x, y) = 0\}$. We shall show that *$X/\rho$ is a metric space.* First, observe that if $x, x' \in X$ and if $y \in q(x)$, $y' \in q(x')$ then $d(y, y') = d(x, x')$. ⟦$d(y, y') - d(x, x') \le d(y, x) + d(y', x') = 0$ using Sec. 2.2, Problem 3.⟧ Thus we may define $D[q(x), q(x')] = d(x, x')$ and this well-defines D. ⟦If $q(x) = q(y)$, $q(x') = q(y')$ then $d(y, y') = d(x, x')$ as just proved.⟧ It is obvious that D is a semimetric; but more, D is a metric. ⟦If $D[q(x), q(x')] = 0$, then $d(x, x') = 0$, thus $q(x) = q(x')$.⟧ Finally we observe that D actually yields the quotient topology for X/ρ ⟦Sec. 6.5, Example 2⟧. ▮

Problems

1. A product of T_0 or T_1 spaces is T_0, T_1, respectively.
2. Let X be a set with subsets A, B. Show that A, B are disjoint if and only if $A \times B$ does not meet the diagonal in $X \times X$.
3. X/ρ can be indiscrete without the equivalence classes being dense in X. (Compare Theorem 6.7.5.) ⟦Let $X = \{a, b, c\}$ with topology $\{\varnothing, \{a, b\}, \{c\}, X\}$, $Y = \{a, b\}$, $f: X \to Y$ given by $a \to a$, $b \to b$, $c \to b$. See Theorem 6.5.5. The same idea can be used to give an example in which $X = \mathbf{R}$.⟧
4. Let $f(x) = 1/x$ for $x \neq 0$, $f(0) = 2$. Show that $f: \mathbf{R} \to \mathbf{R}$ has closed graph.
5. Let $f: X \to Y$ be continuous. Show that the graph of f is a retract of $X \times Y$. ⟦$(x, y) \to (x, fx)$.⟧
6. A continuous map from any space to a Hausdorff space has closed graph. ⟦Example 2. If both spaces are T_2, the result is immediate from Problem 5, and Sec. 4.2, Problem 27.⟧
7. If f has closed graph, so does f^{-1}, if f is one-to-one and onto.
8. Deduce the fact that if X is T_2, the diagonal in $X \times X$ is closed from Problem 6.
9. Let $x, y \in X$ with $y \in \overline{\{x\}}$. Show that $(x, y) \in V$ where V is any neighborhood of the diagonal in $X \times X$. ⟦Let N be a neighborhood of Y with $N \times N \subset V$. Then $x \in N$, hence $(x, y) \in N \times N \subset V$.⟧
10. X is a *US* space if and only if the diagonal in $X \times X$ is sequentially closed. (Compare Example 1.)

---

101. Prove Theorem 6.7.4 for the corresponding spaces of complex-valued functions.
102. If $\prod X_\alpha$ has any one of the properties: connected, Lindelöf, compact, countably compact, locally compact, T_1, Hausdorff, regular, completely regular, normal, first countable, second countable, metrizable; then each X_α has the same property. ⟦For some of these use the fact that the projection on X_α is continuous; see, for example, Theorem 5.4.4. For others use the fact that X_α is homeomorphic with a retract of the product by Sec. 6.6, Problem 3; see, for example Sec. 4.1, Problem 6; Sec. 4.3, Problem 7.⟧
103. A function f is called a *KC function* if the image of every compact set is closed. Show that a function with closed graph is *KC*, hence any continuous function from any space to a Hausdorff space is *KC* ⟦Problem 6⟧. Apply this to the identity map to obtain the result that every compact set in a Hausdorff space is closed.
104. Let X be a *KC* non-Hausdorff space. Show that the identity map on X is a *KC* map without closed graph ⟦Example 1⟧. (Such a space will be shown in Sec. 8.1, Problem 205.)

105. Suppose that each of two sets X, Y has two topologies $S_1, S_2; T_1, T_2$, respectively, and that $S = S_1 \cap S_2$, $T = T_1 \cap T_2$. Show that the product topologies for $(X, S) \times (X, T)$ need not be the intersection of the product topologies for $(X, S_i) \times (Y, T_i)$, $i = 1, 2$.
106. Let S be the product topology for $\mathbf{R}^2$ when $\mathbf{R}$ is given the RHO topology. Let T be the product topology when $\mathbf{R}$ is given the LHO topology. Show that $S \cap T$ is strictly larger than the Euclidean topology.
107. Give an example of a set X with topologies S, T and a subset such that the relative topologies of S, T are discrete on this subset, but that of $S \cap T$ is not. The subset may have two points.
108. The quotient map in Example 4 is open and closed.
109. In Example 4, if X is regular or normal, X/ρ is T_3, T_4 respectively. ⟦For T_3, choose one point in the point of X/ρ which has to be separated from a closed set. See Problem 108.⟧
110. Prove the converse of Problem 109.
111. In Example 4, call X *semidiscrete* if X/ρ is discrete. Show that a symmetric space is semidiscrete if and only if the closure of every singleton is open. (Hence the closure of every set is open.) Deduce that a finite symmetric space is semidiscrete.
112. Let $f: X \to Y$. Show that f has closed graph if X, Y can be given smaller topologies with which f has closed graph. ⟦Closed sets in $X \times Y$ remain closed when the topology is enlarged.⟧ In particular, this is true if f is continuous with the smaller topologies, and that given to Y is Hausdorff ⟦Problem 6⟧.
113. Let T, T' be topologies for a set. Show that (i) $\Rightarrow$ (ii) $\Rightarrow$ (iii) where (i) $T \cap T'$ is Hausdorff; (ii) $i: (X, T) \to (X, T')$ has closed graph; (iii) $T \vee T'$ is Hausdorff. This generalizes Example 2 which is the special case $T = T'$.
114. Let (T_α) be a family of topologies on a set X. Let D be the diagonal in $\prod (X, T_\alpha)$. Show that D (with the product topology) is homeomorphic with $(X, \bigvee T_\alpha)$ under the map $P_\alpha: D \to X$, α being any fixed member of the index set. ⟦This is immediate from Theorems 6.4.1 and 6.2.1.⟧ (Compare Sec. 6.5, Problem 108.)
115. If a topological property is productive and hereditary, it is suping ⟦Problem 114⟧. (A *suping* property is one such that its possession by each of a family of topologies implies its possession by their supremum; similarly for *productive* and product.)
116. Write the details of the following proof of Sec. 4.2, Problem 7. Define $u: X \to Y \times Y$ by $u(x) = (fx, gx)$. Then $S = u^{-1}D$. (See Example 1 and Theorem 6.6.1.)
117. Let T, T' be topologies for a set X. Suppose X contains a proper

subset S which is T-closed and T'-open. Show that $T \vee T'$ is not connected.

118. Use Problem 117 to construct two connected topologies whose sup is not connected. Make them compact metric for good measure. ⟦Let f: $[0, 3] \to [0, 3]$ be one-to-one and onto and carry $[1, 2]$ onto $(1, 2)$. Let T be the Euclidean topology for $[0, 3]$, and T' the weak topology by f, the range of f having T. Then $[1, 2]$ is T-closed and T'-open.⟧
119. If a function from a Hausdorff space has compact graph it is continuous. ⟦$f = P_2 \circ P_1^{-1}$, where P_1, P_2 are the restrictions to the graph of the projections; P_1^{-1} is continuous by Theorem 5.4.9.⟧
120. The graph of f: $X \to Y$ is homeomorphic with $(X, T \vee wf)$. Problem 114 (with two topologies) is a special case.
121. The diagonal in a product of discrete spaces is discrete. Compare Sec. 6.4, Problem 101 ⟦Problem 114⟧.
122. (RHO) $\vee$ (LHO) is discrete ⟦see Problem 114; the diagonal in $\mathbf{R} \times \mathbf{R}$ is discrete, as in Example 3⟧, hence not separable.
123. A countable, locally compact, regular space is second countable ⟦Example 4; Sec. 5.4, Problem 202⟧.

---

201. Let $X = [0, 1]^{\mathbf{R}}$. Then X is separable ⟦consider polynomials with rational coefficients or step functions⟧ but has a nonseparable subspace, namely the set S of characteristic functions of singletons. Show that S is discrete, uncountable, and the cofinite filter on S converges to 0.
202. $\omega^{\mathbf{R}}$ is not normal. ⟦See [Ross and Stone, Theorem 5], for a direct construction. Alternatively, see [Mycielski], where it is shown (with a certain set-theoretical assumption) that this space is not D-separable; with Sec. 5.3, Example 3, this yields the result.⟧
203. If X contains an infinite discrete closed subset, $X^{\mathbf{R}}$ is not normal. ⟦Trivial from Problem 202.⟧ In particular $\mathbf{R}^{\mathbf{R}}$ is not normal.
204. In Problem 113, neither implication can be reversed.
205. Give an explicit constructive description of the topologies in Problem 118.
206. Show that f: $\mathbf{R} \to \mathbf{R}$ is continuous if its graph is closed and connected.
207. Problem 206 fails for more general spaces. ⟦Make f^{-1}: $Y \to X$ continuous with X, T_2, and Y connected; f need not be continuous.⟧
208. If f: $X \to Y$ is continuous must the projection P_1: $G \to X$ be open, where G is the graph of f? (Compare Theorem 6.4.5.)
209. Let X be a compact T_2 space such that the diagonal Δ is a G_δ in $X \times X$. Show that X is second countable. ⟦For each neighborhood G of Δ, cover Δ by open subsets of G of the form $V \times V$, V an open set in X. Reduce this to a finite cover. (Δ is compact by Problem 114.) Do this

for each G_n if $\Delta = \bigcap G_n$. Use an argument like that of Sec. 5.4, Problem 201 to show that the set of all such is a base.⟧

210. In Problem 209, "compact" cannot be replaced by "Lindelöf." (The same procedure yields a countable family, but it need not be a base.) ⟦Let X be a countable space.⟧

211. Must a $T_{3\frac{1}{2}}$ topology be the sup of the set of all smaller metrizable topologies?

Compactness

7.1 Countable and Sequential Compactness

For convenience of presentation we have taken a definition of countable compactness different from the one used by the earliest investigators, and suggested by the classical Bolzano–Weierstrass theorem. In all significant cases these definitions are equivalent; see Theorem 7.1.2.

Historically, the concept of compactness arose from considerations of accumulation points of infinite sets, and from the possibility of selecting convergent subsequences of sequences of functions. Hence, applications to classical analysis will depend on such formulations, and we turn our attention to these in this and the next section.

Compact sets are the most well-behaved members of the family of objects studied by analysts. All finite sets are compact, and all compact sets retain many of the properties of finite sets. A statement about finite sets often remains true of compact sets when some analytic hypothesis is added. For example, every real function on a finite set is bounded; every continuous real function on a compact set is bounded. Some other examples are Theorem 5.4.6; Sec. 11.1, Problem 105. An extended essay on this theme is [Hewitt (c)].

Recall that an accumulation point of a set S is a point every deleted neighborhood of which meets S; that is, a point x, every neighborhood of which contains a point of S other than x itself. We now say that x is an *ω-accumulation point* (pronounced: omega accumulation point) of S if every neighborhood of x contains infinitely many points of S. If X is an indiscrete space, S a nonempty finite subset, and $x \notin S$, then x is an accumulation point of S but not an ω-accumulation point. However this situation is unusual, as we now see.

THEOREM 7.1.1. *In a T_1 space, a point x is an accumulation point of a set S if and only if it is an ω-accumulation point of S.*

Half of this is trivial. Conversely, suppose that x is not an ω-accumulation point of S. Then x has a neighborhood N which meets $S \setminus \{x\}$ in a finite set F. (Perhaps F is empty; note that $x \notin F$.) Then $N \cap \tilde{F}$ is a neighborhood of x which meets S in no point other than, possibly, x itself. Thus x is not an accumulation point of S. ∎

A point x is said to be a *cluster point* of a filterbase F if $x \in \bar{S}$ for all $S \in F$.

★EXAMPLE 1. The filterbase $\{(a, \infty): a \in \mathbf{R}\}$ has no cluster point. The filter consisting of all neighborhoods of $\{0, 1\}$ in $\mathbf{R}$ has two cluster points, 0, and 1. [Indeed 0 and 1 belong to the sets of this filter, not just their closures.] Finally, the filter base $\{(0, a): a > 0\}$ in $\mathbf{R}$ has 0 as its only cluster point.

★EXAMPLE 2. *If $\mathscr{F} \to x$, x is a cluster point of $\mathscr{F}$.* Let N be a neighborhood of x, and $S \in \mathscr{F}$. Since $\mathscr{F} \to x$, $N \supset A$ for some $A \in \mathscr{F}$. Thus $N \cap S \supset A \cap S \neq \varnothing$ proving that $x \in \bar{S}$.

A point is called a *cluster point of a net* if it is a cluster point of the associated filter.

LEMMA 7.1.1. *A point x is a cluster point of a net $(x_\delta: D)$ if and only if for each neighborhood N of x, $x_\delta \in N$ frequently.*

First suppose that x is a cluster point. Fix $\delta_0 \in D$ and let $T = \{x_\delta: \delta \geq \delta_0\}$. Then T belongs to the filter associated with x_δ and so $x \in \bar{T}$. It follows that each neighborhood N of x meets T, in other words $x_\delta \in N$ for some $\delta \geq \delta_0$. Conversely, let A belong to the filter $\mathscr{F}$ associated with x_δ. Then $x_\delta \in A$ eventually and so each neighborhood N of x meets A. It follows that $x \in \bar{A}$; thus x is a cluster point of $\mathscr{F}$. ∎

As a warning against an obvious conjecture, we note Sec. 3.1, Problem 108, or Sec. 8.3, Problem 3, which show a sequence, $\{n\}$, with a cluster point, 0, but with no convergent subsequence.

THEOREM 7.1.2. *The following conditions are equivalent for a topological space X:*

(i) *X is countably compact.*
(ii) *Every sequence in X has a cluster point.*
(iii) *Every infinite subset of X has an ω-accumulation point.*

If X is a T_1 space, these are equivalent to:
(iv) *Every infinite subset of X has an accumulation point.*

Part (iv) is taken care of by part (iii), and Theorem 7.1.1.

Proof. (i) *implies* (ii). Let $\{x_n\}$ be a sequence in a countably compact space. Let $T_n = \{x_n, x_{n+1}, x_{n+2}, \ldots\}$ for $n = 1, 2, \ldots$. Then $\{T_n\}$ is a countable filterbase, and $\{\overline{T}_n : n = 1, 2, \ldots\}$ has nonempty intersection ⟦Theorem 5.4.3⟧. Any point of this intersection is a cluster point of the filterbase $\{T_n\}$, hence of the filter it generates ⟦Problem 1⟧, hence of $\{x_n\}$.

(ii) *implies* (iii). Let S_1 be an infinite set in X. Then S_1 has a countably infinite subset S. Let $S = \{x_1, x_2, \ldots\}$ where $x_i \neq x_j$ if $i \neq j$. By hypothesis the sequence $\{x_n\}$ has a cluster point x. For any neighborhood N of x, $x_n \in N$ for infinitely many values of n ⟦that is (see Lemma 7.1.1), frequently⟧; since all these x_n are different, they constitute an infinite subset of S_1. Thus x is an ω-accumulation point of S_1.

(iii) *implies* (ii). Let $\{x_n\}$ be a sequence and $S = \{x_n : n = 1, 2, \ldots\}$. If S is finite, there must be an index k such that $x_n = x_k$ for infinitely many n, making x_k a cluster point of the sequence. If S is infinite, let x be an ω-accumulation point of S. It follows immediately from Lemma 7.1.1 that x is a cluster point of $\{x_n\}$.

(ii) *implies* (i). Let $\{F_n\}$ be a sequence of closed sets with the finite intersection property. For each n, choose $x_n \in \bigcap \{F_k : k = 1, 2, \ldots, n\}$. By hypothesis, the sequence $\{x_n\}$ has a cluster point x. Then $x \in \bigcap F_k$. ⟦For each k, $x_n \in F_k$ for $n \geq k$ and so each neighborhood of x, containing, as it does, x_n for infinitely many values of n, must meet F_k. Thus $x \in \overline{F}_k = F_k$.⟧ The result follows from Theorem 5.4.3. ∎

REMARK. In applying Theorem 7.1.2 to a subset S of a topological space, we must be careful to add "in S" where appropriate. For example, condition (ii) reads: "Every sequence in S has a cluster point in S." Of course it is possible to discuss sets S in X for which every sequence has a cluster point (in X), and some of this is carried out in the problems.

★EXAMPLE 3. The classical Bolzano–Weierstrass theorem says that a closed bounded interval of real numbers obeys condition (iv) of Theorem 2, thus is countably compact. Conversely a countably compact set of real numbers is compact ⟦Theorem 5.4.1⟧. For sets of real numbers, then, countable compactness is equivalent to compactness. We shall see ⟦Theorem 7.2.1⟧ that this equivalence extends to all semimetric spaces; the Bolzano–Weierstrass theorem will follow as a corollary of this and the Heine–Borel theorem, Sec. 5.4, Example 1. (The theorem was proved by B. Bolzano, a mathematical amateur, in 1851. It was essentially known to Cauchy much earlier. The name of K. Weierstrass is associated through his extensive use of the theorem in 1886.)

We now turn to a third form of compactness. A topological space is called *sequentially compact* if each sequence in it has a convergent subsequence. For analysts this is the most important form of compactness; several classical selection theorems have exactly the conclusion that a certain collection of functions is (pointwise, uniformly) sequentially compact.

In general the only implications among the three main kinds of compactness are that a space which is either compact or sequentially compact must be countably compact. The Tables may be consulted for all the relevant counterexamples. There are other relationships in important special cases ⟦Theorems 7.1.3, 7.2.1, and 13.4.3⟧.

THEOREM 7.1.3. *A first countable space is sequentially compact if and only if it is countably compact.*

Half of this is easy (without first countability) ⟦Problem 4⟧. Conversely, let X be first countable and countably compact, and let $\{x_n\}$ be a sequence in X. Then $\{x_n\}$ has a cluster point x ⟦Theorem 7.1.2⟧. Let $\{N_k\}$ be a shrinking basic sequence of neighborhoods of x ⟦Sec. 3.1, Problem 7⟧. For each k, we know that $x_n \in N_k$ frequently ⟦Lemma 7.1.1⟧; thus there exist n_1 with $x_{n_1} \in N_1$, $n_2 > n_1$ with $x_{n_2} \in N_2, \ldots, n_i > n_{i-1}$ with $x_{n_i} \in N_i, \ldots$. Then $x_{n_i} \to x$ ⟦Sec. 3.1, Problem 5⟧. ▮

We conclude this section with an important phrasing of compactness in terms of cluster points, similar to Theorem 7.1.2.

THEOREM 7.1.4. *The following conditions are equivalent for a topological space X:*

(i) *X is compact.*
(ii) *Every filter in X has a cluster point.*
(iii) *Every net in X has a cluster point.*

Proof. (i) *implies* (ii). If $\mathscr{F}$ is a filter, $\{\bar{S}: S \in \mathscr{F}\}$ is a collection of closed sets with the finite intersection property. Hence it is fixed ⟦Theorem 5.4.3⟧, and each point in its intersection is a cluster point ⟦by definition of cluster point⟧.

(ii) *implies* (iii). Given a net, its associated filter has a cluster point; this is a cluster point of the net, by definition.

(iii) *implies* (ii). Interchange net and filter in the preceding part. ⟦Use Problem 3.⟧

(ii) *implies* (i). Let C be a collection of closed sets with the finite intersection property. Then C is included in a filter $\mathscr{F}$ ⟦Sec. 5.4, Problem 8⟧. Let x be a cluster point of $\mathscr{F}$. Then $x \in \bigcap\{\bar{S}: S \in \mathscr{F}\} \subset \bigcap\{\bar{S}: S \in C\} = \bigcap\{S: S \in C\}$. Thus C is fixed, and X is compact ⟦Theorem 5.4.3⟧. ▮

Problems

★1. A point is a cluster point of a filter if and only if it is a cluster point of every filterbase which generates the filter ("every" may be replaced by "some").

★2. Prove the converse of Lemma 7.1.1. In particular, a limit of a convergent subsequence is a cluster point of the sequence.

★3. If x_δ is any net association with a filter $\mathscr{F}$, every cluster point of x is a cluster point of $\mathscr{F}$. (See Problem 105.)

★4. A sequentially compact space is countably compact ⟦Problem 2 and Theorem 7.1.2 (ii)⟧.

★5. A sequence in a finite space must have a cluster point.

★6. Let $\{x_n\}$ be a sequence in a semimetric space for which there exists $\varepsilon > 0$ with $d(x_m, x_n) \geq \varepsilon$ for all m, n. Show that $\{x_n\}$ has no cluster point. ⟦There would have to be two terms very near the cluster point.⟧

7. Give an example of a T_0 space with a subset which has an accumulation point which is not an ω-accumulation point. ⟦Take a finite subset in Sec. 4.1, Example 2.⟧

8. Every uncountable set in a second countable space has an accumulation point in the space.

★9. A continuous map f carries a cluster point of a filter $\mathscr{F}$ into a cluster point of $f(\mathscr{F})$. The same is true for nets.

★10. Let x be a cluster point of a filter $\mathscr{F}$. Show that there is a filter $\mathscr{F}' \supset \mathscr{F}$ with $\mathscr{F}' \to x$. ⟦Consider $\{S \cap N: S \in \mathscr{F}, N$ a neighborhood of $x\}$.⟧ Conversely, if such $\mathscr{F}'$ exists, x must be a cluster point of $\mathscr{F}$. ⟦Every neighborhood of x belongs to $\mathscr{F}'$, hence meets every member of $\mathscr{F}'$, *a fortiori* every member of $\mathscr{F}$.⟧

11. A set S is compact if and only if every filter on S is included in a filter which converges to a point of S ⟦Problem 10 and Example 2⟧.

★12. A set S is sequentially compact if and only if every infinite subset contains a convergent sequence. ⟦If a sequence has infinite range, apply this condition. If it has finite range, it automatically has a convergent sequence.⟧

★13. Let $\mathscr{F}$ be a filter in a compact space X, such that $\mathscr{F}$ has only one cluster point x. Show that $\mathscr{F} \to x$. Also if a countable set S in a compact space has only one accumulation point, any one-to-one function from ω onto S is a sequence converging to x. ⟦If not, let G be an open neighborhood of x which does not belong to $\mathscr{F}$. Then $\{S \setminus G: S \in \mathscr{F}\}$ is a filter in the compact space $X \setminus G$.⟧

---

101. Give an example of a free filter on $[0, 1]$.

102. Give an example of a nonconvergent filter in $\mathbf{R}$ which has exactly one cluster point. ⟦The filter associated with the sequence $\{[1 - (-1)^n] \cdot n\}$.⟧

103. Show that an ω-accumulation point of the range of a sequence must be a cluster point of the sequence, but a cluster point of a sequence need not be an accumulation point of its range.
104. The following conditions on a T_1 space X are equivalent: (i) X is countably compact, (ii) every countable closed subset of X is countably compact, (iii) every countable closed subset of X is compact. ⟦For (ii) $\Rightarrow$ (iii), use Theorem 5.4.1. For (iii) $\Rightarrow$ (i) use Theorem 7.1.2 (ii).⟧
105. If a point is a cluster point of a filter then it is a cluster point of the canonical net of the filter, but need not be a cluster point of every net associated with the filter. ⟦For example we might have $x_\delta = x$ for all δ, where x is one of a collection of cluster points.⟧
106. Solve Sec. 6.7, Problem 103 ($CG \Rightarrow KC$) by means of Problem 11. ⟦Let $\mathscr{F}$ be a filter in $f[K]$ with $\mathscr{F} \to y$. Then $\mathscr{F} = f(\mathscr{F}_1)$, $\mathscr{F}_1$ a filter in K. $\mathscr{F}_1 \subset \mathscr{F}_2 \to x \in K$ by Problem 11; $f(\mathscr{F}_2) \to y$ since it is larger than $\mathscr{F}$. Hence $y = f(x) \in f(K)$.⟧
107. If $f: X \to Y$ has closed graph and $K \subset Y$ is compact, then $f^{-1}[K]$ is closed ⟦like Problem 106⟧.
108. Let f be a map from a space X to a compact space Y. Then if f has closed graph, it is continuous ⟦Problem 107; Sec. 4.2, Problem 4⟧.
109. Deduce Theorem 5.4.9, from Problem 108 ⟦Sec. 6.7, Problems 6 and 7⟧.
110. Use Problem 11 to show that a compact set in a Hausdorff space must be closed. ⟦Let $\mathscr{F}$ be a filter on K, $\mathscr{F} \to x$. Then $\mathscr{F} \subset \mathscr{F}_1 \to k \in K$ by Problem 11; $\mathscr{F}_1 \to x$ since it is larger than $\mathscr{F}$, hence $x = k$ by Theorem 4.1.2.⟧
111. A metric space can be isometric with a proper subspace. ⟦A subspace of **R** can be found with this property.⟧ However, no such space can be found which is compact ⟦Problem 6⟧.
112. A compact metric space can be homeomorphic with a proper subspace.
113. Find a bounded subset of $\mathbf{R}^2$ which is isometric with a proper subset of itself.
114. A countably compact space is pseudocompact ⟦Example 3 and Theorem 5.4.4⟧.
115. A pseudocompact T_2 space need not be countably compact ⟦Sec. 5.2, Examples 7 and 8; Sec. 5.4, Problem 2⟧.
116. Pseudocompact is not F-hereditary. ⟦A T_1 space which is not countably compact must contain an infinite discrete closed subset. Now see Problem 115.⟧
117. The following condition is necessary and sufficient that d be a u-semimetric: Let $\{x_n\}$ be a sequence with the property that there exists a sequence $\{y_n\}$ satisfying $d(x_n, y_n) > 0$ for all n, $d(x_n, y_n) \to 0$. Then $\{x_n\}$ has a cluster point. (In particular if no such sequence $\{x_n\}$ exists, as in the case of the discrete metric, d is a u-semimetric.)

118. Let X be a compact metric space. Let $\{f_n\}$ be a sequence of homeomorphisms of X onto itself and let $f_n \to f$ uniformly, where also $f: X \to X$. Show that f is onto X. ⟦For $y \in X$ let $f_n(x_n) = y$ and x a cluster point of $\{x_n\}$.⟧
119. The assumption "compact" cannot be omitted in Problem 118.
120. Replace "homeomorphisms" by "continuous maps" in Problem 118.

---

201. Show that if S is relatively compact, then every filterbase in S has a cluster point (not necessarily in S).
202. In a regular space, the converse of the result of Problem 201 holds.
203. Problem 202 becomes false if "regular" is replaced by "Hausdorff." ⟦Give $[0, 1]$ the simple extension of the Euclidean topology by $\mathbf{Q}$. $\mathbf{Q}$ is not relatively compact, by Theorem 5.4.10.⟧
204. A net $(y_\beta: B)$ is called a *subnet* of the net $(x_\alpha: A)$ if there exists a finalizing map $u: B \to A$ such that $y_\beta = x_{u(\beta)}$ for all $\beta \in B$. Show that a subsequence of a sequence is a subnet of the sequence.
205. Give an example of a subnet of a sequence which is not, itself, a sequence. ⟦Let $u: \mathbf{R} \to \omega$ be the largest integer function.⟧
206. Every subnet of a convergent net converges to the same limit.
207. If x is a cluster point of the net $(x_\alpha: A)$; the net has a subnet which converges to x. ⟦Let N be the neighborhood filter of x, $B = \{(\alpha, U): \alpha \in A, U \in N, x_\alpha \in U\}$; $u(\alpha, U) = \alpha$.⟧
208. Does an analogue of Theorem 6.4.1 hold for cluster points?
209. A map is called *perfect* if it is continuous, closed, and the inverse image of every singleton is compact. Show that if f is a perfect map of a regular space onto a space, the latter is regular.

7.2 Compactness in Semimetric Space

Suppose that X is a compact semimetric space. The set of all cells of any fixed radius is an open cover of X, and thus can be reduced to a finite cover. We are thus led easily to a definition and a theorem. A semimetric space X is called *totally bounded* if for each $\varepsilon > 0$, X is covered by a finite collection of cells of radius ε. Put in another way, this says that X contains a finite subset F such that every member of X is within ε of some member of F. The remark just made shows that a compact semimetric space is totally bounded. Lemma 7.2.1 improves this, but only seemingly ⟦see Theorem 7.2.1⟧ and Theorem 9.1.4 gives a result in the converse direction. Problem 4 shows that total boundedness is not topological.

LEMMA 7.2.1. *A countably compact semimetric space is totally bounded.*

Suppose that X is not totally bounded. Then there exists $\varepsilon > 0$ such that no finite collection of cells of radius ε covers X. Let

$$x_1 \in X,$$
$$x_2 \in X \setminus N(x_1, \varepsilon),$$
$$x_3 \in X \setminus [N(x_1, \varepsilon) \cup N(x_2, \varepsilon)], \ldots,$$
$$x_n \in X \setminus \bigcup \{N(x_i, \varepsilon): i = 1, 2, \ldots, n - 1\}, \ldots.$$

This process never terminates ⟦by the defining property of ε⟧. The sequence $\{x_n\}$ has the property, $d(x_m, x_n) \geq \varepsilon$ if $m \neq n$, hence has no cluster point ⟦Sec. 7.1, Problem 6⟧, and so X is not countably compact ⟦Theorem 7.1.2⟧. ▌

LEMMA 7.2.2. *A totally bounded semimetric space is second countable (hence separable and Lindelöf).* p. 77 Th 5.3.4

For $n = 1, 2, \ldots$, let X be covered by finitely many cells of radius $1/n$. Let B be the collection of all such cells; then B is countable; moreover B is a base for the topology. ⟦Let $x \in X$ and let N be a neighborhood of x. Then $N \supset N(x, \varepsilon)$ for some $\varepsilon > 0$. Choose $n > 2/\varepsilon$, and a member S of B containing x and of radius $1/n$. For every $b \in S$, $d(b, x) < 2/n < \varepsilon$; thus $b \in N$. This proves that $S \subset N$.⟧ Thus X is second countable, hence separable and Lindelöf ⟦Theorems 5.3.1 and 5.3.2⟧. ▌

We now prove an important theorem which greatly simplifies the study of compactness in semimetric spaces. In fact, all three of the main kinds of compactness are identical!

THEOREM 7.2.1. *A semimetric space is compact if and only if it is countably compact, and if and only if it is sequentially compact.*

A compact space is countably compact, as is trivial from the definitions. For any first countable space, countable and sequential compactness are the same ⟦Theorem 7.1.3⟧. Finally, let X be countably compact. By Lemmas 7.2.1 and 7.2.2, X is a Lindelöf space. Thus X is compact ⟦Theorem 5.4.1⟧. ▌

Problems on Semimetric Space

1. Total boundedness is hereditary.
2. **R** is not totally bounded.
3. (0, 1) is a totally bounded set in **R**.
4. Total boundedness is not topological. ⟦The spaces of the preceeding two problems are homeomorphic.⟧
5. Give an example of a totally bounded set which is not countably compact.

6. Let $x \notin S$. If S is compact, it has a closest point to x. It is not sufficient that S be closed. ⟦Take $X = \{0\} \cup (1, \infty)$, $S = (1, \infty)$.⟧

---

101. A uniformly continuous function preserves total boundedness, whereas ⟦Problem 4⟧ a continuous function need not. ⟦For $f: X \to Y$, write $X = \bigcup S_i$, with the diameter of each $S_i < \delta$. Then the diameter of each $f[S_i] < \varepsilon$.⟧
102. If a space has a dense totally bounded subspace, it is itself totally bounded.
103. Let Φ be a family of functions from a topological space X to a semimetric space Y. We call Φ *equicontinuous* at $x \in X$ if for every $\varepsilon > 0$, x has a neighborhood N such that the diameter of $f[N] < \varepsilon$ for all $f \in \Phi$; and equicontinuous if it is equicontinuous at x for all x. Show that every member of an equicontinuous family is continuous, and that a finite set of functions is equicontinuous at x if each member is continuous at x.
104. A subset of an equicontinuous family is equicontinuous.
105. A finite union of equicontinuous families is equicontinuous.
106. Let $f, f_n \in C(X)$ for $n = 1, 2, \ldots$. Suppose that $f_n \to f$ pointwise and that $\{f_n(x)\}$ is decreasing for each x. Show that $\{f_n\}$ is equicontinuous. ⟦Assume $f = 0$. If $n \geq m$ implies $f_n(x) < \varepsilon$, then for some neighborhood V of x, $n \geq m$, $y \in V$ imply $f_n(y) < \varepsilon$. Choose a smaller neighborhood W of x such that $|f_n(x) - f_n(y)| < \varepsilon$ for $y \in W$, $n = 1, 2, \ldots, m - 1$.⟧
107. Let $\{f_n\}$ be an equicontinuous sequence of real functions with $f_n \to 0$ pointwise. For each $\varepsilon > 0$, ϕ is continuous, where $\phi(x) = \sum (|f_n(x)| - \varepsilon) \vee 0$. ⟦For each z, there is a neighborhood V of z on which the series is finite, namely, make $|f_n(z)| < \varepsilon/2$ for $n > m$, and $|f_n(x) - f_n(z)| < \varepsilon/2$ for $x \in V$.⟧
108. (*The Dini lemma.*) If X is pseudocompact, monotone convergence (the conditions of Problem 106) implies uniform convergence. ⟦Let $\varepsilon > 0$. Then ϕ, given in Problem 107, is continuous, by Problems 106, 107. Thus $\phi < M$ on X. Now $n \geq M/\varepsilon$ implies $f_n(x) < 2\varepsilon$ for all x, for if $f_n(x) \geq 2\varepsilon$, $(f_k(x) - \varepsilon) \vee 0 \geq \varepsilon$ for $k = 1, 2, \ldots, n$; hence $M > \phi \geq n\varepsilon$. This argument is due to J. Marik.⟧
109. The converse of Problem 108 also holds, namely the given condition implies that X is pseudocompact. ⟦Consider arctan $(|f|/n)$ for unbounded f.⟧

---

201. Let $\{f_\delta\}$ be a net which is equicontinuous at t (Problem 103). Suppose that $f_\delta(x) \to f(x)$ for each x. Show that f is continuous at t. ⟦Use the inequality given in the proof of Theorem 4.2.10 except that x is chosen near t first, and δ is chosen later.⟧

202. We say that $f_\delta \to f$ *continuously*, where $f_\delta, f: X \to Y$ with X, Y topological spaces, if for every net x_δ in X with $x_\delta \to x$ we have $f_\delta(x_\delta) \to f(x)$. Show $f_\delta \to f$ continuously if Y is a semimetric space, $f_\delta \to f$ pointwise, and either f_δ is equicontinuous or $f_\delta \to f$ uniformly. ⟦$d(f_\delta x_\delta, fx) \le d(f_\delta x_\delta, f_\delta x) + d(f_\delta x, fx)$ or $\le d(f_\delta x_\delta, fx_\delta) + d(fx_\delta, fx)$.⟧
203. Let f_δ be an equicontinuous net with $f_\delta \to f$ pointwise. Suppose f_δ, $f: X \to Y$ where X, Y are semimetric spaces and X is compact. Show that $f_\delta \to f$ uniformly. ⟦First show that f_δ is uniformly equicontinuous. Then assume, without loss of generality that X has very small diameter, for it is totally bounded.⟧
204. Compactness cannot be omitted in Problem 203, even if f is a uniformly equicontinuous sequence. ⟦Consider x/n on $\mathbf{R}$.⟧
205. In Sec. 4.2, Problem 202, suppose that Φ is equicontinuous. Show that u is continuous.
206. Let X be a semimetric space and for each $t \in X$ define $f_t: X \to \mathbf{R}$ by $f_t(x) = d(t, x)$. Show that $\{f_t: t \in X\}$ is equicontinuous.
207. Deduce from Problems 205 and 206 that $d(x, S)$ (the distance from x to S) is continuous in x, S being a fixed set in a semimetric space.

7.3 Ultrafilters

We can no longer postpone an explicit statement of an axiom which is used widely in the development of mathematical theories. This axiom will be taken as an unstated hypothesis in all further developments of this book.

AXIOM. *Every nonempty partially ordered set includes a maximal chain.*

For finite or countable posets, this can be proved by induction ⟦Sec. 3.3, Problem 11⟧.

There are several other axioms which would serve the same purpose as the one given here. Eight of them are listed in [Kelley (a), p. 33, Theorem 25], and some are given here in Problems 102, 103, 201, 204; and Theorem 10.2.1.

A maximal filter is called an *ultrafilter*. The following important result will illustrate the use of the axiom just given. We put the reader on notice that this proof will contain the last explicit mention of the axiom. From that point on we shall merely say: "Take a maximal chain," leaving the reader to recall the justification for this.

THEOREM 7.3.1. *Every filterbase is included in an ultrafilter.*

Let $\mathscr{B}$ be a filterbase on a set X. Let P be the collection of all filters $\mathscr{F}$ on X such that $\mathscr{F} \supset \mathscr{B}$; P is not empty, ⟦the filter generated by $\mathscr{B}$ belongs to P⟧, and is a partially ordered set under containment ⟦Sec. 3.3, Example 2⟧. We now appeal to the axiom given above and let C be a maximal chain in P. Thus

C is a collection of filters; of each pair of members of C, one includes the other; moreover every member of C includes $\mathscr{B}$. Let $U = \bigcup \{\mathscr{F} : \mathscr{F} \in C\}$. We shall now show that U is an ultrafilter. (That $U \supset \mathscr{B}$ is trivial.) We begin by checking the four parts of the definition of a filter. First, $\varnothing \notin U$ since $\varnothing \notin \mathscr{F}$ for $\mathscr{F} \in C$; $X \in U$ since $X \in \mathscr{F}$ for some (indeed, all) $\mathscr{F} \in C$. Next suppose $A \in U$, $B \in U$. Then $A \in \mathscr{F}_1$, $B \in \mathscr{F}_2$ for some $\mathscr{F}_1, \mathscr{F}_2 \in C$. Since C is a chain, either $\mathscr{F}_1 \supset \mathscr{F}_2$ or $\mathscr{F}_2 \supset \mathscr{F}_1$; say the latter, for definiteness. Then $A \in \mathscr{F}_2$, $B \in \mathscr{F}_2$ and so $A \cap B \in \mathscr{F}_2$, hence $A \cap B \in U$. Finally, let $A \in U$, $B \supset A$. Then $A \in \mathscr{F}$ for some $\mathscr{F} \in C$. Hence $B \in \mathscr{F}$, and so $B \in U$. Thus U is a filter. It is also an ultrafilter. ⟦If V is a filter and $V \supset U$, it follows that $V \supset \mathscr{F}$ for every $\mathscr{F} \in C$. Thus $C \cup \{V\}$ is a chain. Since C is maximal, it follows that $V \in C$, and so $V \subset U$. Thus $V = U$.⟧ ∎

THEOREM 7.3.2. *If $\mathscr{F}$ is an ultrafilter on X and $S \subset X$, then either $S \in \mathscr{F}$ or $\tilde{S} \in \mathscr{F}$.*

Suppose that $\tilde{S} \notin \mathscr{F}$. Then S meets every member of $\mathscr{F}$. ⟦If not, suppose $A \in \mathscr{F}$, $A \not\cap S$. Then $\tilde{S} \supset A$ and so $\tilde{S} \in \mathscr{F}$.⟧ Thus $\mathscr{F} \cup \{S\}$ has the finite intersection property. ⟦A finite intersection takes the form, on rearrangement, of $(\bigcap A_i) \cap S$, where $A_1, A_2, \ldots, A_n \in \mathscr{F}$. Now $\bigcap A_i \in \mathscr{F}$ and so it meets S in a nonempty set.⟧ Thus the set of all finite intersections of $\mathscr{F} \cup \{S\}$ is a filterbase which generates a filter $\mathscr{F}'$ ⟦Sec. 5.4, Problem 8⟧. Clearly $\mathscr{F} \subset \mathscr{F}'$ and so $\mathscr{F} = \mathscr{F}'$. ⟦$\mathscr{F}$ is an ultrafilter.⟧ But $S \in \mathscr{F}'$ ⟦$S = X \cap S$ and $X \in \mathscr{F}$⟧, and so $S \in \mathscr{F}$. ∎

THEOREM 7.3.3. *Let C be a collection of subsets of a set X with the finite intersection property. Then there exists an ultrafilter $\mathscr{F}$ with $\mathscr{F} \supset C$.*

This is immediate from Theorem 7.3.1 and Sec. 5.4, Problem 8. ∎

THEOREM 7.3.4. *Let X, Y be sets and $f: X \to Y$ onto. Then if $\mathscr{F}$ is an ultrafilter in X, $f[\mathscr{F}]$ is an ultrafilter in Y.*

By Sec. 3.2, Problem 5, $f(\mathscr{F})$ is a filter. If $\mathscr{G}$ is a filter in Y with $\mathscr{G} \supset f[\mathscr{F}]$, let $S \in \mathscr{G}$. Then $f^{-1}[S] \in \mathscr{F}$. ⟦Let $T = (f^{-1}[S])^{\sim}$. Then $f[T] = \tilde{S}$ so that $f[T] \notin \mathscr{G}$. Hence $T \notin \mathscr{F}$ since $f[\mathscr{F}] \subset \mathscr{G}$. By Theorem 7.3.2, $\tilde{T} \in \mathscr{F}$.⟧ Thus $S \in f[\mathscr{F}]$ and so $\mathscr{G} = f[\mathscr{F}]$. ∎

THEOREM 7.3.5. *Let $\mathscr{F}$ be an ultrafilter in a topological space X. Then if x is a cluster point of $\mathscr{F}$, $\mathscr{F} \to x$.*

Let N be a neighborhood of x. Then $\tilde{N} \notin \mathscr{F}$. ⟦$x$ is in the closure of each member of $\mathscr{F}$ while $x \notin \tilde{N}$.⟧ Thus, by Theorem 7.3.2, $N \in \mathscr{F}$. ∎

THEOREM 7.3.6. *A space is compact if and only if every ultrafilter is convergent.*

If X is compact, every ultrafilter has a cluster point ⟦Theorem 7.1.4⟧, hence converges ⟦Theorem 7.3.5⟧. Conversely, if every ultrafilter on X is convergent, let $\mathscr{F}$ be a filter on X. Let $\mathscr{F}'$ be an ultrafilter which includes $\mathscr{F}$. By hypothesis $\mathscr{F}'$ is convergent, say $\mathscr{F}' \to x$. Then x is a cluster point of $\mathscr{F}$ ⟦Sec. 7.1, Problem 10⟧, and so X is compact ⟦Theorem 7.1.4⟧. ∎

Problems

1. D_x is an ultrafilter. ⟦Since $\{x\} \in D_x$, trying to adjoin even one more set will lead to a pair of sets with empty intersection.⟧
2. ★ Let $\mathscr{F}$ be an ultrafilter on X and let $A_1, A_2, \ldots, A_n$ be subsets of X such that $\bigcup A_i \in \mathscr{F}$. Show that at least one $A_i \in \mathscr{F}$. ⟦If not, Theorem 7.3.2 shows that $(\bigcup A_i)^\sim = \bigcap \tilde{A}_i \in \mathscr{F}$.⟧
3. Problem 2 fails if infinitely many A_i are allowed. ⟦An ultrafilter $\mathscr{F}$ containing the cofinite filter can contain no singleton since $\mathscr{F}$ contains its complement.⟧
4. A filter $\mathscr{F}$ is an ultrafilter if and only if $A \cup B \in \mathscr{F}$ implies $A \in \mathscr{F}$ or $B \in \mathscr{F}$. ⟦Problem 2. Conversely, if $A \notin \mathscr{F}$, $\tilde{A} \in \mathscr{F}$ since $A \cup \tilde{A} \in \mathscr{F}$. So no larger filter can contain A.⟧

---

101. Give an example of an ultrafilter other than D_0 which converges to 0 in $\mathbf{R}$. ⟦Apply Theorem 7.3.1 and Sec. 5.4, Problem 8 to $\{(0, 1)\} \cup N_0$.⟧
102. Let P be a poset with the property that every chain in P has an upper bound in P. Show that P contains a maximal member. (This is the famous *Zorn's lemma*.) ⟦Namely, any upper bound for a maximal chain.⟧
103. A property p of certain subsets of a set X is said to be *of finite character* if it is true that a set A has property p if and only if every finite subset of A has property p. Fix $S \subset X$. Show that "does not meet S" and "is a subset of S" are properties of finite character, and that "finite" is not a property of finite character.

---

201. Let p be a property of certain subsets of a set X which is of finite character. Show that if X has a subset with property p, it has a maximal one. (This result is due to J. W. Tukey.) ⟦In the poset of all sets with property p, take the union of a maximal chain.⟧
202. Verify that "linearly independent" is a property of finite character for sets in a linear space.
203. Deduce the existence of a Hamel base for every linear space over a field (a) from the Axiom of this section; (b) from Problem 202. (A *Hamel base* is a set of which each point is a unique (finite) linear combination.)

204. Let Φ be a collection of nonempty sets. Then $\prod \{X: X \in \Phi\}$ is not empty. (This is the famous *Axiom of Choice*, also referred to as Zermelo's multiplicative axiom.) ⟦Let P be the set of all $f: S \to \bigcup \{X: X \in \Phi\}$ with $S \subset \Phi$ and $f(X) \in X$ for each $X \in S$. Order P by $f \geq g$ if f is an extension of g. A maximal member of P is a member of the product.⟧
205. For real x, y, write $x \sim y$ if $x - y$ is rational. Show that the existence of a set A containing exactly one member of each equivalence class follows from Problem 204. Show that $(A + r) \not\cap A$ if r is rational and not zero. (Hence $A \cap [0, 1]$, having infinitely many disjoint translations modulo 1, cannot be measurable.)
206. A *universal net* in X is one such that for each $S \subset X$, either the net is in S eventually, or it is in $\tilde{S}$ eventually. Show that a net is a universal net if and only if its associated filter is an ultrafilter.
207. A universal sequence must be eventually constant.
208. If a universal net is frequently in a set, it is eventually in it. Hence a universal net converges to all its cluster points.
209. Let $\mathscr{G}$ be the filter associated with the net x_δ and $\mathscr{F}$ a filter with $\mathscr{F} \supset \mathscr{G}$. The x_δ has a subnet whose associated filter is $\mathscr{F}$. ⟦[Bartle, Proposition 4.]⟧
210. Every net has a universal subnet ⟦Problems 206 and 209⟧.
211. A space is compact if and only if every net has a convergent subnet ⟦Problems 208 and 210⟧.
212. A compact Hausdorff division ring must be finite. ⟦If the space is not discrete, let x_δ be a universal net of nonzero elements converging to 0. By Problem 208, $x_\delta^{-1} \to y$ and $1 = x_\delta \cdot x_\delta^{-1} \to 0 \cdot y = 0$.⟧
213. Let C be a chain of topologies on a set X and suppose that $f: (X, T) \to Y$ is almost open for every $T \in C$. Then $f: (X, \bigvee C) \to Y$ is almost open.
214. Let X be a set, (Y, T) a topological space and $f: X \to Y$ onto. Then X has a topology M which is maximal among those for which f is almost open. ⟦f is almost open with $f^{-1}[T]$. Take the sup of a maximal chain of such topologies and apply Problem 213.⟧
215. For topologies T and T' on X, say that $T > T'$ if $i: (X, T) \to (X, T')$ is continuous and almost open. Call T *W-maximal* (in honor of J. D. Weston) if it is maximal with respect to this relation. Show that T is W-maximal if and only if every set S with $S \subset (\bar{S})^i$ is open. Hence a W-maximal space is *irresolvable* (that is, a set and its complement cannot both be dense. See [Anderson].)
216. Every topology is included in a W-maximal topology ⟦Problem 214⟧.
217. If $T > T'$ (Problem 215) and (X, T') is T_1 and has no isolated points, neither does (X, T).
218. In any W-maximal T_2 space without isolated points, sequential convergence is trivial. ⟦See [Wilansky (c), p. 21].⟧

7.4 Products

The form of compactness most used by the classical analysts is sequential compactness; however, it is easier to work with compactness itself, and the most usual procedure is to deal with compact spaces where possible, bringing in countable or sequential compactness only where appropriate. There are several reasons for this preference (for example, Theorems 5.4.5 and 5.4.7, both of which fail for countable and sequential compactness), the most important one being the remarkable result (Theorem 7.4.1) proved by A. Tychonoff in 1930. This result fails for countable compactness, even for the product of two spaces. ⟦See the Remark following Example 2.⟧ It also fails for sequential compactness ⟦Example 2⟧, but here Theorem 7.4.2 offers a consolation result.

THEOREM 7.4.1 (Tychonoff's theorem). *A product of compact spaces is compact.*

Let $\mathscr{F}$ be an ultrafilter in $X = \prod \{X_\alpha : \alpha \in A\}$. For each α, $P_\alpha[\mathscr{F}]$ is an ultrafilter in X_α ⟦Theorem 7.3.4⟧, hence has a cluster point x_α ⟦Theorem 7.1.4⟧. Let x be defined by $P_\alpha x = x_\alpha$ for all $\alpha \in A$. Then $\mathscr{F} \to x$. ⟦$P_\alpha[\mathscr{F}] \to x_\alpha$ for each α by Theorem 7.3.5, and so $\mathscr{F} \to x$ by Theorem 6.4.1.⟧ The fact that X is compact now follows from Theorem 7.3.6. ∎

REMARK. If the spaces in Theorem 7.4.1 are Hausdorff, some brief proofs of the theorem fall out of other theories. These proofs are sketched in Sec. 8.6, Problem 102, and Sec. 11.4, Problem 122.

EXAMPLE 1. *The n sphere is compact*, for it is defined to be the subset S_n of $\mathbf{R}^{n+1}$ given by $S_n = \{x : \|x\| = 1\}$, where $x = (x_1, x_2, \ldots, x_{n+1})$ and $\|x\|^2 = \sum |x_i|^2$. Thus $S_n \subset [-1, 1]^{n+1}$, a compact space, by Theorem 7.4.1, moreover S_n is closed since $\|\cdot\|$ is continuous

EXAMPLE 2. *A product of sequentially compact spaces need not be sequentially compact.* Indeed a product of compact spaces need not be sequentially compact. Since such a product is compact ⟦Theorem 7.4.1⟧ this gives an example of *a compact space which is not sequentially compact.* (Another example is given in Sec. 8.3, Problem 3.) Our example is I^U, where $I = [0, 1]$ and U is any set with cardinality $\geq c$ (continuum). Let A be the collection of all strictly increasing sequences $x = \{x_k\}$ of positive integers, and let $\phi : A \to U$ be one-to-one. For $n = 1, 2, \ldots$, let $E_n = \{t \in U : t = \phi(x)$ for some $x \in A$ such that $n = x_k$ with k odd$\}$, and let f_n be the characteristic function of E_n. Clearly each $f_n \in I^U$, and we conclude the example by showing that $\{f_n\}$ has no convergent subsequence. Let $\{f_{x_k}\}$ be a subsequence. Then $x = \{x_k\} \in A$; let $t = \phi(x)$. Now, if k is any positive integer, $t \in E_{x_{2k+1}}$

⟦take $n = x_{2k+1}$ in the definition of E_n⟧ and so $f_{x_{2k+1}}(t) = 1$. Similarly $f_{x_{2k}}(t) = 0$. Thus $\{f_{x_k}(t)\}$ is not convergent, and so, neither is $\{f_{x_k}\}$ ⟦Theorem 6.4.1⟧. ▌ See also Problem 203.

REMARK. Z. Frolik [Frolik (b)] has constructed spaces X_n, $1 \leq n \leq \aleph_0$, such that $(X_n)^K$ is countably compact if and only if $K < n$.

Although the result of Theorem 7.4.1 cannot be extended to sequentially compact spaces, the classical "diagonal selection" argument allows us to prove a consolation result.

THEOREM 7.4.2. *A countable product of sequentially compact spaces is sequentially compact.*

Let $\{x_k\}$ be a sequence in $\prod \{X_n: n = 1, 2, \ldots\}$. Then $\{P_1 x_k\}$ has a convergent subsequence since X_1 is sequentially compact, say $P_1 x_{u(1,k)} \to y_1 \in X_1$ where $\{u(1, 1), u(1, 2), u(1, 3), \ldots\}$ is an increasing sequence of positive integers. Next, $\{P_2 x_{u(1,k)}\}$ has a convergent subsequence, say $P_2 x_{u(2,k)} \to y_2 \in X_2$, where $\{u(2, 1), u(2, 2), u(2, 3), \ldots\}$ is an increasing sequence of integers selected from $\{u(1, k)\}$. Next, $\{P_3 x_{u(2,k)}\}$ has a convergent subsequence, say $P_3 x_{u(3,k)} \to y_3 \in X_3$. Continuing in this way we obtain $u(n, k)$ for all n, k, such that $P_n x_{u(n,k)} \to y_n \in X_n$ as $k \to \infty$ and with $\{u(n, k)\}$ a subsequence of $\{u(n-1, k)\}$. Now we select the diagonal of the matrix thus: let $v(k) = u(k, k)$. Then $x_{v(k)} \to y$ where y is the member of $\prod X_n$ whose nth term is y_n for each n. ⟦For each k, $P_1 x_{v(k)} = P_1 x_{u(k,k)} = P_1 x_i$ for some $i \geq k$. Thus $P_1 x_{v(k)} \to y_1$. For each $k \geq 2$, $P_2 x_{v(k)} = P_2 x_i$ for some $i \geq k$, thus $P_2 x_{v(k)} \to y_2$. In general, for $k \geq n$, $P_n x_{v(k)} = P_n x_i$ for some $i \geq k$ and so $P_n x_{v(k)} \to y_n$. The result follows from Theorem 6.4.1.⟧ ▌

▲ EXAMPLE 3. The following deduction of the axiom of choice from Theorem 7.4.1 is due to J. L. Kelley. (Our proof of Theorem 7.4.1, on the other hand, uses in its proof a condition equivalent to the axiom of choice; namely, the existence of ultrafilters.) Let $\{X_\alpha: \alpha \in A\}$ be a collection of nonempty sets. Let u be an object not in $\bigcup X_\alpha$. Let $Y_\alpha = X_\alpha \cup \{u\}$ and give Y_α the topology which, on X_α is cofinite, and makes u isolated. In $\prod \{Y_\alpha: \alpha \in A\}$, let $S_\alpha = P_\alpha^{-1}[X_\alpha]$. Then $\{S_\alpha: \alpha \in A\}$ is a collection of closed sets ⟦each X_α is closed⟧ with the finite intersection property. ⟦Let F be a finite subset of A. For $\alpha \in F$ choose $x_\alpha \in X_\alpha$, for $\alpha \notin F$ choose $x_\alpha = u$. Then $x \in \bigcap \{S_\alpha: \alpha \in F\}$.⟧ Hence $\bigcap S_\alpha \neq \varnothing$. ⟦Theorem 5.4.3. We are using the fact that each Y_α is compact.⟧ We have thus found a nonempty subset of $\prod X_\alpha$.

Problems

1. Give an example of a noncompact set in $\mathbf{R}^2$ whose projections on the X and Y axes are both compact.

2. If every projection of a closed set in a product is compact, the set is compact.

---

101. Prove Theorem 7.4.1 by means of universal nets instead of ultrafilters ⟦Sec. 7.3, Problems 208–211⟧.
102. Show that $\mathbf{2}^U$ is not sequentially compact; $\mathbf{2}$ is a discrete space with two points, and $|U| \geq c$. ⟦Imitate Example 2.⟧
103. For each n, let $f_n: \mathbf{Q} \to [0, 1]$. Show that $\{f_n\}$ has a pointwise convergent subsequence ⟦Theorem 7.4.2⟧.
104. A product of compact spaces need not be compact in the box topology ⟦Sec. 6.4, Problem 102⟧.
105. Deduce from Theorems 5.4.4 and 7.4.1 that the Cantor discontinuum is compact.
106. A compact Hausdorff space need not be separable. Compare Lemmas 7.2.1 and 7.2.2. ⟦$\mathbf{2}^A$. Use Sec. 5.3, Problem 104.⟧
107. An infinite product of noncompact spaces cannot be locally compact, indeed every compact set in the product must have empty interior ⟦Theorems 6.4.4 and 5.4.4⟧.
108. A product $\prod X_\alpha$ is locally compact if and only if every X_α is locally compact and all but a finite number are compact. ⟦See Problem 107 and Sec. 6.7, Problem 102.⟧
109. Let $\sum a_n$ be an absolutely convergent real series. A subseries is any series gotten by replacing some of the a_n by 0. Show that the set of all sums of subseries of $\sum a_n$ is closed. ⟦Associate each subseries with a point $x \in \mathbf{2}^\omega$ thus: $x_n = 1$ if a_n is retained, $x_n = 0$ if a_n is replaced by 0. The map $x \to$ sum of the subseries is continuous, and $\mathbf{2}^\omega$ is compact by Theorem 7.4.1.⟧
110. In Problem 109, let $a_n = 2 \cdot 3^{-n}$. Show that the resulting set is the Cantor discontinuum.

---

201. In Example 2, is I^U countably compact?
202. Which sets in $\mathbf{R}$ can be obtained from some $\sum a_n$ as in Problem 110?
203. Prove that I^U (Example 2) is not sequentially compact thus: let S be the set of all "sawtooth" functions f on $[0, 1]$, each f being continuous, piecewise linear, and alternately 0 and 1 on a finite set of rationals. By Lebesgue's bounded convergence theorem, no subsequence of S converges to 0. (See *Amer. Math. Monthly* 75 (1968), 1098–1099.)

Compactification

8.1 The One-Point Compactification

Here and in Section 8.3 we shall describe two specific compactifications. In Section 8.4 we shall discuss the *concept* of compactification.

There is a sense in which compact spaces are universal among topological spaces, namely, every topological space is homeomorphic with a subspace of a compact space ⟦Theorem 8.1.1⟧. In its full generality, such a remark is hardly more than witticism, which can be extended also to other properties. (See Problem 1 and Sec. 5.3, Problem 4.) However, the device is often very useful. Here is a simple example: it will be seen ⟦Theorem 8.1.2⟧ that every locally compact Hausdorff space is homeomorphic with a subspace of a compact Hausdorff space; the latter is known to be completely regular ⟦Theorem 5.4.7⟧; complete regularity is hereditary ⟦Sec. 4.3, Problem 2]; and so, without further ado, *every locally compact Hausdorff space is completely regular*. Another application lies in the construction of examples to show that certain properties are not hereditary. The very fact that every locally compact Hausdorff space is homeomorphic with a subset of a compact Hausdorff space shows immediately that the Lindelöf property is not hereditary, since a locally compact Hausdorff space need not be a Lindelöf space. ⟦Consider a discrete uncountable space.⟧ The same type of argument shows that many other properties are not hereditary. (See Problems 1, 2.)

We now turn to the one-point compactification. This may be looked at in two ways. Either, any space X can be made into a compact space Y by adjoining a point to the space (just as $\mathbf{R}$ is made compact by "adding a point at ∞"), or, X is homeomorphic with the complement of a singleton in some compact space. The first point of view is more traditional, and is formalized thus. Let X be a topological space, let ∞ be a point not in X. (Think of X as a subset of a set which contains exactly one point not in X.) Let $X^+ = X \cup \{\infty\}$. A set G in X^+ is called open if either G is an open subset of X, or $X^+ \setminus G$ is a closed compact subset of X. Whenever X is a Hausdorff space we shall omit the word "closed" in this definition, since it is redundant ⟦Theorem 5.4.5⟧. After a few examples, we shall show in Theorem 8.1.1 that X^+ has properties which justify its being called *the one-point compactification* of X.

★EXAMPLE 1. Let $X = (0, 1]$, $Y = [0, 1]$. Then $Y = X^+$ with $\infty = 0$. ⟦For open $G \subset Y$; if $0 \in G$, $Y \setminus G$ is a compact subset of X; if $0 \notin G$, G is an open subset of X. Conversely if G is an open subset of X, G is an open subset of Y, and if K is a compact subset of X, $Y \setminus K$ is an open subset of Y.⟧

▲EXAMPLE 2. What is the point at infinity? In any system of set theory there is no universe set, that is, a set which contains everything. Hence for any X there exists $y \notin X$. For example, $X \notin X$ so X itself may be taken as the point at infinity. History records one lecturer who declaimed to his class "I am the point at infinity." A most reasonable candidate for this honor is the set of all complements of closed compact subsets of X—analogous with the completion process for metric spaces in which the Cauchy sequences are used rather than their fictitious limits.

THEOREM 8.1.1. *X^+ is a compact topological space, and X is a topological subspace. If X is compact, ∞ is an isolated point of X^+. If X is not compact, X is dense in X^+.*

Suppose that G_1, G_2 are open sets in X^+. Then $G_1 \cap G_2$ is open. ⟦If G_1, G_2 are open subsets of X, then so is $G_1 \cap G_2$. If $K_i = X^+ \setminus G_i$ are closed and compact subsets of X for $i = 1, 2$; then so is $X^+ \setminus (G_1 \cap G_2)$ since this is $X^+ \cap (G_1 \cap G_2)^\sim = X^+ \cap (K_1 \cup K_2) = K_1 \cup K_2$. If G_1 is an open subset of X, and $K = X^+ \setminus G_2$ a closed compact subset of X, then $G_1 \cap G_2 = (G_1 \cap X) \cap G_2 = G_1 \cap (X \cap G_2) = G_1 \cap (X \setminus K)$ which is an open subset of X.⟧ Let C be a collection of open sets in X^+. Then $U = \bigcup \{G: G \in C\}$ is open. ⟦If $\infty \notin G$ for all $G \in C$ this is clear. If $\infty \in G_0$ for some $G_0 \in C$, then $X^+ \setminus G_0$ is a closed compact subset of X, and $X^+ \setminus U = X \setminus U = \bigcap \{X \setminus G: G \in C\}$ is closed and compact in X since it is the intersection of a collection of closed sets, at least one of which is compact.⟧

Next, X is a topological subspace of X^+, for if G is an open set in X, G is relatively open ⟦$G = G \cap X$⟧, while if G is relatively open, say $G = O \cap X$ with O an open set in X^+, then G is open in X. ⟦If $\infty \notin O$, $G = O$, if $\infty \in O$,

$G = X \setminus (X^+ \setminus O)$ and $X^+ \setminus O$ is closed in X.]] A special case of this, which we shall now use, is that if F is a closed subset of X^+, $F \cap X$ is a closed subset of X. Next, X^+ is compact. [[Let C be a family of closed sets with the finite intersection property. If $\infty \in \bigcap \{F: F \in C\}$, then C is fixed. If $\infty \notin \bigcap \{F: F \in C\}$, say $\infty \notin F_0 \in C$; then F_0 is a compact subset of X. Let $C_0 = \{F \cap F_0: F \in C\}$. As was just remarked, each $F \cap F_0$ is closed in X, moreover C_0 has the finite intersection property and so C_0 is fixed since F_0 is compact. Hence C is fixed.]] If X is compact, $\{\infty\} = X^+ \setminus X$ is open. If X is not compact, $\{\infty\}$ is not open since its complement is X, hence every neighborhood of ∞ meets X. Thus $\infty \in \overline{X}$ and so $\overline{X} = X^+$. ∎

REMARK. From now on we shall consider compactifications of noncompact spaces only. Theorem 8.1.1 gives an adequate reason for this.

In the following example, X^+ is identified for certain X. This may be justified by direct computation from the definition, or by the uniqueness result, Theorem 8.1.3.

EXAMPLE 3. $(0, 1)^+$ *is the one-sphere* since removal of one point from the latter leaves a copy of $(0, 1)$. [[Let $f(x) = e^{2\pi i x}$.]] ω^+ *is a subset of* $\mathbf{R}$, *indeed it is homeomorphic with* $(0, 1, \frac{1}{2}, \frac{1}{3}, \ldots)$ [[Sec 5.1, Problem 4]].

The degree of separation of the one-point compactification is now solved in its most important special case. See also Problems 4, 113, 123, 203, and 207.

THEOREM 8.1.2. *Let X be a noncompact regular (or T_2) space. Then X^+ is regular (or T_2) if and only if X is locally compact.*

Suppose first that X is locally compact; let $x \in X^+$ and let N be a neighborhood of x in X^+. If $x \in X$, N includes a compact neighborhood K of x in X, and K includes a closed neighborhood F of x in X. Since F is closed and compact [[Theorem 5.4.2]], F is a closed neighborhood of x in X^+. Next suppose that $x = \infty$; N includes an open neighborhood G of x in X^+, and $\tilde{G}$ is included in K^i for a certain compact closed subset K of x, where K^i refers to the interior of K in X. [[Each $y \in \tilde{G}$ has a compact closed neighborhood in X. Since $\tilde{G}$ is compact, its cover by these neighborhoods can be reduced to a finite cover. Let K be the union of the sets in this cover.]] Then $X^+ \setminus K^i$ is a subset of G, hence of N, and is a closed neighborhood of ∞ in X^+ [[it includes $X^+ \setminus K$ which is an open set in X^+ containing ∞.]] Thus in both cases, each neighborhood of x includes a closed neighborhood of x, and so X^+ is regular. Conversely, suppose that X^+ is a regular space. Then X is locally compact by Corollary 5.4.1. Next, if X is a locally compact T_2 space and $x \in X$, x has a compact neighborhood N in X. Then N^i and $X^+ \setminus N$ separate x and ∞ [[Theorem 5.4.5]]. Thus X^+ is T_2. Conversely, if X^+ is T_2, it is regular [[Theorem 5.4.7]] and so X is locally compact, as above. ∎

We now consider the uniqueness of the one-point compactification. We are not so much concerned with the obvious fact that the construction yields a unique object. ⟦The only freedom allowed is what to choose for ∞, and the resulting spaces are identical as to neighborhoods of ∞ since these are defined as $S \cup \{\infty\}$, where S is described completely in terms of X, the space to be compactified.⟧ Rather, we shall consider the question of whether X^+ is the only compact space containing X as a complement of a singleton.

LEMMA 8.1.1. *Let H be a Hausdorff space, K a compact space, and $f: H \to K$ one-to-one and onto. Suppose that for a certain point $h \in H$, $f^{-1}: X \to H$ is continuous, where $X = K \setminus \{f(h)\}$. Then f is continuous at h.*

Let h_δ be a net in H with $h_\delta \to h$. We may assume that $h_\delta \neq h$ for all δ ⟦Sec. 4.2, Problem 33⟧. Let $y_\delta = f(h_\delta)$. By Theorem 7.1.4, y_δ has a cluster point y. It follows that $y = f(h)$ ⟦if not y_δ and y both lie in X and so $f^{-1}(y)$ must be a cluster point of h_δ, by Sec. 7.1, Problem 9. But h is the only cluster point of this net, hence $h = f^{-1}(y)$.⟧ Thus y_δ has exactly one cluster point $f(h)$, and so $y_\delta \to f(h)$ ⟦Sec. 7.1, Problem 13⟧. ▮

The essence of the next result is that the construction of X^+ is the only way to T_2-compactify a locally compact Hausdorff space by adding one point.

THEOREM 8.1.3. *Let Y be a compact Hausdorff space, $h \in Y$ a nonisolated point, and $X = Y \setminus \{h\}$. Then $Y = X^+$ with $h = \infty$.*

Define $f: Y \to X^+$ by $f(x) = x$ for $x \in X$, $f(h) = \infty$. Then f is continuous at h by Lemma 8.1.1. ⟦Take $H = Y$, $K = X^+$.⟧ Hence f is continuous on Y. The same argument shows that f^{-1} is continuous on X^+. Thus f is a homeomorphism. ▮

An abrupt proof of Theorem 8.1.3 runs thus: identifying Y and X^+ as sets with $h = \infty$, an open neighborhood G of ∞ in X^+ has $\tilde{G}$ compact in X, hence in Y; so $\tilde{G}$ is closed in Y and G is a neighborhood of 0 in Y. The same argument shows that every open neighborhood of h in Y is a neighborhood of ∞ in X^+.

★EXAMPLE 4. *σ-properties.* If a topological space is a countable union of sets, each of which has a certain property, P, we say that the space is σ-P. For example, $\mathbf{R}$ is σ-compact. ⟦$\mathbf{R} = \bigcup_{n=1}^{\infty} [-n, n]$, and each $[-n, n]$ is compact.⟧ An uncountable discrete space is not σ-compact since its compact subsets are finite. A σ-finite space is merely a countable space. Of course, a compact space is σ-compact, and a similar statement is true for any other property. If a set in a topological space is σ-closed, we call it an F_σ. For example, $(0, 1]$ is an F_σ in $\mathbf{R}$. ⟦$(0, 1] = \bigcup \{[1/n, 1]: n = 1, 2, \ldots\}$.⟧

Problems

1. Every topological space is a subspace of a normal space. ⟦Add one point, letting its only neighborhood be the whole space. Of any two disjoint closed sets, one is empty.⟧ (Not every space is a subspace of a T_4 space, however.)
2. σ-compactness is not hereditary. ⟦Let D be discrete. Then D^+ is compact; D need not be σ-compact.⟧
3. Let X be the set in the complex plane consisting of $\{z: |z| < 1\}$. Show that X^+ is the two-sphere.
★4. If X is T_1, so is X^+.
5. If X is a topological subspace of Y, X^+ need not be a topological subspace of Y^+. ⟦$Y = [0, 1)$, $X = (0, 1)$.⟧
6. $\mathbf{Q}^+$ is compact, but not locally compact. ⟦$\mathbf{Q}$ is an open subset; now see Theorem 5.4.12.⟧
7. A locally compact KC space must be T_2. ⟦It has a base of compact, hence closed, sets. Thus it is regular.⟧ (We shall see in Problem 205 that a compact KC space need not be T_2.)
8. Any countable subset of a T_1 space is an F_σ.
9. Find all F_σ subsets of (a) a cofinite space; (b) an indiscrete space.
10. Every σ-compact space is a Lindelöf space. ⟦Reduce a cover to a finite cover of every compact set in the sequence.⟧

---

101. Lemma 8.1.1 becomes false if the assumption that K is compact is dropped. ⟦Let $K = [0, 1)$, $H =$ one-sphere, $X = (0, 1)$.⟧
102. Lemma 8.1.1 becomes false if "Hausdorff" is replaced by "T_1." ⟦Let $K = [0, 1]$, $H = K$ with cofinite topology.⟧
103. Lemma 8.1.1 becomes false if "Hausdorff" is replaced by "regular." ⟦Let $K = \{0, 1, \frac{1}{2}, \frac{1}{3}, \ldots\}$, $H = \{0', 0, \frac{1}{2}, \frac{1}{3}, \frac{1}{4}, \ldots\}$, where 0, $0'$ have exactly the same neighborhoods. (A trivial example has H indiscrete.)⟧
104. In the hints to Problems 101, 103, $f^{-1}|X$ is a homeomorphism. Solve Problem 102 with this extra restriction. ⟦Same K; $H = K$ with fewer neighborhoods of 1.⟧
105. Give examples to compare $(A \times B)^+$ with $A^+ \times B^+$. ⟦For example $(\mathbf{R} \times \mathbf{R})^+$ is the two-sphere and $\mathbf{R}^+ \times \mathbf{R}^+$ is the torus.⟧
106. Resolve this contradiction. Let X be a T_1 space. Then X^+ is T_1 ⟦Problem 4⟧. Hence $\{\infty\}$ is closed, and X is open in X^+. By Theorem 5.4.12, X is locally compact. But X is an arbitrary T_1 space!
107. A T_1 space may have two different T_1 compactifications obtained by adding one point. ⟦Let X be cofinite. Consider X^+, and $X \cup \{\infty\}$ with the cofinite topology.⟧
108. $(\mathbf{R}^n)^+ = S_n$.

109. Let $f: X \to \mathbf{R}$. We say that $\lim_{x \to \infty} f(x) = a$ if, for every $\varepsilon > 0$, $\{x : |f(x) - a| > \varepsilon\}$ is included in a compact closed set; and $\lim_{x \to \infty} f(x) = \infty$ if for every m, $\{x : |f(x)| < m\}$ is included in a compact closed set. Let $f^+ : X^+ \to \mathbf{R}$ be defined by $f^+ = f$ on X and $f^+(\infty) = a$. Show that f^+ is continuous at ∞ if and only if $\lim_{x \to \infty} f(x) = a$. Obtain a similar result for a map to $\mathbf{R}^+$ with $f^+(\infty) = \infty$.
110. Let F be a family of subsets of a set X. We call C a *cobase* for F if $C \subset F$ and every member of F is included in some member of C. Show that the family of closed subsets is a cobase for the family of all subsets of a topological space.
111. A filterbase $\mathscr{B}$ generates a filter $\mathscr{F}$ if and only if $\{A : \tilde{A} \in \mathscr{B}\}$ is a cobase for $\{A : \tilde{A} \in \mathscr{F}\}$.
112. A space is called *hemicompact* if its family of compact closed subsets has a countable cobase. Show that a hemicompact T_1 space is σ-compact, but not conversely. For example, if X is hemicompact T_2 and first countable at x, x must have a compact neighborhood. In particular, $\mathbf{Q}$ is not hemicompact. ⟦If $\{G_n\}$ is a shrinking base at x and $\{K_n\}$ is a sequence of compact sets, choose $x_n \in G_n \setminus K_n$; let $K = (x, x_1, x_2, \ldots)$.⟧
113. X is hemicompact if and only if X^+ is first countable at ∞, and is σ-(compact-closed) if and only if $\{\infty\}$ is a G_δ. If X is T_2 or regular, the word "closed" may be omitted. ⟦Sec. 5.4, Problem 9 and Theorem 5.4.5.⟧ (Thus $\mathbf{Q}^+$ is another example of a countable space which is not first countable. See Problem 112.)
114. "Regular" cannot be omitted in Sec. 5.4, Problem 208. ⟦∞ is a G_δ point in $\mathbf{Q}^+$. See Problem 113.⟧
115. A subset of a topological space is called a *k-test set* if its intersection with every compact set K is closed in K. A space is called a *k space* if every k-test set is closed. Show that a T_1 pseudofinite k space must be discrete; hence show an example of a non-k space. ⟦Sec. 5.4, Problem 106. T_2 examples are given in Sec. 3.1, Problem 108; Sec. 6.2, Problem 110; Sec. 8.3, Problem 103.⟧
116. Let X be a space such that, in X, whenever $x \in \bar{S}$, there exists a compact set K such that $x \in K \cap \overline{K \cap S}$. Show that X must be a k space. ⟦If A is a k-test set and $x \in \bar{A}$, then $x \in \overline{A \cap K} \cap K \subset A$.⟧
117. A space in which each point has a compact neighborhood is a k space ⟦Problem 116⟧.
118. Every compact space, and every locally compact space is a k space ⟦Problem 117⟧.
119. A first countable space must be a k space ⟦Problem 116 and Sec. 5.4, Problem 1⟧.
120. Let X be a k space, Y a topological space, and let $f: X \to Y$ have the

property that $f|K$ is continuous for all compact K. Show that f must be continuous.

121. Fix $x \in X$. The *no-point compactification of* X at x, written X_x, is $(X \setminus \{x\})^+$. If we take $\infty = x$ we have $X_x = X$ but with a smaller topology if X is a T_1 space.
122. Every no-point compactification of **R** is a figure 8.
123. A regular space is locally compact if and only if every compact set is included in the interior of a compact set. ⟦Let K be compact. We may assume K is closed by Sec. 5.4, Problem 9. Then $\widetilde{K}$ is a neighborhood of ∞, thus includes a closed neighborhood of ∞.⟧
124. A locally compact Lindelöf space must be σ-compact. ⟦Apply Sec. 5.3, Problem 105 to X^+.⟧
125. A locally compact regular Lindelöf space must be hemicompact. ⟦By Problems 113 and 124, $\{\infty\}$ is a G_δ. Now see Problem 113 and Sec. 5.4, Problem 201.⟧
126. Call X an *a* space if every point has a compact neighborhood, a *b* space if every point has a base of compact closed neighborhoods, a *c* space if every point has a base of relatively compact neighborhoods. Show that for T_2 or regular spaces these definitions are all equivalent to local compactness. Which of these properties does $\mathbf{Q}^+$ have? What of the space in Sec. 5.4, Problem 115? Further discussion is given in [Gross] and [Schnare].
127. X^+ is connected if and only if X has no compact open components. Thus ∞ is a *dispersion point* of $\mathbf{Q}^+$. (This means that $\mathbf{Q}^+$ is connected and $\mathbf{Q}^+ \setminus \{\infty\}$ is totally disconnected.)
128. Let X be locally connected. Then X^+ is connected if and only if X has no compact components.
129. A hemicompact first countable *KC* space must be an *a* space (Problem 126). Hence, if T_2 or regular, it is locally compact ⟦Problem 112⟧.
130. Let X be a not relatively compact subspace of a hemicompact space Y. Show that in X^+, ∞ is a sequential limit point of X. ⟦By Problem 113, X contains a sequence converging to ∞ in Y^+.⟧
131. Let D be an infinite discrete space. Show that the cofinite filter on D^+ converges to ∞.
132. Let D be an uncountable discrete space. Then D^+ is closure-sequential but not first countable ⟦Problems 131 and 113⟧.

---

201. Give an example of a Hausdorff space with a point which has no linearly ordered neighborhood base ⟦Problem 132⟧.
202. $\mathbf{R}^\mathbf{R}$ is not a k space. ⟦See [Kelley, 7J].⟧
203. Let X be a *KC* space. Then X^+ is a *KC* space if and only if X is a k space. ⟦See [Wilansky (b), Theorem 5].⟧
204. Let X be a Hausdorff space. Then X is respectively, locally compact, k,

if and only if X^+ is respectively T_2, KC. ⟦Problem 203 and Theorem 8.1.2.⟧

205. $\mathbf{Q}^+$ is KC, compact, and not T_2 ⟦Problems 204 and 119⟧.
206. The product of two closure-sequential spaces need not be sequential, hence ⟦Sec. 3.1, Problem 204⟧ need not be closure-sequential. ⟦The diagonal in $\mathbf{Q}^+ \times \mathbf{Q}^+$ is sequentially closed by Problems 130, 205, and Sec. 6.7, Problem 10, but is not closed by Sec. 6.7, Example 1. See Problem 130 or 209.⟧
207. Give an example of a US space X such that X^+ is not US. ⟦See [Wilansky (b)].⟧
208. Give an example of a nonhomogeneous metric space X such that $X_x = X_y$ for all $x, y \in X$. (See Problem 121.) ⟦Try $X = \omega \cup Y$ with $Y = \bigcup_{n=1}^{\infty} (-n - 1, -n)$.⟧
209. If X is a metric space, X^+ is closure-sequential.

8.2 Embeddings

In this section we discuss some embeddings which lead to compactifications of various types. Let X be a Tychonoff space and C the set of all continuous maps of X into $I = [0, 1]$. Then the topology of X is equal to the weak topology by C ⟦Theorem 6.7.4; the proof obviously covers this case⟧, and so Theorem 6.6.2 shows that X is homeomorphic with a subspace of I^C. Moreover, the map described in Theorem 6.6.2 is $x \to \hat{x}$ (pronounced x-hat), where $\hat{x}(f) = f(x)$. The map $x \to \hat{x}$ is called the *evaluation* from X to I^C. The range of the evaluation is written $\hat{X}$; thus $\hat{X} = \{\hat{x} : x \in X\}$. We state these remarks formally.

THEOREM 8.2.1. *Let X be a Tychonoff space and C the set of all continuous maps of X into $I = [0, 1]$. Then the evaluation is a homeomorphism of X into I^C.*

Any space of the form I^A, where $I = [0, 1]$ and A is a set, is called a *cube*. We now see that cubes are universal among Tychonoff spaces.

THEOREM 8.2.2. *A topological space is a Tychonoff space if and only if it is homeomorphic with a subspace of a cube.*

Half of this is Theorem 8.2.1. Conversely each cube is a Tychonoff space by Theorem 6.7.3 (or because it is compact T_2), and hence, so is each subspace by Sec. 4.3, Problem 2. ∎

▲ EXAMPLE 1. A certain embedding, suggested by M. Fréchet in 1909, has some interesting applications. One of these is an alternative construction of the completion of a metric space in Sec. 9.2, Example 3. Another is the construction of a universal separable metric space. (See [Banach, p. 187, Theorem 10].) Let X be a topological space and let $Y = C^*(X)$. For $f \in Y$ define $\|f\| = \sup\{|f(x)| : x \in X\}$, ($\|f\|$ is pronounced norm-f), and for f, $g \in C$, define $n(f, g) = \|f - g\|$. Then (Y, n) is a metric space. Now, assuming that (X, d) is a semimetric space, fix $z \in X$ and define $\rho: X \to Y$ by $\rho(x)(t) = d(x, t) - d(t, z)$, for all $x, t \in X$. (Note, $\rho(x)$ is that function on X whose value at t is $\rho(x)(t)$.) Then ρ is a distance-preserving map, that is for $x, y \in X$, $n[\rho(x), \rho(y)] = d(x, y)$.

$$
\begin{aligned}
[\![n[\rho(x), \rho(y)] &= \|\rho(x) - \rho(y)\| \\
&= \sup\{|\rho(x)(t) - \rho(y)(t)| : t \in X\} \\
&= \sup|d(x, t) - d(t, z) - d(y, t) + d(t, z)| \\
&= \sup|d(x, t) - d(y, t)| \leq d(x, y) \\
&= |d(x, y) - d(y, z) - d(y, y) + d(y, z)| \\
&\leq \sup|d(x, t) - d(t, z) - d(y, t) + d(t, z)| \\
&= n[\rho(x), \rho(y)].
\end{aligned}
$$

(The third expression from the end is just the value of the succeeding expression when $t = y$.)$]\!]$ Thus if X is a metric space we see that X is isometric with a subspace of (Y, n). Some refinements are suggested in Problems 103, 104, and 105.

Problems

1. Let A be a nonempty collection of continuous functions from a topological space X into $I = [0, 1]$. For $x \in X$, define $\hat{x} \in I^A$ by $\hat{x}(f) = f(x)$ for $f \in A$. Let the map $x \to \hat{x}$ be called $e: X \to I^A$. Show that e is continuous.
2. The map e of Problem 1 is one-to-one if and only if A is separating over X.
3. The map e of Problem 1 is a homeomorphism if and only if A is *determining*. (This means that A is separating and $f[\mathscr{F}] \to f(x)$ for all $f \in A$ implies $\mathscr{F} \to x$, where $\mathscr{F}$ is a filter.)
4. If X is a T_0 space it is redundant, in the definition of "determining," to assume that A is separating.
5. Give an example of a separating family of continuous functions on $\mathbf{R}$ which has only one member. Prove that such a family must be determining.

---

101. Let $f_k(x) = x/(k + x^2)$ for $k = 1, 2$; $x \in \mathbf{R}$. Let $A = \{f_1, f_2\}$. Show that the map e of Problem 1 goes from $\mathbf{R}$ to I^2 (a square in $\mathbf{R}^2$) and carries $\mathbf{R}$ onto a simple closed curve. This is an example in which e is one-to-one but not a homeomorphism.

102. Show directly that A, in Problem 101, is not determining.
103. If X is a semimetric space, the embedding of Example 1 is distance preserving but need not be isometric.
104. Let D be a dense subset of a semimetric space (X, d), and let $Y = C^*(D)$. For $f \in Y$, let $\|f\| = \sup\{|f(x)| : x \in D\}$. Define $\rho: X \to Y$ by $\rho(x)(t) = d(x, t) - d(t, z)$, z being a fixed member of X. Show that ρ is distance preserving ⟦as in Example 1⟧.
105. Every separable metric space is isometric with a subset of the space of all bounded real sequences metrized by $d(x, y) = \sup|x_n - y_n|$ ⟦Problem 104⟧.
106. Show that the evaluation cannot be onto I^C.

---

201. Give an example for Problem 1 in which $X = \mathbf{R}$, A has two members and e is onto I^2.

8.3 The Stone–Cech Compactification

We are going to embed a space X in a compact Hausdorff space with remarkable properties. Since a compact Hausdorff space is $T_{3\frac{1}{2}}$ ⟦Theorem 5.4.7⟧ and this property is hereditary ⟦Sec. 4.3, Problem 2⟧, only Tychonoff spaces can be so embedded; as we shall see, this assumption is also sufficient. The main result (Theorem 8.3.1) was proved by M. H. Stone and E. Cech, both in 1937. The special case $Y = \mathbf{R}$ had been given by A. Tychonoff in 1930.

The compactification will be carried out by means of the evaluation of Section 8.2, and for this purpose we set up our notation: X is a $T_{3\frac{1}{2}}$ space, C is the set of all continuous maps from X to I, the unit interval $[0, 1]$. As before, the evaluation is the map $x \to \hat{x}$ where $\hat{x}(f) = f(x)$ for all $f \in C$; it is a homeomorphism of X into I^C, the range being X ⟦Theorem 8.2.1⟧.

LEMMA 8.3.1. *Let $g: \hat{X} \to I$ be continuous. Then g has a continuous extension $G: I^C \to I$.*

Define $g_0 \in C$ by $g_0(x) = g(\hat{x})$. ⟦g_0 is continuous since $g_0 = g \circ \rho$, where ρ is the evaluation; g, ρ being continuous.⟧ Define $G: I^C \to I$ by $G(h) = h(g_0)$ for all $h \in C$. Then G is continuous. ⟦Let $h_\delta \to h$. Then $h_\delta(g_0) \to h(g_0)$ since convergence is pointwise. Thus $G(h_\delta) \to G(h)$.⟧ Finally $G \mid \hat{X} = g$. ⟦Let $\hat{x} \in \hat{X}$. Then $G(\hat{x}) = \hat{x}(g_0) = g_0(x) = g(\hat{x})$.⟧ ∎

The space I^C is a compact Hausdorff space and hence, so is the closure of $\hat{X}$ in I^C. Let this closure be called βX; it is a compact Hausdorff space with a dense subspace homeomorphic with X. If we identify X with $\hat{X}$ we may say that *we have embedded X as a dense subspace of βX in such a way that every continuous function $g: X \to I$ has a continuous extension $G: \beta X \to I$.* It is in this form that we shall use the result of Lemma 8.3.1. Note that if X is compact, $\beta X = X$. (More accurately, $\beta X = \hat{X}$.)

Before reading the proof of Lemma 8.3.2, the reader may be guided by a glance over the remark which follows it.

LEMMA 8.3.2. *Let $g\colon X \to Y$ be continuous, where Y is a compact Hausdorff space. Then g has a continuous extension $G\colon \beta X \to Y$.*

Let $C(Y)$ be the set of all continuous functions from Y into I, and $\hat{Y}$ the image of the natural embedding of Y in $I^{C[Y]}$. Since Y, $\hat{Y}$ are homeomorphic ⟦Theorems 8.2.1 and 5.4.7⟧, it will be sufficient to prove Lemma 8.3.2 under the assumption that $g\colon X \to \hat{Y}$. Fix $u \in C[Y]$, and let $g_u\colon X \to I$ be defined thus: for each $x \in X$, let $g_u(x) = g(x)(u)$, the value of $g(x)$, a member of $\hat{Y}$, at u. Then g_u is continuous. ⟦Let q be the natural embedding of Y in $I^{C(Y)}$; a homeomorphism, as just mentioned. Then $g_u = u \circ q^{-1} \circ g$, as is seen thus: let $x \in X$, $g(x) = \hat{y}$. Then $u\{q^{-1}[g(x)]\} = u[q^{-1}(\hat{y})] = u(y) = \hat{y}(u) = g(x)(u)$.⟧ By Lemma 8.3.1, g_u has a continuous extension $G_u\colon \beta X \to I$.

We now reverse the process which led from g to g_u, "pasting together" the G_u's to make G. Define $G\colon \beta X \to I^{C(Y)}$ as follows: for each $x \in X$ let $G(x)$ be that member of $I^{C(Y)}$ whose value at $u \in C(Y)$ is $G_u(x)$; thus $G(x)(u) = G_u(x)$. Now $G|X = g$. ⟦For $x \in X$, $G(x)(u) = G_u(x) = g_u(x) = g(x)(u)$ for all $u \in C[Y]$, hence $G(x) = g(x)$.⟧ G is continuous. ⟦Let z_δ be a net in βX with $z_\delta \to z$. Then $G(z_\delta) \to G(z)$ since for each $u \in C(Y)$, $G(z_\delta)(u) = G_u(z_\delta) \to G_u(z) = G(z)(u)$, each G_u being continuous on βX.⟧ Finally $\hat{Y}$ is compact, hence a closed subset of $I^{C(Y)}$. ⟦The latter space is T_2 since it is a product of intervals, and a compact subset is closed, by Theorem 5.4.5.⟧ It follows that the range of G is included in $\hat{Y}$. ⟦$G[\beta X] = G[\bar{X}] \subset \hat{Y}$ by Theorem 4.2.4.⟧ ∎

REMARK. The proof of Lemma 8.3.2 can be conceptually simplified in the following way. Let $X' = C(X)$, $X'' = I^{X'}$, $Y' = C(Y)$, $Y'' = I^{Y'}$, $g\colon X \to Y$. Define the so-called *dual map* $g'\colon Y' \to X'$ by $g'(u) = u \circ g$ for $u \in Y'$. The *second dual map* $g''\colon X'' \to Y''$ is similarly defined by $g''(h) = h \circ g'$ for $h \in X''$. We may identify X with $\hat{X}$, Y with $\hat{Y}$, $g''|\hat{X}$ with g. ⟦Actually $g''|\hat{X} = \rho_Y \circ g \circ \rho_X^{-1}$, where ρ_Y, ρ_X are the evaluations.⟧ Now the problem is to extend g''. For each $u \in Y'$ we consider the map $\hat{x} \to g''(\hat{x})(u)$. This is a map to I and can be extended, by Lemma 8.3.1. The extension then yields a map $G\colon X'' \to Y''$ whose value at u is $g''(h)(u)$ for $h \in X''$. Then the restriction of G to $\bar{\hat{X}}$ is the required extension of g''.

THEOREM 8.3.1. *Let X be a Tychonoff space. Then there exists a compact Hausdorff space βX such that X is a dense subspace of βX and such that every continuous function from X into a compact Hausdorff space Y can be extended to a continuous function from βX into Y.*

This is the content of Lemma 8.3.2. ∎

The space βX is called the Stone–Cech compactification of X. A construction of βX by the process of completion of a uniform space is given in Sec. 11.5, Example 1; a construction by the techniques of Banach algebra, due to M. H. Stone, I. Gelfand, and A. Kolmogoroff, may be found in Section 14.3 of [Wilansky (a)]; a construction by means of ultrafilters, due in part to P. Samuel, M. H. Stone, and H. Wallman, may be found in Chapter 6 of [Gillman and Jerison]; still other constructions have been given by S. Kakutani, G. E. Silov and others.

Theorem 8.5.2 is a uniqueness theorem for βX.

★EXAMPLE 1. Let $X = (0, 1]$, $f(x) = \sin 1/x$, and let F be the extension of f to βX. (Here the range space is $[-1, 1]$.) Let $-1 \leq t \leq 1$. Then there exists a sequence $\{x_n\}$ in X such that $x_n \to 0, f(x_n) = t$. The sequence $\{x_n\}$ has a cluster point u in βX ⟦Theorem 7.1.4⟧ and $F(u) = t$. ⟦$F(u)$ is a cluster point of $\{F(x_n)\}$ by Sec. 7.1, Problem 9, and $\{F(x_n)\}$ is a constant sequence in $[-1, 1]$.⟧ Thus $\beta X \setminus X$ has, at least, a point for each number between -1 and 1. This shows that βX is large for even very small spaces. The next result shows its pathology in a different way.

THEOREM 8.3.2. *Let X be a T_4 space. Then βX is not first countable at any point of $\beta X \setminus X$. No point of $\beta X \setminus X$ is a sequential limit point of X.*

In Sec. 10.2, Problem 112 and Sec. 14.6, Problem 1, we shall see that the first result holds, and the second result fails, for $T_{3\frac{1}{2}}$ spaces.

The first conclusion of Theorem 8.3.2 follows from the second, and Theorem 3.1.1. Suppose that $\{x_n\}$ is a sequence in X and $x_n \to y \in \beta X \setminus X$. Then $\{x_{2n}\}$ and $\{x_{2n+1}\}$ are disjoint closed subsets of X; they can be completely separated. When the separating function f is extended to βX, we have $f(y) = \lim f(x_{2n}) = \lim f(x_{2n+1})$ which is impossible by the definition of f. ∎

A point in X may be a sequential limit point of $\beta X \setminus X$ ⟦Problem 110⟧.

COROLLARY 8.3.1. *βX cannot be metrizable unless X is already compact.*

If βX is metrizable, X is also; hence X is T_4 ⟦Theorem 4.3.3⟧ and the result follows from Theorem 8.3.2. ∎

REMARK. The Stone–Cech compactification is a useful tool. Like the one-point compactification, it can be used to show that certain properties are not hereditary ⟦Problem 101⟧. It can be used to construct examples of measures, integrals, and generalized limits; see [Wilansky (a), Section 14.3, Application 3]. It can be used to prove a form of Tychonoff's product theorem; see Sec. 8.6, Problem 102. It serves as a readily available source of examples, (see Problem 103 and many other places throughout this book), and can even be used to construct a set of real numbers which is not Lebesgue measurable; see [Semadeni]. We put aside its study with reluctance, referring the reader to an excellent discussion in [Gillman and Jerison, Chapter 6].

Problems

In this list X is a Tychonoff space, f a continuous function defined on X, and F the continuous extension of f to βX.

★1. Every bounded continuous real function on X can be extended to a continuous real function on βX. ⟦In Theorem 8.3.1, take $Y = \overline{f[X]}$.⟧

★2. No unbounded real function on X can be extended to a continuous real function on βX ⟦Sec. 5.4, Example 2⟧.

3. $\beta\omega$ is not sequentially compact. ⟦Consider $\{n\}$ and Theorem 8.3.2.⟧ (This also follows from Theorem 14.1.5.)

4. Let $X = (0, 1) \cup (1, 2)$. Let f be the characteristic function of $(0, 1)$. Find $F[\beta X]$.

5. Let F, K be disjoint with F closed and K compact. Show that they can be completely separated. ⟦Apply Urysohn's lemma (Theorem 4.2.11) to the normal space βX. Note that use of X^+ would yield a weaker result.⟧

---

101. T_4 is not hereditary ⟦Theorem 8.3.1; Sec. 6.7, Example 3 and Theorem 6.7.3⟧.

102. Let X be the subspace of $\beta\omega$ consisting of ω and one more point t. Show that X is normal ⟦Sec. 4.1, Problem 119⟧.

103. The space X of Problem 102 is countable and (a) not first countable, (b) pseudofinite, (c) not locally compact, (d) not a k space. ⟦For (a) use the second conclusion of Theorem 8.3.2. For (b), if S is an infinite compact subset, there must be a sequence converging to t, by Sec. 7.1, Problem 13. This contradicts Theorem 8.3.2. For (c), argue directly that no neighborhood of t is closed in $\beta\omega$ or use Sec. 5.4, Problem 202. For (d), a pseudofinite k space must be discrete.⟧ Also ω is sequentially closed but not closed in X ⟦Theorem 8.3.2⟧.

104. βX is connected if and only if X is ⟦Lemma 8.3.2, Lemma 5.2.1, and Theorem 5.2.3⟧.

105. X is locally compact if and only if it is open in βX ⟦Theorems 5.4.12 and 5.4.13.⟧

106. X is σ-compact if and only if it is an F_σ in βX.

107. The points of ω are isolated in $\beta\omega$. ⟦Each is open in ω which, by Problem 105, is open in $\beta\omega$.⟧

108. The cardinality of $\beta\omega$ is $2^{\mathfrak{c}}$. ⟦Let D be a dense countable subset of

$[0, 1]^{\mathbf{R}}$ (Sec. 6.7, Problem 201), and $f: \omega \to D$ onto. Then $F[\beta\omega] = [0, 1]^{\mathbf{R}}$ since it is compact and dense. See Sec. 4.1, Problem 204.⟧

109. βX is zero-dimensional, hence totally disconnected, if X is countable ⟦Sec. 4.3, Problem 108⟧.
110. Each point of $\mathbf{Q}$ is a sequential limit point of $\beta\mathbf{Q}\backslash\mathbf{Q}$. ⟦Choose x_n in the closure in $\beta\mathbf{Q}$ of $(x - 1/n, x + 1/n)$.⟧
111. Open intervals of $\mathbf{R}$ are open in $\beta\mathbf{R}$ ⟦Problem 105⟧.
112. If X is T_4, no point of $\beta X \setminus X$ is a G_δ in βX. ⟦Theorem 8.3.2; Sec. 5.4, Problem 201.⟧ (The remark on Theorem 8.3.2 shows that T_4 may be replaced by $T_{3\frac{1}{2}}$.) It follows that no point can be a zero-set. ⟦Every zero-set is a G_δ.⟧
113. A locally compact separable metric space must be hemicompact ⟦Sec. 8.1, Problem 125⟧, but a locally compact separable Hausdorff space need not even be σ-compact. ⟦Remove one point from $\beta\omega$ and apply Problem 112 or Sec. 5.3, Problem 105; or Sec. 14.1, Example 2.⟧
114. Let X be pseudocompact and $f \in C(X)$. Then F, f have the same range. ⟦$F[X] = f[X]$ is pseudocompact, hence compact, hence closed. It thus includes $F[\beta X]$ by Theorem 4.2.4.⟧
115. The following condition is necessary and sufficient that X be pseudocompact. If $f \in C^*(X)$ and $f(x) \neq 0$ for all $x \in X$, then $F(t) \neq 0$ for all $t \in \beta X$. ⟦Problem 114; conversely, if X is not pseudocompact, let g be unbounded, and $f = 1/(1 \vee |g|)$.⟧
116. A Hausdorff space is called *absolutely closed* if it is a closed subspace of any Hausdorff space in which it can be embedded. Show that a compact space is absolutely closed, and that an absolutely closed Tychonoff space is compact. ⟦It is closed in βX.⟧
117. Let X be first countable at x. Show that βX is first countable at x. ⟦Take the closure in βX of the X-neighborhoods of x. These will form a base for the closed neighborhoods of x in the regular space βX.⟧
118. In contrast with Problem 117, a point may be a G_δ in X but not in βX. ⟦Consider t in Problem 102 and use Problem 112.⟧
119. Let X, Y be first countable and such that βX, βY are homeomorphic. Show that X, Y are homeomorphic ⟦Theorem 8.3.2 and Problem 117⟧. ("First countable" cannot be omitted, nor can it be replaced by the assumption that all points are G_δ's. See Sec. 8.5, Problem 110.)
120. $C^*(X) = C(\beta X)$.
121. Take as known the result: If $C(X)$, $C(Y)$, are ring isomorphic, with X, Y compact Hausdorff, then X, Y are homeomorphic. (See [Wilansky (a), Sec. 14.1, Corollary 1].) Show that if X, Y are first countable Tychonoff spaces with $C^*(X)$, $C^*(Y)$ ring isomorphic, then X, Y are homeomorphic ⟦Problems 120 and 119⟧.
122. Let $t \in \beta\omega \setminus \omega$. Let V be a neighborhood of (t, t) in $\beta\omega \times \beta\omega$. Show that V must contain a point (m, m) and a point (m, n) with $m \in \omega$

$n \in \omega$, $m \neq n$. ⟦$V \supset U \times U$ and $U \cap \omega$ has more than one member.⟧

123. Must two sequences in $\mathbf{Q}$ which converge to $\sqrt{2}$ have the same accumulation points in $\beta\mathbf{Q}$?
124. Two different topologies may have the same compact connected sets ⟦Problem 103 and the discrete topology⟧.
125. There exists a convergent net with no one-to-one subnet. ⟦In the space of Problem 103, let $x_\delta \in \omega$, $x_\delta \to t$. A one-to-one subnet would be a sequence.⟧

---

201. Criticize this argument for showing that $\beta(0, 1]$ is not metrizable. With $f(x) = \sin 1/x$, F would be uniformly continuous ⟦Sec. 5.4, Problem 108⟧. But $F \mid (0, 1] = f$ is not uniformly continuous. What does this argument actually prove?
202. Call an open set G *saturated* if there exists no point $x \notin G$ with $G \cup \{x\}$ open. Find the saturated open sets in $\mathbf{R}$ and $\mathbf{R}^2$. Show that ω is saturated in $\beta\omega$ ⟦Problem 112⟧. (More generally, this shows if X is σ-compact and locally compact, it is saturated in βX.) Express this result in terms of isolated points of $\beta X \setminus X$.
203. Let X be separable, and first countable at each point of a dense countable subset. Then X has a countable pseudobase but need not be second countable ⟦$\beta\omega$⟧.
204. βX need not be a continuous image of $I^{C(X)}$. ⟦Problem 104; Theorem 6.6.3; Theorem 5.2.2.⟧ In particular, it need not be a retract.
205. In Lemma 8.3.1, g was extended all the way to I^C. This is not in general possible in Lemma 8.3.2. ⟦If we could extend the inclusion map: $X \to \beta X$ to $r\colon I^C \to \beta X$; then r is onto, since its range includes a dense set. This contradicts Problem 204.⟧
206. For a T_2 space X, these are equivalent: (a) X is absolutely closed, (b) every filterbase of open sets has a cluster point, (c) X is closed in every T_2 space Y such that $Y = X \cup$ singleton. ⟦For $c \Rightarrow b$, add a new point whose neighborhoods are the sets of the filterbase. For $b \Rightarrow a$, form a filterbase from the open neighborhoods of some $y \in \overline{X}$.⟧
207. Let A and $\tilde{A}$ be dense in a compact T_2 space (X, T). Show that the simple extension of T by A is absolutely closed and not compact. ⟦Use Theorem 5.4.10.⟧
208. An absolutely closed T_3 space is compact.
209. The character of x in X is the same as in βX. ⟦Imitate Problem 117.⟧ This is not true for the weight of x ⟦Problem 118⟧.

8.4 Compactifications

DEFINITION 1. *A compactification of a topological space X, is a pair (Y, f),* *where Y is a compact space and f is a homeomorphism from X onto a dense subspace X_0 of Y. If Y is a T_2 space, (Y, f) is called a T_2 compactification.*

Thus if Y is compact and X is a dense subspace of Y, then (Y, i) is a compactification of X, where $i: X \to Y$ is the inclusion map, $i(x) = x$. The statement that X^+, βX are compactifications of X is to be interpreted to mean that (X^+, i) and $(\beta X, i)$ are compactifications, where $i: X \to X$ is the identity map. (In the case of βX, our construction yields $(\beta X, \rho)$ as the compactification before any identifications are made, where ρ is the evaluation map of X in I^C.)

There are two "compactifications" which do not fit Definition 1. One is the important Bohr compactification (see, for example, [Hewitt and Ross]), and another is the less important no-point compactification introduced in Sec. 8.1, Problem 121. In each of these cases, a space is not a topological subspace of its compactification, but is continuously embedded in it; that is, has a larger topology. (In the case of the Bohr compactification, the new topology has the same convergent sequences as the old, see [Reid]; while the no-point compactification of X has the same topology as X on all but one point of X.)

THEOREM 8.4.1. *A locally compact T_2 space is open in any T_2 compactification: that is, if (Y, f) is a compactification of X, then $f[X]$ is an open subset of Y.*

This theorem is merely a restatement of Theorem 5.4.13. ▮

We now introduce a partial ordering among T_2 compactifications of a fixed space X. If (Y, f), (Z, g) are compactifications of X we say that $(Y, f) \geq (Z, g)$ if there is a continuous function h from Y onto Z with $g = h \circ f$.

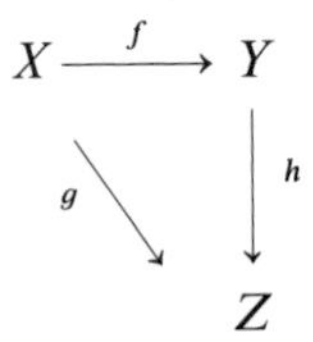

The diagram shown is *commutative* (that is $g = h \circ f$) and $u \equiv h \mid f[X]$ is a homeomorphism onto $g[X]$. ⟦$u^{-1} = f \circ g^{-1}$ where we consider $g^{-1}: g[X] \to X$.⟧ The function h has the additional property that it carries $Y \setminus f[X]$ onto $Z \setminus g[X]$ ⟦Lemma 4.3.1⟧.

★EXAMPLE 1. Let $X = (0, 1)$, $Y = [0, 1]$, S_1 = one-sphere, $f(x) = x$, $g(x) = e^{2\pi i x}$. Then (Y, f), (S_1, g) are compactifications of X, and $(Y, f) \geq (S_1, g)$ since if we set $h(x) = e^{2\pi i x}$, then h maps Y continuously onto S_1, and $h \circ f = g$. It is false that $(S_1, g) \geq (Y, f)$, since if $k: S_1 \to Y$ satisfies $k \circ g = f$ we have $k[g[X]] = f[X] = (0, 1)$, while $S_1 \setminus g[X]$ has only one point, so that k cannot possibly map S_1 onto Y, which has two points outside of $(0, 1)$.

EXAMPLE 2. Let Y, Z be compact T_2 spaces such that $X \equiv Y \cap Z$ is dense in both Y and Z, and the relative topologies of Y and Z are the same on X. Then (Y, i) and (Z, i) are compactifications of X, and if $(Y, i) \geq (Z, i)$ the requisite map $h\colon Y \to Z$ must be the identity map on X, and must carry $Y \setminus X$ onto $Z \setminus X$ ⟦Lemma 4.3.1⟧.

We now see that the one-point and Stone–Cech compactifications are minimum and maximum, respectively. X is assumed noncompact to avoid trivialities.

THEOREM 8.4.2. (a) *Let X be a locally compact T_2 space, and (Y, f) any T_2 compactification. Then $(Y, f) \geq (X^+, i)$.* (b) *Let X be a $T_{3\frac{1}{2}}$ space and Y any T_2 compactification, then $(\beta X, i) \geq (Y, f)$.*

To prove (a), we define $h(y) = f^{-1}(y)$ for $y \in f[X]$, and $h(y) = \infty$ for $y \in Y \setminus f[X]$. It is clear that $h \circ f = i$. ⟦For $x \in X$, $h[f(x)] = f^{-1}[f(x)] = x$.⟧ To see that h is continuous, let $y \in Y$. If $y \in f[X]$ and y_δ is a net converging to y, then $y_\delta \in f[X]$ eventually ⟦$f[X]$ is open, by Theorem 8.4.1⟧ and so, eventually, $h(y_\delta) = f^{-1}(y_\delta) \to f^{-1}(y) = h(y)$, so that h is continuous at y. If $y \in Y \setminus f[X]$ so that $h(y) = \infty$, let N be an open neighborhood of ∞ in X^+. Then $X^+ \setminus N = X \setminus N$ is a compact set, so that $h^{-1}[N] = Y \setminus f[X^+ \setminus N]$ is open, being the complement of a compact set. Thus h is continuous at y.

To prove (b), we note that $f\colon X \to Y$ is continuous and that Y is compact. Thus by Theorem 8.3.1, f can be extended to a continuous function $h\colon \beta X \to Y$. Moreover $h \circ i = f$. ⟦For $x \in X$, $h(x) = f(x)$.⟧ ∎

THEOREM 8.4.3. *The partial ordering of T_2 compactifications of a fixed space is antisymmetric, in the sense that if (Y, f), (Z, g) are T_2 compactifications of X, and $(Y, f) \geq (Z, g) \geq (Y, f)$, then (Y, f) and (Z, g) are equivalent in the sense that there exists a homeomorphism h from Y onto Z such that $h \circ f = g$. (In particular h carries $f[X]$ onto $g[X]$ "in the right way.")*

There exists a continuous $h\colon Y \to Z$ with $h \circ f = g$, and continuous $k\colon Z \to Y$ with $k \circ g = f$. Now for $y \in f[X]$, say $y = f(x)$, $k[h(y)] = k[g(x)] = f(x) = y$. Thus $k \circ h$ is the identity on $f[X]$. Since $k \circ h\colon Y \to Z$ and the identity map $i\colon Y \to Z$ agree on a dense subset of Y, they agree on Y ⟦Sec. 4.2, Problem 7⟧. This shows that $k = h^{-1}$, so that h is a homeomorphism onto. ∎

REMARK. The parenthesized phrase in the statement of Theorem 8.4.3 means that for each $x \in X$, $f(x)$ is carried to $g(x)$ by h. In particular if $f = g = i$, h leaves X pointwise fixed.

Since it is customary to identify any two homeomorphic spaces, we shall henceforth consider a space X as a dense subspace of any compactification, and shall refer to Y as a compactification of X (instead of Y and a map), thinking of X as a dense subspace of Y. Thus, as in Example 1, S_1 is a compactification of (0, 1). If $Y \setminus X$ has n points we call Y an *n-point compactification*. If $Y \setminus X$ is infinite, or countable, we call Y an *infinite*, or *countable*, *compactification*, respectively. For an interesting remark on finite compactifications, see [Sanderson].

▲ EXAMPLE 3. **R** has a one-point T_2 compactification, S_1, and a two-point T_2 compactification [0, 1], but no three-point T_2 compactification as we now prove. Let Y be a compact T_2 space and assume that Y contains three points y_1, y_2, y_3 such that $Y \setminus \{y_1, y_2, y_3\} = \mathbf{R}$. Let G_1, G_2, G_3 be disjoint open neighborhoods of y_1, y_2, y_3 respectively, and $G = G_1 \cup G_2 \cup G_3$. Then $Y \setminus G$ is a closed subset of Y hence is compact. But $Y \setminus G \subset \mathbf{R}$ hence $Y \setminus G \subset [a, b]$, a closed interval in **R**. It follows that $(-\infty, a) \cup (b, \infty) \subset G$ and so one of G_1, G_2, G_3 must fail to meet $(-\infty, a) \cup (b, \infty)$ since the latter has only two components. Suppose it is G_1. Then $G_1 \setminus \{y_1\} = G_1 \cap [a, b]$, from which it follows that y_1 is an isolated point of Y; **R** is not dense in Y; Y is not a compactification of **R**. ∎

Problems

1. Let X be a subspace of a compact space Y. Show that the closure of X in Y, with the inclusion map, is a compactification of X.
2. The subspace $\{z : |z| < 1\}$ of the complex plane has the compactifications, $\{z : |z| \leq 1\}$ and S^2. Are they comparable?
3. All one-point T_2 compactifications of a given space are equivalent ⟦Theorem 8.1.3⟧.
4. A compact T_2 space has only one T_2 compactification: itself. ⟦It cannot be a dense proper subset by Theorem 5.4.5.⟧ Compare Problem 103.

---

101. **R** has an infinite metrizable compactification. ⟦It may be a subset of $\mathbf{R}^2$; consider the graph of $y = \sin 1/x$, $0 < x \leq 1$, and its closure.⟧
102. All n-point T_2 compactifications of a given space X for a given value of n are equivalent. True or false?
103. An infinite cofinite space is a compactification of any infinite subspace. Such a subspace is compact and dense.
104. Construct a subspace of **R** which has a one- and a two-point compactification which are homeomorphic with each other. (See [Sanderson].)
105. A space is called *C-complete* (or Cech-complete) if it can be embedded

as a G_δ in some compact T_2 space. Show that a locally compact T_2 space is C-complete. ⟦Consider X^+ or βX.⟧

106. If X is C-complete, it must be a G_δ in βX. ⟦If $X \subset Y$, let $Z = \bar{X}$. Let $f: \beta X \to Z$ satisfy $f^{-1}[x] = x$. Use Lemma 4.3.1.⟧
107. C-complete is F-hereditary.
108. A G_δ in a C-complete space is C-complete.

8.5 C- and C^*-Embedding

A set S in a topological space X is said to be *C^*-embedded in X* if every bounded continuous real function on S can be extended to a continuous real function on X, and *C-embedded in X* if every continuous real function on S can be extended to a continuous real function on X. (See Problem 3.) The importance of βX stems partly from the fact that X is C^*-embedded in it.

EXAMPLE 1. *Let X be a Tychonoff space. Then every finite subset S is C-embedded in X.* (This result is also a special case of Corollary 8.5.1, below.) Let $S = (x_1, x_2, \ldots, x_n)$. For each $i = 1, 2, \ldots, n$, there exists a continuous $f_i: X \to \mathbf{R}$ with $f_i(x_i) = 1, f_i(x_j) = 0$ for $j \neq i$. ⟦Separate x_i from the closed set $\{x_j: j \neq i\}$.⟧ Now for any $g: S \to \mathbf{R}$ let

$$f(x) = \sum_{i=1}^{n} g(x_i)f_i(x) \quad \text{for } x \in X.$$

Then f is continuous ⟦it is a finite sum of continuous functions⟧ and $f|S = g$. ⟦For $j = 1, 2, \ldots, n, f(x_j) = \sum g(x_i)f_i(x_j) = g(x_j)$.⟧ ∎

EXAMPLE 2. A C-embedded set is certainly C^*-embedded, but not conversely. For example, X is always C^*-embedded in βX, but need not be C-embedded. Indeed an unbounded function could not have a continuous extension to βX ⟦Sec. 8.3, Problem 2⟧.

Let us say that a subspace S of X is *K-embedded* in X if for every compact Hausdorff space Y, every continuous $f: S \to Y$ has a continuous extension $F: X \to Y$. A K-embedded subspace is C^*-embedded ⟦take $Y = \overline{f[S]}$⟧, and we have the following important partial converse.

THEOREM 8.5.1. *A dense C^*-embedded subspace of a Tychonoff space is K-embedded.*

In the proof of Lemma 8.3.2, only the fact that X is C^*-embedded and dense in βX was used; thus the same proof applies. ∎

(The density of X was used in the last step of that proof. It cannot be omitted; see Problem 5.)

THEOREM 8.5.2. *βX is the only T_2 compactification of X in which X is C^*-embedded. More precisely, if (Y, f) is a T_2 compactification of X such that $f[X]$ is C^*-embedded in Y, then (Y, f) is equivalent to $(\beta X, i)$.*

(We are thinking of X as a subspace of βX.) Consider the map $f^{-1}: f[X] \to \beta X$. This has an extension $h: Y \to \beta X$ ⟦Theorem 8.5.1⟧. Now $h \circ f = i$. ⟦For $x \in X$, $h[f(x)] = f^{-1}[f(x)] = x$⟧ and so $(Y, f) \geq (\beta X, i)$. Since also $(\beta X, i) \geq (Y, f)$ ⟦Theorem 8.4.2⟧ the result follows from Theorem 8.4.3. ∎

EXAMPLE 3. Let X be a Tychonoff space and $X \subset Y \subset \beta X$. Then $\beta Y = \beta X$. The use of the "equals" sign indicates not only that βY is homeomorphic with βX, but that Y is actually a dense C^*-embedded subspace of βX. ⟦Y is dense since X is. Also if $g: Y \to \mathbf{R}$, let $f = g \mid X$. Extend f to $F: \beta X \to \mathbf{R}$. Then $F \mid Y = g$; since this is true on X, a dense subset of Y; by Sec. 4.2, Problem 7.⟧

The following result, known as Tietze's extension theorem, was given for metric spaces by H. Tietze in 1923, and extended by P. Urysohn in 1925. A very elegant proof by use of linear operators is given in [McCord].

THEOREM 8.5.3. *Every closed subspace S of a normal space X is C-embedded.*

Let $f: S \to \mathbf{R}$ be continuous.

Case I. Suppose that $f[S] \subset [-2, 2]$. Let $f_1 = f$,

$$A_1 = \{x \in S: f_1(x) \leq -\tfrac{2}{3}\}, \quad B_1 = \{x \in S: f_1(x) \geq \tfrac{2}{3}\}.$$

There exists continuous $g_1: X \to [-\frac{2}{3}, \frac{2}{3}]$ with $g_1 = -\frac{2}{3}$ on A_1, $g_1 = \frac{2}{3}$ on B_1. ⟦By Urysohn's lemma (Theorem 4.2.11) since A_1, B_1 are disjoint, and are closed in X since they are closed in S which is closed in X. (If A_1 is empty, take $g_1 = \frac{2}{3}$ everywhere; if $B_1 = \varnothing$, $g_1 = -\frac{2}{3}$; if $A_1 = B_1 = \varnothing$, take $g_1 = 0$. These remarks apply also to the rest of the proof.)⟧ Now define f_2 on S by $f_2(x) = f_1(x) - g_1(x)$ for $x \in S$. Then $f_2[S] \subset [-\frac{4}{3}, \frac{4}{3}]$. ⟦For $x \in A_1$, $f_1(x) \in [-2, -\frac{2}{3}]$, $g_1(x) = -\frac{2}{3}$; for $x \in B_1$, $f_1(x) \in [\frac{2}{3}, 2]$, $g_1(x) = \frac{2}{3}$; for other $x \in S$, $f_1(x) \in (-\frac{2}{3}, \frac{2}{3})$, $g_1(x) \in [-\frac{2}{3}, \frac{2}{3}]$.⟧ Next, let

$$A_2 = \{x \in S: f_2(x) \leq -(\tfrac{2}{3})^2\}, \quad B_2 = \{x \in S: f_2(x) \geq (\tfrac{2}{3})^2\},$$

and continuous $g_2: X \to [-(\frac{2}{3})^2, (\frac{2}{3})^2]$ with $g_2 = -(\frac{2}{3})^2$ on A_2, $g_2 = (\frac{2}{3})^2$ on B_2, and define $f_3: S \to [-\frac{8}{9}, \frac{8}{9}]$ by $f_3(x) = f_2(x) - g_2(x)$ for $x \in S$. ⟦For $x \in A_2$, $f_2(x) \in [-\frac{4}{3}, -\frac{4}{9}]$ and $g_2(x) = -\frac{4}{9}$, etc.⟧ Continuing in this way we get, for $n = 1, 2, \ldots$, continuous

$$g_n: X \to [-(\tfrac{2}{3})^n, (\tfrac{2}{3})^n], \qquad f_n: S \to [-3(\tfrac{2}{3})^n, 3(\tfrac{2}{3})^n],$$

and, on $S, f_{n+1} = f_n - g_n$. Finally, define $F: X \to [-2, 2]$ by

$$F(x) = \sum_{n=1}^{\infty} g_n(x).$$

⟦The series converges, since for each x, and n, $|g_n(x)| \leq (\frac{2}{3})^n$, so $|F(x)| \leq \sum(\frac{2}{3})^n = 2$.⟧ F is continuous ⟦Theorem 4.2.10 and Sec. 4.2, Problem 15.⟧ Finally, $F \mid S = f$. ⟦For $x \in S$,

$$\sum_{n=1}^{k-1} g_n(x) = \sum_{n=1}^{k-1} [f_n(x) - f_{n+1}(x)] = f_1(x) - f_k(x) \to f(x)$$

since $|f_k(x)| \leq 3(\frac{2}{3})^k \to 0$ as $k \to \infty$.⟧

REMARK ON CASE I. Suppose $f: S \to (-2, 2)$. Then we can choose F so that also $F: X \to (-2, 2)$. ⟦Note that F, chosen in Case I satisfies $F: X \to [-2, 2]$. Let $T = \{x : |F(x)| = 2\}$. Then T is closed and disjoint from S. If T is empty there is nothing to prove. Otherwise, choose a continuous $h: X \to [0, 1]$ with $h = 1$ on S, $h = 0$ on T. Then $h \cdot F$ maps X continuously into $(-2, 2)$ and agrees with f on S.⟧

Case II. For arbitrary continuous $f: S \to \mathbf{R}$, let $u: \mathbf{R} \to (-2, 2)$ be any homeomorphism onto ⟦Sec. 4.2, Example 5⟧. Then $u \circ f$ is a continuous map of S into $(-2, 2)$. By the remark on Case I, we may extend $u \circ f$ to a continuous $G: X \to (-2, 2)$. Then $u^{-1} \circ G$ is a continuous map of G into $\mathbf{R}$ and $(u^{-1} \circ G) \mid S = f$. ⟦For $x \in S$, $u^{-1}[G(x)] = u^{-1}[u(fx)] = f(x)$.⟧ ∎

COROLLARY 8.5.1. *Every compact subset K of a Tychonoff space X is C-embedded.*

K is C-embedded in βX ⟦Theorem 8.5.3⟧, hence it is C-embedded in X. ⟦Extend $f: K \to \mathbf{R}$ to $F: \beta X \to \mathbf{R}$ and consider $F|X$.⟧ ∎

NOTE. The problem of extending metrics and complete metrics is treated in [Bacon].

Problems

In this list, wherever βX is mentioned, it is assumed that X is a Tychonoff space.

1. With S, X, f, F as in Theorem 8.5.3, if $f[S] \subset (a, b)$ then F can be so chosen that $F[X] \subset (a, b)$.
2. In Problem 1, replace (a, b) by (i) $[a, b]$, (ii) $[a, b)$.
3. ★ If S is C^*-embedded in X and $f \in C^*(S)$, f can be extended to $F \in C^*(X)$. ⟦If $f[S] \subset (a, b)$, replace F by $(a \vee F) \wedge b$.⟧
4. $(0, 1]$ is not C^*-embedded in $[0, 1]$.

5. "Dense" cannot be omitted in Theorem 8.5.1. ⟦$S = Y = \{0, 1\}$. $X = [0, 1]$, $g(x) = x$.⟧
6. A C-embedded subspace need not be K-embedded. ⟦See the hint for Problem 5.⟧
7. Let X be a T_1 space. Then every finite subspace is C-embedded if and only if every two points can be completely separated. (Compare Example 1.)
8. Give an example of a locally compact Hausdorff space Y such that $\beta Y = Y^+$. ⟦Remove one point from βX. Use Example 3 and Theorem 8.1.3.⟧

★9. Let X be a dense and C-embedded subspace of Y. Let $F \in C(Y)$ and suppose that $F(y) = 0$ for some $y \in Y$. Show that $F(x) = 0$ for some $x \in X$. ⟦If $F(x) \neq 0$ for all $x \in X$, let $g(x) = 1/|F(x)|$ for $x \in X$; $g \in C(X)$ by Theorems 4.2.6 and 4.2.8. Let G be the extension of g to all of Y. Then $(G \cdot |F|)(x) = 1$ for all $x \in X$, hence for all $y \in Y$ since X is dense. In particular $F(y)$ can never be 0.⟧ By Problem 12, C cannot be replaced by C^*.

10. Under the assumptions of Problem 9, suppose also that Y is a Tychonoff space and that $y \in Y \setminus X$. Then $F(x) = 0$ for at least two values of x. ⟦Suppose $F(x) = 0$ for $x = x_0 \in X$ and for no other $x \in X$. Choose $G \in C(Y)$ with $G(x_0) = 1$, $G(y) = 0$. Then $|F| + |G|$ vanishes at y but not anywhere in X. This contradicts Problem 9.⟧
11. In Problem 10, $F(x) = 0$ for infinitely many values of $x \in X$ ⟦similar proof⟧.
12. $\beta\omega \setminus \omega$ is a zero-set in $\beta\omega$. ⟦Consider $f(n) = 1/n$.⟧
13. A K-embedded subset of a connected space is connected ⟦Lemma 5.2.1⟧. In particular this applies to a dense C^*-embedded subset ⟦Theorem 8.5.1⟧. Here K cannot be replaced by C^*. ⟦Consider a subspace with two points.⟧

---

101. A Tychonoff space X is C-embedded in βX if and only if it is pseudocompact.
102. Prove this converse of Theorem 8.5.3. If every closed subspace of X is C-embedded, X is normal. ⟦Extend the characteristic function of one of two disjoint closed sets.⟧
103. For normal spaces, pseudocompact and countably compact are equivalent. ⟦Sec. 7.1, Problem 114. If X is not countably compact, it includes a closed copy of ω; now apply Theorem 8.5.3.⟧
104. $\omega \times \omega$ is not C^*-embedded in $\beta\omega \times \beta\omega$. ⟦The characteristic function of the diagonal cannot be extended, by Sec. 8.3, Problem 122.⟧
105. $\beta(X \times Y) \geq \beta X \times \beta Y$ in the sense of Section 8.4, and they need not be equivalent compactifications ⟦Problem 104 and Theorem 8.4.2(b)⟧. (I. Glicksberg has proved that they are equivalent if and only if

$X \times Y$ is pseudocompact. See [Isbell]. See also Problem 106, and Sec. 14.1, Problem 106.)

106. The compactifications in Problem 105 are equivalent if and only if $X \times Y$ is C^*-embedded in $\beta X \times \beta Y$.
107. Let X be a T_1 space. Which subsets S of X have the property: (a) Every bounded real function on S can be extended to a bounded real function on X. (b) Every bounded real function on S can be extended to a bounded continuous real function on X. (c) Every continuous real function on S can be extended to a bounded continuous real function on X?
108. Let S be C^*-embedded in X and such that if C is any closed set not meeting S, S and C can be completely separated. Show that if $f: S \to (-1, 1)$ is continuous, f has a continuous extension $F_0: X \to (-1, 1)$. ⟦Let $C = \{x: |F(x)| \geq 1\}$. Imitate the remark on Case I of Theorem 8.5.3.⟧ Hence show that S is C-embedded ⟦Case II of Theorem 8.5.3⟧.
109. Express Sec. 4.2, Problem 203 as a sufficient condition that a dense subspace of a space be C-embedded. ⟦Take $Y = \mathbf{R}$.⟧
110. With X as in Sec. 8.3, Problem 102, show that $\beta X = \beta\omega$ but X, ω are not homeomorphic ⟦Example 3; Sec. 8.3, Problem 103⟧.
111. X is called an *absolute retract* if for every T_4 space Y, every closed subspace of Y homeomorphic with X is a retract of Y. (In other words, X is a retract in every T_4 space in which it is embedded as a closed subset.) Show that a retract of a cube is an absolute retract. ⟦Let $S \subset Y$, Y normal; f: $S \to X \subset I^A$ a homeomorphism onto. For each $a \in A$, the map $s \to f(s)(a)$ is a continuous map from S to I. By Tietze's extension theorem, this map may be extended to g_a: $Y \to I$. This yields g: $Y \to I^A$. The required retraction is $f^{-1} \circ r \circ g$, where r is the retraction on X.⟧ In particular, a cube is an absolute retract.
112. $\mathbf{R}^n$ is an absolute retract. ⟦Use Urysohn's lemma and imitate the hint for Problem 111.⟧
113. A compact Hausdorff space is an absolute retract if and only if it is a retract of a cube ⟦Problem 111 and Theorem 8.5.3⟧.
114. Let A be a topological space such that for every normal space X and closed subspace S, every continuous f: $S \to A$ can be extended to continuous F: $X \to A$. Show that A is an absolute retract. (Theorem 8.5.3 shows that $\mathbf{R}$ has this property.) ⟦If A is a closed subspace of a normal space X, extend the identity i: $A \to A$ to r: $X \to A$.⟧ See Problem 201.
115. Give an example of a compact connected subspace of $\mathbf{R}^2$ which is not an absolute retract ⟦Sec. 6.5, Problem 210⟧.
116. Let (X, d) be a semimetric space and Y a metric space which is an absolute retract. Show that every continuous function from X to Y is

uniformly continuous if and only if d is a u-semimetric ⟦Sec. 4.2, Problems 115 and 112; Sec. 4.3, Problem 203⟧. In particular, if every bounded continuous real function is uniformly continuous, d is a u-semimetric. ⟦Take $Y = [0, 1]$.⟧

---

201. A normal space A is an absolute retract if and only if it has the property expressed in Problem 114. ⟦If A has this property and $f: S \to A$, with S a closed subspace of the normal space X, let $Z = (X + A)/\rho$ where $X + A$ is the free union (Sec. 6.4, Problem 106) and ρ is the relation $s \, \rho \, f(s)$ for $s \in S$ and for $u, v \in (X \setminus S) \cup (A \setminus f[S])$ $u \, \rho \, v$ means $u = v$; Z is normal, A is a closed subspace so there exists $r: Z \to A$. For $x \notin S$, define $F(x) = r(x)$.⟧
202. Let S be a subset of a Tychonoff space X. Let $A = \{f \in C(S): f$ has a continuous extension to all of $X\}$. Then S is C-embedded in X if and only if $A = C(S)$. Show that A is a subalgebra of $C(S)$, contains all constant functions, and separates points of S; that is if $x \neq y$, A contains f with $f(x) \neq f(y)$.
203. Deduce from Tietze's extension theorem the result of Sec. 5.3, Example 3 that a separable normal space is D-separable. ⟦If X is not D-separable, $C(X)$ will have 2^c members. See Sec. 5.3, Problem 203.⟧

8.6 Realcompact Spaces

The results of this section will be used only in Section 11.4 (Examples 3 and 4 and Problems). They have been developed from foundations laid in 1948 by Edwin Hewitt. Our treatment is taken from [Gillman and Jerison], with modifications so as to avoid the use of algebraic tools. We begin with a Tychonoff space X. For every $f \in C(X)$ we may also consider f as a map into $\mathbf{R}^+$; as such, it has a continuous extension $F: \beta X \to \mathbf{R}^+$. For any $t \in \beta X$ we have either $F(t) = \infty$, or $F(t) \in \mathbf{R}$; in the latter case the function $f: X \to \mathbf{R}$ has a continuous extension to $F: X \cup \{t\} \to \mathbf{R}$. (In the former case $\mathbf{R}$ would have to be replaced by $\mathbf{R}^+$ here.) It follows that if t has the curious property that $F(t) \in \mathbf{R}$ for every $f \in C(X)$, where $F: \beta X \to \mathbf{R}^+$ is the extension of $f: X \to \mathbf{R}^+$, then every $f \in C(X)$ has a continuous extension $F: X \cup \{t\} \to \mathbf{R}$, in other words, X is C-embedded in $X \cup \{t\}$. The set of all such t is written υX (pronounced: upsilon X), and called the *realcompactification* of X. Of course υX is never empty since it includes X; $X \subset \upsilon X \subset \beta X$, and

X is C-embedded in υX, and in no subspace of βX which is not included in υX. (8.6.1)

⟦If $t \notin \upsilon X$, there exists $f \in C(X)$ with $F(t) = \infty$, moreover F is uniquely determined by f since X is dense in βX, hence in υX.⟧ We call X *realcompact* if $\upsilon X = X$. Thus a compact Hausdorff space is realcompact, and a Tychonoff space X is realcompact if and only if, for every $t \in \beta X \setminus X$, there exists $f \in C(X)$ such that $F(t) = \infty$ where $F\colon \beta X \to \mathbf{R}^+$ is the extension of $f\colon X \to \mathbf{R}^+$.

REMARK ON NOTATION. In this section we shall denote members of $C(X)$ by lower case $f, g, h, \phi, \psi, \ldots$, and their extensions as maps from βX to $\mathbf{R}^+$ by the corresponding capitals, $F, G, H, \Phi, \Psi, \ldots$; X is always a Tychonoff space.

The following result justifies the name realcompactification.

THEOREM 8.6.1. *υX is realcompact.*

Let $Y = \upsilon X$. Then $\beta Y = \beta X$ ⟦Sec. 8.5, Example 3⟧. Moreover X is C-embedded in υY. ⟦Every $f \in C(X)$ has an extension $F_1\colon \upsilon X \to \mathbf{R}$, in other words $F_1\colon Y \to \mathbf{R}$. Then F_1 has an extension $F_2\colon \upsilon Y \to \mathbf{R}$ and F_2 clearly extends f.⟧ By (8.6.1) $\upsilon Y \subset \upsilon X = Y$, hence Y is realcompact. ▮

An example of a Tychonoff space which is not realcompact is given in Sec. 14.1, Example 2. In fact, by Problem 1, it is sufficient to give a countably compact, noncompact space. Since υX is always realcompact, it follows that realcompactness is not hereditary.

THEOREM 8.6.2. *Let $f \in C(X)$ and suppose that $F(t) = 0$ for some $t \in \upsilon X$. Then $f(x) = 0$ for some $x \in X$.*

Thus $\upsilon X \setminus X$ includes no zero-set of υX. The result is immediate from Sec. 8.5, Problem 9. ▮

COROLLARY 8.6.1. *No point of $\upsilon X \setminus X$ can be a G_δ in υX.*

For such a point would be a zero-set ⟦Sec. 4.3, Problem 5⟧, contradicting Theorem 8.6.2. ▮

The corresponding result for βX seems much more difficult to obtain. (See Sec. 8.3, Problem 112.)

DEFINITION 1. *Let $t \in \upsilon X$. Let $E_t = \{f \in C(X)\colon F(t) = 0\}$. Let $Z_t = \{f^\perp\colon f \in E_t\}$.*

Thus Z_t is a collection of subsets of X.

LEMMA 8.6.1. *Fix $t \in \upsilon X$. Then Z_t has the countable intersection property.*

This result extends Theorem 8.6.2 in that it says that if the intersection of a countable collection of zero-sets of υX is not empty, it must meet X. A countable collection of members of Z_t is a sequence $\{f_n^\perp\}$ with $f_n \in C(X)$, $F_n(t) = 0$ for each n. Let $g = \sum (|f_n| \wedge 2^{-n})$. Then $g \in C(X)$ by Theorems 4.2.8, 4.2.9, and 4.2.10. Also $\bigcap f_n^\perp = g^\perp \neq \varnothing$ by Theorem 8.6.2. ▮

LEMMA 8.6.2. *Let $t \in \upsilon X \setminus X$. Then the collection Z_t has empty intersection.*

Given $x \in X$, choose $F \in C(\upsilon X)$ with $F(t) = 0$, $F(x) = 1$. Let $f = F \mid X$. Then $f^{\perp} \in Z_t$, but $f(x) = 1$ so that $x \notin f^{\perp}$. Thus x does not belong to the intersection of the members of Z_t; but x is an arbitrary member of X. ▌

THEOREM 8.6.3. *Every Lindelöf $T_{3\frac{1}{2}}$ space is realcompact.*

If $t \in \upsilon X$, the collection Z_t has the countable intersection property. Hence it has nonempty intersection. ⟦Sec. 5.3, Problem 9. Note that each member of Z_t is a closed subset of X.⟧ By Lemma 8.6.2, $t \in X$. ▌

COROLLARY 8.6.2. *Every separable metric space and every countable T_3 space are hereditarily realcompact.*

They are hereditarily Lindelöf, and completely regular, by Theorems 4.3.3, 5.3.4, and 5.3.5. The result follows from Theorem 8.6.3. ▌

Let us say that a subspace S of X is *RK-embedded in X* if for every realcompact space Y, every continuous $f: S \to Y$ has a continuous extension $F: X \to Y$. An *RK*-embedded subspace is *C*-embedded ⟦take $Y = \mathbf{R}$⟧, and we have the following partial converse. (Compare Theorem 8.5.1.)

THEOREM 8.6.4. *A dense C-embedded subspace S of a Tychonoff space X is RK-embedded.*

We may assume that $S \subset X \subset \beta S = \beta X$. ⟦$S$ is C^*-embedded in X hence in βX. It is also dense in βX. By Theorem 8.5.2, we may declare that βX *is* the Stone–Cech compactification of S.⟧ Now let $f: S \to Y$, with Y realcompact. We may consider $f: S \to \beta Y$. Let $F_0 : \beta S \to \beta Y$ be the continuous extension of f ⟦Theorem 8.3.1⟧. Finally, let $F = F_0 \mid X$. We shall show that F is the required extension. First, F extends f, ⟦$F|S = F_0|S = f$⟧, and the result will follow when we show that $F[X] \subset Y$. To this end, let $t \in F[X]$, say $t = F(x)$, $x \in X$. To show that $t \in Y$, the definitions suggest showing that every $g \in C(Y)$ has $G(t) \in \mathbf{R}$ where $G: \beta Y \to \mathbf{R}^+$ is the extension of g. ⟦Let $h = g \circ f$. Then $h \in C(S)$. By hypothesis h has a continuous extension $H: X \to \mathbf{R}$. Now $H = G \circ F$ on S, since this equation merely says $h = g \circ f$ on S, hence $H = G \circ F$ since S is dense in X. Thus $G(t) = G(Fx) = H(x) \in \mathbf{R}$.⟧ ▌

LEMMA 8.6.3. *A retract of a realcompact space is realcompact.*

Let $r: X \to S$ be a retraction from the realcompact space X onto S. Let $i: S \to X$ be the inclusion map; it has an extension $I: \upsilon S \to X$ ⟦Theorem 8.6.4⟧. Then $r \circ I$ is a retraction of υS onto S ⟦for $s \in S$, $r(Is) = r(s) = s$⟧, hence S is closed in υS ⟦Sec. 4.2, Problem 27⟧, but it is also dense. Thus $S = \upsilon S$ and S is realcompact. ∎

THEOREM 8.6.5. *A product of realcompact spaces is realcompact.*

Let $X = \prod \{X_\alpha : \alpha \in A\}$ with each X_α realcompact; X is a Tychonoff space ⟦Theorem 6.7.3⟧. We may extend each projection $P_\alpha : X \to X_\alpha$ to a continuous map $F_\alpha : \upsilon X \to X_\alpha$ ⟦Theorem 8.6.4⟧. Form the map $F: \upsilon X \to X$ by defining $(Ft)_\alpha = F_\alpha(t)$ for $t \in \upsilon X$; then F is continuous ⟦Theorem 6.6.1 (iv); note that $P_\alpha \circ F = F_\alpha$⟧, and indeed F is a retraction of υX onto X. ⟦For $x \in X$, $F(x) = x$ since $(Fx)_\alpha = F_\alpha(x) = P_\alpha(x) = x_\alpha$ for all α.⟧ The result follows by Lemma 8.6.3. ∎

REMARK. An application of β and υ is to the classification problem. For example, if βX and βY are not homeomorphic, then X, Y are certainly not homeomorphic. The converse problem is more interesting. For example, if βX and βY are homeomorphic, and if X, Y are first countable, then X, Y are homeomorphic ⟦Sec. 8.3, Problem 119⟧. However, β is not sensitive enough to distinguish between ω and $X = \omega \cup \{t\}$, $t \in \beta\omega \setminus \omega$ ⟦Sec. 8.5, Problem 110⟧, even though these are both G_δ *spaces*. (Every point is a G_δ.) On the other hand, υ is at least as sensitive as β since if υX, υY are homeomorphic, it follows that βX, βY are ⟦$\beta X = \beta(\upsilon X)$ by Sec. 8.5, Example 3⟧, and it is strictly more sensitive in that it does distinguish between G_δ spaces ⟦Problem 107⟧. However, even υ cannot distinguish between all spaces; for example, let X be a nonrealcompact space. Then $\upsilon X = \upsilon(\upsilon X)$ ⟦Theorem 8.6.1⟧ but X, υX are not homeomorphic.

Problems on Tychonoff Space

In this set we follow the remark on notation given before Theorem 8.6.1.

1. $\upsilon X = \beta X$ if and only if X is pseudocompact. ⟦If X is pseudocompact, apply (8.6.1). If $\upsilon X = \beta X$, X is C-embedded in βX; every member of $C(\beta X)$ is bounded.⟧ Hence a realcompact pseudocompact space is compact; so also is a realcompact countably compact space ⟦Sec. 7.1, Problem 114⟧.
2. Deduce that $\mathbf{R}$ is a realcompact from Lemma 4.3.1. ⟦Let i be the identity map: $\mathbf{R} \to \mathbf{R}$, f its extension: $\beta\mathbf{R} \to \mathbf{R}^+$.⟧
3. State Corollary 8.6.1 in the language of Sec. 8.5, Problem 9 (with no mention of υX).
4. Let X be a realcompact space such that every point is a G_δ, then X is hereditarily realcompact ⟦Corollary 8.6.1⟧. Compare Corollary 8.6.2.
5. υX cannot be first countable at any point of $\upsilon X \setminus X$ ⟦Corollary 8.6.1⟧.
6. "Dense" cannot be omitted in Theorem 8.6.4 ⟦Sec. 8.5, Problem 5⟧.
7. ($\mathbf{R}$, RHO) is realcompact ⟦Sec. 5.3, Example 2⟧.

8. A realcompact space need not be normal. ⟦Theorem 8.6.5; Problem 7; Sec. 6.7, Example 3. Another example is Sec. 6.7, Problem 203.⟧

---

101. An RK-embedded subspace is K-embedded but not conversely ⟦Theorem 8.5.1; Sec. 8.5, Example 2⟧.
102. Prove Tychonoff's theorem (Theorem 7.4.1), in the special case of the product of compact T_2 spaces, by a close imitation of the proof of Theorem 8.6.5. ⟦Use Theorem 8.5.1.⟧
103. Prove an analogue of Theorem 8.5.2 for realcompact spaces. ⟦Use Theorem 8.6.4.⟧
104. Let X, Y be first countable and such that υX, υY are homeomorphic. Show that X, Y are homeomorphic ⟦Problem 5; Sec. 8.3, Problem 117⟧.
105. Every singleton $\{x\}$ which is a zero-set in X is also a zero-set in υX. ⟦Let $f \in C(X)$ with $f^{\perp} = \{x\}$. Then $F(t) \neq 0$ for $t \in \upsilon X \setminus X$ by Sec. 8.5, Problem 10. Thus $F^{\perp} = \{x\}$.⟧
106. Let x be a G_δ point in X, then it is a G_δ point in υX ⟦Problem 105; Sec. 4.3, Problem 5⟧. Compare Sec. 8.3, Problem 118.
107. Let X, Y be G_δ spaces (each point is a G_δ point.) Then if υX, υY are homeomorphic, it follows that X, Y are homeomorphic ⟦Problem 106 and Corollary 8.6.1⟧.
108. Take as known the result: If $C(X)$, $C(Y)$ are ring isomorphic with X, Y realcompact, then X, Y are homeomorphic. (See [Gillman and Jerison, Theorem 8.3].) Show that if X, Y are G_δ spaces with $C(X)$, $C(Y)$ ring isomorphic, then X, Y are homeomorphic. (Compare Sec. 8.3, Problem 121.) ⟦Problem 107, with $C(X) = C(\upsilon X)$.⟧
109. Every intersection X of realcompact subspaces X_α of a Tychonoff space Y is realcompact. ⟦Let $i: X \to Y$ be the inclusion map. We may consider $i: X \to X_\alpha$. Then i has a continuous extension $I: \upsilon X \to X_\alpha$ for each α by Theorem 8.6.4, the same I for each α! Thus $I: \upsilon X \to X$. Hence X is a retract of υX. Apply Lemma 8.6.3.⟧
110. Realcompact is F-hereditary. ⟦Let F be a closed subset of X. Extend the inclusion map to $I: \upsilon F \to X$ by Theorem 8.6.4. Then $I^{-1}[F] = F$ by Lemma 4.3.1. Hence F is closed in υF; it is also dense.⟧
111. Any discrete space with the cardinality of the continuum or less is realcompact. ⟦Sec. 6.7, Example 3, or Sec. 6.7, Problem 202; consider any subset of D. It is discrete and realcompact by Problem 110 and Theorem 8.6.5.⟧
112. Let a subset S of X be called *pseudobounded* if every $f \in C(X)$ is bounded on S. Every subset of a pseudocompact space is pseudobounded. Every closed pseudobounded subset of a normal space is pseudocompact. (But this does not imply normality, see the next problem.)

113. Let X be pseudocompact. Then every closed pseudobounded subset of X is pseudocompact if and only if X is countably compact.

114. A set S in a Tychonoff space X is pseudobounded if and only if $\text{cl}_{\beta X} S \subset \upsilon X$. ⟦If $t \notin \upsilon X$ choose f with $f(t) = \infty$. Conversely, the inclusion implies that S is relatively compact in υX.⟧

115. A Tychonoff space is called an *NS* space (in honor of L. Nachbin and T. Shirota) if every closed pseudobounded set is compact. Show that every realcompact space is an *NS* space ⟦Problem 114⟧. The following (trivial) result improves Problem 1. A pseudocompact *NS* space is compact.

116. Every metric space is an *NS* space ⟦Problem 112, Sec. 8.5, Problem 103⟧.

117. The collection Z_t (Definition 1) may not be a filter ⟦it contains only closed sets⟧, but it behaves like an ultrafilter in the following respect. (Compare Sec. 7.3, Problem 2.) If $f_1, f_2, \ldots, f_n \in C(X)$ with $\bigcup f_i^\perp \supset h^\perp$ for some $h \in E_t$, then at least one $f_i \in E_t$. ⟦If not, select $G_i \in C(\upsilon X)$ with $G_i(t) = 0$, $G_i(x) = 1$ for $x \in F_i^\perp$. Let $g_i = G_i | X$ and $q = |h| + \sum_{i=1}^n |g_i|$. Then $q \in E_t$ but $q^\perp = \varnothing$ contradicting Theorem 8.6.2.⟧

118. Let $X = \prod X_\alpha$, $t \in \upsilon X$, $E_t^\alpha = \{h \in C(X_\alpha)\colon\ P_\alpha^{-1}[h^\perp] \supset f^\perp$ for some $f \in E_t\}$, $Z_t^\alpha = \{h^\perp : h \in E_t^\alpha\}$. Show that for each α, Z_t^α has the countable intersection property. ⟦For any $\{h_n\}$ and associated $\{f_n\}$ choose $x \in \bigcap f_n^\perp$ by Lemma 8.6.1. Then $x_\alpha \in$ all $h_n^\perp$.⟧

119. In Problem 118, let $t \in \upsilon X \setminus X$, then Z_t^α has empty intersection for some α. ⟦If not, let $x_\alpha \in \bigcap Z_t^\alpha$, $x = (x_\alpha)$. Let $H(t) = 0$, $H(x) = 1$, $h = H | X$. By Sec. 4.3, Problem 6, $x \in \bigcap_{i=1}^n P_{\alpha_i}^{-1}[V_i]$, where $\tilde{V}_i = g_i^\perp$, $g_i \in C(X_{\alpha_i})$. Then $h^\perp \subset \bigcup (g_i \circ P_{\alpha_i})^\perp$. By Problem 117, some $g_i \in E_t^{\alpha_i}$. But $x_{\alpha_i} \notin g_i^\perp$.⟧

120. Deduce from Problem 119 that every product of Lindelöf $T_{3\frac{1}{2}}$ spaces is realcompact. (This is also a special case of Theorems 8.6.3 and 8.6.5.)

121. X is realcompact if and only if $\beta X \setminus X$ is the union of all its subsets each of which is a zero set of βX; while X is pseudocompact if and only if $\beta X \setminus X$ has no subset which is a zero set of βX. This strengthens the result (Problem 1) that for noncompact spaces, these two properties are incompatible. ⟦For X realcompact and $t \notin X$, let $f = (|g| \vee 1)^{-1}$, where $g \in C(X)$, $g(t) = \infty$. Then $F^\perp \not\pitchfork X$. For X not realcompact, apply Theorem 8.6.2. The rest is Sec. 8.3, Problem 115.⟧

122. Imitate Section 8.3, embedding X in $\mathbf{R}^{C(X)}$ (rather than I^C). Show that $\hat{X}$ is C-embedded ⟦as in Lemma 8.3.1⟧, and its closure is realcompact ⟦Problems 109 and 110⟧. Thus a space is realcompact if and only if it is a closed subspace of a product of copies of $\mathbf{R}$.

---

201. Study υX, obtaining, or disproving, where possible, analogues of Theorem 5.4.5; Sec. 8.3, Problems 104, 105, 106, 117, 121 and 209; Sec. 8.5, Problems 8, 104, 105 and 106.
202. Examine the definition and properties of realcompactness if $\mathbf{R}^+$ is replaced by the two-point compactification of $\mathbf{R}$ (add $\pm\infty$), or if $\mathbf{R}$ is replaced by $[0, 1)$.

Complete Semimetric Space

9.1. Completeness

In order to test a sequence $\{x_n\}$ to see if it is convergent, it seems necessary to look "outside" the sequence, in the sense that some proposed limit t—not a term of the sequence—is put forth as a candidate, and $|x_n - t|$ is examined for large n. In the early part of the nineteenth century it was recognized that it would be desirable to have a purely internal characterization of convergence, that is, one involving only the terms of the sequence. An example of such an internal criterion is the sufficient condition, $\sum |a_n| < \infty$, for the convergence of the series $\sum a_n$ of real numbers. A solution of the problem was given at that time by A. Cauchy, who proved that a sequence of real numbers is convergent if and only if it satisfies the condition (9.1.1) which is given below. A sequence satisfying this condition is called a Cauchy sequence in his honor. We shall first give a corresponding definition for filterbases. Let $\mathscr{F}$ be a filterbase in a semimetric space; $\mathscr{F}$ is called a *Cauchy filterbase* if for every $\varepsilon > 0$, $\mathscr{F}$ contains a set of diameter $\leq \varepsilon$. This concept will be extended to uniform spaces in Section 11.3. Example 1, below, shows that it cannot be extended to topological spaces in general. The same remarks refer to the concept of completeness, as given below.

THEOREM 9.1.1. *Every convergent filterbase is a Cauchy filterbase.*

Suppose $\mathscr{F} \to x$ and $\varepsilon > 0$. Then $N(x, \varepsilon/2)$ is a neighborhood of x, hence belongs to $\mathscr{F}$. It has diameter $\leq \varepsilon$. ∎

The converse of Theorem 9.1.1 is not true ⟦Example 1⟧, but those spaces for which it is true, called *complete spaces*, play an important role in classical analysis. This importance may be judged from the fact that the Lebesgue integral displaced the Riemann integral, almost entirely because of the completeness of various spaces of Lebesgue integrable functions.

★EXAMPLE 1. Let $X = (-1, 1)$ with the Euclidean metric. Let $\mathscr{F} = \{(n/(n+1), 1): n = 1, 2, \ldots\}$, then $\mathscr{F}$ is a Cauchy filterbase. ⟦The diameter of $(n/(n+1), 1)$ is $1/n$.⟧ But $\mathscr{F}$ is not convergent. ⟦For any x, we can choose n so large that $n/(n+1) > x$. Then $x \notin [n/(n+1), 1) = (n/(n+1), 1)$. Thus x is not a cluster point of $\mathscr{F}$ and so $\mathscr{F} \not\to x$.⟧ We may use this example to illustrate the nontopological character of these concepts; let $f(x) = x/(1 - |x|)$, so that f is a homeomorphism of $(-1, 1)$ onto $\mathbf{R}$. Then $f[\mathscr{F}]$ is the filterbase $\{(n, \infty): n = 2, 3, \ldots\}$ which is not Cauchy since it consists entirely of sets of infinite diameter. Since a homeomorphism of semimetric spaces fails to preserve Cauchy filterbases, there is no topological concept which specializes to Cauchy filterbase for semimetric space.

A net x_δ in a semimetric space is said to be a *Cauchy net* if for every $\varepsilon > 0$, there exists δ_0 such that $\delta \geq \delta_0$, $\delta' \geq \delta_0$ implies $d(x_\delta, x_{\delta'}) \leq \varepsilon$. Thus a *Cauchy sequence*, $\{x_n\}$, is one which satisfies:

$$\text{For every } \varepsilon > 0, \text{ there exists } k \text{ such that } m \geq k, n \geq k \text{ implies } d(x_m, x_n) \leq \varepsilon. \tag{9.1.1}$$

THEOREM 9.1.2. *x_δ is a Cauchy net if and only if the associated filter $\mathscr{F}$ is a Cauchy filter.*

Suppose that x_δ is a Cauchy net and let $\varepsilon > 0$. Choosing δ_0 as above, we see that $\{x_\delta: \delta \geq \delta_0\}$ has diameter $\leq \varepsilon$. Since this set belongs to $\mathscr{F}$, $\mathscr{F}$ is Cauchy. Conversely, suppose that $\mathscr{F}$ is Cauchy and let $\varepsilon > 0$. Choose $S \in \mathscr{F}$ with diameter $S \leq \varepsilon$. Since $x_\delta \in S$ eventually, there exists δ_0 such that $\delta \geq \delta_0$ implies $x_\delta \in S$. But then $\delta \geq \delta_0$, $\delta' \geq \delta_0$ implies $d(x_\delta, x_{\delta'}) \leq$ diameter $S \leq \varepsilon$. ▌

A semimetric space is called *complete* if every Cauchy filterbase is convergent. A clearly equivalent condition is that every Cauchy filter is convergent. Example 1 shows that $(-1, 1)$ is not complete. It is also customary to refer to the semimetric as being complete, as in the sentence: on $(-1, 1)$, the Euclidean metric is not complete.

EXAMPLE 2. *On any space, the discrete metric is complete.* Let $\mathscr{F}$ be a Cauchy filter; then $\mathscr{F}$ contains a set S of diameter less than 1. This implies that S has only one point, say $S = \{x\}$. Since every neighborhood of x includes a member of $\mathscr{F}$ ⟦namely S⟧, it follows that $\mathscr{F} \to x$. ▌ Note that *a metric space with the discrete topology may not be complete*! For example the subspace $(1, \frac{1}{2}, \frac{1}{3}, \frac{1}{4}, \ldots)$ of $\mathbf{R}$ is discrete, but the Cauchy filterbase

$$\left\{\left(\frac{1}{n}, \frac{1}{n+1}, \frac{1}{n+2}, \cdots\right): n = 1, 2, \ldots\right\}$$

is not convergent. The point is that this space does not have the discrete metric, but rather an equivalent but different metric.

THEOREM 9.1.3. *The following conditions on a semimetric space X are equivalent.*

(i) *X is complete.*
(ii) *Every Cauchy net is convergent.*
(iii) *Every Cauchy sequence is convergent.*

Proof. (i) *implies* (ii). Let x_δ be a Cauchy net and $\mathscr{F}$ the associated filter. Then $\mathscr{F}$ is a Cauchy filter ⟦Theorem 9.1.2⟧ hence convergent. The net is convergent by Sec. 3.4, Example 5.

(ii) *implies* (iii). This is trivial since every sequence is a net.

(iii) *implies* (i). Let $\mathscr{F}$ be a Cauchy filterbase. For each $n = 1, 2, \ldots$, choose $S_n \in \mathscr{F}$ with diameter $\leq 1/n$, and let $x_n \in S_n$. Then $\{x_n\}$ is a Cauchy sequence. ⟦Let $\varepsilon > 0$. Let $n_0 > 2/\varepsilon$ be an integer. Now if $n \geq n_0$, $n' \geq n_0$ we may choose $y \in S_n \cap S_{n'}$ since the latter set is not empty, and obtain $d(x_n, x_{n'}) \leq d(x_n, y) + d(y, x_{n'}) \leq 1/n + 1/n' \leq 2/n_0 < \varepsilon$.⟧ By hypothesis $x_n \to x$, for some $x \in X$. Finally, $\mathscr{F} \to x$. ⟦Let N be a neighborhood of x. Then $N \supset N(x, \varepsilon)$ for some x. There exists $n > 2/\varepsilon$ with $d(x_n, x) < \varepsilon/2$. Then for every $s \in S_n$ we have $d(s, x) \leq d(s, x_n) + d(x_n, x) < 1/n + \varepsilon/2 < \varepsilon$ so that $s \in N$. This shows that $S_n \subset N$ and so N includes a member of $\mathscr{F}$, namely S_n.⟧ ∎

We now consider relations between completeness and compactness. The next two results will be repeated for uniform space in Section 11.3.

LEMMA 9.1.1. *If a Cauchy filterbase $\mathscr{F}$ in a semimetric space has a cluster point x, then $\mathscr{F} \to x$.*

Let N be a neighborhood of x. Then for some $\varepsilon > 0$, $N \supset N(x, \varepsilon)$. Let S be a member of $\mathscr{F}$ with diameter $< \varepsilon/2$. Now S contains a point y with $d(x, y) < \varepsilon/2$. ⟦Our assumption is that $x \in \bar{S}$.⟧ It follows that $S \subset N$. ⟦Let $s \in S$. Then $d(s, x) \leq d(s, y) + d(y, x) \leq \varepsilon/2 + \varepsilon/2 = \varepsilon$ so that $s \in N(x, \varepsilon)$.⟧ Thus N includes a member of $\mathscr{F}$ and so $\mathscr{F} \to x$. ∎

THEOREM 9.1.4. *A semimetric space is compact if and only if it is totally bounded and complete.*

That a compact space is totally bounded was noted at the beginning of Section 7.2. To prove completeness, let $\mathscr{F}$ be a Cauchy filter. Then $\mathscr{F}$ has a cluster point ⟦every filter does, by Theorem 7.1.4⟧, hence is convergent ⟦Lemma 9.1.1⟧. Conversely let X be a totally bounded complete semimetric space, and $\mathscr{F}$ an ultrafilter in X. It will be sufficient to prove $\mathscr{F}$ convergent ⟦Theorem 7.3.6⟧. It will be sufficient to prove that $\mathscr{F}$ is a Cauchy filter. To this end, let $\varepsilon > 0$. Then X is the union of finitely many subsets of diameter $<\varepsilon$; but $X \in \mathscr{F}$ hence $\mathscr{F}$ contains a set of diameter $<\varepsilon$ ⟦Sec. 7.3, Problem 2⟧. Thus $\mathscr{F}$ is a Cauchy filter. ▮

REMARK. Theorem 9.1.4 yields one of many remarkable examples of nontopological properties whose conjunction is topological. A homeomorph of a totally bounded space need not be totally bounded and the same is true for a complete space. ⟦Example 1. Indeed $(-1, 1)$ is totally bounded and **R** is complete!⟧ However "totally bounded and complete" is topological by Theorem 9.1.4. Again, we often get topological conclusions from nontopological assumptions, as for example Lemma 7.2.2, and nontopological conclusions from topological assumptions as in Lemma 7.2.1.

A set S in a semimetric space X is called *complete* if, as a semimetric subspace of X, S is complete.

LEMMA 9.1.2. *If a Cauchy filterbase contains a complete set, it is convergent.*

Let $\mathscr{F}$ be a Cauchy filterbase, and $S \in \mathscr{F}$, a complete set. Let $\mathscr{F}_1 = \{S \cap A : A \in \mathscr{F}\}$. Then $\mathscr{F}_1$ is a Cauchy filterbase on S, hence converges to some point $x \in S$. It follows that $\mathscr{F} \to x$. ⟦If $\mathscr{F}'$ is the filter generated by $\mathscr{F}$, it is clear that $\mathscr{F}' \supset \mathscr{F}_1$. Hence $\mathscr{F}' \to x$.⟧ ▮

LEMMA 9.1.3. *If a Cauchy filterbase contains a compact set, it is convergent.*

This follows from Lemma 9.1.2 and Theorem 9.1.4. ▮

★EXAMPLE 3. **R** *is complete.* A Cauchy filter $\mathscr{F}$ must contain a bounded set A; then $\bar{A} \in \mathscr{F}$. But $\bar{A}$ is compact ⟦Sec. 5.4, Example 2⟧, and the result follows by Lemma 9.1.3. ▮ Thus **R** has the property that a sequence is convergent if and only if it is a Cauchy sequence ⟦Theorems 9.1.1, 9.1.2, and 9.1.3⟧. This is Cauchy's result, mentioned at the beginning of this section. An indication of the way Cauchy himself proved it is given in Problem 110.

★EXAMPLE 4. *$[C^*(X), d]$ is complete, where X is any topological space, and $d(f, g) = \sup\{|f(x) - g(x)| : x \in X\}$.* Let $\{f_n\}$ be a Cauchy sequence. For each x, $\{f_n(x)\}$ is a Cauchy sequence of real numbers ⟦$|f_n(x) - f_m(x)| \le d(f_n, f_m)$⟧, hence is convergent ⟦Example 3⟧. Let $f(x) = \lim f_n(x)$ for each $x \in X$. Then $f_n \to f$ uniformly. ⟦Given $\varepsilon > 0$, choose k such that $m, n > k$ implies $d(f_m, f_n) < \varepsilon$. Now let $n > k$, $x \in X$; then $|f(x) - f_n(x)| =$

$\lim_m |f_m(x) - f_n(x)| \le \varepsilon$ since $|f_m(x) - f_n(x)| < \varepsilon$ as soon as $m > k$.⟧ Hence f is continuous ⟦Theorem 4.2.10⟧ and $f \in C^*(X)$; that is, f is bounded. ⟦Choose n such that $|f(x) - f_n(x)| < 1$ for all x. Then $|f(x)| \le |f(x) - f_n(x)| + |f_n(x)| < 1 + \sup\{|f_n(t)|: t \in X\}$ for all x.⟧ Finally, the statement $f_n \to f$ uniformly, means, since $f \in C^*(X)$, that $d(f_n, f) \to 0$, that is $f_n \to f$ in $[C^*(X), d]$. ∎

★EXAMPLE 5. *In Sec. 6.7, Example 5, $(X/\rho, D)$ is complete if (X, d) is.* Let $\{q(x^n)\}$ be a Cauchy sequence. Then $\{x^n\}$ is a Cauchy sequence in X. ⟦$d(x^m, x^n) = D[q(x^m), q(x^n)$.⟧ Say $x^n \to x$. Then $q(x^n) \to q(x)$. ⟦$D[q(x^n), q(x)] = d(x^n, x) \to 0$.⟧ ∎

Problems on Semimetric Space

★1. Let $\mathscr{F}$, $\mathscr{F}_1$ be filterbases which generate the same filter; then if $\mathscr{F}$ is Cauchy, so is $\mathscr{F}_1$. In particular the filter generated by $\mathscr{F}$ is Cauchy if and only if $\mathscr{F}$ is. ⟦Every member of $\mathscr{F}$ includes a member of $\mathscr{F}_1$.⟧
2. If $\mathscr{F}$ is a Cauchy filter, $\mathscr{G}$ is a convergent filter, and $\mathscr{F} \subset \mathscr{G}$, then $\mathscr{F}$ is convergent. ⟦Lemma 9.1.1. If $\mathscr{G} \to x$, x is a cluster point of $\mathscr{F}$.⟧
★3. A Cauchy sequence which has a convergent subsequence must be convergent ⟦Problem 2⟧.
4. A Cauchy net with a cluster point must converge to it.
5. A complete set in a metric space must be closed. This is not true for semimetric space. ⟦Consider finite or, more generally, compact sets in, for example, the indiscrete semimetric.⟧
6. Completeness is F-hereditary.
7. A subset of a complete metric space is complete if and only if it is closed ⟦Problems 5 and 6⟧.
8. The indiscrete semimetric is complete. ⟦Every filter converges!⟧
9. (The range of) a Cauchy sequence is metrically bounded.
10. A finite semimetric space is complete ⟦Theorem 9.1.4⟧.
★11. If $\{x_n\}$ is a Cauchy sequence in a semimetric space it has a subsequence $\{y_n\}$ satisfying $\sum d(y_n, y_{n+1}) < \infty$. ⟦Having chosen $y_n = x_a$, let $y_{n+1} = x_b$ with $b > a$ and $d(y_n, x_b) < 2^{-n}$.⟧

---

101. A locally compact metric space need not be complete. (But see Sec. 9.2, Problem 101 and Sec. 12.2, Problem 127.)
102. A semimetric space is complete if every bounded closed subset is complete ⟦Lemma 9.1.2⟧.
103. A space is called *BTB* if every bounded set is totally bounded. Show that *BTB* is not a topological property. ⟦Try $d(x, y) = |x - y|/(1 + |x - y|)$; consider a covering of $\mathbf{R}$ by spheres of radius 1/2.⟧ (*Note:* A complete *BTB* space has the classical property that a

subset is complete if and only if it is closed and bounded. In [Busemann], it is proved that every locally compact separable space has an equivalent metric with this property.)

104. A countable product of complete spaces is complete. (See Theorem 6.4.2. For larger products we have, as yet, no definition for completeness; see Sec. 11.4, Problem 7.)
105. A uniformly continuous function preserves Cauchy filterbases. (This means that $f(\mathscr{F})$ is a Cauchy filterbase if $\mathscr{F}$ is. In Example 1, it is shown that this is false for continuous functions.)
106. A uniform homeomorphism f preserves completeness. (f and f^{-1} are both uniformly continuous.) Indeed, it is sufficient that f be continuous, and f^{-1} uniformly continuous.
107. A retract of a complete space is complete. ⟦If $x_n \to x$, then $x_n = r(x_n) \to r(x)$.⟧
108. Obtain a necessary and sufficient condition that a space Y be isometric with a retract of a space X similar to that of Sec. 4.2, Problem 32.
109. State and prove a result about completeness of the factors of a complete product similar to Sec. 6.7, Problem 102 ⟦Problem 107⟧.
110. Prove that $\mathbf{R}$ is complete by showing that if $\{x_n\}$ is a Cauchy sequence, $x_n \to \sup\{\inf\{x_k : k > n\} : n = 1, 2, \ldots\}$.
111. Prove that $\mathbf{R}$ is complete thus: a Cauchy sequence is metrically bounded ⟦Problem 9⟧, hence contained in a closed interval, hence has a convergent subsequence ⟦Sec. 5.4, Example 2; Theorem 7.2.1⟧, hence is convergent ⟦Problem 3⟧. Note that the local compactness of $\mathbf{R}$ seems to enter, and compare Problem 101.
112. Let $f: X \to X$ be continuous, X a complete semimetric space. Suppose there exists t, $0 < t < 1$, such that $d(fx, fy) \le td(x, y)$; show that f must have a fixed point. ⟦Fix $x_1 \in X$. Let $x_n = f(x_{n-1})$. Then $\{x_n\}$ is a Cauchy sequence.⟧ (For an application of this result to the study of differential equations see [Kelley and Namioka, 5E].
113. Give an example of an ultrafilter which is not a Cauchy filter. ⟦Apply Theorem 7.3.1 to $\{(a, \infty) : a \in \mathbf{R}\}$.⟧ (This spoils any chance of using Lemma 9.1.1 to deduce Theorem 7.3.5.)
114. A u-semimetric must be complete. ⟦Consider $\{x_{2n}\}$, $\{x_{2n+1}\}$ where $\{x_n\}$ is a Cauchy sequence. Use either the definition or Sec. 7.1 Problem 117.⟧
115. A space is called *locally complete* if every neighborhood of any point x includes a complete neighborhood of x. Show that every open or closed subspace of a locally complete space is locally complete, and that every complete space is locally complete. ⟦Compare Theorems 5.4.11 and 5.4.12. Note that a semimetric space is regular.⟧
116. A dense locally complete subspace of a complete metric space is open. (Compare Theorem 5.4.13.)

117. A locally compact semimetric space is locally complete ⟦Theorem 9.1.4⟧. (Compare Problem 101.)
118. Let $f: X \to Y$ be one-to-one, continuous and almost open, where X is a complete metric space and Y a T_2 space. Show that f is a homeomorphism. ⟦Let U_n^i be an abbreviation for $f[N(x_n^i, \varepsilon/2^n)]$. Define x_n^i for $i = 1, 2$; $n \in \omega$; inductively, thus, x_1^1 arbitrary; let $y \in \overline{U_1^1} \cap f[X]$, $y = f(x_1^2)$; $x_n^i \in N(x_{n-1}^i, \varepsilon/2^{n-1})$; $f(x_n^1) \in \overline{U_{n-1}^2}$; $f(x_n^2) \in \overline{U_n^1}$. Choose in this order, $x_1^1, x_1^2, x_2^1, x_2^2, x_3^1, x_3^2, \ldots$. Then each $\{x_n^i\}$ is a Cauchy sequence hence $x_n^i \to a^i$; $a^1 = a^2$ by Sec. 4.1, Problem 201; hence $d(x_1^1, x_1^2) \leq 2\varepsilon$ so $y \in f[N(x_1^1, 2\varepsilon)]$ and f is an open map. This argument is due to J. D. Weston.⟧
119. In Problem 118, "complete" cannot be omitted.

$$⟦X = (\mathbf{Q} \cap [-1, 0] \cup (\mathbf{J} \cap [0, 1]). \qquad Y = [0, 1], \qquad f(x) = \pm x.⟧$$

120. In Problem 118, we cannot conclude that f is onto. ⟦$i: \mathbf{R} \to \mathbf{R}^+$; or $i: \mathbf{J} \to \mathbf{R}$. (See Problem 205.)⟧

---

201. Let X be the set of all sequences of complex numbers which converge to 0. Define $d(x, y) = \sup|x_n - y_n|$. Prove that X is complete.
▲202. Prove this converse of part of Theorem 9.1.4. If X is a noncompact semimetric space, it has an equivalent noncomplete semimetric. ⟦See [Niemytzki and Tychonoff].⟧
203. A topological space is called *topologically complete* if its topology can be given by a complete semimetric. Show that every open set in a complete semimetric space is topologically complete. ⟦Let $f(x) = 1/d(x, \tilde{G})$. See Sec. 4.2, Problem 12 and Theorem 4.2.6. Let $d'(x, y) = d(x, y) + |f(x) - f(y)|$.⟧
204. Every G_δ in a complete semimetric space is topologically complete. ⟦Say $S = \bigcap G_n$, each G_n complete, by Problem 203. Map S onto D, the diagonal of $\prod G_n$. In the metric case D is closed, hence complete. In general, D is complete by a direct argument, for example, Problem 107.⟧
205. $\mathbf{J}$ is topologically complete. ⟦Problem 204; note that $\tilde{\mathbf{J}} = \mathbf{Q}$ is countable, hence an F_σ; that is, a countable union of closed sets.⟧ (See also Sec. 6.4, Problem 203.)
206. As a converse to Problem 204, prove that every topologically complete space X is an *absolute* G_δ. This means that if X is homeomorphic with a subset S of a metric space Y, S is a G_δ in Y. ⟦Let $G_n = \{x \in \bar{S}$: there exists r such that the diameter of $f[N(x, r)]$ is less than $1/n\}$.⟧ (See also Sec. 8.3, Problem 111, which shows that "metric space Y" cannot be replaced by "Hausdorff space Y.")
207. Let $\{f_\delta\}$ be an equicontinuous net of maps from one semimetric space

to another. Show that $\{x: \{f_\delta(x)\}$ is a Cauchy net$\}$ is closed. ⟦$d(f_\alpha x, f_\beta x) \le d(f_\alpha x, f_\alpha y) + d(f_\alpha y, f_\beta y) + d(f_\beta y, f_\beta x)$.⟧

208. A metric space is complete if and only if it is Cech-complete. ⟦See [Cech, p. 838].⟧

209. Let $X = \{0\} \cup (1, \infty)$ and suppose that the Euclidean topology for X is given by a complete metric. Must $(1, \infty)$ contain a closest point to 0?

9.2. Completion

It is often possible to apply results about complete spaces to those which are not complete by the device of embedding spaces in complete spaces. This process, called completion, should be compared with the compactification of Chapter 8. (Compactifications may be obtained as special cases of completions; see Sec. 11.5, Example 1.)

A *completion* of a semimetric space X is a pair (Y, f), where Y is a complete semimetric space and f is an isometry from X onto a dense subspace of Y. A completion (Y, f) will be called a *metric completion* if Y is a metric space. (Obviously, only a metric space could have a metric completion.) Just as in Chapter 8, we shall usually think of a space as being a subspace of its completion.

Before constructing completions we shall consider their uniqueness. This is covered by three remarks. Theorem 9.2.1 will show that a metric completion is unique. Example 1 will show that more general completions are not unique, and Example 2 will show that a metrizable space may have different metric completions depending on the metric chosen to induce its topology.

We give here a direct proof of Theorem 9.2.1. In Section 11.5, we shall show how to deduce it from the extension theorem for uniformly continuous functions.

THEOREM 9.2.1. *Let (Y, f) and (Z, g) be metric completions of a metric space X. Then there exists an isometry h from Y onto Z such that $h \circ f = g$. (In particular h carries $f[X]$ onto $g[X]$ "in the right way.")*

(The parenthesized phrase is explained in the remark following Theorem 8.4.3.) Define $u: f[X] \to g[X]$ by $u = g \circ f^{-1}$. Then u is an isometry of $f[X]$ onto $g[X]$. For any $y \in Y$, let $\{t_n\}$ be a sequence of points in $f[X]$ with $t_n \to y$. Then $\{t_n\}$ is a Cauchy sequence and so $\{u(t_n)\}$ is a Cauchy sequence in Z. Let $z = \lim u(t_n)$. ⟦z exists and is unique since Z is a complete metric space.⟧ We now show that z is uniquely determined by y, although the sequence $\{t_n\}$ need not be. ⟦Let $t'_n \to y$. Then $d(t_n, t'_n) \to 0$ and so $d[z, u(t'_n)] \le d[z, u(t_n)] + d[u(t_n), u(t'_n)] = d[z, u(t_n)] + d(t_n, t'_n) \to 0$. Thus $u(t'_n) \to z$ also.⟧ Now define $h: Y \to Z$ by $h(y) = z$. Then $h \mid f[X] = u$, ⟦for $y \in f[X]$, take $t_n = y$ for all n. Then $h(y) = \lim u(t_n) = u(y)$⟧, and so $h \circ f = u \circ f = g$. Finally h is an isometry. ⟦Let $y, y' \in Y$. Let $\{t_n\}, \{t'_n\}$ be sequences of points in $f[X]$ with $t_n \to y$, $t'_n \to y'$. Then $d[h(y), h(y')] = \lim d[u(t_n), u(t'_n)] = \lim d(t_n, t'_n) = d(y, y')$, using the definition of h and the fact that u is an isometry.⟧ ∎

EXAMPLE 1. Let $Y = \mathbf{R}^2$ with the semimetric d given by

$$d[(x, y), (x', y')] = |x - x'|.$$

The X axis, $X = \{(x, 0): x \in \mathbf{R}\}$, is dense in Y. ⟦For any $P \in \mathbf{R}^2$ and $\varepsilon > 0$, say $P = (x, y)$, we have $(x, 0) \in X \cap N(P, \varepsilon)$ since $d[(x, 0), (x, y)] = |x - x| = 0 < \varepsilon$.⟧ Moreover Y is complete. ⟦Let $\{(x_n, y_n)\}$ be a Cauchy sequence. Then $\{x_n\}$ is a Cauchy sequence in $\mathbf{R}$ since $|x_m - x_n| = d[(x_m, y_m), (x_n, y_n)]$. Let $x = \lim x_n$ in $\mathbf{R}$. Then $(x_n, y_n) \to (x, 0)$ in Y.⟧ Thus (Y, j) is a completion of $\mathbf{R}$, j being the inclusion map $j(x) = (x, 0)$. But $(\mathbf{R}, i)$ is also a completion of $\mathbf{R}$, i being the identity map, and so Theorem 9.2.1 fails for completions not assumed to be metric. ⟦$\mathbf{R}$ and Y are not isometric, indeed not homeomorphic, since $\mathbf{R}$ is a T_2 space and Y is not.⟧

EXAMPLE 2. Let S_1 be the one-sphere, $X = (0, 1)$, $Y = [0, 1]$, $Z = [0, \infty)$ with the Euclidean metric in each case. Then S_1, Y, Z are complete metric spaces, no two homeomorphic, and all three have dense subspaces homeomorphic with X. ⟦Remove any point from S_1; remove 0 and 1 from Y; and remove 0 from Z.⟧ Now (Y, i) is the metric completion of X, and X can be given different metrics (inducing the same topology) to make S_1 or Z its metric completion. Finally X can be given a complete metric. ⟦X is homeomorphic with $\mathbf{R}$.⟧ These ideas are continued in Problems 104, 105, 203, and 204.

▲EXAMPLE 3. We saw in Sec. 8.2, Example 1 (and Sec. 9.1, Problem 201), that every metric space X is isometric with a subspace of a complete metric space Y. Let $Z = \overline{\rho[X]}$ in Y where $\rho: X \to Y$ is the isometry. Then (Z, ρ) is a completion of X. Thus *every metric space has a completion.*

In spite of the availability of Example 3 we shall spell out another construction of a completion. There are two reasons for doing this. We can complete an arbitrary semimetric space in this way; and Example 3 presupposes the availability and completeness of $\mathbf{R}$ (in proving that Y is complete) and so this method cannot be used to construct $\mathbf{R}$ from $\mathbf{Q}$ by a completion process. It should also be pointed out that the construction of Example 3 is, in a certain sense, external, while the construction now to be given is internal.

We begin by proving two results which will reduce the labor of proving that our final product is complete.

LEMMA 9.2.1. *Let S be a dense subset of a semimetric space Y, and suppose that each Cauchy sequence of points in S converges in Y. Then Y is complete.*

Let $\{y_n\}$ be a Cauchy sequence in Y. For each n, choose $s_n \in S$ with $d(s_n, y_n) < 1/n$. Then $\{s_n\}$ is a Cauchy sequence.

$$[\![d(s_m, s_n) \leq d(s_m, y_m) + d(y_m, y_n) + d(y_n, s_n) < \frac{1}{m} + d(y_m, y_n) + \frac{1}{n} < \varepsilon$$

if
$$m > \frac{3}{\varepsilon}, \quad n > \frac{3}{\varepsilon}, \quad d(y_m, y_n) < \frac{\varepsilon}{3}.]\!]$$

Thus by hypothesis $s_n \to y$ for some $y \in Y$. But then $y_n \to y$.

$$[\![d(y_n, y) \leq d(y_n, s_n) + d(s_n, y) < \frac{1}{n} + d(s_n, y) \to 0.]\!]$$ ∎

LEMMA 9.2.2. *Let $\{x_n\}$ be a Cauchy sequence in a semimetric space X. Then* $\lim_{k\to\infty} \lim_{n\to\infty} d(x_k, x_n) = 0$.

For each $a \in X$, $\{d(a, x_n)\}$ is a Cauchy sequence of real numbers, $[\![|d(a, x_n) - d(a, x_m)| \leq d(x_n, x_m)]\!]$, hence convergent. This shows that for each k, $\lim_{n\to\infty} d(x_k, x_n)$ exists. Now let $\varepsilon > 0$. Choose M so that $n > M$, $k > M$ implies $d(x_k, x_n) < \varepsilon$. Then $k > M$ implies that $\lim_{n\to\infty} d(x_k, x_n) \leq \varepsilon$. [\![For all such k, $d(x_k, x_n) < \varepsilon$ when n is large enough hence its limit $\leq \varepsilon$.]\!] This is precisely the statement of Lemma 9.2.2. ∎

THEOREM 9.2.2. *Every semimetric space has a completion.*

Let (X, d) be a semimetric space and let Y be the set of all Cauchy sequences in X. We shall define a complete semimetric on Y and show that X is isometric with a dense subset. If $u = \{u_n\}$, $v = \{v_n\}$ are Cauchy sequences in X, then $\{d(u_n, v_n)\}$, is a Cauchy sequence of real numbers $[\![|d(u_n, v_n) - d(u_m, v_m)| \leq d(u_n, u_m) + d(v_n, v_m)$ by Sec. 2.2, Problem 3.]\!] We define $D(u, v) = \lim_{n\to\infty} d(u_n, v_n)$. Clearly

$$D(u, v) = D(v, u), \qquad D(u, u) = 0,$$

and

$$D(u, w) = \lim d(u_n, w_n) \leq \lim[d(u_n, v_n) + d(v_n, w_n)] = D(u, v) + D(v, w)$$

so that (Y, D) is a semimetric space. Now define $f: X \to Y$ by $f(x) = \{x, x, x, \ldots\}$; that is, $f(x)$ is the sequence each of whose terms is x, surely a Cauchy sequence. Then f is an isometry. [\![It is obviously one-to-one, and $D[f(x), f(y)] = \lim_{n\to\infty} d(x, y) = d(x, y)$.]\!] Also, $f[X]$ is dense in Y. [\![Let $u \in Y$ and $\varepsilon > 0$. Choose M so that $m \geq M$, $n \geq M$ implies $d(u_m, u_n) < \varepsilon$. Now $u_M \in X$ and $D[f(u_M), u] = \lim_{n\to\infty} d(u_M, u_n) < \varepsilon$ by choice of M. Hence $f(u_M) \in N(u, \varepsilon)$. Thus $f[X]$ meets every open set.]\!] It remains only to show that Y is complete. Let $\{u^k\}$ be a Cauchy sequence in Y: by Lemma 9.2.1 we may assume $u^k \in f[X]$ for each k; say $u^k = f(x_k)$. Now $\{x_n\}$ is a Cauchy sequence $[\![d(x_n, x_m) = D(u^n, u^m)$ since f is an isometry]\!], in other words, setting

$v = \{x_n\}$ we have $v \in Y$. Finally $u^k \to v$. ⟦$D(u^k, v) = \lim_{n\to\infty} d(u_n^k, x_n) = \lim_{n\to\infty} d(x_k, x_n) \to 0$ as $k \to \infty$ by Lemma 9.2.2. We used here the fact that u_n^k in the nth term of the constant sequence $f(x_k)$.⟧ ▮

THEOREM 9.2.3. *Every metric space has a metric completion.*

Let X be a metric space and Y a completion ⟦Theorem 9.2.2⟧. There is no loss of generality in assuming that $X \subset Y$ ⟦since if f is an isometry of X onto a dense subspace of Y, $f[X]$ is a metric space isometric with X, (recall that f must be one to one), and we may discuss $f[X]$ instead of X⟧. We know that there is a complete metric space Z and a function q from Y onto Z which preserves distances. ⟦Namely $Z = Y/\rho$ as in Sec. 6.7, Example 5, and Sec. 9.1, Example 5.⟧ Then $q: X \to Z$ is an isometry. ⟦X is a metric space so q, as a distance-preserving map, must be one-to-one.⟧ Moreover $q[X]$ is dense in Z. ⟦Let $z \in Z$. Then $z = q(y)$ for some $y \in Y$ since q is onto. Since X is dense in Y there is a sequence $\{x_n\}$ of points of X with $x_n \to y$. Then $q(x_n) \in q[X]$ and $q(x_n) \to z$ since $d[q(x_n), z] = d[q(x_n), q(y)] = d(x_n, y) \to 0$.⟧ It follows that (Z, q) is a metric completion of X. ▮

Problems on Semimetric Space

1. If X has more than one point, the space Y constructed in Theorem 9.2.2 cannot be a metric space.
2. The completion of X is compact if and only if X is totally bounded ⟦Theorem 9.1.4; Sec. 7.2, Problem 102⟧.

---

101. A locally complete metric space has an equivalent complete metric ⟦Theorem 9.2.3; Sec. 9.1, Problems 116 and 204⟧. The same is true for a locally compact space ⟦Sec. 9.1, Problem 117⟧.
102. A space is locally complete if and only if it is an open subset of its completion. (Compare Theorem 8.4.1.)
103. A noncomplete metric space may be homeomorphic with its completion. ⟦The first quadrant in $\mathbf{R}^2$ together with the positive X axis.⟧
104. Call (Y, f) an *h completion* of X if Y is complete and $f: X \to Y$ is a homeomorphism of X onto a dense subspace of Y. Show that an h completion of (X, d) is a completion of (X, D), where D is a semimetric equivalent with d.
105. Two one-point h completions of a metric space need not be homeomorphic. ⟦Consider $[0, \infty)$ and S^1 as h completions of $\mathbf{R}$.⟧
106. The following are equivalent for a metric space X: (a) X is separable; (b) X has an equivalent totally bounded metric; (c) X has a compact h completion; (d) X has a metrizable compactification. ⟦(b) ⇒ (a) by Lemma 7.2.2, and (a) ⇒ (d) by Sec. 6.6, Problem 203, and Theorem 6.4.2. See also Problem 2.⟧ Compare Sec. 10.1, Problem 112.

---

201. A metric space may be isometric with a proper subspace obtained by deleting one nonisolated point.

$[\![\bigcup\{[-2n-1,-2n):n=0,1,2,\ldots\}\cup(\bigcup\{(2n-1,2n):n=1,2,\ldots\})$.

Remove -1 and translate.$]\!]$ Completion of such a space must always add infinitely many points. (Compare Sec. 8.4, Problem 104.)

202. Show that (0, 1), [0, 1), and **J** can all be taken as h completions of **Q** by metrizing their topologies suitably and defining embeddings of **Q** into them.

203. Show that **R** has a three-point h completion. (Compare Sec. 8.4, Example 3.) See *Amer. Math. Monthly, 76* (1969), 569.

204. Define an ordering among h completions of a given space analogous to that among compactifications. Compare, in this ordering, **R**, [0, 1), [0, 1], and S_1 as h completions of **R**.

9.3 Baire Category

The category theorem (Theorem 9.3.5), given by R. Baire in 1899 is one of the principal avenues through which applications of completeness are made in classical and functional analysis. The most important of these arise through the uniform boundedness principle, as well as other category theorems in functional analysis. See Theorem 12.4.4. Other applications are indicated here in Example 3, Problems 114, 115, 121, 202, 203. The Baire theorem and its applications are not trivial, even for **R**.

We begin with a nineteenth-century result of G. Cantor which was at one time in great vogue for discussing the structure of subsets of **R**. Cantor proved it for **R**, but his proof extends easily.

THEOREM 9.3.1. *Let X be a complete semimetric space, and $\{F_n\}$, a sequence of nonempty closed sets such that $F_{n+1} \subset F_n$ for all n, and $d_n \to 0$, where d_n is the diameter of F_n. Then $\bigcap F_n$ is not empty.*

For each n, choose $x_n \in F_n$. The sequence $\{x_n\}$ is a Cauchy sequence. $[\![$Given $\varepsilon > 0$, choose N so that $d_N < \varepsilon$. Let $m > N$, $n > N$. Then $x_m \in F_N$, $x_n \in F_N$, so that $d(x_m, x_n) \le d_N < \varepsilon.]\!]$ Thus $\{x_n\}$ is convergent, say $x_n \to x$. It follows that $x \in \bigcap F_n$. $[\![$Let n be arbitrary. For $m > n$, $x_m \in F_n$, hence $x \in F_n$ since F_n is closed and $x_m \to x.]\!]$ ∎

A topological space is called a *Baire space* if the intersection of every sequence of dense open sets is dense.

THEOREM 9.3.2. *Every complete semimetric space is a Baire space.*

Let $\{G_n\}$ be a sequence of dense open sets. We shall show that their intersection meets an arbitrary nonempty open set G. Since G_1 is dense and open,

$G \cap G_1$ must have interior, hence includes a disc $D(x_1, r_1)$, $r_1 > 0$. For the same reason $G_2 \cap N(x_1, r_1)$ includes a disc, say $D(x_2, r_2)$ and we may take $0 < r_2 < \frac{1}{2}r_1$. Continuing in this way, we obtain a sequence $\{D_n\}$ with $D_n = D(x_n, r_n)$, $D_{n+1} \subset D_n$, and $r_{n+1} < \frac{1}{2}r_n$. Each D_n is closed and has diameter $\leq 2r_n < 2^{-n+2}r_1 \to 0$. By Theorem 9.3.1,

$$G \cap \left(\bigcap_{n=1}^{\infty} G_n\right) = (G \cap G_1) \cap \left(\bigcap_{n=2}^{\infty} G_n\right) \supset \bigcap_{n=1}^{\infty} D_n \neq \varnothing. \blacksquare$$

Note that the definition of Baire space is purely topological and so, for example, (0, 1) is a Baire space since it is homeomorphic with the complete metric space **R**.

EXAMPLE 1. **Q** *is not a Baire space.* Let $\mathbf{Q} = \{r_n\}$ (an enumeration of the rationals as a sequence), and let $G_n = \{r_n, r_{n+1}, r_{n+2}, \ldots\}$. Each G_n is open ⟦since it is the complement of a finite set⟧ and dense ⟦since each Euclidean neighborhood of a rational contains infinitely many rationals⟧. But $\bigcap G_n = \varnothing$. On the other hand, **J** is a Baire space by Problem 118. ⟦**J** is a G_δ since **Q** is an F_σ.⟧

The concepts discussed above can be expressed in terms of the concept of Baire category which we now define. A set S in a topological space X is said to be *nowhere dense in* X if the interior of $\bar{S}$ is empty.

EXAMPLE 2. Let X be the X axis in $\mathbf{R}^2$, thus $X = \{(x, 0): x \in \mathbf{R}\}$; and let S be the set of rational points in X, $S = \{(x, 0): x \in \mathbf{Q}\}$. Then S is nowhere dense in $\mathbf{R}^2$, but S is not nowhere dense in X since the closure of S in X is all of X so that, in X, $(\bar{S})^i = X$.

A set S in a topological space X is said to be *of first category* in X if it is σ-nowhere dense; that is, if S is the union of countably many sets each of which is nowhere dense in X. For example, **Q** is of first category in **R** because it is the union of countably many singletons, each of which is nowhere dense in **R**. In Example 2, X is nowhere dense in $\mathbf{R}^2$, hence of first category in $\mathbf{R}^2$⟦$X = \bigcup X_n$ where $X_n = X$ for all n⟧, but X is not of first category in itself as is shown by Theorem 9.3.5, below. A set which is not of first category in X is said to be *of second category in* X. The abbreviations $S \in$ Cat I or $S \in$ Cat II are sometimes used; and, when no confusion is likely, a space is said to be of first or second category, with "in itself" understood. As an exception to these definitions, we shall adopt the convention that the empty space is of second category in itself; although, of course, it is nowhere dense in every space.

THEOREM 9.3.3. *A topological space X is a Baire space if and only if each nonempty open set is of second category in X.*

Let X be a Baire space and G a nonempty open set. Let $\{S_n\}$ be a sequence of sets each of which is nowhere dense in X, and let $G_n = X \setminus \bar{S}_n$. Each G_n is a dense open subset of X, hence $\bigcap G_n$ is dense, in particular $\bigcap G_n$ meets G. But this means that G contains a point not in $\bigcup \tilde{G}_n = \bigcup \bar{S}_n$, *a fortiori*, G contains a point not in $\bigcup S_n$, in particular $G \neq \bigcup S_n$ and so G is of second category in X. Conversely, if X is not a Baire space, let $\{G_n\}$ be a sequence of dense open sets whose intersection is not dense. There then exists a nonempty open set G with $G \not\pitchfork \bigcap G_n$. Let $S_n = G \setminus G_n$. Then each S_n is nowhere dense $[\![S_n \subset X \setminus G_n]\!]$, and $G = G \setminus \bigcap G_n = \bigcup (G \setminus G_n) = \bigcup S_n$ so that G is of first category in X. ∎

THEOREM 9.3.4. *A Baire space is of second category in itself.*

Apply Theorem 9.3.3 to the whole space (if it is not empty). ∎

THEOREM 9.3.5. *Every complete semimetric space is of second category in itself.*

For it is a Baire space, by Theorem 9.3.2. ∎

THEOREM 9.3.6. *Every locally compact regular space is a Baire space.*

Let $\{G_n\}$ be a sequence of dense open sets. We shall show that the intersection meets an arbitrary nonempty set G. Since G_1 is dense and open, $G \cap G_1$ must have interior, hence includes a closed compact set K_1 with nonempty interior $[\![$Sec. 5.4, Problem 9$]\!]$. For the same reason $K_1^i \cap G_2$ includes a closed compact set K_2 with nonempty interior. Continuing in this way we obtain a sequence $\{K_n\}$ of closed compact nonempty sets with $K_{n+1} \subset K_n$. Since K_1 is compact it follows that $\bigcap K_n \neq \varnothing$ $[\![$Theorem 5.4.3$]\!]$. Thus $G \cap (\bigcap G_n) = (G \cap G_1) \cap (\bigcap_{n=2}^{\infty} G_n) \supset \bigcap K_n \neq \varnothing$. ∎

EXAMPLE 3. Two interesting uses of Baire category may be found in [Brunk] and [Fort]. In the first of these it is proved that if for each $x \in \mathbf{R}$, there exists k such that the kth derivative of f vanishes at x, then f is a polynomial. (It is assumed that f has derivatives of all orders.) The other shows that if a real function is discontinuous on a dense set and differentiable on a dense set, it must be continuous and nondifferentiable on a residual set.

Problems

1. In Theorem 9.3.1, the diameter of $\bigcap F_n$ must be 0.
2. Determine all the sets constructed in the proof of Theorem 9.3.2 if $X = \{0\} \cup [1, 2]$, $G_n = X \setminus \{2 - 1/k : k = 1, 2, \ldots, n\}$, $G = \{0\}$.

3. Every locally countably compact regular space is a Baire space. ⟦The proof of Theorem 9.3.6.⟧
4. Every locally compact semimetric space is a Baire space. ⟦Theorem 9.3.6. It also follows from Sec. 9.1, Problem 115 and Theorem 8.4.1.⟧
5. Let $\{G_n\}$ be a sequence of nonempty open sets in a Baire space such that G_n is dense for all sufficiently large n. Show that $\bigcap G_n \neq \varnothing$.
6. Let $S \subset X$. If S is of first category in itself, it must be of first category in X. This is false for second category. ⟦Consider ω in $\mathbf{R}$.⟧
7. Let $A \subset G \subset X$, where G is an open subset of X. Show that A is of first category in G if and only if it is of first category in X.
8. In Problem 7, replace "open" by "dense."
9. In Problem 7, "open" cannot be omitted. ⟦$G = \omega$, $X = \mathbf{R}$.⟧
10. A set is called *residual* if its complement is of first category. Show that a dense G_δ is residual. ⟦$(\bigcap G_n)^{\sim} = \bigcup \tilde{G}_n$, and each $\tilde{G}_n$ is nowhere dense since it is closed and has dense complement.⟧
11. $\mathbf{Q}$ is not a G_δ in $\mathbf{R}$. ⟦Problem 10. $\mathbf{Q}$ is not residual since it is of first category in $\mathbf{R}$.⟧
12. Express $[0, 1]$ as the union of two subsets, neither of which is a Baire space. ⟦One of them is $[\mathbf{Q} \cap (0, \frac{1}{2})] \cup [\mathbf{J} \cap (\frac{1}{2}, 1)]$.⟧
13. Being a Baire space is not hereditary; it is not even F-hereditary. However it is G-hereditary, and a retract of a Baire space need not be Baire. ⟦Consider $\mathbf{R}^2$ with all irrational points on the X axis removed; $\mathbf{Q}$ is a closed subspace. For G hereditary, see Problem 7 and Theorem 9.3.3. Finally, an example of a retraction of a Baire space onto a non-Baire subspace is given in a correction in Volume 77 of the *American Mathematical Monthly* to Example 2.2 of my article "Life Without T_2."⟧ See also Problem 117.
14. Every finite space is a Baire space. ⟦Sec. 2.5, Problem 110 with "finitely many" instead of "two."⟧

---

101. In Theorem 9.3.1 the condition "$d_n \to 0$" can be dropped if at least one F_n is compact, but cannot be dropped in general; nor can "closed" be dropped. ⟦Easy examples in $\mathbf{R}$ can be given.⟧
102. State and prove a converse of Theorem 9.3.1, namely that the condition on closed subsets of X implies that X is complete.
103. A metric space X may be of second category and have a nonempty open subset which is of first category in itself, hence ⟦Problem 6⟧ in X. ⟦$\mathbf{Q} \cup [0, 1]$.⟧
104. A countable connected T_1 space must be of first category in itself ⟦Sec. 5.2, Problem 9⟧.
105. A countably infinite locally compact regular space must have infinitely many isolated points. ⟦Otherwise, removal of the isolated points yields a contradiction to Theorem 9.3.6.⟧ "Regular" cannot be re-

placed by "T_1" ⟦cofinite⟧. "Locally compact regular" can be replaced by "locally compact KC" but not by "compact KC" ⟦$\mathbf{Q}^+$⟧.

106. The space in Problem 105 may also have infinitely many nonisolated points. ⟦In **R**, make a sequence of convergent sequences converging to $1, \frac{1}{2}, \frac{1}{3}, \ldots$⟧

107. Let X be a countable T_2 space with exactly one nonisolated point x. Then if X is compact it is homeomorphic with ω^+. In particular, X is a subspace of **R**. ⟦If N is an open neighborhood of x, $\tilde{N}$ is discrete and compact, hence finite. Thus $x_n \to x$ where $X = \{x_n\}$. See Sec. 5.1, Problem 4.⟧

108. With X as in Problem 107 suppose that X is locally compact and not compact. Show that X is a subspace of **R**, indeed X is homeomorphic with $(0, \frac{1}{2}, \frac{1}{3}, \frac{1}{4}, \ldots) \cup \boldsymbol{\omega}$. ⟦If N is a compact neighborhood of x, $N = \boldsymbol{\omega}^+$ by Problem 107; and $\tilde{N}$ is discrete, open, and closed.⟧

109. The first result of Example 3 is false if the parenthesized phrase is omitted. ⟦Consider $x \cdot |x|$.⟧

110. Is **J** σ-compact? ⟦Problem 11.⟧

111. A perfect set in a complete metric space cannot be countable. ⟦It is a complete metric space with nowhere dense points.⟧ Generalize.

112. A cofinite space is a Baire space if and only if it is either finite or uncountable.

113. A compact T_2 space must be a Baire space ⟦Theorem 9.3.6⟧, but neither a locally compact T_1 space nor a compact KC space need be. ⟦Problem 112. Also $\mathbf{Q}^+$ is countable and self-dense; see Sec. 8.1, Problem 205.⟧

114. Theorem 9.3.5 implies that **R** is uncountable. Write the proof in full assuming that **R** is self-dense and complete metric. Derive a contradiction from the assumption that **R** is countable, following the proof of Theorem 9.3.5.

115. (**R**, RHO) is not σ-compact. ⟦An RHO compact set is Euclidean compact and nowhere dense. Hence **R** would be of (Euclidean) first category.⟧

116. Find a residual set in [0, 1] which has (Lebesgue) measure 0. ⟦Let G_n be open, $G_n \supset \mathbf{Q} \cap [0, 1]$, $|G_n| < 1/n$, $S = \bigcap G_n$. Use Problem 10.⟧ (This set is large and with small complement in the sense of category, but small and with large complement in the sense of measure.)

117. A residual subset S of a Baire space X is a Baire space. ⟦Let $\tilde{S} \subset \bigcup N_k$, each N_k nowhere dense. Let $\{G_n\}$ be dense open sets in S, say $G_n = H_n \cap S$, H_n open in X. Let $V_n = H_n \setminus \bigcup \{\overline{N}_k : k = 1, 2, \ldots, n\}$. Each V_n is dense in X, partly by Sec. 2.5, Problem 109. Then $\bigcap V_n = \bigcap (G_n \cap S)$.⟧

118. A dense G_δ in a Baire space is a Baire space ⟦Problems 117 and 10⟧. "Dense" cannot be omitted. ⟦First example in the hint for Problem

13. $\mathbf{Q} = \bigcap \{(x, y): |y| < 1/n\}$.⟧ However a G_δ in a complete semi-metric space must be a Baire space ⟦Sec. 9.1, Problem 204⟧. See also Problem 13.

119. Prove the (paradoxical) fact that in $[0, 1]^\omega$, the set S is residual, where $S = \{x: x_n = 1 \text{ for at least one } n\}$. In $[0, 1]^A$, for finite A, the corresponding set is nowhere dense. ⟦Each $[0, a]^\omega$, $0 < a < 1$, is nowhere dense, by Theorem 6.4.4. Take $a = 1 - 1/n$.⟧
120. Every Cech-complete space is a Baire space ⟦Problem 118⟧.
121. $\mathbf{Q}$ cannot be given a smaller T_2 topology which makes it locally compact ⟦Problem 105⟧. However $\mathbf{Q}$ has a smaller compact KC topology ⟦Sec. 8.1, Problems 203, 121, and 119⟧.
122. Every locally compact T_2 space X is residual in βX. ⟦Theorem 8.4.1. Indeed, $\tilde{X}$ is nowhere dense.⟧
123. $\mathbf{J}$ is residual in $\beta\mathbf{J}$ ⟦Problem 10; Sec. 9.1, Problems 205, 116, and 115⟧. $\mathbf{Q}$ is not residual in $\beta\mathbf{Q}$. ⟦Indeed $\tilde{\mathbf{Q}}$ is residual.⟧
124. Let $X = \bigcup \{S_\alpha: \alpha \in A\}$ where each S_α is open and closed and of first category in X. Show that X is of first category in itself. ⟦$S_\alpha = \bigcup \{N_\alpha^n: n = 1, 2, \ldots\}$, $X = \bigcup \{\bigcup \{N_\alpha^n: \alpha \in A\}; n = 1, 2, \ldots\}$.⟧

---

201. Use the category theorem to show that the space of Sec. 5.3, Problem 201 is not normal. ⟦$\mathbf{Q}$ and $\mathbf{J}$, subsets of X, cannot be separated: If G_1, G_2 separated them; for each $x \in \mathbf{J}$ choose one circle C_x, tangent at x and in G_2. Let $S_n = \{x \in \mathbf{J}: \text{radius } C_x > 1/n\}$. The Euclidean closure F_n of each S_n is a subset of $\mathbf{J}$, and $\mathbf{J} = \bigcup F_n$, an F_σ. This contradicts Problem 11.⟧
202. Let G be an unbounded open set in $\mathbf{R}$. Let $A_n = \{x \in \mathbf{R}: k > n \text{ implies } kx \notin G\}$. Show that each A_n is nowhere dense. Hence there exists a number not in any A_n. What does this say about G?
203. Let $\{N_n\}$ be a disjoint sequence of cells included in a disc D in $\mathbf{R}^2$ such that their union, G, is dense. Show that $D \setminus G$ is not homogeneous. ⟦Theorem 9.3.5 shows that it contains two kinds of points, those on the circumference of some N_n, and those not on any such.⟧
204. Let T, T' be topologies for a set X with $T' \supset T$, and let $S \subset X$. Give an example to illustrate each of the following possibilities: (a) S is T-dense and T'-nowhere dense. (b) S is T-nowhere dense but T'-somewhere dense. ⟦A simple extension by S.⟧ (c) X is of T category I and of T' category II. ⟦Take T' discrete.⟧ (d) X is of T category II and of T' category I. Prove in addition that: (e) if S is T'-dense it is T-dense, (hence (b) cannot be improved by omitting "somewhere.") (f) If X is T'-self-dense it is T-self-dense. (g) If X is countable and of T category II, and if T' is a T_1 topology, then X is of T' category II ⟦Problem 105⟧. (h) "Countable" cannot be omitted in (g). ⟦See (d).⟧

205. (See [Oxtoby].) Call a space *pseudocomplete* if it has a sequence $\{B_n\}$ of pseudobases such that whenever $V_n \in B_n$ and $V_n \supset \overline{V}_{n+1}$ it follows that $\bigcap V_n \neq \varnothing$. A pseudocomplete regular space is a Baire space. ⟦Imitate the proof of Theorem 9.3.2.⟧ The following are pseudocomplete: (a) A complete semimetric space. ⟦Let $B_n = \{N(x, \varepsilon): 0 < \varepsilon < 1/n\}$.⟧ (b) A locally compact regular space. ⟦Let B_n be the set of relatively compact open sets for all n.⟧ (c) Any product of pseudocomplete regular spaces.
206. Every product of complete semimetric spaces is a Baire space ⟦Problem 205(c)⟧.
207. There are only three countably infinite metric spaces with exactly one nonisolated point; all are subspaces of **R** ⟦Problems 107, 108; also Sec. 6.6, Problem 205⟧.

Metrization

10.1 Separable Spaces

In the late nineteenth and early twentieth century search for a sufficiently general basis for classical analysis, proposals were made that the metric space is the natural domain of analysis. That there is virtue in this proposal is attested to by the fact that many texts in advanced calculus, at the present time, do develop their subject matter for metric spaces. The inadequacy of metric and semimetric space shows up chiefly in the fact that product topologies are seldom semimetrizable ⟦Sec. 6.4, Problem 6⟧, and so, important function spaces lead to nonmetrizable (pointwise) topologies. However, a metric space is a very general object, and those who proposed its study attempted to show that certain types of spaces could actually be metrized. Such a metrization theorem is valuable since the knowledge that one is dealing with a metric space gives the investigator a great deal of information, for example, that the space is first countable and normal. The principal result of this section was given by P. Urysohn in 1925.

The key idea in the metrization theorems is the construction of a family of maps from a space X to metric spaces, such that the topology of X is precisely equal to the weak topology by the family of maps. Then, when this family is countable, the topology of X is semimetrizable, by Theorem 6.3.4.

THEOREM 10.1.1. *A second countable regular space is semimetrizable.*

Let B be a countable base. For each pair V, W of members of B which satisfy $\overline{W} \subset V$, we choose a continuous function $f: X \to [0, 1]$ with $f = 0$

on $W, f = 1$ on $\tilde{V}$. ⟦Such a function exists, for X is normal by Theorems 5.3.2 and 5.3.5; Urysohn's lemma (Theorem 4.2.11) may be applied to $\overline{W}$ and $\tilde{V}$.⟧ Let Φ be the family of all these functions f, then Φ is countable. ⟦There is one member of Φ for each pair, V, W of members of B with $\overline{W} \subset V$, and the set of such pairs is a subset of $B \times B$.⟧ We are going to show that $w(\Phi)$, the weak topology by Φ, is identical with the topology T of X. This will complete the proof. ⟦$w(\Phi)$ is semimetrizable by Theorem 6.3.4.⟧ Surely $T \supset w(\Phi)$. ⟦$w(\Phi)$ is the smallest topology making all members of Φ continuous.⟧ Conversely, let F be a T-closed subset of X. We shall show that F is $w(\Phi)$-closed. ⟦Let $x \notin F$. Then $\tilde{F}$ is a T neighborhood of x, hence there exists $V \in B$ with $x \in V \subset \tilde{F}$. Since X is regular, there exists $W \in B$ with $x \in W$, $\overline{W} \subset V$. By definition of Φ, there exists $f \in \Phi$ with $f = 0$ on $W, f = 1$ on $\tilde{V}$. In particular $f(x) = 0, f = 1$ on F and so, since f is $w(\Phi)$-continuous, x cannot belong to the $w(\Phi)$ closure of F by Theorem 4.2.4.⟧ Hence $T \subset w(\Phi)$, and so finally $T = w(\Phi)$. ∎

For a certain class of topological spaces, the metrization problem is completely solved.

COROLLARY 10.1.1. *A separable topological space is semimetrizable if and only if it is second countable and regular.*

Half of this is Theorem 10.1.1. The other half is Theorem 5.3.1, and Theorem 4.3.3. ∎

We now show how Theorem 10.1.1 may be interpreted in terms of embedding.

COROLLARY 10.1.2. *A second countable T_3 space X is homeomorphic with a subspace of* $[0, 1]^\omega$.

It follows that X is metrizable ⟦Theorem 6.4.2⟧, but we know this already from Theorem 10.1.1. To prove the corollary, note that the family Φ constructed for the proof of Theorem 10.1.1 is now separating ⟦Theorem 6.3.2⟧. Thus by Theorem 6.6.2, X is homeomorphic into $[0, 1]^\Phi$. ⟦$Y_f = [0, 1]$ for each $f \in \Phi$.⟧ ∎

REMARK. We have chosen Theorem 10.1.1 as our metrization theorem, instead of the more usual Corollary 10.1.2, because of its applicability to non-T_1 spaces, as well as what appears to be a slight gain in simplicity of the presentation. It is possible for a non-T_1 topology to be the weak topology by maps into metric spaces, whereas an embedding into a product would be impossible.

Theorem 10.1.1 is quite good in some ways. There are directions in which it cannot be improved. For example "second countable" cannot be replaced by the related weaker condition "Lindelöf and separable." A separable Lindelöf T_3 space need not be metrizable; indeed we can go further and state that a separable compact T_3 space need not be metrizable. (Of course, for compact spaces, T_2, T_3, T_4 are equivalent.) As an example we may cite $\beta\omega$. ⟦It is separable, since ω is dense; and not metrizable by Theorem 8.3.2.⟧ Moreover, a converse of Theorem 10.1.1 also holds, in the sense that a topological space is a second countable T_3 space if and only if it is a separable metric space ⟦Theorems 10.1.1 and 5.3.1⟧. A third remark is that "regular" cannot be replaced by "T_2" in Theorem 10.1.1 ⟦Problem 101⟧.

However Theorem 10.1.1 is also so special that it applies only to separable metric spaces. In Section 10.3, we shall give a result which is free from this defect, and is a "best possible" one.

Problems

1. A compact Hausdorff space is metrizable if and only if it is second countable.
2. Deduce from Theorem 10.1.1 the result (Sec. 4.1, Problem 113) that a finite space is semimetrizable if and only if it is regular. (Of course a finite T_3 space is discrete.)

---

101. A second countable normal space need not be semimetrizable ⟦Sec. 4.1, Problem 8⟧. A second countable T_4 space is metrizable ⟦Theorem 10.1.1⟧. A second countable T_2 space need not be metrizable ⟦Sec. 5.2, Problem 113⟧.
102. Deduce the fact that RHO is not second countable from Theorem 10.1.1, and the fact that if RHO were metric, $\mathbf{R} \times \mathbf{R}$ would be normal, contradicting Sec. 6.7, Example 3.
103. A free union of semimetric (metric) spaces is semimetrizable (metrizable). ⟦$d(x, y) = 1$ for x, y in different spaces. $d(x, y) = d \wedge 1$ for x, y in the same space, a distance d from each other.⟧
104. A topological space is semimetrizable if it can be written as the disjoint union of finitely many closed subspaces, each of which is semimetrizable in its relative topology ⟦Problem 103⟧.
105. In Problem 104, "finitely" may not be replaced by "countably." ⟦Every space is a union of singletons.⟧
106. In Problem 104, "closed" may not be omitted. More specifically, a T_4 space which is not metrizable may be the union of two disjoint metrizable subspaces, one of which is compact. ⟦Sec. 8.3, Problem 103. One subspace is countable and discrete, and the other is a singleton!⟧ (Note also Sec. 5.3, Problem 201, which shows a non-normal space which is the union of a Euclidean and a discrete subspace.)

107. A countable regular space need not be semimetrizable. ⟦Sec. 8.3, Problem 103; a T_4 space!⟧
108. A countable regular space is semimetrizable if it is first countable ⟦Theorem 10.1.1⟧. "Regular" cannot be replaced by "T_2" ⟦Sec. 5.2, Problem 113⟧. Other metrization theorems for countable spaces are given in Sec. 9.3, Problems 107, 108; and in the following problem.
109. A countable regular space is semimetrizable if it is locally compact ⟦Theorem 10.1.1, and Sec. 5.4, Problem 202⟧. "Regular" may be replaced by "T_2" ⟦Sec. 5.4, Problem 101⟧.
110. A set S in a T_2 space X is sequentially closed if and only if $S \cap C$ is closed for every compact metrizable subset C of X. ⟦Note that every point in $\overline{S \cap C}$ is a sequential limit point of S. Conversely, if $x_n \in S$ and $x_n \to x$, then $(x, x_1, x_2, \ldots)$ is compact and metrizable by Sec. 9.3, Problem 107.⟧
111. A Hausdorff space is sequential if and only if every subset S which satisfies the following condition is closed: $S \cap C$ is closed for every compact metrizable subset C ⟦Problem 110⟧. (Compare k space in Sec. 8.1, Problem 115.)
112. X^+ is metrizable if and only if X is locally compact, separable, metrizable. Compare Sec. 9.2, Problem 106. ⟦Theorem 8.1.2; Theorem 5.3.1. Note that compact implies Lindelöf, and second countability is hereditary. Conversely, if X is locally compact, separable, metrizable, X^+ is first countable at ∞ by Sec. 8.1, Problems 113 and 124, and Theorems 5.3.1 and 5.3.2. Since X is second countable, this makes X^+ second countable. The result follows from Theorem 10.1.1.⟧
113. A locally compact separable metric space can be given a smaller compact metrizable topology ⟦Problem 112; Sec. 8.1, Problem 121⟧. A locally compact T_2 (or regular) space can be given a smaller compact T_2 (or regular) topology ⟦Sec. 8.1, Problems 121 and 123⟧. Compare Sec. 9.3, Problem 121.
114. Let $S \subset \mathbf{R}^2$ have the property that each vertical line meets it exactly once. Show that S can be given a smaller compact metrizable topology. ⟦Apply Problem 113 to the weak topology by the projection on the X axis.⟧ (Such sets can be quite pathological, e.g. totally disconnected; connected and dense in $\mathbf{R}_2$. See [May, pp. 101, 118].)
115. Let X be a compact regular space such that the diagonal Δ is a G_δ in $X \times X$. Show that X is semimetrizable ⟦Sec. 6.7, Problem 209⟧. The converse holds if X is also assumed Hausdorff ⟦Δ is closed, and $X \times X$ is metric⟧, but not otherwise ⟦X indiscrete⟧.
116. In Problem 115, "regular" may not be dropped or replaced by "T_1" ⟦ω with the cofinite topology⟧.

---

201. A first countable T_0 space X must be the continuous open image of a metric space A. ⟦Let A be the collection of all shrinking countable local bases at every point. For α, $\beta \in A$ let $d(\alpha, \beta) = 1/n$, where $n = \min\{k: \alpha_k \neq \beta_k\}$. (*Note*: $\alpha = \{\alpha_k\}$.) Define $f: A \to X$ by $f(\alpha) = x$, where x is the point at which α is a local base. To see that f is continuous, let N be a neighborhood of x; let $\alpha_k \subset N$; then $d(\alpha, \beta) < 1/(k + 1)$ implies $\alpha_k = \beta_k$ hence $f(\beta) \in \beta_k \subset N$. To see that f is open, fix α, k. We shall show that $f[N(\alpha, 1/k)] \supset \bigcap_{i=1}^{k+1} \alpha_i$. Let $y \in \bigcap_{i=1}^{k+1} \alpha_i$ and $\beta \in A$ a base at y with $\beta_1 \subset \alpha_{k+1}$. Let $\gamma = (\alpha_1, \alpha_2, \ldots, \alpha_{k+1}, \beta_1, \beta_2, \ldots)$. Then $f(\gamma) = y$ and $d(\gamma, \alpha) < 1/k$.⟧ This result is due to S. Hanai and V. I. Ponomarev.
202. The space A of Problem 201 is zero-dimensional.
203. If X^+ is metrizable (Problem 112) it is an h completion of X. Is it a minimum h completion in the sense of Sec. 9.2, Problem 204?

10.2 Local Finiteness

In this section we shall need to make use of the *well-ordering principle.* This is expressed in Theorem 10.2.1. A totally ordered set $(X, \leq)$ is said to be *well ordered* if every nonempty subset has a minimum (or first) member; that is every nonempty S has a member x such that $x \leq y$ for all $y \in S$.

THEOREM 10.2.1. *Every nonempty set can be well ordered.*

Let X be a nonempty set and let P be the set of all well-ordered sets D for which $D \subset X$. Then P is not empty. ⟦Let $x \in X$. Let $D = \{x\}$ and order D by equality. Then $D \in P$.⟧ We make P into a poset by defining $D' \succ D$ to mean that $D' \supset D$ (as sets), that their orderings agree on D (that is, if $x, y \in D$ then $x \leq y$ in D if and only if $x \leq y$ in D'), and finally that, in the ordering of D', $y \geq x$, wherever $y \in D' \setminus D$ and $x \in D$. Let C be a maximal chain in P, and let $U = \bigcup \{D: D \in C\}$. We first give U a total order; namely, for x, $y \in U$ we have $x, y \in D$ for some $D \in C$ since C is a chain. Define $x \leq y$ in U to mean $x \leq y$ in D. This definition makes sense since if $x, y \in D'$ for some other $D' \in C$, the orderings agree by definition of P and the fact that C is a chain. Next we show that this is actually a well-ordering for U: Let S be a nonempty subset of U. Then $S \cap D$ is not empty for some $D \in C$. Let x be the smallest member in D of $S \cap D$. Then x is also the smallest member in U of S ⟦for if $y \in S$, and $y \leq x$ in U, then $y \leq x$ in D' for some $D' \in C$, by definition of the ordering of U. Now if $y \in D$ we shall have $y \geq x$ by definition of x; while if $y \notin D$, we have $y \in D' \setminus D$ and $D' \succ D$ since C is a chain, hence $y \geq x$. Thus in all cases $y \geq x$ and so $y = x$ since the ordering of each member of C is antisymmetric.⟧ Having seen that U is well ordered, that is, that $U \in P$, we note that U is maximal in P. ⟦If $U' \succ U$ and $U' \neq U$, we could adjoin U'

to C making a strictly larger chain.⟧ Finally $U = X$. ⟦If not, let $y \in X \setminus U$ and declare $y \geq x$ for all $x \in U$. Then $U \cup \{y\}$ is well ordered and is strictly larger than U in the order of P, contradicting the maximality of U.⟧ ▮

Let S be a family of sets in a topological space X and let $x \in X$. We shall say that x is an *ω-accumulation point* of S if every neighborhood of x meets infinitely many members of S.

EXAMPLE 1. Let S be the family of open intervals $(1/n, 1 + 1/n)$, $n = 1, 2, \ldots$, in $\mathbf{R}$. Then 0 is an ω-accumulation point of S, as are all points in the interval $[0, 1]$. However, -1 is not an ω-accumulation point since $(-2, -\frac{1}{2})$ does not meet any members of S; and $\frac{5}{4}$ is also not an ω-accumulation point, since for example $(\frac{9}{8}, 2)$ meets only 7 members of S.

EXAMPLE 2. Let S be a family of singletons, and let $A = \{x: \{x\} \in S\} = \bigcup \{W: W \in S\}$. Then a point is an ω-accumulation point of S if and only if it is an ω-accumulation point of A in the sense of Section 7.1. (But see Problem 8.)

A family of sets is called *locally finite* if it has no ω-accumulation point. Thus a family S of sets in X is locally finite if and only if every point in X has a neighborhood which meets only finitely many members of S. Neither of the families in Examples 1, 2 is locally finite. Note that local finiteness is a property of the family of sets *and* the space in which it is located. See Problem 4 for an illustration of this important point.

EXAMPLE 3. Let S be the set of intervals $[n, n + 2]$, $n = 1, 2, \ldots$, in $\mathbf{R}$. Then S is locally finite since every real number lies in an interval meeting at most 3 members of S.

EXAMPLE 4. Any finite family is locally finite, as is a finite union of locally finite families. Thus for example if S is a locally finite family in the space X, so is $S \cup \{V\}$ for any subset V of X.

LEMMA 10.2.1. *Let S be a locally finite family. Then $\{\overline{W}: W \in S\}$ is also locally finite.*

Let x be any point and N an open neighborhood of x meeting only finitely many $W \in S$. Then N meets only finitely many $\overline{W}$. ⟦$N \not\pitchfork W$ implies $N \not\pitchfork \overline{W}$ since N is open; the argument being: $W \subset \tilde{N}$ hence $\overline{W} \subset \tilde{N}$.⟧ ▮

Locally finite families have some interesting permanence properties (Corollary 10.2.1 and Theorem 10.2.3) due to the fact that any action taken by a locally finite family can usually be attributed to one of its members. For example, Theorem 10.2.2 says that any point in the closure of such a family is in the closure of one of its members. This result can also be expressed in this way: for a locally finite family, the closure of the union of its members is equal to the union of the closures of its members.

THEOREM 10.2.2. *Let S be a locally finite family of sets and $A = \bigcup\{W: W \in S\}$. Then $\bar{A} = \bigcup\{\bar{W}: W \in S\}$.*

For each $W \in S$, $W \subset A$ hence $\bar{W} \subset \bar{A}$. Conversely, let $x \in \bar{A}$ and let N be a neighborhood of x meeting only finitely many members of S; say that N meets only $W_1, W_2, \ldots, W_k \in S$. Then for some i, $x \in \bar{W}_i$. ⟦If not, let $N_1 = \bigcap \tilde{\bar{W}}_i$. Then N_1 is a finite intersection of neighborhoods of x, hence $N \cap N_1$ is a neighborhood of x not meeting A. Thus $x \notin \bar{A}$.⟧ ▮

COROLLARY 10.2.1. *The union of a locally finite family of closed sets is closed.*

We have previously considered whether continuity of a function could be deduced from the fact of its continuity on certain subsets of a space. Examples are Sec. 8.1, Problem 120; Sec. 4.2, Problem 203; and the easily proved fact that if $f \mid G$ is continuous for every G in some open cover of X, then f is continuous on X. The next result is of this type.

THEOREM 10.2.3. *Let $f: X \to Y$ where X, Y are topological spaces, and suppose that S is a locally finite cover of X such that $f \mid \bar{A}$ is continuous for each $A \in S$. Then f is continuous.*

We may assume that the members of S are closed since $\{\bar{A}: A \in S\}$ may be considered instead of S ⟦Lemma 10.2.1⟧. Let $x \in X$. Let S' be the family of those sets in S which contain x. There exists a neighborhood N of x meeting only a finite collection of members of S', say $A_1, A_2, \ldots, A_k \in S'$, and meeting no members of $S \setminus S'$. ⟦Let $U = \bigcup\{W: W \in S \setminus S'\}$. Then U is closed, by Corollary 10.2.1, and $x \notin U$. Thus $\tilde{U}$ is a neighborhood of x not meeting any member of $S \setminus S'$ and we may choose $N \subset \tilde{U}$.⟧ Let V be a neighborhood of $f(x)$. Then for each $i = 1, 2, \ldots, k$, there exists a neighborhood W_i of x such that $f[W_i \cap A_i] \subset V$ ⟦since $f \mid A_i$ is continuous⟧. Let $W = N \cap W_1 \cap W_2 \cdots \cap W_k$. Then W is a neighborhood of x and $f[W] \subset V$. ⟦Let $w \in W$. Then $w \in N$ hence $w \in A_i$ for some i. Then $w \in W_i \cap A_i$ and so $f(w) \in V$.⟧ Thus f is continuous at x. ▮

Some light is thrown on various forms of compactness by the concept of local finiteness (Theorem 10.2.5). We begin the discussion with a sufficient condition for C-embedding which often allows the extension to completely regular spaces of theorems given for normal spaces. See, for example, Problem 112.

THEOREM 10.2.4. *Let X be completely regular. Suppose that $\{G_n\}$ is a locally finite disjoint sequence of open sets, and that $\{x_n\}$ is a sequence of points with $x_n \in G_n$ for each n. Then $\{x_n\}$ is C-embedded (and C*-embedded) in X.*

The parenthesized phrase follows from Sec. 8.5, Example 2. Let g be a continuous real function on $\{x_n\}$, say $g(x_n) = t_n$; thus $\{t_n\}$ is a sequence of real numbers. For each n there exists a continuous $f_n \colon X \to [0, 1]$ with $f_n(x_n) = 1, f_n = 0$ on $\tilde{G}_n$. ⟦By definition of complete regularity.⟧ For $x \in X$, let $f(x) = \sum_{n=1}^{\infty} t_n f_n(x)$. ⟦This sum is well defined since for all but at most one value of n, $x \in \tilde{G}_n$ and so $f_n(x) = 0$.⟧ We have $f | \{x_n\} = g$. ⟦For each k, $f(x_k) = \sum t_n f_n(x_k) = t_k = g(x_k)$, since $x_k \in \tilde{G}_n$ if $n \neq k$, implying $f_n(x_k) = 0$.⟧ Finally we shall use the criterion of Theorem 10.2.3 to show that f is continuous. Let $G_0 = X \setminus \bigcup \{G_n : n = 1, 2, \ldots\}$. Then $\{G_n : n = 0, 1, 2, \ldots\}$ is locally finite and covers X. Now $f | \bar{G}_0 = f | G_0 = 0$ is continuous, while for $k = 1, 2, \ldots, f | \bar{G}_k = t_k f_k$. ⟦Let $x \in \bar{G}_k$. If $n \neq k$, $G_n \not\pitchfork \bar{G}_k$ since G_n is open, hence $x \notin G_n$ and so $f_n(x) = 0$.⟧ Hence $f | \bar{G}_k$ is continuous for $k = 0, 1, 2, \ldots$ and so f is continuous, by Theorem 10.2.3. ∎

THEOREM 10.2.5. *A topological space is countably compact if and only if every locally finite disjoint family of subsets is finite. A completely regular space is pseudocompact if and only if every locally finite disjoint family of open subsets is finite.*

(A normal space is countably compact if and only if it is pseudocompact ⟦Sec. 8.5, Problem 103⟧, hence for normal spaces the two finiteness conditions cited are equivalent.) Suppose that X is countably compact and that S is a locally finite disjoint family of subsets of X. Let F be a set gotten by choosing exactly one point from each nonempty set in S. Then F has no ω-accumulation point. ⟦If x is an ω-accumulation point of F, every neighborhood of x meets infinitely many members of S.⟧ Thus F is finite. Since S is a disjoint family, it too must be finite. Conversely, if X is not countably compact it has an infinite subset with no ω-accumulation point, that is, a locally finite infinite family of singletons. Next, let X be not pseudocompact. Then there exists a real continuous function f which is unbounded above. ⟦If f is unbounded, either f or $-f$ is unbounded above.⟧ Choose x_1 with $f(x_1) > 1$, x_2 with $f(x_2) > f(x_1) + 1$, and so on, by induction, obtaining a sequence $\{x_n\}$ such that $f(x_n) > f(x_{n-1}) + 1$ for all n. Let

$$V_n = \{x : f(x_n) - \tfrac{1}{4} < f(x) < f(x_n) + \tfrac{1}{4}\}.$$

Then $\{V_n\}$ is an infinite family. ⟦No V_n is empty since $x_n \in V_n$.⟧ Also each V_n is open, no two V_n meet, and $\{V_n\}$ is locally finite. ⟦For each x, choose N so that $|f(t) - f(x)| < \frac{1}{4}$ for $t \in N$. Then N meets at most one V_n.⟧ Finally, assume that there exists an infinite locally finite disjoint family of open sets in the completely regular space X. Then there exists a countably infinite subfamily, which we may denote by $\{G_n\}$. Clearly $\{G_n\}$ is locally finite. For each n, let $x_n \in G_n$, and denote $\{x_n: n = 1, 2, \ldots\}$ by S. Now S is discrete ⟦in its relative topology; for each singleton $\{x_k\}$ is $G_k \cap S$⟧ and so an arbitrary real function defined on S is continuous. Let $f(x_n) = n$ for each n. Since S is C-embedded in X ⟦Theorem 10.2.4⟧, f has an extension to all of X. This is unbounded. ∎

REMARK. A few other facts may be gleaned from the proof of Theorem 10.2.5. (a) In the first half of the theorem "subsets" may be replaced by "singletons" or, if X is T_1, by "closed subsets." (b) The condition in the second half implies that X is pseudocompact without the assumption of complete regularity.

A collection S is called *σ-locally finite* if $S = \bigcup S_n$ with each S_n locally finite.

EXAMPLE 5. Let S be the collection of all intervals $(k - 1/n, k + 1/n)$, $k, n = 1, 2, 3, \ldots$. Then S is not a locally finite family in $\mathbf{R}$ since, for example, every neighborhood of 0 meets infinitely many $(-1/n, 1/n)$. However, S is σ-locally finite since we may set $S_n = \{(k - 1/n, k + 1/n): k = 1, 2, \ldots\}$ for each n, and it is clear that each S_n is locally finite and that $S = \bigcup S_n$.

The interest of the following result, given by A. H. Stone in 1948, lies in the fact that the given property actually characterizes semimetric spaces. This statement is made precise and proved in Theorem 10.3.1.

THEOREM 10.2.6. *Every semimetric space X has a σ-locally finite base for its topology.*

We begin by well-ordering X ⟦Theorem 10.2.1⟧. Thus $(X, \leq)$ is a well-ordered set. Take $x < y$ to mean $x \leq y$ and $x \neq y$. For each $x \in X$, and positive integers k, n, let

$$U(k, n, x) = N\left(x, \frac{1}{k} - \frac{1}{n}\right) \setminus \overline{\bigcup \left\{D\left(y, \frac{1}{k} - \frac{1}{n+1}\right): y < x\right\}}.$$

Then each $U(k, n, x)$ is open, (possibly empty, possibly equal to $N(x, 1/k - 1/n)$). For each fixed k, the set of all $U(k, n, x)$ is a cover of X. ⟦Let $z \in X$, and let x be the first member of X such that $z \in N(x, 1/k)$. Then $z \in U(k, n, x)$ if $1/n < 1/k - d(x, z)$, since if $y < x$, $z \notin N(y, 1/k) \supset D(y, 1/k - 1/(n+1))$.⟧ The set of all $U(k, n, x)$ is a base for the topology. ⟦Let $z \in X$ and $\varepsilon > 0$. Choose $k > 2/\varepsilon$ and n, x such that $z \in U(k, n, x)$. Then

$$U(k, n, x) \subset N\left(x, \frac{1}{k} - \frac{1}{n}\right) \subset N\left(x, \frac{\varepsilon}{2}\right) \subset N(z, \varepsilon)$$

since $z \in N(x, \varepsilon/2)$.⟧ Finally we have to show that the set of all $U(k, n, x)$ is σ-locally finite. Let $B(k) = \{U(k, n, x): n = 1, 2, \ldots; x \in X\}$. It is sufficient to show that each $B(k)$, $k = 1, 2, \ldots$, is σ-locally finite since the original set is $\bigcup B(k)$. Fix k; let $S_n = \{U(k, n, x): x \in X\}$ for $n = 1, 2, \ldots$. It is sufficient to show that S_n is locally finite for each n. We first note that for any x, y with $x \neq y$, $d[U(k, n, x), U(k, n, y)] \geq 1/(n(n + 1))$. ⟦We may assume $y < x$. Then $U(k, n, x) \not\pitchfork D(y, 1/k - 1/(n + 1))$, while $U(k, n, y) \subset N(y, 1/k - 1/n)$.⟧ It follows that if any cell has radius less than $1/(2n(n + 1))$ it could meet $U(k, n, x)$ for no more than one value of x. Thus S_n is locally finite. ∎

We saw ⟦Theorem 5.3.5⟧ that a regular Lindelöf space is normal. Lemma 10.2.2 is of this type, and has a similar proof, although it is not a generalization ⟦Sec. 10.3, Problem 2⟧. The reader should avoid memorizing the statement of Lemma 10.2.2 until he has read that of Theorem 10.3.1.

LEMMA 10.2.2. *If a regular space X has a σ-locally finite base $\mathscr{B}$, then X is normal.*

We shall use the criterion of Lemma 5.3.1. Let A, B be disjoint closed sets. For each x in A choose $V \in \mathscr{B}$ with $x \in V$ and $\bar{V} \not\pitchfork B$. The set of all such V, one for each $x \in A$, is an open cover, O, of A and since $O \subset \mathscr{B}$ we may write $O = \bigcup S_n$ where each S_n, $n = 1, 2, \ldots$, is a locally finite subset of $\mathscr{B}$. Let $V_n = \bigcup \{W: W \in S_n\}$ for each n. Then V_n is open and $\bar{V}_n \not\pitchfork B$. ⟦$\bar{V}_n = \bigcup \{\bar{W}: W \in S_n\}$, by Theorem 10.2.2, and each $\bar{W} \not\pitchfork B$.⟧ Repeating the argument with A replaced by B yields the sufficient condition of Lemma 5.3.1. ∎

Problems

In this list, X is a topological space.

1. Write out a proof that P, in Theorem 10.2.1, is a poset and that U is totally ordered.
2. $\{(n, n + 5): n = 1, 2, \ldots\}$ is a locally finite collection in $\mathbf{R}$.
3. Is the collection of all open intervals with integer end-points a locally finite collection in $\mathbf{R}$?
4. $\{(1/n, 2/n): n = 1, 2, \ldots\}$ is not a locally finite collection in $\mathbf{R}$, but is a locally finite collection in $(0, \infty)$.
5. $\mathbf{R}$, with its natural order, is not well ordered, but ω is.
6. "Locally finite" cannot be omitted in Theorem 10.2.3. ⟦Let S be a family of singletons.⟧
7. The criterion of Theorem 10.2.3 would not be sufficient if $\bar{A}$ were replaced by A, even if S has only two members! ⟦Consider the characteristic function of one of the two members.⟧

8. Describe a family S of sets in $\mathbf{R}$ and a number x such that every neighborhood of x meets a member of S, yet x is not an ω-accumulation point of S. (Thus Theorem 7.1.1 has no analogue in this setting.)
9. Let T, T' be topologies for a set X such that on each member V of a locally finite family of closed sets in (X, T) the relative topology of T is larger than that of T'. Show that $T \supset T'$. ⟦Apply Theorem 10.2.3 to the identity map.⟧
10. Let $Y \subset X$ and let S be a locally finite family in Y. If Y is closed, S must be a locally finite family in X.
11. "Closed" cannot be omitted in Problem 10. ⟦$X = \omega^+$, $Y = \omega$, S = family of all singletons in Y.⟧

---

101. A family S of sets in X is called *discrete* if every point of X has a neighborhood which meets at most one member of S. Show that if S is a discrete family of singletons in a T_1 space X, the union of the members of S is closed in X, and discrete (in its relative topology).
102. Show that a family of closed sets is discrete if and only if it is locally finite and disjoint.
103. A disjoint family S of open and closed sets in X need not be locally finite in X ⟦Problem 11⟧, but is discrete (hence locally finite) in $\bigcup\{W: W \in S\}$.
104. If X is locally connected, the family of components is discrete (hence locally finite) ⟦Problem 103 and Theorem 5.2.5⟧. The converse fails ⟦Sec. 5.2, Problem 115⟧.
105. If X is the union of a disjoint family S of open and closed sets and $f: X \to Y$ has the property that $f \mid A$ is continuous for each $A \in S$, then f is continuous ⟦Theorem 10.2.3 and Problem 103⟧.
106. Let X be locally connected, and assume that $f: X \to Y$ has the property that $f \mid C$ is continuous for each component C of X. Show that f is continuous ⟦Problems 104, 105, and Theorem 5.2.5⟧.
107. "Locally connected" cannot be omitted in Problem 104 ⟦Problem 6⟧.
108. Let X be pseudocompact and Y a subset which is the closure of its interior. Then Y is pseudocompact ⟦Theorem 10.2.5 and Problem 10⟧. It is not sufficient to assume Y closed ⟦Sec. 7.1, Problem 116⟧.
109. A T_1 space has a locally finite base if and only if it is discrete.
110. Let X be a topological space which is not pseudocompact. Show that X contains a discrete disjoint infinite family of open subsets. ⟦See the proof of Theorem 10.2.5.⟧
111. A family of sets is called *σ-discrete* if it is a countable union of discrete families. Show that a semimetric space has a σ-discrete base. ⟦See the proof of Theorem 10.2.6.⟧
112. Let X be a $T_{3\frac{1}{2}}$ space. Show that βX cannot be first countable at any $t \in \beta X \setminus X$. ⟦If $\{G_n\}$ is a base of open neighborhoods of t, let

$x_n \in G_n \cap X$ and let $f \in C^*(X)$ with $f(x_n) = (-1)^n$. This is possible by Theorem 10.2.4. Clearly f has no continuous extension to t.⟧

113. Let S be a locally finite cover of X and let $A \subset X$. Show that if $A \cap \overline{W}$ is closed for each $W \in S$, then A is closed. ⟦$\{A \cap \overline{W}: W \in S\}$ is locally finite by Lemma 10.2.1. Hence, by Corollary 10.2.1, $\bigcup \{A \cap \overline{W}\}$ is closed.⟧
114. In Problem 113 it is sufficient to assume about S that it is locally finite and covers a dense subset of X ⟦Theorem 10.2.2⟧.
115. If X has a locally finite cover consisting of compact closed sets, X must be a k space ⟦Problem 113⟧. (Compare, also, Theorem 10.2.3 with Sec. 8.1, Problem 120.)
116. Let X be completely regular. Show that X is pseudocompact if and only if whenever $\{G_n\}$ is a shrinking sequence of nonempty open sets, $\bigcap \overline{G}_n \neq \varnothing$. ⟦If f is unbounded, let $G_n = (|f| > n)$. If $\bigcap \overline{G}_n = \varnothing$, apply Theorem 10.2.5 to $\{G_n \setminus \overline{G}_{n+1}\}$, for each x, there exists n such that $\tilde{\overline{G}}_n$ is a neighborhood of x.⟧
117. A T_2 compactification of a pseudocompact space cannot be first countable at any added point ⟦Problem 116⟧.
118. Let X be a locally compact, σ-compact T_2 space. Then if X is pseudocompact it is compact. ⟦X^+ is first countable at ∞ by Sec. 8.1, Problems 125 and 113; now apply Problem 117. Another proof is to cite Sec. 8.5, Problem 103; Theorem 5.3.5 and 5.4.1.⟧
119. The neighborhood system of a point x in a T_1 space is well ordered under inclusion if and only if x is isolated.
120. Let C_1, C_2 be covers of X. We say that C_2 *refines* C_1 or C_2 is a *refinement* of C_1 if every member of C_2 is included in some member of C_1. In particular a cobase for a cover is refined by the cover. Show that a refinement of a base for a topology need not be a base. ⟦Let C_2 be all intervals of unit length in **R**.⟧
121. Let X be a semimetric space and $A_n = \{N(x, 1/n): x \in X\}$ for $n = 1, 2, \ldots$. For each n let B_n be a refinement of A_n. Show that $\bigcup B_n$ is a base for X. ⟦Keep in mind that each B_n is a cover.⟧
122. A topological space is called *paracompact* if it is regular and has the property that every open cover has a locally finite refinement which is also an open cover. Show that a compact regular space is paracompact.
123. Let a topological space X be a disjoint union of paracompact subspaces each of which is open and closed in X. Show that X is paracompact. ⟦Take a refinement on each of the subspaces.⟧ In particular a discrete space is paracompact.
124. Paracompact is F-hereditary. ⟦Add $\tilde{F}$ to an open cover of F, refine, and remove $\tilde{F}$.⟧
125. Every open cover C of a paracompact space has a locally finite refine-

ment which is a closed cover; that is, a cover by closed sets. ⟦For each $x \in U \in C$, let $x \in V$, $\bar{V} \subset U$ with V open. Refine the set of all V, and use Lemma 10.2.1.⟧

---

201. Let $\{G_n\}$ be a locally finite disjoint sequence of open sets in X. Suppose that for each n there is given continuous $f_n: X \to [0, 1]$ with $f_n = 0$ on $\tilde{G}_n$ and $f_n = 1$ somewhere. Let $g_n = \sum \{f_k: k = 1, 2, \ldots, n\}$. Show that $\{g_n\}$ is equicontinuous.
202. Every set X can be well ordered in such a way that for every $a \in X$, $S_a = \{x: x < a\}$ has smaller cardinality than X. In such a case, X is said to be *initially ordered.* ⟦Well-order X. Let y be the smallest member of X such that $|S_y| = X$. Then S_y is initially ordered.⟧ (The applications given in the next two problems assume the continuum hypothesis.)
203. Well-order $[0, 1]$ in such a way that for every $a \in [0, 1]$, $\{x: x < a\}$ has Lebesgue measure 0 ⟦Problem 202⟧.
204. $\mathbf{R}^3$ is a union of disjoint copies of S_1. ⟦Problem 202. At each stage a circle may be drawn avoiding countably many points.⟧ (Such a decomposition of $\mathbf{R}^3$ must be very pathological. See [Jones].)
205. True or false? A subset of $\mathbf{R}$ is well ordered in the usual ordering of $\mathbf{R}$ if and only if it is RHO-discrete.

10.3. Metrization

In this section we give a characterization of those topological spaces which are semimetrizable; this is the Nagata–Smirnov theorem, which was given by S. Nagata in 1950 and Y. M. Smirnov in 1951.

In Section 10.1 we metrized a space by mapping it into a countable product of spaces, all of these spaces being $[0, 1]$. They could just as well be taken to be $\mathbf{R}$. Since such a product is separable, this particular technique is restricted to separable spaces. We might try to generalize the result of Section 10.1 by repeating that construction with an arbitrary base, not necessarily countable. This would indeed give a description of the topology of X as the weak topology by maps into $\mathbf{R}^A$ for some A, (or, as in Corollary 10.1.2, an embedding in $\mathbf{R}^A$), but would not yield metrizability since A need not be countable. What is actually done is very nice indeed! In outline, we replace $\mathbf{R}^A$ by a countable product in this way. Write $A = \bigcup_{n=1}^{\infty} S_n$, (each S_n may be uncountable), then consider $\prod_{n=1}^{\infty} L[S_n]$, where each $L[S_n]$ is a space resembling $\mathbf{R}^{S_n}$, (actually a subset of $\mathbf{R}^{S_n}$), with a (metrizable) topology. It stands to reason that the metrizable space $\prod L[S_n]$ should be something like a subspace of $\prod \mathbf{R}^{S_n} = \mathbf{R}^{\cup S_n} = \mathbf{R}^A$.

This ends the preliminary description and we turn to the details. Let S be a nonempty set and $u: S \to \mathbf{R}$. The *support* of u is defined to be

$\{x \in S: u(x) \neq 0\}$, and u will be said to have *countable support* if its support is countable. (This includes finite and empty support as special cases.) A function $u: S \to \mathbf{R}$ will be called *summable* if it has countable support and $\sum \{|u(x)|: u(x) \neq 0\} < \infty$. Let $L[S]$ be the set of all summable functions from S to $\mathbf{R}$, and for $u \in L[S]$ set $\|u\| = \sum\{|u(x)|: u(x) \neq 0\}$, $\|u\| = 0$ if $u = 0$. ⟦The sum of a countable set of positive numbers is independent of the way in which these numbers are arranged in a sequence.⟧ ($\|u\|$ is pronounced norm u.) It is clear that $L[S]$ contains all functions with finite support. Finally, for $u, v \in L[S]$, set $d(u, v) = \|u - v\|$. [Clearly $u - v$ has countable support.] Then d is a metric for $L[S]$. Future references to $L[S]$ will be to the metric space $(L[S], d)$. A final remark: $0 \in L[S]$ stands for the (constant) function taking every point in S to $0 \in \mathbf{R}$; $\|0\| = 0$.

LEMMA 10.3.1. *Let S be a locally finite family of sets in a topological space X. Suppose that for each $V \in S$ there exists a continuous $f_V: X \to \mathbf{R}$ with $f_V = 0$ on $\tilde{V}$. For $x \in X$, define $h(x): S \to \mathbf{R}$ by $h(x)(V) = f_V(x)$ for all $V \in S$. Then h is a continuous map from X into $L[S]$.*

For each x, $h(x)$ has finite support. ⟦If $h(x)(V) \neq 0$, then $f_V(x) \neq 0$ and so $x \in V$. This is true for only finitely many V.⟧ Thus $h(x) \in L[S]$ and so $h: X \to L[S]$. Also h is continuous. ⟦Let $x_\delta \to x \in X$. Let N be a neighborhood of x meeting only finitely many $V \in S$. If N meets none, then $h = 0$ on N and so $h(x_\delta) = 0 = h(x)$ eventually. If N meets only $V_1, V_2, \ldots, V_k$, then, as soon as $x_\delta \in N$,

$$d[h(x_\delta), h(x)] = \sum_{i=1}^{k} |h(x_\delta)(V_i) - h(x)(V_i)|$$

$$= \sum_{i=1}^{k} |f_{V_i}(x_\delta) - f_{V_i}(x)| \to 0$$

since k is fixed and each f_{V_i} is continuous.⟧ ∎

The reader will not fail to notice the great similarity in the proofs of Theorem 10.3.1 and 10.1.1.

THEOREM 10.3.1. *A topological space is semimetrizable if and only if it is regular and has a σ-locally finite base.*

Half of this is Theorem 10.2.6. Conversely, suppose that X is a regular space and that B is a σ-locally finite base; say $B = \bigcup B_n$, each B_n locally finite. Fix positive integers m, n. We shall set up a continuous map $h_{mn}: X \to L[B_n]$ in the following way. Let $V \in B_n$. Let

$$A = \bigcup \{W: W \in B_m, \overline{W} \subset V\}.$$

Then $\bar{A} \subset V$ [[Theorem 10.2.2]], and since X is normal [[Lemma 10.2.2]], Urysohn's lemma (Theorem 4.2.11) supplies a continuous $f: X \to [0, 1]$ with $f = 1$ on A, $f = 0$ on $\tilde{V}$. (If A is empty, take $f = 0$.) Let $h_{mn}(x)(V) = f(x)$ for $x \in X$. In Lemma 10.3.1, we proved that h_{mn} is a continuous map of X into $L[B_n]$. Let H be the family of all h_{mn}, $m, n = 1, 2, \ldots$; then H is countable. We are going to show that $w(H)$, the weak topology by H is identical with the topology T of X. This will complete the proof. [[$w(H)$ is semimetrizable by Theorem 6.3.4.]] Surely $T \supset w(H)$. [[$w(H)$ is the smallest topology making all members of H continuous.]] Conversely, let F be a T-closed subset of X. We shall show that F is $w(H)$ closed. [[Let $x \notin F$. Then $\tilde{F}$ is a T-neighborhood of x, hence there exists $V \in B$ with $x \in V \subset \tilde{F}$. Say $V \in B_n$. Since X is regular, there exists $W \in B$ with $x \in W$, $\overline{W} \subset V$. Say $W \in B_m$. (Possibly $m = n$.) Then for any $y \in F$,

$$d[h_{mn}(y), h_{mn}(x)] = \|h_{mn}(y) - h_{mn}(x)\|$$

$$\geq |h_{mn}(y)(V) - h_{mn}(x)(V)| = |h_{mn}(x)(V)| = 1$$

since $x \in \bigcup \{W: W \in B_m, \overline{W} \subset V\}$, the set on which $f = f_V$ is 1. Thus $d[h_{mn}[F], h_{mn}(x)] \geq 1$ and so, since h_{mn} is $w(H)$-continuous, x cannot belong to the $w(H)$ closure of F.]] Hence $T \subset w(H)$, and so, finally $T = w(H)$. ∎

COROLLARY 10.3.1. *A T_3 space with a σ-locally finite base is homeomorphic with a subspace of $\prod \{L_\alpha: \alpha \in A\}$ for some countable set A, where each $L_\alpha = L(B_\alpha)$ for some set B_α.*

This is proved exactly as Corollary 10.1.2. ∎

Problems

1. Write out the proof of Corollary 10.3.1.
2. Let X be a T_4 space. Then X may be Lindelöf without having a σ-locally finite base and conversely [[Theorems 10.3.1 and 10.2.6]].
3. Describe $L[S]$ if S is finite.
4. Deduce the metrization theorem for second countable spaces from Theorem 10.3.1.

---

101. If a space has a σ-locally finite base it is first countable. (For regular spaces this follows from Theorem 10.3.1.)

---

201. Show that $L[S]$ is complete.
202. Can "disjoint" be omitted in Sec. 10.1, Problem 104? [[MR *13*, 264, Nagata.]]

Uniformity

11.1 Uniform Space

The concept of uniform space lies between those of semimetric and topological space, in the sense that every semimetric space is a uniform space, and every uniform space is a topological space. In the context of uniform space we can deal with certain nontopological concepts of classical analysis such as completeness and uniform convergence. This yields a setting for this type of analysis which is significantly wider than that of semimetric space, including, for example, topological groups. There is also the advantage over semimetric space, which is purely a matter of taste, that uniformity may be introduced without reference to the real number system, a fact that upset some widely held beliefs to the contrary, when, in 1937, the concept of uniformity was introduced by André Weil.

A crucial property of a semimetric space X is that it is possible to talk about the "same" neighborhood of two different points: we may refer to a cell of radius r and center x, and another cell of radius r and center y. This is essentially the reason that it is possible to define a Cauchy filter, namely for every $\varepsilon > 0$, there is a set in the filter lying in some cell of radius ε. What happens is that each positive real number ε designates a neighborhood for each point in the space, namely, with x we associate $N(x, \varepsilon)$. We phrase this formally:

DEFINITION 1. *Let X be a set. A connector for X is a map $U: X \to 2^X$ (that is, for each x, $U(x)$ is a subset of X), such that for each x, $x \in U(x)$. (But see the important remark following Lemma* 11.1.1, *below.)*

Thus, in semimetric space, we may, for each $\varepsilon > 0$, speak of the connector N_ε, namely $N_\varepsilon(x) = N(x, \varepsilon)$, the cell of center x, radius ε. Each connector U for X defines a subset, which we may call U also, of $X \times X$, namely $U = \{(x, y): y \in U(x)\}$; moreover, $U \supset \Delta$, the diagonal of $X \times X$, $\Delta = \{(x, x): x \in X\}$. Conversely, every subset U of $X \times X$ which includes Δ leads to a connector by the formula $U(x) = \{y: (x, y) \in U\}$. Thus it is a matter of indifference whether the theory is set up by postulating the existence of connectors or of certain distinguished subsets of $X \times X$. We shall follow the latter route. We begin with some notation. Given a connector U; equivalently, a subset of $X \times X$ which includes the diagonal; let

$$U^{-1} = \{(x, y): (y, x) \in U\} = \{(x, y): x \in U(y)\}.$$

Thus $y \in U^{-1}(x)$ if and only if $x \in U(y)$. For a set S in X and connector U, $U[S] = \{U(x): x \in S\}$. For connectors U, V, the connector $U \circ V$ is defined by $(U \circ V)(x) = U[V(x)]$. Thus $y \in (U \circ V)(x)$; equivalently, $(x, y) \in U \circ V$; if and only if $y \in U(a)$ for some $a \in V(x)$; equivalently, $(a, y) \in U$, $(x, a) \in V$ for some a; the latter formulation being harder to remember. Some simple properties of these notations are listed in Problems 1–6, and we shall use these immediately without citation.

DEFINITION 2. *Let X be a set. A uniformity for X is a collection $\mathscr{U}$ of subsets of $X \times X$ satisfying*

(i) *$\mathscr{U}$ is a filter;*
(ii) *$\Delta \subset U$ for every $U \in \mathscr{U}$, where Δ is the diagonal;*
(iii) *$U \in \mathscr{U}$ implies $U^{-1} \in \mathscr{U}$;*
(iv) *for each $U \in \mathscr{U}$ there exists $V \in \mathscr{U}$ with $V \circ V \subset U$.*

A base for a uniformity is a collection $\mathscr{B}$ of subsets of $X \times X$ satisfying

(i′) *$\mathscr{B}$ is a filterbase;*
(ii′) *$\Delta \subset U$ for every $U \in \mathscr{B}$;*
(iii′) *for each $U \in \mathscr{B}$, there exists $V \in \mathscr{B}$ with $V^{-1} \subset U$,*
(iv′) *for each $U \in \mathscr{B}$, there exists $V \in \mathscr{B}$ with $V \circ V \subset U$.*

A pair $(X, \mathscr{U})$, where $\mathscr{U}$ is a uniformity for X is called a *uniform space*.

LEMMA 11.1.1. *Let $\mathscr{B}$ be a base for a uniformity on X. Let $\mathscr{U}$ be the filter on $X \times X$ generated by $\mathscr{B}$. Then $\mathscr{U}$ is a uniformity, and $\mathscr{U}$, $\mathscr{B}$ are related by the conditions.*

$$\mathscr{B} \subset \mathscr{U}, \text{ and every member of } \mathscr{U} \text{ includes a member of } \mathscr{B}. \quad (11.1.1)$$

Conversely, if $(X, \mathscr{U})$ is a uniform space and $\mathscr{B}$ is a collection of subsets satisfying condition (11.1.1), *then $\mathscr{B}$ is a base for a uniformity.*

By definition, $\mathscr{U} = \{U: U \supset S \text{ for some } S \in \mathscr{B}\}$. Then $\mathscr{U}$ is a filter and Conditions (ii), (iii), (iv) of Definition 2 are easy to check. ⟦Use Problem 3.⟧ To prove the converse proposition, we first check that $\mathscr{B}$ is a filterbase. First

$\varnothing \notin \mathscr{B}$ since $\varnothing \notin \mathscr{U}$; $\mathscr{B} \neq \varnothing$, [[by (condition 11.1.1)]], also if $U, V \in \mathscr{B}$, choose $W \in \mathscr{U}$ with $W \subset U \cap V$. It follows from (11.1.1) that $U \cap V$ includes a member of $\mathscr{B}$. Thus $\mathscr{B}$ is a filterbase. Condition (ii′) of Definition 2 is trivial; Condition (iii′) is easy. To check Condition (iv′), let $U \in \mathscr{B}$. Choose $W \in \mathscr{U}$ with $W \circ W \subset U$, and $V \in \mathscr{B}$ with $V \subset W$. Then $V \circ V \subset U$. ∎

When $\mathscr{U}, \mathscr{B}$ are related as in Lemma 11.1.1 we say that $\mathscr{U}$ is the uniformity *generated by* $\mathscr{B}$ and that $\mathscr{B}$ is a *base for* $\mathscr{U}$.

In Definition 2, Condition (ii) is equivalent to the condition $x \in U(x)$, in other words each member of a uniformity is a connector.

REMARK. When a uniform space $(X, \mathscr{U})$ is specified, the phrase: "let U be a connector" shall mean "let $U \in \mathscr{U}$."

In Definition 2, Condition (iv) will appear in situations where in classical analysis, inequalities are chosen using $\varepsilon/2$ instead of ε. A classical argument ending: "$A + B \leq \varepsilon/2 + \varepsilon/2 = \varepsilon$" will appear in uniform spaces as "$(x, y) \in V$, $(y, z) \in V$ hence $(x, z) \in V \circ V \subset U$." Compare for example the proofs of Lemma 11.3.1, and Lemma 9.1.1. In the following Example, it will be seen that Condition (iv) is related to the triangle inequality, which is the essential tool in the solution of Problem 6; also in Lemma 12.1.1 it will be shown that Condition (iv) is intimately related to the continuity of binary operations, such as addition.

★EXAMPLE 1. Let (X, d) be a semimetric space. For each $\varepsilon > 0$, let $U_\varepsilon = \{(x, y): d(x, y) < \varepsilon\}$, and let $\mathscr{B} = \{U_\varepsilon: \varepsilon > 0\}$. Then $\mathscr{B}$ is a base for a uniformity. To prove this we check first that $\mathscr{B}$ is a filterbase. [[In Sec. 3.2, Definition 2, only Condition (iii) is nontrivial. Let U_α, $U_\beta \in \mathscr{B}$, and let $\varepsilon = \frac{1}{2} \cdot \min(\alpha, \beta)$. Then, as in Problem 6, $U_\varepsilon \circ U_\varepsilon \subset U_{2\varepsilon} \subset U_\alpha \cap U_\beta$ since $2\varepsilon \leq \alpha$, $2\varepsilon \leq \beta$.]] In Definition 2, Conditions (ii′) and (iii′) are trivial [[$U_\varepsilon^{-1} = U_\varepsilon$]], and Condition (iv′) follows from $U_{\varepsilon/2} \circ U_{\varepsilon/2} \subset U_\varepsilon$ [[Problem 3]]. The uniformity generated by $\mathscr{B}$ is called the *semimetric uniformity*, also the uniformity *induced* by the semimetric d. If a uniformity has the property that there exists a semimetric which induces it, it is said to be *semimetrizable*. We shall see many examples of uniformities which are not semimetrizable in Section 11.4.

REMARK. We shall refer to certain spaces as having their natural uniformity. This will be the uniformity induced by the natural metric (Section 2.4). For example, when **R** is named without comment, it is assumed to be a uniform space with the uniformity induced by d where $d(x, y) = |x - y|$. This is the *Euclidean uniformity*.

Let $(X, \mathscr{U})$ be a uniform space. Then $\mathscr{U}$ *induces* a topology in the following way. (Compare Section 2.3.) We declare the empty set to be open; moreover, if $G \subset X$, G is called open if, for every $x \in G$, there exists $U \in \mathscr{U}$ such that $U(x) \subset G$. ⟦Certainly $\varnothing$ and X are open. Next suppose that G_1, G_2 are open sets and $x \in G_1 \cap G_2$. For $i = 1, 2$ choose $U_i \in \mathscr{U}$ with $U_i(x) \subset G_i$. Let $V = U_1 \cap U_2$; $V \in \mathscr{U}$ since $\mathscr{U}$ is a filter, also $V(x) \subset U_1(x) \cap U_2(x) \subset G_1 \cap G_2$. (See Problem 3.) Thus $G_1 \cap G_2$ is open. Finally, let Σ be a collection of open sets and let $x \in \bigcup \{G: G \in \Sigma\}$. Then $x \in S$ for some $S \in \Sigma$, and so there exists $U \in \mathscr{U}$ with $U(x) \subset S$. Then $U(x) \subset \bigcup \{G: G \in \Sigma\}$.⟧

We shall now apply topological concepts to a uniform space $(X, \mathscr{U})$ with the understanding that they apply to (X, T), where T is the topology induced by $\mathscr{U}$. In certain cases this would be ambiguous; for example if we say that X is metrizable, this could mean that X has a metrizable topology or a metrizable uniformity; these are not the same thing. In such cases we shall exercise a little more care in specifying exactly which is meant.

★EXAMPLE 2. Let d_1 be the Euclidean metric for $\mathbf{R}$, $d_2 = 2d_1$, and let $d_3(x, y) = |x/(1 + |x|) - y/(1 + |y|)|$. These metrics are all equivalent. The equivalence between d_1, d_3 follows from the fact that with $f(x) = x/(1 + |x|)$, $f: \mathbf{R} \to (-1, 1)$ is a homeomorphism onto ⟦Sec. 4.2, Example 5⟧, and so $d_1(x_n, x) \to 0$ if and only if $d_3(x_n, x) \to 0$. ⟦$d_3(x, y) = d_1(fx, fy)$.⟧ Moreover, d_1, d_2 induce the same uniformity since each cell in one is a cell in the other; we say that they are *uniformly equivalent*. However d_1, d_3 do not induce the same uniformity. Let the respective uniformities be $\mathscr{U}_1, \mathscr{U}_3$. First $\mathscr{U}_1 \supset \mathscr{U}_3$ ⟦Problem 9⟧. Next, let $U = \{(x, y): d_1(x, y) < 1\}$. Then $U \in \mathscr{U}_1$, but $U \notin \mathscr{U}_3$. ⟦Let $r > 0$ and $V_r = \{(x, y): d_3(x, y) < r\}$. Let x be a large positive number, and $y = x + 2$. Then $(x, y) \notin U$, and $(x, y) \in V_r$ if x is large enough since $d_3(x, y) = d_3(x, x + 2) = 2/(3 + x)(1 + x) < r$ if $x^2 > 2/r$.⟧ We say that d_1 is *uniformly stronger* than d_3 since $\mathscr{U}_1 \supset \mathscr{U}_3$, and *strictly* so, since $\mathscr{U}_1 \neq \mathscr{U}_3$. Since d_1 and d_3 are equivalent, $\mathscr{U}_1$ and $\mathscr{U}_3$ induce the same topology. For this reason we call them *equivalent uniformities*. As we shall see in Section 11.2, these concepts can all be phrased in terms of uniform continuity of the identity map. The important points to be noticed are (a) *the same topology may be induced by two different uniformities*; (b) *the same uniformity may be induced by two different semimetrics.*

If $\mathscr{U}, \mathscr{V}$ are uniformities for a set X, and $T_{\mathscr{U}}, T_{\mathscr{V}}$ are the induced topologies, we say that $\mathscr{U}$ is *stronger* than $\mathscr{V}$ if $T_{\mathscr{U}} \supset T_{\mathscr{V}}$, and $\mathscr{U}$ is *equivalent* to $\mathscr{V}$ if $T_{\mathscr{U}} = T_{\mathscr{V}}$.

It is tiresome to remember whether $x \in U(y)$ or $y \in U(x)$ if $(x, y) \in U$. The burden of remembering the distinction is lifted by means of Theorem 11.1.1. We call a connector U *symmetric* if $U = U^{-1}$; thus $(x, y) \in U$ if and only if $(y, x) \in U$, and $y \in U(x)$ if and only if $x \in U(y)$. Working with symmetric connectors will avoid such pathology as is expressed in Problem 116.

THEOREM 11.1.1. *Every uniformity $\mathscr{U}$ has a base of symmetric connectors.*

Let $\mathscr{B}$ be the collection of all symmetric connectors in $\mathscr{U}$. Then $\mathscr{B}$ is the required base. To see this we check that $\mathscr{B}$ satisfies (11.1.1) in Lemma 11.1.1. That $\mathscr{B} \subset \mathscr{U}$ is by definition; next, given $U \in \mathscr{U}$, let $V = U \cap U^{-1}$. Then $V \in \mathscr{B}$ ⟦Problems 1 and 5⟧ and $V \subset U$. ▮

THEOREM 11.1.2. *Let $(X, \mathscr{U})$ be a uniform space, let $\mathscr{B}$ be a base for $\mathscr{U}$ and let $x \in X$. Then $\{U(x): U \in \mathscr{B}\}$ is a base at x (for the neighborhood system of x in the topological space X).*

We must first show that each $U(x)$ is a neighborhood of x. Fix $U \in \mathscr{U}$ and let $G = \{y: V(y) \subset U(x) \text{ for some } V \in \mathscr{U}\}$. We shall show that $x \in G \subset U(x)$ and that G is open. First $x \in G$. ⟦$x \in U(x) \subset U(x)$.⟧ Next $G \subset U(x)$. ⟦For $y \in G$, $y \in V(y) \subset U(x)$ for some V.⟧ Finally, to show that G is open, let $g \in G$. Then there exists V such that $V(g) \subset U(x)$. Choose $W \in \mathscr{U}$ with $W \circ W \subset V$. Then $W(g) \subset G$. ⟦For $y \in W(g)$, $W(y) \subset U(x)$ since $z \in W(y)$ implies $z \in W \circ W(g) \subset V(g) \subset U(x)$. Thus $y \in G$.⟧ Next we must show that every neighborhood N of x includes $U(x)$ for some $U \in \mathscr{B}$. By definition of neighborhood and open set in a uniform space, there exists a connector U_1 with $U_1(x) \subset N$; then, since $\mathscr{B}$ is a base, there exists $U \in \mathscr{B}$ with $U \subset U_1$, so that $U(x) \subset N$. ▮

We now obtain an important formula for the closure of a set in a uniform space $(X, \mathscr{U})$.

THEOREM 11.1.3. *Let A be a set in a uniform space $(X, \mathscr{U})$. Then $\bar{A} = \bigcap \{V(A): V \in \mathscr{B}\}$, where $\mathscr{B}$ is any base for $\mathscr{U}$.*

Let $x \in \bar{A}$. For each $V \in \mathscr{B}$ we may, by Theorem 11.1.1, choose a symmetric connector $W \subset V$. Then $W(x)$ meets A. ⟦It is a neighborhood of x by Theorem 11.1.2.⟧ Let $y \in A \cap W(x)$. Then $x \in W(y) \subset V(y) \subset V(A)$. Conversely, suppose that $x \notin \bar{A}$. By Theorems 11.1.1 and 11.1.2, there exists a symmetric connector U with $U(x) \not\cap A$; and $\mathscr{B}$ contains a connector V with $V \subset U$. It follows that $x \notin V(A)$. ⟦If $x \in V(A)$, then $x \in U(A)$; say $x \in U(a)$, $a \in A$; then $a \in U(x) \cap A$.⟧ Thus $x \notin \bigcap \{V(A): V \in B\}$. ▮

EXAMPLE 3. There are reasons for picturing a connector for X as a union of vertical "line segments" rather than as an amorphous "neighborhood of the diagonal." For a connector U, let $L_x = x \times U(x)$ for $x \in X$. Then in the usual diagram of $X \times X$, L_x is a subset of a vertical line (a copy of X), also $U = \bigcup \{L_x : x \in X\}$. ⟦$(x, y) \in U$ implies $y \in U(x)$ hence $(x, y) \in L_x$, and conversely.⟧ As an example, let $X = [0, 1]$, let v be the vertical segment $\{(0, y): \frac{1}{2} < y \leq 1\}$ and let h be the horizontal segment $\{(x, 0): \frac{1}{2} < x \leq 1\}$. Let U be the connector $(X \times X) \setminus (v \cup h)$. Then U is dense in $I \times I$ with the product topology. Next, let P be the point $(0, \frac{3}{4})$. We shall see that there exists a connector V such that $P \notin V \circ U$, a somewhat paradoxical fact if we view V as a neighborhood of Δ; thinking instead of V as a union of vertical lines we see that if we choose $V = \{(x, y): |y - x| < \frac{1}{8}\}$ we shall have $P \notin V \circ U$. Indeed $V \circ U = (X \times X)\setminus v'$, where $v' = \{(0, y): \frac{5}{8} < y \leq 1\}$ as we see by adding on to U all possible vertical line segments of length $\frac{1}{8}$, each one meeting U. Further properties of this Example are given in Problems 113–117. (It would also be possible to have made the definition so as to introduce a horizontal, rather than a vertical bias, or to have maintained a two-sided point of view.)

THEOREM 11.1.4. *Every uniform space* $(X, \mathscr{U})$ *is regular. Moreover the following conditions are equivalent.*
(i) *X is a T_0 space.*
(ii) *X is a T_3 space.*
(iii) $\bigcap \{U: U \in \mathscr{U}\} = \Delta$.

Let N be a neighborhood of a point x. Choose connectors U with $U(x) \subset N$ and V with $V \circ V \subset U$. Then $\overline{V(x)} \subset V[V(x)]$ ⟦Theorem 11.1.3⟧ $\subset U(x) \subset N$. Thus N includes a closed neighborhood of x, hence X is regular. It is now clear that (i) ⇒ (ii) in the statement of the theorem. Next, let X be T_3, and $y \neq x$. Then $y \notin \overline{\{x\}}$, hence, by Theorem 11.1.3 (with $A = \{x\}$), $y \notin U(x)$ for some symmetric connector U ⟦Theorem 11.1.1⟧. But then $(x, y) \notin U$. This proves (iii). Finally, assuming (iii), let $x \neq y$. Then there exists U with $(x, y) \notin U$. Thus $y \notin U(x)$ and so X is a T_0 space. ▮

A uniform space satisfying the three conditions of Theorem 11.1.4 is said to be *separated.* We shall speak of a *separated uniform space*, and a *separated uniformity*. In particular a separated uniform space is a Hausdorff space. We shall see in Theorem 11.5.2 that a uniform space is always completely regular so that Condition (ii) of Theorem 11.1.4 can be raised to $T_{3\frac{1}{2}}$.

REMARK. Theorem 11.1.3 expresses $\bar{A}$ as the intersection of certain neighborhoods of A. The intersection of *all* neighborhoods of A is equal to $\bar{A}$ if A is compact ⟦Theorem 11.1.4; Sec. 5.4, Problem 113⟧, but in general is smaller than $\bar{A}$ ⟦Sec. 4.1, Problem 114⟧.

Since a connector U for a uniform space X is a subset of $X \times X$, it is possible to discuss $\bar{U}$, the closure of U in the product topology. If $\bar{U} = U$, we call U a *closed connector*.

THEOREM 11.1.5. *Let U be a connector in a uniform space* $(X, \mathscr{U})$ *and let* $\mathscr{B}$ *be a base for* $\mathscr{U}$. *Then* $\bar{U} = \bigcap \{V \circ U \circ V: V \in \mathscr{B}\}$.

Suppose first that $(a, b) \notin \bar{U}$. There is a neighborhood N of (a, b) in the product topology which does not meet U, and this neighborhood includes $W(a) \times W(b)$ for some symmetric connector W. ⟦$N \supset P_1^{-1}N_1 \cap P_2^{-1}N_2$ with N_1, N_2 neighborhoods of a, b by definition. By Theorem 11.1.2, there exist connectors W_1, W_2 with $W_1(a) \subset N_1$, $W_2(b) \subset N_2$; and by Theorem 11.1.1, there exists symmetric W with $W \subset W_1 \cap W_2$.⟧ It follows that $(a, b) \notin W \circ U \circ W$. ⟦Otherwise $x \in W(a)$, $y \in U(x)$, $b \in W(y)$ hence $(x, y) \in [W(a) \times W(b)] \cap U \subset N \cap U$.⟧ Now choose $V \in \mathscr{B}$ with $V \subset W$ and obtain $(a, b) \notin V \circ U \circ V$. Conversely, let $(a, b) \in \overline{U}$ and $V_1 \in \mathscr{B}$. Choose symmetric $V \subset V_1$. Now $V(a) \times V(b)$ is a neighborhood of (a, b) in the product topology; hence, it meets U; say $(x, y) \in [V(a) \times V(b)] \cap U$. Then $a \in V(x)$, $x \in U(y)$, $y \in V(b)$ and so $(a, b) \in V \circ U \circ V \subset V_1 \circ U \circ V_1$. ∎

THEOREM 11.1.6. *Every uniformity $\mathscr{U}$ has a base of symmetric closed connectors.*

Let $\mathscr{B}$ be the collection of all symmetric closed connectors in $\mathscr{U}$. We check Condition (11.1.1) in Lemma 11.1.1. That $\mathscr{B} \subset \mathscr{U}$ is by definition. Next, let $U \in \mathscr{U}$. Choose symmetric V with $V \circ V \circ V \subset U$ ⟦Problem 10⟧. Then $\overline{V} \subset V \circ V \circ V$ ⟦Theorem 11.1.5⟧ $\subset U$ and the proof is concluded by showing that $\overline{V} \in \mathscr{B}$. Since $\overline{V}$ is closed, it is sufficient to show that it is symmetric. Now by Theorem 11.1.5, $\overline{V} = \bigcap \{W \circ V \circ W : W \in S\}$, where S is the set of symmetric connectors ⟦Theorem 11.1.1⟧. Each $W \circ V \circ W$ is symmetric ⟦Problem 5⟧ and since any intersection of symmetric connectors is symmetric the result follows. ∎

Problems

In this list X is a uniform space, U, V are connectors (belonging to the uniformity), A, B, G, K are subsets of X.

★1. $(U^{-1})^{-1} = U$.
★2. $U \subset V$ implies $U(x) \subset V(x)$ for all x.
★3. $U \subset V$ implies $U^{-1} \subset V^{-1}$.
★4. If $W = U \cap V$, $W(x) = U(x) \cap V(x)$ for all x.
★5. $(U \cap V)^{-1} = U^{-1} \cap V^{-1}$ and $(U \circ V)^{-1} = V^{-1} \circ U^{-1}$.
★6. Let (X, d) be a semimetric space and $U_\varepsilon = \{(x, y) : d(x, y) < \varepsilon\}$ for $\varepsilon > 0$. Show that $U_\varepsilon \circ U_\delta \subset U_{\varepsilon + \delta}$.
★7. $U \subset V \circ U$ and $U \subset U \circ V$.
★8. If $\mathscr{B}$ is a base for uniformities $\mathscr{U}_1$ and $\mathscr{U}_2$, then $\mathscr{U}_1 = \mathscr{U}_2$. ⟦Each is the unique filter generated by the filterbase $\mathscr{B}$.⟧
★9. If d_1, d_2 are semimetrics for a set X and $d_1 \geq d_2$, then $\mathscr{U}_1 \supset \mathscr{U}_2$, where $\mathscr{U}_1$, $\mathscr{U}_2$ are the semimetric uniformities. ⟦Let $U \in \mathscr{U}_2$. Then $U \supset V_\varepsilon^2$ in the notation of Example 1. The superscript refers to d_2. Now $V_\varepsilon^2 \supset V_\varepsilon^1$ since $d_1(x, y) < r$ implies $d_2(x, y) < r$. Hence $U \in \mathscr{U}_1$.⟧
★10. For each connector U, there exists a symmetric closed connector V with $V \circ V \circ V \subset U$. ⟦Apply Condition (iv) of Definition 2 twice with Theorem 11.1.6; then use Problem 7.⟧

11. Given a set X, let $\mathscr{U}$ be the collection of all subsets of $X \times X$ which include the diagonal, Δ; let $\mathscr{V} = \{X \times X\}$. Show that $\mathscr{U}$, $\mathscr{V}$ are uniformities. They are called the *discrete* and *indiscrete* uniformities, respectively. Show that they generate the discrete and indiscrete topologies. Show also that $\{\Delta\}$ is a base for the discrete uniformity.

★12. A subset S of a uniform space $(X, \mathscr{U})$ is made into a uniform space by means of the *relative uniformity*: For $U \in \mathscr{U}$, let

$$U_S = \{(a, b): a, b \in S, (a, b) \in U\}.$$

Show that $\{U_S: U \in \mathscr{U}\}$ is a uniformity and induces the relative topology.

13. Express Theorem 11.1.3 as a fact about semimetric space.

★14. Let f_δ be a net of functions from a set S to a uniform space Y, and let $f: S \to Y$. We say that $f_\delta \to f$ *uniformly on* S if for each connector U, there exists δ_0 such that $\delta \geq \delta_0$ implies $(f_\delta x, fx) \in U$ for all $x \in S$. Show that the limit of a uniformly convergent net of continuous functions from a topological space to a uniform space must be continuous. ⟦Imitate Theorem 4.2.10.⟧

15. Exhibit a uniformity for ω which is not discrete but induces the discrete topology. ⟦Consider the natural uniformity of $\{1/n\}$ in **R**.⟧

16. If a set A has the property that $U[A] \subset A$ for some connector U, then A is open and closed. ⟦Actually $U[A] = A$. For $x \in A$, $U(x) \subset A$; also $\bar{A} \subset U[A] \subset A$.⟧

17. For any set S and connector U, $\bigcup \{U^n[S]: n = 1, 2, \ldots\}$ is open and closed, where $U^n = U \circ U \circ \cdots \circ U$ with n terms ⟦Problem 16⟧.

---

101. In a semimetric space, $U_\varepsilon \cap U_\delta = U_{\varepsilon \wedge \delta}$.

102. Show that $U_\varepsilon \circ U_\delta \neq U_{\varepsilon + \delta}$ is possible in a semimetric space; indeed, in a subspace of **R**.

103. If U, V are symmetric, $U \circ V$ need not be. ⟦A brother's friend need not be a friend's brother.⟧ Compare Sec. 3.3, Problem 205.

104. A *uniform neighborhood* of a set S is a set which includes $U[S]$ for some connector U. Show that every uniform neighborhood of S is a neighborhood of S, and that (0, 1) is not a uniform neighborhood of its subset of rational points (in the Euclidean uniformity).

105. Every neighborhood of a compact set is a uniform neighborhood. ⟦Say $K \subset N$. For each $x \in K$ let $(U_x \circ U_x)(x) \subset N$. Reduce $\{U_x(x): x \in K\}$ to a finite cover $\{U_{x_i}(x_i)\}$ and set $V = \bigcap U_{x_i}$.⟧ This implies, by Theorem 11.1.3, that $\bar{K} \subset N$ if N is a neighborhood of K; a result which was obtained in Sec. 5.4, Problem 113.

106. For any connectors U, V and $x \in X$, $U(x) \times V(x) \subset V \circ U^{-1}$. In particular, if V is symmetric, $V(x) \times V(x) \subset V \circ V$ for all x. Since

every connector U includes a connector of the form $V \circ V$ it follows that *U is a neighborhood of Δ in the product topology of $X \times X$.*

107. Let (X, T) be a topological space and $\mathcal{N}$ the set of all neighborhoods of Δ, the diagonal, in $X \times X$. Show that $\mathcal{N}$ satisfies Conditions (i), (ii), and (iii) of Definition 2. Thus $\mathcal{N}$ is a uniformity if and only if (iv) is satisfied. If $\mathcal{N}$ is a uniformity T need not be regular ⟦see the next two problems⟧ but it must be normal. ⟦For disjoint closed A, B choose symmetric $V \in \mathcal{N}$ with $V \circ V \subset (A \times B)^{\sim}$. Then $V(A)$, $V(B)$ are neighborhoods of A, B since, for example, $V(a) = j^{-1}V$ where $j(x) = (a, x)$. Also, they are disjoint since if $V(a) \cap V(b) \neq \varnothing$, then $(a, b) \in (V \circ V) \cap (A \times B)$.⟧
108. Use $\mathcal{N}$ (Problem 107) to induce a topology $T_{\mathcal{N}}$ in the same way that a uniformity induces a topology: G is open if for each $x \in G$ there exists $U \in \mathcal{N}$ with $U(x) \subset G$. Show that $T_{\mathcal{N}} \subset T$.
109. In Problem 108, $T_{\mathcal{N}}$ may be strictly smaller than T even if $\mathcal{N}$ is a uniformity ⟦Sec. 4.1, Problem 8; $\mathcal{N}$ is the indiscrete uniformity⟧, and $T_{\mathcal{N}} = T$ if and only if T is symmetric. In particular, $T_{\mathcal{N}} = T$ if T is regular or T_1. ⟦If T is symmetric and $G \in T$, let $x \in G$. Then $U = (\overline{\{x\}} \times \tilde{G})^{\sim} \in T$ by Sec. 4.1, Problem 13. Also $U(x) \subset G$. Conversely, if $T_{\mathcal{N}} = T$ and $x \notin \overline{\{y\}}$, there exists $U \in \mathcal{N}$ with $U(x) \subset \widetilde{\overline{\{y\}}}$. As in Theorem 11.1.3, $\overline{\{x\}} \subset \widetilde{\overline{\{y\}}}$ and so $y \notin \overline{\{x\}}$.⟧
110. If $\mathcal{N}$ (Problem 107) is a uniformity, $T_{\mathcal{N}} = T$ if and only if T is regular ⟦Problem 107; Sec. 4.1, Problems 9 and 12⟧. In particular, if (X, T) is a T_1 space and $\mathcal{N}$ is a uniformity, then $T_{\mathcal{N}} = T$.
111. Let $I = \bigcap \{U: U \in \mathcal{B}\}$, where $\mathcal{B}$ is any base for the uniformity of X. Then $\Delta \subset I \subset G$ for every neighborhood G of Δ. ⟦If $(x, y) \in I$, then $y \in U(x)$ for all U so that $y \in \overline{\{x\}}$. Now see Sec. 6.7, Problem 9.⟧
112. The uniformity $\mathcal{U}$ of a compact uniform space X is precisely the set of neighborhoods of Δ in $X \times X$. ⟦Problem 106. Conversely, let G be an open neighborhood of Δ. Then $\tilde{G}$ is compact and does not meet I, Problem 111. (Take $\mathcal{B}$ to be the set of closed connectors; see Problem 10.) Hence $\tilde{G}$ fails to meet a finite intersection of members of $\mathcal{B}$ and so G includes a finite intersection of members of $\mathcal{B}$, hence $G \in \mathcal{U}$.⟧
113. Let vU, the *vertical closure* or *v closure* of a connector U be defined by $vU(x) = \overline{U(x)}$. Call U *v-closed* if $vU = U$. Show that a closed connector is v-closed, but not conversely. ⟦Example 3; $vU = (X \times X)\backslash v$⟧.
114. For a connector U, $vU = \bigcap \{V \circ U: V \in \mathcal{B}\}$, where $\mathcal{B}$ is any base for the uniformity ⟦Theorem 11.1.3⟧. Hence $U \subset vU \subset V \circ U$ for all V.
115. Every uniformity has a base of symmetric v-closed connectors ⟦Problems 10 and 113⟧.
116. The v closure of a symmetric connector need not be symmetric ⟦Example 3⟧.

117. Is it possible to have U v-closed and U^{-1} not v-closed? ⟦Consider vU in Example 3.⟧
118. For a connector U and base $\mathscr{B}$, $(vU)^{-1} = \bigcap \{U^{-1} \circ V: V \in \mathscr{B}\}$, $v(U^{-1}) = \bigcap \{V \circ U^{-1}: V \in \mathscr{B}\}$, hence $\bigcap \{U \circ V: V \in \mathscr{B}\} = [v(U^{-1})]^{-1}$ and $U \subset [v(U^{-1})]^{-1} \subset U \circ V$ for all V. Compare Problem 114.)
119. If $S \times S \subset U$, then $S \times \bar{S} \subset vU$ and $\bar{S} \times \bar{S}$ need not be included in vU.

---

201. In Problem 107, let $X = \mathbf{R}$. Then $\mathscr{U}$ is a uniformity.
202. Construct a *transitive* uniformity for $\mathbf{Q}$; that is, with a base of *transitive* connectors. ($U \circ U \subset U$.)
203. A transitive uniformity must induce a zero-dimensional topology ⟦Problem 17⟧.
204. A uniform space has an equivalent transitive uniformity if and only if it is zero-dimensional. (A. F. Monna. See [Banaschewski].)

11.2 Uniform Continuity

Let $(X, \mathscr{U})$, $(Y, \mathscr{V})$ be uniform spaces. Then $f: X \to Y$ is called *uniformly continuous* if for each $V \in \mathscr{V}$, there exists $U \in \mathscr{U}$ such that $(a, b) \in U$ implies $(fa, fb) \in V$; phrased differently, such that $b \in U(a)$ implies $f(b) \in V(fa)$; briefly $f[U(x)] \subset V(fx)$ for all $x \in X$.

★EXAMPLE 1. Let (X, d_X), (Y, d_Y) be semimetric spaces and $f: X \to Y$ a uniformly continuous function when X, Y are given the uniformities $\mathscr{U}$, $\mathscr{V}$, induced by their semimetrics. Let $\varepsilon > 0$ be given. Then $V = \{(y, z): d_Y(y, z) < \varepsilon\} \in \mathscr{V}$. Choose $U \in \mathscr{U}$ as in the above definition. There exists $\delta > 0$ such that $\{(a, b): d_X(a, b) < \delta\} \subset U$ ⟦Sec. 11.1, Example 1⟧. Thus f satisfies the condition:

$$\text{For every } \varepsilon > 0, \text{ there exists } \delta > 0 \text{ such that } d_X(a, b) < \delta \text{ implies } d_Y(fa, fb) < \varepsilon. \tag{11.2.1}$$

This is true since with the δ just chosen, $d_X(a, b) < \delta$ implies $(a, b) \in U$; this implies that $(fa, fb) \in V$ whence $d_Y(fa, fb) < \varepsilon$. It is left to the reader ⟦Problem 1⟧ to show that conversely (11.2.1) implies that f is uniformly continuous.

Several topological results have uniform analogues. Problem 5 is similar to the result that f is continuous if and only if $f^{-1}[G]$ is open when G is open. Theorem 11.2.1 is an analogue of Theorem 4.2.5. (Of course, some analogues fail: see Problem 103.)

THEOREM 11.2.1. *Let $\mathscr{U}, \mathscr{V}$ be uniformities for a set X. Then $\mathscr{U} \supset \mathscr{V}$ if and only if the identity map $i: (X, \mathscr{U}) \to (X, \mathscr{V})$ is uniformly continuous; and $\mathscr{U}$ is stronger than $\mathscr{V}$ if and only if i is continuous.*

The second statement is merely Theorem 4.2.5. If $\mathscr{U} \supset \mathscr{V}$, let $V \in \mathscr{V}$. Let $U = V$. Then $(a, b) \in U$ implies $(ia, ib) \in V$ since $ia = a, ib = b, U = V$. Thus i is uniformly continuous. Conversely, let i be uniformly continuous and $V \in \mathscr{V}$. There exists U such that $(a, b) \in U$ implies $(ia, ib) \in V$; this merely says $U \subset V$, hence $V \in \mathscr{U}$ since $\mathscr{U}$ is a filter. ∎

Since there are equivalent uniformities which are not equal ⟦Sec. 11.1, Example 2⟧, it follows from Theorem 11.2.1 that uniform continuity is not topological, in the sense that a uniformly continuous function can be made not uniformly continuous without altering the topologies of its domain and range. ⟦Let $\mathscr{U}, \mathscr{V}$ be equivalent uniformities for a set X with $\mathscr{U} \not\supset \mathscr{V}$. Then $i: (X, \mathscr{U}) \to (X, \mathscr{U})$ is uniformly continuous, while $i: (X, \mathscr{U}) \to (X, \mathscr{V})$ is not.⟧

A *uniform homeomorphism* is a homeomorphism f such that both f and f^{-1} are uniformly continuous. (Here $f^{-1}: f[X] \to X$ with the relative uniformity; see Sec. 11.1, Problem 12.)

Certain covers of a uniform space are of such a nature that every point is covered by a "nonsmall" set. For example the set of all intervals of unit length covers **R** in such a way that each point is included in an interval of unit length. In contrast let $C = \{(x - 1/x, x]: x > 1\}$. This is a cover of $(0, \infty)$ and there is no lower bound to the smallness of members of C required; for example, if $\varepsilon > 0$ choose $x > 1/\varepsilon$; then every member of C which contains x has diameter $< \varepsilon$. The first kind of cover is called uniform; we proceed to define this concept. A cover C of a uniform space X is called a *uniform cover* if there exists a connector U such that for every $x \in X$ some member of C includes $U(x)$; that is the cover $\{U(x): x \in X\}$ refines C. It may always be assumed that U is symmetric ⟦Theorem 11.1.1⟧. Uniform continuity can be phrased in terms of uniform covers.

LEMMA 11.2.1. *Let $f: (X, \mathscr{U}) \to (Y, \mathscr{V})$ and for each $V \in \mathscr{V}$, let $C_V = \{f^{-1}[V(y)]: y \in Y\}$ then f is uniformly continuous if and only if C_V is a uniform cover of X for every $V \in \mathscr{V}$.*

Suppose that C_V is always a uniform cover, and let $W \in \mathscr{V}$. Choose symmetric $V \in \mathscr{V}$ with $V \circ V \subset W$. Then C_V is a uniform cover. Choose U as in the definition of uniform cover, and let $(a, b) \in U$. There exists y such that $f^{-1}[V(y)] \supset U(a)$. It follows that $f(a) \in V(y)$ and $f(b) \in V(y)$ ⟦both $a, b \in U(a)$⟧, hence $(fa, fb) \in V \circ V \subset W$. Conversely, suppose that f is uniformly continuous, and let $V \in \mathscr{V}$. Choose U as in the definition of uniform continuity, let $x \in X$ and set $y = f(x)$. Then $f[U(x)] \subset V(fx)$ so that $U(x) \subset f^{-1}[V(fx)]$. Thus C_V is a uniform cover. ∎

THEOREM 11.2.2. *Every open cover C of a compact uniform space X is a uniform cover.*

For each $x \in X$, choose $G_x \in C$ with $x \in G_x$, a connector V_x with $V_x(x) \subset G_x$ ⟦G_x is open⟧, and a symmetric connector U_x with $U_x \circ U_x \subset V_x$. Now $\{[U_x(x)]^i : x \in X\}$ is an open cover of X, hence can be reduced to a finite cover $\{[U_{x_k}(x_k)]^i : k = 1, 2, \ldots, n\}$. Let $U = \bigcap U_{x_k}$ and we shall check that U satisfies the requirements that make C a uniform cover. Let $x \in X$. Then $x \in U_{x_k}(x_k)$ for some k, hence

$$U(x) \subset U \circ U_{x_k}(x_k) \subset U_{x_k} \circ U_{x_k}(x_k) \subset V_{x_k}(x_k) \subset G_{x_k} \in C. \quad \blacksquare$$

We shall require a slight modification of Theorem 11.2.2 in which we replace the assumption that C is an open cover by the assumption that $\{S^i : S \in C\}$ is a cover of X; that is, that for each x, some member of C is a neighborhood of x. Since this latter cover is open, it is uniform and it follows immediately that C is a uniform cover.

COROLLARY 11.2.1. *Every continuous function f from a compact uniform space $(X, \mathscr{U})$ to a uniform space $(Y, \mathscr{V})$ is uniformly continuous.*

Let $V \in \mathscr{V}$. For each $x \in X, f^{-1}[V(fx)]$ is a neighborhood of x, since f is continuous and $V(fx)$ is a neighborhood of $f(x)$. The result is now immediate from Lemma 11.2.1 and the modification of Theorem 11.2.2 just given. $\blacksquare$

COROLLARY 11.2.2. *The uniformity of a compact space is unique; that is, a compact topological space has at most one uniformity which induces its topology.*

If two uniformities are given, an application of Corollary 11.2.1 and Theorem 11.2.1 to the identity map shows that they are equal. $\blacksquare$

Corollary 11.2.2 is also a special case of Sec. 11.1, Problem 112.

Problems

★1. Prove the condition (11.2.1) to be equivalent to uniform continuity of a map between semimetric spaces.

★2. A uniformly continuous function is continuous. ⟦For $x \in X$, every neighborhood of $f(x)$ includes some $V[f(x)]$ hence, choosing U with $(fa, fb) \in V$ for $(a, b) \in U$, we have $f[U(x)] \subset V[f(x)]$.⟧

★3. A bounded continuous real function on a uniform space X need not be uniformly continuous, even if X is a bounded subset of $\mathbf{R}$. ⟦Consider $\sin 1/x$ on $(0, 1]$.⟧

4. Give $(\mathbf{R}, d)$, d being the Euclidean metric, an equivalent uniformity $\mathscr{U}$ such that $f: (\mathbf{R}, \mathscr{U}) \to (\mathbf{R}, d)$ is uniformly continuous, where $f(x) = x^2$; and show that $f: (\mathbf{R}, d) \to (\mathbf{R}, d)$ is not uniformly continuous.

★5. Given $f: (X, \mathscr{U}) \to (Y, \mathscr{V})$, define $f_2: X \times X \to Y \times Y$ by $f_2(a, b) = (fa, fb)$. Show that f is uniformly continuous if and only if $f_2^{-1}[V] \in \mathscr{U}$ for all $V \in \mathscr{V}$.

6. A cover is uniform if it has a refinement which is uniform.

7. Let $\mathscr{U}$, $\mathscr{V}$ be uniformities for a set X inducing topologies $T_{\mathscr{U}}$, $T_{\mathscr{V}}$. Assume that $T_{\mathscr{U}} \supset T_{\mathscr{V}}$ and that $T_{\mathscr{U}}$ is compact. Show that $\mathscr{U} \supset \mathscr{V}$ ⟦Corollary 11.2.1⟧. See Problem 108.

8. All semimetrics which induce the same compact topology on a set are uniformly equivalent ⟦Corollary 11.2.2⟧.

★9. The composition of two uniformly continuous functions is uniformly continuous.

10. A retraction of $\mathbf{R}^2$ onto $\mathbf{R}$ need not be uniformly continuous. ⟦$r(x, y) = x^{1+|y|}$.⟧

---

101. Show that d and $d \wedge 1$ are uniformly equivalent semimetrics.

102. The product of two bounded uniformly continuous real functions is uniformly continuous.

103. The product of two uniformly continuous real functions need not be uniformly continuous even if one is bounded ⟦x and $\sin x$ on $\mathbf{R}$⟧.

104. The square of a uniformly continuous real function need not be uniformly continuous.

105. Prove, directly from the definition, that $\sqrt{x}$ is uniformly continuous for $0 \leq x \leq 1$.

106. If f, g are uniformly continuous real functions, $af + bg$, $|f|$, $f \vee g$, $f \wedge g$ are uniformly continuous. (Compare Theorems 4.2.6, 4.2.8, and 4.2.9.)

107. Is $\{(x, 1/x): 0 < x < 1\}$ a uniform cover of $(0, \infty)$ with the Euclidean uniformity?

108. In Problem 7, the assumption that $T_{\mathscr{U}}$ is compact cannot be omitted ⟦Sec. 11.1, Example 2⟧.

109. Is sgn uniformly continuous on $\mathbf{R} \setminus \{0\}$? (sgn $x = 1$ if $x > 0$; 0 if $x = 0$; -1 if $x < 0$.)

110. Give a reasonable definition for the concept, *uniformly open map*.

111. Write out a statement of Theorem 11.2.2 specialized to a compact metric space X. (This is the Lebesgue covering lemma, given for $X = [a, b]$ by H. Lebesgue before 1900.)

112. The limit of a uniformly convergent net of uniformly continuous functions is uniformly continuous.

---

201. Exhibit $f: \mathbf{R} \to \mathbf{R}$ such that f is unbounded and f^n is uniformly continuous for all $n = 1, 2, \ldots$.

202. Is there any analogue of Lemma 2.5.1 for uniformities?

203. Construct two uniform spaces X, Y which are not uniformly homeo-

morphic and for which there exist uniformly continuous homeomorphisms from X onto Y and from Y onto X.

11.3 Uniform Concepts

In the context of uniformity, nontopological concepts such as completeness and total boundedness may be studied. We begin by defining a *Cauchy filterbase* to be a filterbase $\mathscr{F}$ in a uniform space with the property that for every connector U, $\mathscr{F}$ contains a member S with $S \times S \subset U$; that is $x, y \in S$ imply $(x, y) \in U$. The suggestive word *small* may be used here. Suppose we call a set S *small of order* U if $S \times S \subset U$. (For example, in a semimetric space, U might be $\{(x, y): d(x, y) \leq \varepsilon\}$; then a set is small of order U if and only if it has diameter $\leq \varepsilon$.) A Cauchy filterbase is then one which contains a set which is small of any order. (The set chosen depends on the order.)

THEOREM 11.3.1. *Every convergent filterbase is a Cauchy filterbase.*

Suppose that $\mathscr{F} \to x$. Let U be a connector. Choose a symmetric connector V with $V \circ V \subset U$. Then $V(x)$ is a neighborhood of x, hence includes some $S \in \mathscr{F}$. It follows that $S \times S \subset V(x) \times V(x) \subset V \circ V \subset U$ ⟦Problem 4⟧. ∎

It will come as no surprise that a uniform space in which every Cauchy filterbase is convergent is called *complete*. A uniform space is called *sequentially complete* if every Cauchy sequence is convergent. For a semimetric space these concepts are equivalent ⟦Theorem 9.1.3⟧. We shall see in Sec. 11.4, Example 3, that a sequentially complete uniform space need not be complete. These concepts are, of course, uniform invariants; they have some additional permanence properties which we now show.

THEOREM 11.3.2. *A uniformly continuous map preserves Cauchy filterbases.*

Let $\mathscr{F}$ be a Cauchy filterbase in $(X, \mathscr{U})$ and $f: X \to (Y, \mathscr{V})$. Fix $V \in \mathscr{V}$ and set $U = f_2^{-1} V$ ⟦Sec. 11.2, Problem 5⟧. $\mathscr{F}$ contains a set S which is small of order U. Then $f[S]$ is small of order V. ⟦For $a, b \in S$, we have $(a, b) \in U$, hence $(fa, fb) = f_2(a, b) \in V$.⟧ Since $f[S] \in f[\mathscr{F}]$, it follows that $f[\mathscr{F}]$ is a Cauchy filterbase. ∎

THEOREM 11.3.3. *Let $f: X \to Y$ be a uniformly continuous homeomorphism onto. Then if Y is complete, X is complete also.*

Let $\mathscr{F}$ be a Cauchy filterbase in X. By Theorem 11.3.2 and the hypothesis, $f[\mathscr{F}]$ is convergent in Y. Since f^{-1} is continuous $\mathscr{F}$ is convergent in X. ∎

COROLLARY 11.3.1. *If a uniformity $\mathscr{U}$ for a set X is complete, then any* *equivalent larger uniformity $\mathscr{V}$ is also complete. The same holds for sequentially complete.*

Apply Theorem 11.3.3 to the identity map from $(X, \mathscr{V})$ to $(X, \mathscr{U})$, taking account of Theorem 11.2.1. ∎

EXAMPLE 1. Let e be the Euclidean uniformity for $\mathbf{R}$ and let b be the (smaller) uniformity which makes $\mathbf{R}$ uniformly homeomorphic with $(-1, 1)$ ⟦the uniformity induced by d_3 in Sec. 11.1, Example 2⟧. Then $(\mathbf{R}, e)$ is complete, $(\mathbf{R}, b)$ is not complete ⟦since $(-1, 1)$ is not⟧, and $e \supset b$.

LEMMA 11.3.1. *If a Cauchy filterbase $\mathscr{F}$ in a uniform space has a cluster point x, then $\mathscr{F} \to x$.*

Let N be a neighborhood of x. There exists a closed connector U with $U(x) \subset N$ ⟦Theorem 11.1.6⟧, and $\mathscr{F}$ contains a set S which is small of order U. Then for any $s \in S$, $(x, s) \in \bar{S} \times S \subset U$. ⟦Theorem 6.4.7 implies that $\bar{S} \times S \subset \bar{S} \times \bar{S} = \overline{S \times S} \subset \bar{U} = U$.⟧ Hence $s \in U(x) \subset N$, and so $S \subset N$. This proves that $\mathscr{F} \to x$. ∎

We have seen examples of subspaces which are not C- and C^*-embedded (Section 8.5). The following useful extension theorem reveals, in particular, that every dense subspace allows extension of a real *uniformly* continuous function.

THEOREM 11.3.4. *Let D be a dense subspace of a uniform space $(X, \mathscr{U})$, let $(Y, \mathscr{V})$ be a complete uniform space, and $g: D \to Y$ a uniformly continuous map. Then g has an extension to a uniformly continuous map $G: X \to Y$.*

For $x \in X$ we have to choose some $y \in Y$ and call it $G(x)$. The problem of which point in Y to choose is answered by the requirement that G be continuous, thus a filterbase converging to x is mapped to a filterbase converging to y. The details are as follows. Let $x \in X \setminus D$ and let

$$\mathscr{F}_x = \{U(x) \cap D : U \in \mathscr{U}\};$$

$\mathscr{F}_x$ is a filterbase in X, because D is dense and each $U(x)$ is a neighborhood of x. Moreover $\mathscr{F}_x \to x$. ⟦Every neighborhood of x includes a set of the form $U(x)$, hence a member of $\mathscr{F}_x$.⟧ Thus $\mathscr{F}_x$ is a Cauchy filterbase in X, hence in D; it follows that $g[\mathscr{F}_x]$ is a Cauchy filterbase ⟦Theorem 11.3.2⟧. Since Y is complete, $g[\mathscr{F}_x]$ is convergent. Let y be any limit of $g[\mathscr{F}_x]$ and define $G(x) = y$. (If Y is separated, y is uniquely determined.) If $x \in D$, define $G(x) = g(x)$. By definition $G \mid D = g$, and it remains to show that G is uniformly continuous. Let $V_0 \in \mathscr{V}$. Choose closed symmetric $V \in \mathscr{V}$ with $V \circ V \circ V \subset V_0$ ⟦Sec. 11.1, Problem 10⟧. There exists $U_0 \in \mathscr{U}$ such that $(r, s) \in U_0$, $r, s \in D$ imply $(gr, gs) \in V$ ⟦Sec. 11.1, Problem 12⟧, and symmetric

$U \in \mathscr{U}$ with $U \circ U \circ U \subset U_0$. We shall prove that $(a, b) \in U$ implies $(Ga, Gb) \in V_0$, completing the proof. Let $(a, b) \in U$. Choose $r \in D \cap U(a)$. ⟦Possible because D is dense, and $U(a)$ is a neighborhood of a.⟧ Choose $s \in D \cap U(b)$. Then (r, a), (a, b), (b, s) all belong to U, hence $(r, s) \in U \circ U \circ U \subset U_0$. Hence $(gr, gs) \in V$. If now we show that $(Ga, gr) \in V$ and $(gs, Gb) \in V$ we shall have $(Ga, Gb) \in V \circ V \circ V \subset V_0$, completing the proof. We shall prove the first of these two statements only; that is, $(Ga, gr) \in V$; the other is proved similarly. What we have to prove is that $G(a) \in V(gr)$. Since the latter set is closed it is sufficient to show that every neighborhood of $G(a)$ meets $V(gr)$. Let N be a neighborhood of $G(a)$, then N includes a member of $g[\mathscr{F}_a]$, say $g[W(a) \cap D] \subset N$, $W \in \mathscr{U}$. Now $W(a) \cap U(a)$ is a neighborhood of a in X, hence meets D; say $d \in D \cap W(a) \cap U(a)$. But then $g(d) \in N \cap V(gr)$. ⟦$d \in D \cap W(a)$ implies $g(d) \in N$ by definition of W; also $d \in U(a)$, $a \in U(r)$ implies $(d, r) \in U \circ U \subset U_0$ so that $(gd, gr) \in V$ by definition of U_0.⟧ This proves that N meets $V(gr)$ and concludes the proof of Theorem 11.3.4. ∎

REMARK. In case Y is separated, the extension G is uniquely determined by g, indeed G is the only continuous extension ⟦Sec. 4.2, Problem 7⟧.

A set S in a uniform space is called *totally bounded* if for every connector U, S is covered by a finite collection of sets, each of which is small of order U.

THEOREM 11.3.5. *The following are equivalent for a nonempty set S in a uniform space:*

(i) *S is totally bounded.*
(ii) *For every connector U, there exists a finite subset F of S with $S \subset U[F]$.*
(iii) *For every connector U, there exists a finite subset F of X with $S \subset U[F]$.*

Proof. (i) *implies* (ii). Let U be a connector. We have $S = \bigcup \{S_i: i = 1, 2, \ldots, n\}$ with each S_i small of order U. Assuming, as we may, that each S_i is not empty, choose $x_i \in S_i$ for each i. Then $S_i \subset U(x_i)$, ⟦for $y \in S_i$, $(x_i, y) \in S_i \times S_i \subset U$⟧, and so $S \subset \bigcup U(x_i) = U[F]$, where $F = (x_1, x_2, \ldots, x_n)$.

(iii) *implies* (i): Let V be a connector. Choose a symmetric connector U with $U \circ U \subset V$, and choose F according to (iii), say $F = (x_1, x_2, \ldots, x_n)$. Then $S \subset \bigcup \{U(x_i): i = 1, 2, \ldots, n\}$ and each $U(x_i)$ is small of order V. ⟦$a \in U(x_i)$, $b \in U(x_i)$ imply $(a, b) \in U \circ U \subset V$.⟧ The result is trivial from this. ∎

THEOREM 11.3.6. *A uniform space is totally bounded if and only if every ultrafilter is a Cauchy filter.*

Let X be totally bounded and $\mathscr{F}$ an ultrafilter. Let U be a connector. Then X is the union of finitely many sets each of which is small of order U. One of these sets must belong to $\mathscr{F}$ ⟦Sec. 7.3, Problem 2⟧. Hence $\mathscr{F}$ is a Cauchy filter. Conversely suppose that X is not totally bounded. There exists a symmetric connector U with $X \neq U[A]$ for every finite set A. Then there exists a "uniformly discrete" sequence $\{x_n\}$; that is with $(x_m, x_n) \notin U$ if $m \neq n$. ⟦Inductively, choose x_1 at random, and

$$x_{n+1} \notin U[\{x_i: i = 1, 2, \ldots, n\}].⟧$$

Let A_n be the set $(x_n, x_{n+1}, x_{n+2}, \ldots)$. Then if $\mathscr{F}$ is any filter which includes the filterbase $\{A_n: n = 1, 2, \ldots\}$ $\mathscr{F}$ cannot be Cauchy; indeed it cannot contain a set which is small of order U. ⟦Let $S \in \mathscr{F}$. Then S must contain two different x_i; namely S must meet A_1, say $x_m \in S \cap A_1$, and S must meet A_{m+1}, say $x_n \in S \cap A_{m+1}$. But then $(x_m, x_n) \notin U$ hence $S \times S \not\subset U$.⟧ Since Theorem 7.3.1 assures us that there exists an ultrafilter which includes the above mentioned filterbase, the proof is complete. ∎

THEOREM 11.3.7. *A uniform space is compact if and only if it is totally bounded and complete.*

If X is compact and $\mathscr{F}$ is an ultrafilter, then $\mathscr{F}$ must be convergent ⟦Theorem 7.3.6⟧ hence Cauchy ⟦Theorem 11.3.1⟧. Thus X is totally bounded ⟦Theorem 11.3.6⟧. To see that X is complete, let $\mathscr{F}$ be a Cauchy filter. Then $\mathscr{F}$ has a cluster point ⟦Theorem 7.1.4⟧, hence converges ⟦Lemma 11.3.1⟧. Conversely, let X be totally bounded and complete, and let $\mathscr{F}$ be an ultrafilter. By Theorem 11.3.6, $\mathscr{F}$ is Cauchy, hence converges, and so X is compact ⟦Theorem 7.3.6⟧. ∎

Problems on Uniform Space

★1. Check that the definitions of Cauchy filterbase and completeness given here and in Section 9.1 agree.
2. On any space, the discrete and indiscrete uniformities are complete.
3. Exhibit a noncomplete uniformity for the integers which induces the discrete topology ⟦Sec. 9.1, Example 2⟧.
★4. Let V be a symmetric connector with $V \circ V \subset U$. Then $V(x)$ is small of order U for all x. ⟦If $(a, b) \in V(x) \times V(x)$, then $a \in V(x)$, $x \in V(b)$ so $a \in (V \circ V)(b)$.⟧
5. A space is totally bounded if and only if every filter is included in a Cauchy filter.
★6. A subspace of a uniform space X is totally bounded in its relative uniformity if and only if it is a totally bounded subset of X.
7. Do Problems 1 and 2 in Sec. 9.1, for uniform space.
★8. A closed subset of a complete uniform space is complete (in the relative uniformity), and a subspace of a complete separated uniform space is complete if and only if it is closed. Obtain also analogues of Sec. 9.1,

Problems 7 and 8. ⟦Let $\mathscr{F}$ be a Cauchy filter in $S \subset X$. Then $\mathscr{F}$ is a Cauchy filterbase in X. Thus $\mathscr{F} \to x$. But then $x \in S$.⟧

9. A uniformly continuous function preserves total boundedness. In particular a uniformly continuous real function is bounded on every totally bounded set.
10. If the restriction of a continuous function g to a dense subset is uniformly continuous, g is uniformly continuous. ⟦See the proof of Theorem 11.3.4.⟧
★11. Let X, Y be separated complete uniform spaces and suppose they have uniformly homeomorphic dense subspaces. Then X, Y are uniformly homeomorphic. ⟦By Theorem 11.3.4, extend $f: D_X \to D_Y$ to $F: X \to Y$, and $f^{-1}: D_Y \to D_X$ to $G: Y \to X$. Then $(F \circ G) \mid D_Y$ and $(G \circ F) \mid D_X$ are identity maps. By Sec. 4.2, Problem 7, $G = F^{-1}$.⟧ See Problem 104.
12. Must a bounded continuous real function on a totally bounded set be uniformly continuous?
★13. The closure of a totally bounded set S is totally bounded. ⟦If S is a finite union of sets S_i which are small of order U, $\bar{S}$ is a union of sets $\bar{S}_i$ which are small of order $\bar{U}$, by Theorem 6.4.7. We may assume U closed, by Theorem 11.1.6.⟧
★14. A net x_δ is said to be a *Cauchy net* if for every connector U, there exists δ_0 such that $\delta, \delta' \geq \delta_0$ implies $(x_\delta, x_{\delta'}) \in U$. Show that x_δ is a Cauchy net if and only if its associated filter is a Cauchy filter. ⟦Let $\mathscr{B}_\delta = \{x_{\delta'}: \delta' \geq \delta\}$. Then $\{\mathscr{B}_\delta\}$ is a filterbase which contains small sets if and only if x_δ is a Cauchy net.⟧
★15. Every net associated with a Cauchy filter is a Cauchy net.
★16. A uniform space is complete if and only if every Cauchy net is convergent. ⟦Necessity by Problem 14; sufficiency by Problem 15.⟧
17. Call two uniformities for a set *Cauchy equivalent* if they have the same Cauchy filters; equivalently ⟦Problem 14⟧ if they have the same Cauchy nets. Cauchy equivalent uniformities need not be equal ⟦for example, two equivalent complete uniformities⟧; and equivalent uniformities need not be Cauchy equivalent; ⟦that is, a homeomorphism need not be uniform; consider Sec. 4.2, Example 5⟧. Two complete uniformities are Cauchy equivalent if and only if they are equivalent. See also Problem 110.
18. A uniformity is totally bounded if it is Cauchy equivalent to a totally bounded uniformity ⟦Theorem 11.3.6⟧.
19. Let X be not totally bounded. Then there exists a connector V and a sequence $\{x_n\}$ with $V(x_m) \not\cap V(x_n)$ for $m \neq n$. ⟦Choose U so that X has no finite cover by sets which are small of order U. Choose symmetric V with $V \circ V \circ V \circ V \subset U$. Let $x_1 \in X$. Choose x_2 with $V(x_2) \not\cap V(x_1)$; if this is not possible, $X \subset V \circ V(x_1)$, which is small

of order U. Choose x_3 with $V(x_3) \not\subset [V(x_1) \cup V(x_2)]$. Continue inductively.⟧

20. A countably compact uniform space must be totally bounded ⟦Problem 19; Theorem 10.2.5⟧.
21. A countably compact complete uniform space is compact ⟦Problem 20; Theorem 11.3.7⟧. Thus a countably compact noncompact space is *very incomplete* in the sense that it cannot be given a complete uniformity. Hence a realcompact countably compact space is compact; a result noted in improved form in Sec. 8.6, Problem 1.
22. A pseudocompact uniform space must be totally bounded. ⟦Imitate Problem 19, using the interiors of the sets $V(x_n)$.⟧ This improves Problem 20 ⟦Sec. 7.1, Problem 114⟧. Improve Problem 21 in the same way.

---

101. Prove Lemmas 9.1.2 and 9.1.3 for uniform space.
102. Do Sec. 9.1, Problems 115, 116, and 117 for uniform space.
103. Theorem 11.3.4 becomes false if it is not assumed that D is dense ⟦Sec. 8.5, Problem 5⟧, or if it is not assumed that Y is complete. ⟦The identity map from **Q** to **Q** cannot be extended to **R** by Sec. 4.2, Problem 27.⟧
104. Problem 11 fails if "uniform" and "uniformly" are omitted. (Replace "separated" by "Hausdorff.") ⟦Sec. 9.2, Example 2. $S_1 \setminus \{P\}$ (P any point) and (0, 1) are not uniformly homeomorphic, by Problem 11, although there is a homeomorphism between them which is uniformly continuous. What is it?⟧
105. For a filter $\mathscr{F}$ in a uniform space $(X, \mathscr{U})$, let $\mathscr{B} = \{U[S]: U \in \mathscr{U}, S \in \mathscr{F}\}$. Show that $\mathscr{B}$ is a filterbase and that $\mathscr{B}$ is convergent or Cauchy if and only if $\mathscr{F}$ is. ⟦Since $\mathscr{B} \subset \mathscr{F}$, half of each part is trivial.⟧
106. The closure of a complete set in a uniform space is complete. ⟦For a separated space, see Problem 8. In general use Problem 105.⟧ (In particular, the closure of a compact set is complete, but this is trivial from Sec. 5.4, Problem 9.)
107. A subset $\mathscr{B}$ of a uniformity $\mathscr{U}$ is called a *subbase* for $\mathscr{U}$ if every member of $\mathscr{U}$ includes a finite intersection of members of $\mathscr{B}$. Suppose a set S in $(X, \mathscr{U})$ has the property that for every U in some subbase $\mathscr{B}$ for $\mathscr{U}$, S is the union of finitely many sets which are small of order U. Show that S is totally bounded. ⟦If $S = \bigcup S_{ij}$ where, for each i, S_{ij} is small of order U_j; let $A = \bigcap S_{i_k k}$ and consider the set of all such A.⟧
108. In contrast with Problem 107; in Part (ii) of Theorem 11.3.5, it is not sufficient to assume that the condition holds for all U in a subbase. ⟦Take $X = \mathbf{R}$, let subbasic connectors be $\{(x, y): |x - y| < \varepsilon$ or $x - y$ is an odd integer$\}$ and the same with odd replaced by even. Then ω is not totally bounded.⟧

109. An attempt to extend Lemma 11.3.1 to a nonuniform situation might be made thus: CONJECTURE. If a convergent filterbase $\mathscr{F}$ in a topological space has a cluster point x, then $\mathscr{F} \to x$. Settle this conjecture.
110. Two Cauchy equivalent uniformities are equivalent. ⟦Let $\mathscr{U}_1, \mathscr{U}_2$ be Cauchy equivalent. For any x, the $\mathscr{U}_1$-neighborhood filter N_x must be $\mathscr{U}_2$-Cauchy, hence, by Lemma 11.3.1, converges to x in the topology induced by $\mathscr{U}_2$. (x *belongs* to each member of N_x.) The result follows by Sec. 4.2, Problem 117.⟧

---

201. Let $\mathscr{F}$ be a Cauchy filter. Show that the filter generated by $\mathscr{B}$ in Problem 105 is a minimal Cauchy filter. In particular N_x is a minimal Cauchy filter for each x.
202. Suppose that $\mathscr{F}$ is a filter but not a Cauchy filter. Is it possible to choose a point $x_S \in S$ each $S \in \mathscr{F}$ such that $(x_S: \mathscr{F})$ is not a Cauchy net?
203. A set S in a uniform space is called *bounded* if for every connector U there exists a finite set F and positive integer n such that $S \subset U^n[F]$ where $U^n = U \circ U \circ \cdots \circ U$ with n terms. Show that in a semimetric space, a bounded set must be metrically bounded, but not conversely.
204. The closure of a bounded set is bounded.
205. Every totally bounded set is bounded but not conversely.
206. A uniformly continuous map preserves bounded sets.
207. A set is bounded if and only if every uniformly continuous real function is bounded on it. ⟦See [Nakano, p. 80, Theorem 2].⟧
208. A compact uniform space must be complete ⟦Theorem 11.3.7⟧, but a locally compact uniform space cannot always be given a complete uniformity. Compare Sec. 9.2, Problem 101; Sec. 12.2. Problem 127. ⟦By Theorem 11.3.7 and Problem 20 it is sufficient to give a certain noncompact space.⟧

11.4 Uniformization

In this and the next section we solve two natural converse problems: Which topologies can be uniformized? Which uniformities can be semimetrized? Briefly: a topology is induced by some uniformity if and only if it is completely regular; a uniformity is induced by a semimetric if and only if it has a countable base. In these sentences, if the uniformity is taken to be separated, replace completely regular by $T_{3\frac{1}{2}}$ (Tychonoff), and semimetrizable by metrizable. Some discussion of the limitations of this program is given at the beginning of Section 11.5.

Since every completely regular space has the weak topology by its continuous real functions ⟦Theorem 6.7.4⟧, it is sufficient to show that such a space can be uniformized. (The converse is Theorem 11.5.2.) It is natural to imitate the procedures of Chapter 6. Suppose that a set X has a nonempty collection Φ of uniformities specified for it. Let $\mathscr{B}$ be the collection of all subsets of $X \times X$ of the form $\bigcap \{U_i: i = 1, 2, \ldots, n\}$, where each U_i is a member of $\mathscr{U}$ for some $\mathscr{U} \in \Phi$. We shall show that $\mathscr{B}$ is a base for a uniformity by checking Sec. 11.1, Definition 2. To see that $\mathscr{B}$ is a filterbase, we note that $\mathscr{B}$ is not empty, and that no member of $\mathscr{B}$ is empty. ⟦Every member of every $\mathscr{U} \in \Phi$ includes the diagonal, Δ.⟧ Also it is clear that $\mathscr{B}$ is closed under finite intersection. Conditions (ii′), (iii′), and (iv′) are easily checked. ⟦For example $(\bigcap U_i)^{-1} = \bigcap U_i^{-1}$ since $(x, y) \in \bigcap U_i$ if and only if $(y, x) \in \bigcap U_i^{-1}$.⟧ The uniformity generated by $\mathscr{B}$ is called the *supremum* (or *sup*) of Φ and written $\bigvee \Phi$. (It is not hard to see that $\mathscr{B}$ actually *is* a uniformity.)

THEOREM 11.4.1. *Let Φ be a collection of uniformities for a set X. Then $\bigvee \Phi$ induces the topology $\bigvee \{T_{\mathscr{U}}: \mathscr{U} \in \Phi\}$, where $T_{\mathscr{U}}$ is the topology induced by $\mathscr{U}$.*

This should be compared with the analogous result for semimetrization, Theorem 6.2.2, where it could be obtained for a countable family only. To prove Theorem 11.4.1, let T be the topology induced by $\bigvee \Phi$ and let $S = \bigvee T_{\mathscr{U}}$. Suppose that N is a T neighborhood of x. Then there exist $U_1, U_2, \ldots, U_n$, each U_i belonging to some $\mathscr{U} \in \Phi$, with $(\bigcap U_i)(x) \subset N$ ⟦Theorem 11.1.2⟧. Now $(\bigcap U_i)(x) = \bigcap [U_i(x)]$, and $U_i(x)$ is a $T_{\mathscr{U}}$ neighborhood of x if $U_i \in \mathscr{U}$. Thus $\bigcap [U_i(x)]$ is an S neighborhood of x, hence N is an S neighborhood of x. Conversely, that every S neighborhood of x is a T neighborhood is proved similarly. ∎

THEOREM 11.4.2. *With the notation of Theorem 11.4.1, let $\mathscr{V} = \bigvee \Phi$. Then a filterbase $\mathscr{F}$ in $(X, \mathscr{V})$ is a Cauchy filterbase if and only if it is a Cauchy filterbase in $(X, \mathscr{U})$ for each $\mathscr{U} \in \Phi$. The same result holds for nets.*

The filterbase result implies the net result by Sec. 11.3, Problem 14. Half of Theorem 11.4.2 is trivial since $\mathscr{V} \supset \mathscr{U}$ for each $\mathscr{U} \in \Phi$. Conversely, let $\mathscr{F}$ be $\mathscr{U}$-Cauchy for every $\mathscr{U}$ and let $V \in \mathscr{V}$. There exist $U_1, U_2, \ldots. U_n$, each U_i belonging to some $\mathscr{U} \in \Phi$ with $\bigcap U_i \subset V$. For each i, $\mathscr{F}$ contains S_i with $S_i \times S_i \subset U_i$. Let $S = \bigcap S_i$. Then $S \times S \subset V$. Thus $\mathscr{F}$ is $\mathscr{V}$-Cauchy. ∎

Next, let X be a set, $(Y, \mathscr{V})$ a uniform space, and $f: X \to Y$. We define $u(f)$, the *weak uniformity by f*, to be the uniformity on X generated by $\mathscr{B}$, where $\mathscr{B}$ is the collection of all sets of the form

$$U_V = \{(a, b) \in X \times X: (fa, fb) \in V\},$$

V being a symmetric connector for Y. It is easy to see that $\mathscr{B}$ is a base for a uniformity ⟦for example, each member of $\mathscr{B}$ is symmetric, includes the diagonal, and if $V \circ V \subset W$, then $U_V \circ U_V \subset U_W$, showing that $\mathscr{B}$ satisfies Part (iv′) of Sec. 11.1, Definition 2⟧. Thus $u(f)$ is a well-defined uniformity; moreover $f: X \to Y$ is uniformly continuous when X has the uniformity

$u(f)$. ⟦For any connector V_1 for Y, let V be a symmetric connector with $V \subset V_1$. Then $U_V \in \mathscr{B}$ and so $U_{V_1} \in u(f)$. Comparing the definitions of U_V and of uniform continuity yields the result.⟧ Indeed, *$u(f)$ is the smallest uniformity which makes f uniformly continuous.* ⟦Any such uniformity must include $\mathscr{B}$, as is shown by a glance at the definitions.⟧

Extending the idea of weak uniformity to more than one map, let X be a set and Φ a family of functions f, each f defined on X and with range in some uniform space Y_f. *The weak uniformity by Φ for X*, written $u(\Phi)$ or $u\Phi$, is $\bigvee \{u(f): f \in \Phi\}$, where $u(f)$ is the weak uniformity by f as just defined.

THEOREM 11.4.3. *Let Φ be a collection of maps from a set X to uniform spaces. Then $u(\Phi)$ induces the topology $w(\Phi)$, the weak topology by Φ.*

It will be sufficient to prove this for a single map f. ⟦Theorem 11.4.1 will then yield the general result.⟧ Let $u(f)$ also stand for the topology induced by $u(f)$. Then $u(f) \supset w(f)$ ⟦$w(f)$ is the smallest topology making f continuous, and $u(f)$ does indeed make f continuous⟧. Conversely, let N be a $u(f)$ neighborhood of x. Then there is a connector V for Y such that $x \in U_V(x) \subset N$. Then $N \supset f^{-1}[V(fx)]$. ⟦If $a \in f^{-1}[V(fx)]$, $f(a) \in V(fx)$, hence $(fx, fa) \in V$ which implies $(x, a) \in U_V$ and so $a \in U_V(x) \subset N$.⟧ Since $V(fx)$ is a neighborhood of $f(x)$ in Y, this shows that N is a $w(f)$ neighborhood of x. ∎

As a special case of Theorem 11.4.3, we have

THEOREM 11.4.4. *The product of a family of uniform spaces is a uniform space with the weak uniformity by the family of projections.*

This uniformity is called the *product uniformity*; it is part of the content of Theorem 11.4.3 that the product uniformity induces the product topology.

We show a simple application of the sup and weak uniformities.

EXAMPLE 1. *Every metric space can be given a complete uniformity* (which induces its topology). For separable metric spaces, a possibly better result is given in Example 5, and Corollary 8.6.2. (Of course for *separable* spaces it is an equivalent result.) Let (X, d) be a metric space, and we shall use the letter d to denote its uniformity also. Let υ be the weak uniformity by $C(X)$ and let $\mathscr{U} = d \vee \upsilon$. Then $\mathscr{U}$ induces the metric topology of X ⟦Theorems 11.4.1 and 6.7.4⟧. We shall show that $\mathscr{U}$ is complete. Let $\mathscr{F}$ be a $\mathscr{U}$-Cauchy filter. Then F is a d-Cauchy filter ⟦since $\mathscr{U} \supset d$⟧, hence $\mathscr{F} \to y \in Y$, where Y is the metric completion of (X, d) ⟦Theorem 9.2.3⟧. Now $y \in X$. ⟦If $y \notin X$, let $g(x)$ be the distance from x to y in Y for each $x \in X$. Then g is never 0, so $1/g \in C(X)$, hence $1/g$ is υ-uniformly continuous. But this is a contradiction since $\mathscr{F}$ is a υ-Cauchy filter and $1/g$ is unbounded on each member S of $\mathscr{F}$ because $y \in \bar{S}$ in Y. The contradiction derives from

Theorem 11.3.2.⟧ Thus $\mathscr{F} \to x$ in X, a topological (not just uniform) statement; in particular $\mathscr{F} \to x$ in $(X, \mathscr{U})$. ∎

THEOREM 11.4.5. *Every completely regular topology is induced by a uniformity.*

By Theorem 6.7.4, a completely regular topology for X is precisely the weak topology by $C(X)$. The result follows from Theorem 11.4.3. ∎

REMARK. We shall refer to a space as having a uniformity. It must be clearly understood that the uniformity is to induce the topology of the space. For instance, Example 1 implies that **Q** can be given a complete uniformity—this means, *a complete uniformity which induces the natural (Euclidean) topology.*

THEOREM 11.4.6. *The topology of a compact regular space, and, in particular, a compact Hausdorff space, is given by a unique uniformity.*

Such a space is completely regular ⟦Theorem 5.4.7⟧, hence is a uniform space ⟦Theorem 11.4.5⟧. The uniformity is unique by Corollary 11.2.2. ∎

Because of Theorem 11.4.6 we may speak unambiguously in uniform terms when referring to a compact regular space. For example such a space is complete ⟦Theorem 11.3.7⟧, and it is unnecessary to specify the uniformity with respect to which it is complete. The converse of Theorem 11.4.6 is false in that a noncompact space may have a unique uniformity ⟦Sec. 11.5, Problem 108⟧, but, as a partial converse, a space with a unique uniformity must be pseudocompact ⟦Example 4⟧. See also Sec. 14.1, Problem 111.

★EXAMPLE 2. *The α and β uniformities.* Let X be a Tychonoff space. Then βX has a unique uniformity. The relative uniformity for X of βX will be denoted by β. Thus for example, (X, β) *is always totally bounded* ⟦Sec. 11.3, Problem 6; βX is compact hence totally bounded⟧, and never complete, unless X is already compact. The important identification of β as $u[C^*(X)]$ is spelled out in Problem 110. Next, let X be a locally compact regular space. If X is not compact, X^+ is compact and regular ⟦Theorem 8.1.2⟧ hence has a unique uniformity. The relative uniformity for X of X^+ will be denoted by α. If X is compact and regular, α will denote its (unique) uniformity. As a sample result we have: *For a locally compact Hausdorff space X, $\beta \supset \alpha$.* ⟦Extend $i: X \to X$ to $f: \beta X \to X^+$. By Corollary 11.2.1, f is uniformly continuous, hence so is $i: (X, \beta) \to (X, \alpha)$. The result follows from Theorem 11.2.1.⟧ If X is compact, $\beta = \alpha$ is the unique uniformity of X. For a noncompact space it is still possible that $\beta = \alpha$ ⟦Sec. 8.5, Problem 8⟧. Of course β, α are equivalent; they both induce the topology of X.

EXAMPLE 3. Let X be a noncompact T_4 space. Then X is sequentially closed in βX and not closed ⟦Theorem 8.3.2⟧. This means that (X, β) is sequentially complete, where β is given in Example 2. ⟦βX is complete by Theorem 11.3.7.⟧ However (X, β) is not complete since X is dense ⟦Sec. 11.3, Problem 8⟧, or since X is totally bounded and not compact. Thus, *if X is a noncompact Tychonoff space, β is not metrizable.* ⟦If it were metrizable it would be T_4. This leads, as just shown to the impossible situation of a sequentially complete non-complete space, contradicting Theorem 9.1.3.⟧ As a special case we have: *The topology of every noncompact Tychonoff space can be induced by a nonmetrizable uniformity.* It is interesting to apply this result to any noncompact metric space. Its topology is given by a non-metrizable uniformity, as well as a metrizable one.

The criteria for continuity of functions with ranges in sup, weak, and product spaces (Theorem 6.6.1) have exact analogues for uniform continuity. In Theorem 11.4.7, $\{\mathscr{U}_\alpha : \alpha \in A\}$ is a collection of uniformities on a set Y; $\{f_\alpha : \alpha \in A\}$ is a family of maps, each $f_\alpha : Y \to Z_\alpha$, where each Z_α is a uniform space; X_α is a family of uniform spaces; and X is a uniform space.

THEOREM 11.4.7.

(i) *A function $f: X \to (Y, \bigvee \{\mathscr{U}_\alpha : \alpha \in A\})$ is uniformly continuous if and only if $f: X \to (Y, \mathscr{U}_\beta)$ is uniformly continuous for each $\beta \in A$.*

(ii) *A function $f: X \to (Y, u\{f_\alpha : \alpha \in A\})$ is uniformly continuous if and only if $f_\beta \circ f$ is uniformly continuous for all $\beta \in A$.*

(iii) *A function of $f: X \to \prod\{X_\alpha : \alpha \in A\}$ is uniformly continuous if and only if $P_\beta \circ f$ is uniformly continuous for each $\beta \in A$.*

Half of Part (i) is trivial since $\mathscr{U}_\beta \subset \bigvee \mathscr{U}_\alpha$. Conversely, suppose that $f: (X, \mathscr{V}) \to (Y, \mathscr{U}_\beta)$ is uniformly continuous for each β and let $U \in \bigvee \mathscr{U}_\alpha$. Then $U \supset U_1 \cap U_2 \cap \cdots \cap U_n$ where each $U_i \in \mathscr{U}_{\alpha_i}$ for some $\alpha_i \in A$. For each i, there exists $V_i \in \mathscr{V}$ with the property that $(a, b) \in V_i$ implies $(fa, fb) \in U_i$. Let $V = \bigcap V_i$. Then $V \in \mathscr{V}$ and $(a, b) \in V$ implies $(fa, fb) \in U$. Thus f is uniformly continuous. Part (ii) follows from Part (i) and the special case of Part (ii) in which A has one member, which we now prove. We denote the sole function f_α by g. Half of the result is trivial by Sec. 11.2, Problem 9. Conversely, suppose that $g \circ f$ is uniformly continuous and let $U \in u(g)$. There exists W in the uniformity of Z_α such that $U \supset \{(a, b): (ga, gb) \in W\}$. Since $g \circ f$ is uniformly continuous, there exists $V \in \mathscr{V}$ such that $(x, x') \in V$ implies $(g \circ fx, g \circ fx') \in W$ which implies that $(fx, fx') \in U$. Thus f is uniformly continuous. Finally, Part (iii) follows from Part (ii) and the definition of product uniformity. ▌

EXAMPLE 4. *The uniformity υ.* Let X be a Tychonoff space and let $\upsilon = uC(X)$, the weak uniformity by $C(X)$. Since $C(X) = C(\upsilon X)$ in a natural way ⟦X is dense and C-embedded in υX⟧, the uniformity υ is also defined for υX. We first note that like β, υ *induces the natural topology of* υX; that is, the relative topology of βX ⟦Theorems 11.4.3, and 6.7.4⟧. Since $\beta = uC^*[X]$ ⟦Problem 110⟧, we see that υ bears the same relation to υX that β does to βX. See also Problems 112, 113, and 114. Now υ, β are equivalent uniformities, as just mentioned, but $\beta \subset \upsilon$, ⟦$C^*(X) \subset C(X)$⟧, also $\beta = \upsilon$ if and only if X is pseudocompact. ⟦An unbounded $f \in C(X)$ is υ-uniformly continuous, but not β-uniformly continuous by Problem 111.⟧ It follows that a *non-pseudocompact Tychonoff space always has more than one uniformity*. A pseudocompact, indeed a countably compact space, may have more than one uniformity ⟦Sec. 14.1, Problem 111⟧.

The role of realcompactness in uniformity is illustrated by the following result.

EXAMPLE 5. *A Tychonoff space X is complete in the uniformity υ if and only if it is realcompact.* Suppose first that X is realcompact and that $\mathscr{F}$ is an υ-Cauchy filter. Then $\mathscr{F}$ is a β-Cauchy filter since $\upsilon \supset \beta$. Thus $\mathscr{F} \to t \in \beta X$ since $(\beta X, \beta)$ is complete. ⟦It is compact!⟧ If $t \notin X$, the definitions yield $f \in C(X)$ whose extension to βX is ∞ at t; such f is unbounded on each member of $\mathscr{F}$ ⟦t is an accumulation point of each member of $\mathscr{F}$⟧. This contradicts Theorem 11.3.2. Thus $t \in X$. That $\mathscr{F} \to x$ is a topological statement, as true for υ as for β. Conversely if X is not realcompact, (X, υ) is a dense proper subspace of $(\upsilon X, \upsilon)$ hence is not complete. ∎

Problems

1. If $\mathscr{U}$ is a uniformity for a set X which includes each member of a family of uniformities on X, show that $\mathscr{U}$ includes the supremum of the family.
2. The (Euclidean) topology of **R** can be given by a noncomplete metrizable uniformity ⟦$\mathbf{R} = (0, 1)$⟧, a complete metrizable uniformity, and a sequentially complete noncomplete uniformity ⟦Example 3⟧.
3. The discrete topology of an infinite space can be given by a nonmetrizable uniformity ⟦Example 3⟧.
4. The weak uniformity by a family Φ of maps is the smallest uniformity which makes every member of Φ uniformly continuous.
5. Let $X = \mathbf{R}$. Let α be as in Example 2, let e be the Euclidean uniformity, and let t be the uniformity which makes **R** uniformly homeomorphic with $[(-1, 1), e]$. Thus $(\mathbf{R}, \alpha)$ is a dense uniform subspace of S_1, the unit circumference; $(\mathbf{R}, e)$ is complete; and $(\mathbf{R}, t)$ is a dense uniform subspace of $[-1, 1]$. Show that $e \supset t \supset \alpha$. ⟦These are metrizable uniformities so the inclusions can be checked analytically. Also there

is a continuous (hence uniformly continuous) map from $[-1, 1]$ onto S_1, hence from $(\mathbf{R}, t)$ onto $(\mathbf{R}, \alpha)$. Compare Sec. 11.5, Problem 2.⟧

★6. Every projection of a product is uniformly continuous.

★7. Show that a product of complete uniform spaces is complete with the product uniformity. ⟦Let $\mathscr{F}$ be a Cauchy filter. Then each projection of $\mathscr{F}$ is a Cauchy filterbase by Problem 6 and Theorem 11.3.2. Thus each projection converges and so $\mathscr{F}$ converges by Theorem 6.4.1.⟧

8. A filterbase is Cauchy in $\bigvee \Phi$ (see Theorem 11.4.1) if and only if it is Cauchy in each $\mathscr{U} \in \Phi$.

9. Let $f: (X, \mathscr{U}) \to (Y, \mathscr{V})$ and let $\mathscr{W} = \mathscr{U} \vee u(f)$. Show that $\mathscr{W}$ is equivalent to $\mathscr{U}$ if and only if f is continuous, and $\mathscr{W} = \mathscr{U}$ if and only if f is uniformly continuous.

10. In Problem 9, assume that f is continuous and $\mathscr{U}$ is complete. Show that $\mathscr{W}$ is complete ⟦Corollary 11.3.1⟧.

11. Let X, Y be semimetric spaces with X complete, and let $f: X \to Y$ be continuous. Define $D(a, b) = d(a, b) + d(fa, fb)$ for $a, b \in X$. Show that (X, D) is complete ⟦Problem 10⟧.

12. In Problem 9, assume that f has closed graph, and that $\mathscr{U}$, $\mathscr{V}$ are complete. Show that $\mathscr{W}$ is complete. Specialize as in Problem 11.

13. The weak uniformity $u(f)$ need not be the smallest uniformity which induces $w(f)$.

---

101. Let $(X, \mathscr{U})$ and $(Y, \mathscr{V})$ be complete uniform spaces and $S \subset X$. Let $f: S \to Y$ have the property that its graph is a closed subset of $X \times Y$. Show that $(S, \mathscr{W})$ is complete, where $\mathscr{W} = \mathscr{U} \vee u(f)$. ⟦A Cauchy net s_δ is $\mathscr{U}$-Cauchy, hence $s_\delta \to x \in X$. Also $f(s_\delta)$ is $\mathscr{V}$-Cauchy, hence $f(s_\delta) \to y \in Y$. Then $y = f(x)$, $x \in S$ and $s_\delta \to x$ in X, hence in S.⟧

102. If every continuous real function on a semimetric space is uniformly continuous the space must be complete ⟦Sec. 9.1, Problem 114; Sec. 8.5, Problem 116⟧. This result is false for uniform space ⟦Example 5⟧.

103. Let X be a bounded subspace of $\mathbf{R}$. Then $\beta \supset e$, where e is the Euclidean uniformity, and β is as in Example 2. ⟦Use the argument of Example 2 with $\bar{X}$ in place of X^+. This argument extends to any subspace of a compact space and is taken to its full generality in Sec. 11.5, Problem 104.⟧

104. In contrast to Problem 103; on ω, e is strictly larger than β. ⟦e is discrete; β is not metrizable.⟧

105. If X is an unbounded subset of $\mathbf{R}$, β is not larger than e ⟦Sec. 11.3, Problem 9⟧.

106. Generalize the result of Problem 105.

107. On $\mathbf{R}$, β is neither larger nor smaller than e. ⟦Problem 105. Also β

makes every bounded continuous real function uniformly continuous, while e does not. Hence $\beta \not\subset e$.⟧

108. There are two different equivalent uniformities for $\boldsymbol{\omega}$ each of them making every bounded continuous real function uniformly continuous ⟦Problem 104⟧.
109. Example 3 shows that for noncompact X, β is never metrizable. Deduce this also from the fact that such a metric would be complete by the arguments used in Problem 102; hence compact.
110. The uniformity β of Example 2 is the weak uniformity by $C^*(X)$. ⟦$\beta \supset u(C^*)$ by Problem 4. Conversely, β is the relative uniformity of the product uniformity of I^C. (See the definition of βX in Section 8.3. We are also using Corollary 11.2.2.) A typical $U \in \beta$ is a finite intersection of sets of the form

$$\{(x, y): |\hat{x}(f) - \hat{y}(f)| < \varepsilon\} = \{(x, y): |f(x) - f(y)| < \varepsilon\}$$

with $f \in C$. Since $C \subset C^*(X)$ we have $U \in u(C^*)$.⟧
111. Let $f \in C(X)$. Then f is β-uniformly continuous (Example 2) if and only if it is bounded, that is if and only if $f \in C^*(X)$. ⟦Problems 110 and 4. Conversely if f is uniformly continuous it is bounded, by Sec. 11.3, Problem 9 since, by Example 2, (X, β) is totally bounded.⟧
112. For any subspace X of $\mathbf{R}$, $\upsilon \supset e$. Compare Problems 103 and 104. ⟦e is the weak uniformity by the inclusion map $i: X \to \mathbf{R}$, and $i \in C(X)$.⟧
113. On $\mathbf{R}$, υ is strictly larger than e. Compare Problem 107. ⟦The same proof shows $\upsilon \not\subset e$. Now see Problem 112.⟧
114. On $\boldsymbol{\omega}$, $\upsilon = e =$ discrete uniformity. Compare Problem 104 ⟦Problem 112⟧.
115. Let X be a Tychonoff space and $U \in \upsilon$. Then X is a countable union of sets which are small of order U. ⟦Suppose first

$$U = \{(x, y): |f(x) - f(y)| < \varepsilon\}$$

for some $f \in C(X)$. With $S_n = \{x: \frac{1}{2}n\varepsilon \le f(x) \le \frac{1}{2}(n+1)\varepsilon\}$ we have $X = \bigcup_{-\infty}^{\infty} S_n$. Proceed as in Sec. 11.3, Problem 107.⟧
116. Let X be a noncountable discrete space and d any metric which induces the discrete topology. Then $\upsilon \not\supset d$, and υ is not metrizable. ⟦Problem 115; Sec. 2.3, Problem 109.⟧
117. Let d be the discrete metric for a set X and form υ for the space (X, d). Then $\upsilon \supset d$ if and only if X is countable, in which case $\upsilon = d$. In all other cases, $\upsilon \subset d$ strictly. ⟦d is always the discrete uniformity, than which there is no larger. For an uncountable space, d cannot have the property given in Problem 115.⟧
118. A complete uniform space may be of first category. Compare the Baire category theorem. ⟦Consider $(\mathbf{Q}, \upsilon)$. Note Example 5 and Corollary 8.6.2.⟧

119. State and prove for uniform spaces the results on retraction and completeness for products given in Sec. 6.7, Problem 102 (also Sec. 9.1, Problems 107 and 109). Add a result on total boundedness (Sec. 11.3, Problem 9).
120. The uniformity of a uniform space need not be the weak uniformity by the set of all bounded uniformly continuous real functions ⟦Problem 108⟧. Compare Theorem 6.7.4.
121. A uniform (separated) space need not be normal (T_4) ⟦Sec. 6.7, Example 3⟧.
122. Show that a product of totally bounded spaces is totally bounded in the product uniformity ⟦Sec. 11.3, Problem 107⟧. Deduce Tychonoff's theorem, Theorem 7.4.1, for regular spaces; in particular, for Hausdorff spaces ⟦Theorem 11.4.5; Problem 7; Theorem 11.3.7⟧.
123. Let X be pseudocompact. Show that $\upsilon(X \times X)$ is homeomorphic with $(\upsilon X) \times (\upsilon X)$ if and only if $X \times X$ is pseudocompact. ⟦By Sec. 11.3, Problem 22, they are homeomorphic if and only if $\upsilon(X \times X)$ is compact, that is if and only if $\upsilon(X \times X) = \beta(X \times X)$. Now see Sec. 8.6, Problem 1.⟧
124. Deduce the fact that a countably compact metric space is compact from Example 1, and Sec. 11.3, Problem 21.
125. The result of Sec. 9.2, Problem 106 fails for uniform space; indeed every uniform space has an equivalent totally bounded uniformity.

---

201. Can the topology of **R** be given by a nonmetrizable complete uniformity?
202. Discuss quotients of uniform spaces ⟦MR *35* #972⟧.

11.5 Metrization and Completion

To the concepts of first and second countability for a uniform space, we add another countability concept which lies between them, namely, a uniform space may have a countable base for its uniformity. Suppose we designate a uniformity $\mathscr{U}$ as $2C$ if it induces a second countable topology, FC if it induces a first countable topology, CB if it has a countable base, M_u if it is semimetrizable, and M_t if it induces a semimetrizable topology. A summary of past and future results is contained in the list:

$$M_t \leftarrow 2C,\ CB \leftrightarrow M_u \rightarrow M_t \rightarrow FC,$$

and no arrow can be reversed.

That $2C \rightarrow M_t$ is Theorem 10.1.1. That $M_u \rightarrow CB$ is clear from consideration of the countable base $\{U_n\}$, where $U_n = \{(x, y): d(x, y) < 1/n\}$. That $CB \rightarrow M_u$ is given in Theorem 11.5.1. That $M_u \rightarrow M_t \rightarrow FC$ is trivial. Also $CB \not\rightarrow 2C$ ⟦consider any nonseparable metric space⟧, $M_t \not\rightarrow M_u$ ⟦Sec.

11.4, Example 3⟧, and $FC \not\to M_t$ ⟦Theorem 11.4.5; Sec. 5.3, Examples 1 and 3⟧.

In preparation for the metrization theorem we show how to obtain a function obeying the triangle inequality from an arbitrary nonnegative real function D defined on $X \times X$, where X is an arbitrary nonempty set. For x, y in D define

$$d(x, y) = \inf\left\{\sum_{i=1}^{k} D(x_i, x_{i+1}) : x_1 = x, x_{k+1} = y\right\}. \qquad (11.5.1)$$

In Equation (11.5.1), it is understood that the inf is taken over all finite collections $x_1, x_2, x_3, \ldots, x_{k+1}$ with $x_1 = x$, $x_{k+1} = y$.

LEMMA 11.5.1. *With these definitions $d(x, z) \leq d(x, y) + d(y, z)$.*

Let $\varepsilon > 0$. Choose $x_1, x_2, \ldots, x_{k+1}; y_1, y_2, \ldots, y_{n+1}$ with $x_1 = x$, $x_{k+1} = y_1 = y$, $y_{n+1} = z$, and

$$\sum D(x_i, x_{i+1}) < d(x, y) + \frac{\varepsilon}{2}, \quad \sum D(y_i, y_{i+1}) < d(y, z) + \frac{\varepsilon}{2}.$$

Now set $x_{k+r} = y_r$ for $r = 2, 3, \ldots, n + 1$ and obtain

$$d(x, z) \leq \sum_{i=1}^{k+n} D(x_i, x_{i+1}) = \sum_{i=1}^{k} D(x_i, x_{i+1}) + \sum_{i=1}^{n} D(y_i, y_{i+1}) < d(x, y) + d(y, z) + \varepsilon. \quad \blacksquare$$

THEOREM 11.5.1. *A uniformity with a countable base is semimetrizable.*

As pointed out above, the converse is also true. Let X be a uniform space whose uniformity $\mathscr{U}$ has a countable base $\{U_n\}$. Now let $V_0 = X \times X$. Let V_1 be a symmetric connector with $V_1 \subset U_1$. Let $V_2 \in \mathscr{U}$ be symmetric, $V_2 \circ V_2 \circ V_2 \subset U_2 \cap V_1$. ⟦Possible since $U_2 \cap V_1 \in \mathscr{U}$, and by Sec. 11.1, Problem 10.⟧ Choose symmetric $V_3 \in \mathscr{U}$ with $V_3 \circ V_3 \circ V_3 \subset U_3 \cap V_2$ and, in general, choose symmetric $V_n \in \mathscr{U}$ with $V_n \circ V_n \circ V_n \subset U_n \cap V_{n-1}$. Then $\{V_n\}$ is a base for $\mathscr{U}$ ⟦Lemma 11.1.1; furthermore $V_n \subset U_n$ for all n⟧, and, in addition, all V_n are symmetric, and $V_n \circ V_n \circ V_n \subset V_{n-1}$ for $n = 1, 2, \ldots$. For each $x, y \in X$, set $D(x, y) = 0$ if $(x, y) \in V_n$ for all n, while if $(x, y) \in V_n \setminus V_{n+1}$, set $D(x, y) = 2^{-n}$. This definition makes sense since every $(x, y) \in V_0$ and once $(x, y) \notin V_k$ then $(x, y) \notin V_m$ for all $m > k$. The function D has some of the properties of a semimetric; clearly $D(x, y) \geq 0$, $D(x, y) = D(y, x)$ since each V_n is symmetric, $D(x, x) = 0$. Note that for $n = 0, 1, 2, \ldots$,

$$D(x, y) \leq 2^{-n} \qquad \text{if and only if } (x, y) \in V_n. \qquad (11.5.2)$$

We shall use D to obtain the desired semimetric. We begin by proving

$$D(w, z) \leq 2 \max [D(w, x), D(x, y), D(y, z)] \quad \text{for } w, x, y, z \in X. \quad (11.5.3)$$

⟦Let the right-hand side of Equation (11.5.3) be called $2r$, so that $r = 2^{-n}$ for some n. Then $D(w, x) \leq 2^{-n}$ so that $(w, x) \in V_n$ by Equation (11.5.2). Similarly, $(x, y), (y, z) \in V_n$ and so $(w, z) \in V_n \circ V_n \circ V_n \subset V_{n-1}$. It follows from Equation (11.5.2) that $D(w, z) \leq 2^{-(n-1)} = 2r$.⟧ From Equation (11.5.3) we can deduce

$$D(x_1, x_{k+1}) \leq 2 \sum_{i=1}^{k} D(x_i, x_{i+1}) \qquad \text{for } x_1, x_2, \ldots, x_{k+1} \in X. \quad (11.5.4)$$

⟦If all $D(x_i, x_{i+1})$ are 0, $x_{i+1} \in \overline{\{x_i\}}$ for all i by Theorem 11.1.3, hence $x_{k+1} \in \overline{\{x_1\}}$ and $D(x_1, x_{k+1}) = 0$. In the rest of the proof we shall assume that at least one $D(x_i, x_{i+1}) > 0$. For $k = 0, 1, 2$, Equation (11.5.4) is obvious from Equation (11.5.3), and the fact that $D(x, x) = 0$. We shall prove Equation (11.5.4) by induction. Let the right-hand side of Equation (11.5.4) be called $2b$. Let $k > 2$, and let m be the largest integer such that

$$\sum_{i=1}^{m} D(x_i, x_{i+1}) \leq \frac{b}{2}. \qquad (11.5.5)$$

In case $D(x_1, x_2) > b/2$, take m and the left-hand side of Equation (11.5.5) to be 0. Then $0 \leq m < k$ also

$$\sum_{i=m+2}^{k} D(x_i, x_{i+1}) < \frac{b}{2} \qquad (11.5.6)$$

since $b - \sum_{i=m+2}^{k} = \sum_{i=1}^{m+1} > b/2$ by definition of m. In case $m = k - 1$, the left-hand side of (11.5.6) is taken to be 0. Since the sums in the left-hand sides of Equation (11.5.5) and Equation (11.5.6) contain fewer than k terms (or are 0) we have, by the induction hypothesis,

$$D(x_1, x_{m+1}) \leq 2 \sum_{i=1}^{m} D(x_i, x_{i+1}) \quad \leq b,$$

$$D(x_{m+2}, x_{k+1}) \leq 2 \sum_{i=m+2}^{k} D(x_i, x_{i+1}) \leq b,$$

and, trivially,

$$D(x_{m+1}, x_{m+2}) \qquad \leq b.$$

Applying Equation (11.5.3) to these inequalities yields Equation (11.5.4).⟧ We are now ready to define our semimetric d. It is defined by the formula (11.5.1). The triangle inequality is Lemma 11.5.1; $d(x, x) = 0$, ⟦take $k = 1$ in Equation (11.5.1)⟧; $d(x, y) = d(y, x)$ since $D(x, y) = D(y, x)$, and $d(x, y) \geq 0$ since $D(x, y) \geq 0$. To relate d to the uniformity, we prove

$$d \leq D \leq 2d. \qquad (11.5.7)$$

⟦That $d \leq D$ follows by taking $k = 1$ in Equation (11.5.1). Next, let $x, y \in X$. For any choice of $x_1, x_2, \ldots, x_{k+1}$ with $x_1 = x$, $x_{k+1} = y$ we have, from Equation (11.5.4), $\sum_{i=1}^{k} D(x_i, x_{i+1}) \geq \frac{1}{2} D(x, y)$. Taking the inf of the left-hand side yields $d(x, y) \geq \frac{1}{2} D(x, y)$.⟧ Finally, d induces the uniformity of X. ⟦Let U be a connector. Then, for some n,

$$U \supset V_n = \{(x, y): D(x, y) \leq 2^{-n}\} \supset \{(x, y): d(x, y) \leq 2^{-n-1}\}$$

by Equations (11.5.2) and (11.5.7). Thus U is a d connector. Conversely, let U be a d connector, which we may take to be of the form $\{(x, y): d(x, y) \leq 2^{-n}\}$. Then $U \supset \{(x, y): D(x, y) \leq 2^{-n}\} = V_n$ by Equation (11.5.2) and (11.5.7). Thus U is a connector for the uniformity of X.⟧ ∎

Some details in the metrization of some special uniformities were given in the problems of Section 11.4. See also [Rainwater], [Waterhouse], [Mrowka], and the references given there.

THEOREM 11.5.2. *A uniform space is completely regular.*

Let F be a closed set in the uniform space X, and $x \notin F$. Choose a symmetric connector V with $V(x) \not\cap F$. Inductively we may choose $V_1 = V$, a symmetric connector V_2 with $V_2 \circ V_2 \subset V_1$, and so, in general, a symmetric connector V_n with $V_n \circ V_n \subset V_{n-1}$. Then $\{V_n\}$ is a base for a uniformity, which we may denote by $\mathscr{V}$. By Theorem 11.5.1, $(X, \mathscr{V})$ is semimetrizable, hence, completely regular ⟦Theorem 4.3.3⟧, and so, since $V(x)$ is a neighborhood of x in $(X, \mathscr{V})$, there exists a continuous map f of $(X, \mathscr{V})$ into $[0, 1]$ with $f(x) = 0$, $f(y) = 1$ for $y \notin V(x)$, in particular for $y \in F$. But $\mathscr{V} \subset \mathscr{U}$ where $\mathscr{U}$ is the original uniformity hence f is a continuous function on $(X, \mathscr{U})$. ∎

REMARK. Theorem 11.5.2, together with Theorem 11.4.5, characterizes uniform topologies. *A topology is given by a uniformity if and only if it is completely regular.* (Compare Sec. 6.3, Problem 202.)

LEMMA 11.5.2. *Let V be a symmetric connector in a uniform space $(X, \mathscr{U})$. Then there exists a semimetric D whose uniformity is smaller than $\mathscr{U}$, such that V is a D connector.*

(It is not intended that D induce the topology of X.) Let the uniformity $\mathscr{V}$ be defined exactly as in the proof of Theorem 11.5.2. By Theorem 11.5.1, $\mathscr{V}$ is given by a semimetric D. The uniformity of D is smaller than $\mathscr{U}$ ⟦namely, $\mathscr{V} \subset \mathscr{U}$⟧. Let $V_\varepsilon = \{(x, y): D(x, y) < \varepsilon\}$. Then $\{V_\varepsilon: \varepsilon > 0\}$ is a base for $\mathscr{V}$, thus $V_\varepsilon \subset V$ for some $\varepsilon > 0$. ∎

LEMMA 11.5.3. *Let $(X, \mathscr{U})$ be a separated uniform space and $x, y \in X$, $x \neq y$. Then there exists a semimetric D whose uniformity is smaller than $\mathscr{U}$ such that $D(x, y) \neq 0$.*

(It is not intended that d induce the topology of X.) Choose a symmetric connector V with $(x, y) \notin V$ ⟦Theorem 11.1.4⟧. Choose D as in Lemma 11.5.2. Then $D(x, y) \neq 0$. ⟦For some $\varepsilon > 0$, $D(x, y) < \varepsilon$ implies $(x, y) \in V$, since V is a D connector. Hence $(x, y) \notin V$ implies $D(x, y) \geq \varepsilon$.⟧ ∎

THEOREM 11.5.3. *A uniform space $(X, \mathscr{U})$ is uniformly homeomorphic into a product of semimetric spaces.*

Let $S = \{d: d$ is a semimetric inducing a uniformity which is smaller than $\mathscr{U}\}$. (d need not be equivalent with $\mathscr{U}$.) S is not empty since the indiscrete semimetric ($d = 0$) belongs to S. For each $d \in S$, let $X_d = (X, d)$ and let $v: X \to \prod \{X_d: d \in S\}$ be the "diagonal" map, that is, $v(x)$ is the constant member of the product whose value is always x; precisely: $v(x)_d = x$ for all $d \in S$. We first show that v is uniformly continuous. ⟦For each $d \in S$, $(P_d \circ v)(x) = v(x)_d = x$ so that $P_d \circ v$ is the identity map from X to X_d. Since the d-uniformity is smaller than $\mathscr{U}$, $P_d \circ v$ is uniformly continuous by Theorem 11.2.1. It follows from Theorem 11.4.7 that v is uniformly continuous.⟧ It is obvious that v is one-to-one, thus it remains to prove that for each connector V for X, there exists U in the product uniformity such that $(a, b) \in U$, $a, b \in v[X]$ imply $(v^{-1}a, v^{-1}b) \in V$. Phrased another way, what we have to prove, given V, is the existence of U with the property $x, y \in X$, $(vx, vy) \in U$ imply $(x, y) \in V$. Let $V \in \mathscr{U}$ and choose D as in Lemma 11.5.2. Then $D \in S$, and with $V_\varepsilon = \{(x, y): D(x, y) < \varepsilon\}$, we have $V_\varepsilon \subset V$ for some $\varepsilon > 0$. Now V_ε belongs to the uniformity of the space X_D, so if we set $U = \{(a, b): a, b \in \prod X_d, (P_D a, P_D b) \in V_\varepsilon\}$, U is in the product uniformity, and $(x, y) \in X$, $(vx, vy) \in U$ imply $(x, y) \in V$ as required. ⟦Each of the following statements implies the next: $(vx, vy) \in U$, $(P_D vx, P_D vy) \in V_\varepsilon$, $[(vx)_D, (vy)_D] \in V_\varepsilon$, $(x, y) \in V_\varepsilon$, $(x, y) \in V$.⟧ ∎

Several methods of completing a uniform space are known. The one given here will use Theorem 11.5.3. A *completion* of a uniform space X is a pair (Y, f), where Y is a complete uniform space and f is a uniform homeomorphism of X onto a dense subspace of Y. (When X, Y are semimetric spaces, it was required in Section 9.2 that f be an isometry; where it is necessary to make the distinction we shall call Y a *semimetric completion* when f is an isometry.) A completion in which Y is separated will be called a *separated completion.*

Completion resembles Stone–Cech compactification in that it has the same sort of extension property. Suppose that X is a dense subspace of a complete space Y. Then *every uniformly continuous map of X into a complete space has a uniformly continuous extension to all of Y* ⟦Theorem 11.3.4⟧. We may identify X with a dense subspace of its completion and apply this result.

THEOREM 11.5.4. *Every uniform space X has a completion.*

By Theorem 11.5.3 and the completion theorem for semimetric spaces, Theorem 9.2.2, there exists a product $\prod$ of complete semimetric spaces, and a uniform homeomorphism $v: X \to \prod$. Then $(\overline{v[X]}, v)$ is the required completion; for since $\prod$ is complete ⟦Sec. 11.4, Problem 7⟧, so is any closed subspace ⟦Sec. 11.3, Problem 8⟧. ∎

In order to construct a separated completion for a separated uniform space X, we begin with an appropriate modification of Theorem 11.5.3.

THEOREM 11.5.5. *A separated uniform space $(X, \mathscr{U})$ is uniformly homeomorphic into a product of metric spaces.*

Accepting Theorem 11.5.3 and the notation of its proof we have a uniform homeomorphism $v: X \to \prod X_d$, where $X_d = (X, d)$ is a semimetric space for each $d \in S$. For each $D \in S$ there exists a metric space Y_D and a function q_D from X_D onto Y_D which preserves distances ⟦Sec. 6.7, Example 5⟧, hence is uniformly continuous. Define $q: \prod X_d \to \prod Y_d$ by $q(z)_D = q_D(z_D)$, where $z = (z_d) \in \prod X_d$. Then, denoting P_D, Φ_D, the projections of $\prod X_d$, $\prod Y_d$ on

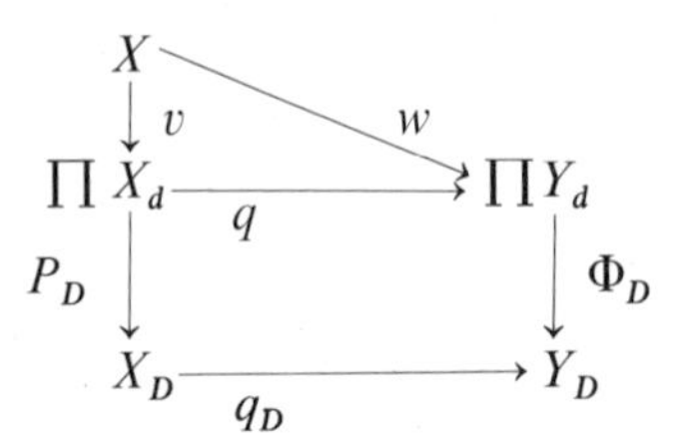

X_D, Y_D respectively, we have $\Phi_D \circ q = q_D \circ P_D$, ⟦$q_D(P_D z) = q_D(z_D) = q(z)_D = \Phi_D(qz)$⟧; that is, the lower part of the diagram shown is commutative. It follows that q is uniformly continuous. ⟦By Theorem 11.4.7, q is uniformly continuous if and only if $\Phi_D \circ q$ is uniformly continuous for all $D \in S$. But this function is $q_D \circ P_D$, the composition of two uniformly continuous functions.⟧ The required embedding is w, defined by $w = q \circ v$; w is uniformly continuous since both q and v are; and it is one-to-one. ⟦If $w(a) = w(b)$, then, for each D, $\Phi_D(wa) = \Phi_D(wb)$. Since q_D is distance preserving, a and b must be carried into two points of X_D whose distance apart is 0; that is, $D[P_D(va), P_D(vb)] = 0$ for all D. But $P_D(va) = a$ by definition of v, and so $D(a, b) = 0$ for all $D \in S$. By Lemma 11.5.3, $a = b$.⟧ It remains to prove that w^{-1} is uniformly continuous. As in the proof of Theorem 11.5.3, this requires, for a given connector V for X, construction of a connector U for the product uniformity of $\prod Y_d$ such that

$$a, b \in X, (wa, wb) \in U \qquad \text{imply} \qquad (a, b) \in V. \tag{11.5.8}$$

Let V be given and, using Lemma 11.5.2, choose $D \in S$ and $\varepsilon > 0$ so that $V \supset V_\varepsilon = \{(a, b): a, b \in X, D(a, b) < \varepsilon\}$. Now D is the semimetric of X_D, and there is no harm in using the same letter, D, to denote the metric of Y_D;

thus $D(q_D a, q_D b) = D(a, b)$ for $a, b \in X_D$, (that is, $a, b \in X$). ⟦q_D is distance preserving.⟧ Let $W_\varepsilon = \{(y, y'): y, y' \in Y_D, D(y, y') < \varepsilon\}$. Then W_ε belongs to the uniformity of Y_D and so, if we set $U = \{(s, t): s, t \in \prod Y_d, (\Phi_D s, \Phi_D t) \in W_\varepsilon\}$, U is a connector for the product uniformity of $\prod Y_d$; and it remains to check the truth of Equation (11.5.8). ⟦Let $a, b \in X$, $(wa, wb) \in U$. Then $(\Phi_D wa, \Phi_D wb) \in W_\varepsilon$, hence

$$\varepsilon > D(\Phi_D wa, \Phi_D wb) = D(q_D P_D va, q_D P_D vb) = D(P_D va, P_D vb) = D(a, b).$$

Hence $(a, b) \in V_\varepsilon \subset V$.⟧ ∎

We shall deduce from Theorem 11.5.5 that every separated uniform space has a separated completion. We saw in Sec. 9.2, Example 1, that completions are not unique. However, uniqueness is restored in the presence of separation.

THEOREM 11.5.6. *Every separated uniform space has a separated completion. Moreover, the completion is unique, in the sense that if (Y, f), (Z, g) are separated completions of X, there exists a uniform homeomorphism h from Y onto Z such that $h \circ f = g$.*

By Theorem 11.5.5 and the completion theorem for metric spaces, Theorem 9.2.3, there exists a product $\prod$ of complete metric spaces, and a uniform homeomorphism $w: X \to \prod$. Then $(\overline{w[X]}, w)$ is the required completion. See the details in the proof of Theorem 11.5.4. For the uniqueness part of the theorem, consider $g \circ f^{-1}: f[X] \to g[X]$. This is a uniform homeomorphism onto ⟦Sec. 11.2, Problem 9⟧, and the result follows from Sec. 11.3, Problem 11 and its hint. ∎

DEFINITION 1. *The* separated completion *of a separated uniform space will be denoted by γX.*

It is unique in the sense of Theorem 11.5.6.

EXAMPLE 1. The Stone–Cech compactification may be constructed by completion. Let X be a Tychonoff space and let $b = uC^*(X)$, the weak uniformity by the bounded continuous real functions. (Ignore the earlier discussions of this uniformity under the name β.) Let bX be the completion of (X, b). It is clear that *X is dense in bX*. Also *X is C^*-embedded in bX*. ⟦Every $f \in C^*(X)$ is b-uniformly continuous by definition of b and the result follows from Theorem 11.3.4.⟧ Next X is b-totally bounded. ⟦For $f \in C^*(X)$ let $a = \inf\{f(x): x \in X\}$, $b = \sup\{f(x): x \in X\}$. Given $\varepsilon > 0$, choose integer $n > (b - a)/\varepsilon$ and let

$$X_k = \{x: a + (k - 1)(b - a)/n \le f(x) \le a + k(b - a)/n\}$$

for $k = 1, 2, \ldots, n$. Then $X = \bigcup X_k$ and each X_k is small of order $\{(x, y): |f(x) - f(y)| < \varepsilon\}$. Since the set of such connectors is a subbase for

b, the result follows by Sec. 11.3, Problem 107.⟧ Thus *bX is compact*. We thus have that X is densely C^*-embedded in a compact Hausdorff space. Such an object is unique ⟦Theorem 8.5.2⟧. That $b = \beta$ as defined in Sec. 11.4, Example 2, is simply by uniqueness; with the earlier construction of βX, this required proof (Sec. 11.4, Problem 110).

COROLLARY 11.5.1. *Let X, Y be uniform spaces with Y separated, and $f: X \to Y$ a continuous function such that $f \mid D$ is uniformly continuous for some dense subspace D of X. Then f is uniformly continuous on X.*

$f \mid D$ has an extension to a uniformly continuous function $g: X \to \gamma Y$ ⟦Theorem 11.3.4⟧. Since γY is separated and $f = g$ on a dense subspace, $f = g$ ⟦Sec. 4.2, Problem 7⟧. ∎

COROLLARY 11.5.2. *Let Y be a set, X a subset and suppose that Y has two equivalent separated uniformities $\mathscr{U}$, $\mathscr{V}$ such that X is dense in Y in the uniform topology, and the relative uniformities of $\mathscr{U}$, $\mathscr{V}$ on X are equal. Then $\mathscr{U} = \mathscr{V}$.*

The identity map $i: (Y, \mathscr{U}) \to (Y, \mathscr{V})$ is continuous, and $i \mid X$ is uniformly continuous. By Corollary 11.5.1, i is uniformly continuous, hence $\mathscr{V} \subset \mathscr{U}$. By symmetry, the result follows. ∎

Problems on Uniform Space

NOTE. All uniformities mentioned are supposed to induce the given (or natural) topology in each case.

1. The completion of X is compact if and only if X is totally bounded ⟦Theorem 11.3.7; Sec. 11.3, Problem 13⟧. (In case X is not separated, the result refers to every completion.) For this reason the word *pre-compact* is also used for totally bounded. Similarly a *pre-realcompact* space is one whose completion is realcompact.
2. Let $\mathscr{U}_1$ and $\mathscr{U}_2$ be equivalent uniformities for a set X, and let X_1, X_2 be the respective completions. Then $\mathscr{U}_1 \supset \mathscr{U}_2$ if and only if there is a uniformly continuous map f from X_1 into X_2 with $f(x) = x$ for $x \in X$ ⟦Theorem 11.2.1⟧.
3. A separable uniform space need not be second countable. ⟦$\beta\omega$.⟧

---

101. A topological space (X, T) is (completely) regular if for every closed set F and $x \notin F$ there exists a (completely) regular topology T' for X such that $T' \subset T$ and $x \notin \mathrm{cl}_{T'} F$. ⟦See the proof of Theorem 11.5.2.⟧
102. There exists a nonmetrizable uniformity for ω with which it has a one-point completion. ⟦Let X be the subspace of $\beta\omega$ consisting of ω and one more point. Then X is realcompact by Theorem 8.6.3, hence υ-complete by Sec. 11.4, Example 5. Let υ' be the relative uniformity

of υ on ω. Then υ' is not metrizable since if it were, its completion would be also by Theorem 11.5.6 and Theorem 9.2.3, whereas X is not metrizable by Sec. 8.3, Problem 103.⟧

103. Obtain υX as a completion similar to Example 1. ⟦Complete (X, υ).⟧
104. Let X be a Tychonoff space. Then β is the largest totally bounded uniformity for X. ⟦Let $(X, \mathscr{U})$ be totally bounded. By Theorem 8.3.1, extend the identity map on X to continuous $I: \beta X \to \gamma(X, \mathscr{U})$; (the latter space is compact, by Problem 1). Then I is uniformly continuous by Corollary 11.2.1; hence $i: (X, \beta) \to (X, \mathscr{U})$ is uniformly continuous.⟧
105. Let X be a locally compact regular space. Then α is the smallest totally bounded uniformity. ⟦We may assume that $(X, \mathscr{U})$ is totally bounded and not compact. Define $f: \gamma(X, \mathscr{U}) \to X^+$ by $f \mid X = i$ and $f[Y \setminus X] = \{\infty\}$. Then f is uniformly continuous by Problem 1 and Corollary 11.2.1. Hence $i: (X, \mathscr{U}) \to (X, \alpha)$ is uniformly continuous.
106. "Totally bounded" cannot be omitted in Problem 104 ⟦Sec. 11.4, Problem 104⟧. (It can, however, be omitted in Problem 105 ⟦see [Gillman and Jerison, 15K1]⟧.)
107. In contrast with Problem 104, υ need not be the largest pre-realcompact uniformity ⟦Sec. 11.4, Problem 116; Sec. 8.6, Problem 111⟧.
108. Let X be a countably compact noncompact space such that $X^+ = \beta X$. (Examples are shown in Section 14.5 and Section 14.6.) Then X has only one uniformity even though it is not compact ⟦Problems 104 and 105; Sec. 11.3, Problem 20⟧.
109. A subset of a Tychonoff space X is pseudobounded if and only if it is υ-totally bounded. ⟦Half is trivial from Sec. 11.3, Problem 9. Conversely if S is pseudobounded, Sec. 8.6, Problem 114 implies that it is a relatively compact set in υX. Another proof of the converse imitates the proof that (X, β) is always totally bounded.⟧
110. A uniform space is called *N-complete* if every closed totally bounded subset is compact. (This is in honor of John von Neumann.) Show that every complete space is N-complete ⟦Problem 1⟧, and every N-complete space is sequentially complete. ⟦A Cauchy sequence is totally bounded.⟧ Thus for semimetric space all these definitions coincide.
111. A sequentially complete first countable space need not be N-complete, hence need not be complete ⟦Sec. 11.4, Example 3⟧.
112. Show that a Tychonoff space X is an NS space if and only if (X, υ) is N-complete ⟦Problem 109⟧. Compare the facts that X is realcompact if and only if (X, υ) is complete ⟦Sec. 11.4, Example 5⟧, and compact if and only if (X, β) is complete (equivalently, N-complete, since it is totally bounded).

201. Compare completions with different uniformly equivalent metrics, for example draw the completions of $(0, 1)$ with d, $d/(1 + d)$, and $d \wedge 1$ if d is the Euclidean metric.
202. Show that for each uniform space there exists a family F of semimetrics such that $\{V_{d,\varepsilon}: d \in F, \varepsilon > 0\}$ is a subbase for the uniformity, where $V_{d,\varepsilon} = \{(x, y): d(x, y) < \varepsilon\}$. ⟦Choose one d for each $\mathscr{U}$ as in Lemma 11.5.2.⟧
203. Can "separated" be omitted in Corollaries 11.5.1 and 11.5.2?

Topological Groups

12

12.1 Group Topologies

The spaces of classical analysis such as **R** and $C(X)$ have, in addition to their topological (uniform, metric) structure, various algebraic properties in the form of certain natural operations which are defined on them. For example, members of **R** or $C(X)$ may be added. One of the simplest algebraic structures is the group. In attempting to combine the studies of groups and topological spaces, one might begin with a set that is simultaneously a group and a topological space. Such an assumption would be too general; for example, it would be impossible to generalize the simple fact about **R** that if $x_n \to 0$, then $x_n + 2 \to 2$; since it is possible to put a topology on **R** which makes this false. ⟦Add $\{2\}$ to the Euclidean topology.⟧ It is natural to study those spaces, each of which is a group and a topological space, and is such that the group operations are continuous. The most interesting and valuable results are those in which hypotheses in one theory imply conclusions in the other, and topological concepts are phrased algebraically or vice versa. See for example Problems 21, 103, and 113; Sec. 12.3, Problems 3 and 4.

Let X be a group in which the group operation is designated by $(x, y) \to xy$. By definition, X has an identity, e, such that $ex = xe = x$ for all x; (in particular, X is not empty), also, for each x, X contains a member x^{-1}, called the inverse of x, such that $xx^{-1} = x^{-1}x = e$. Finally the group operation is associative, $(xy)z = x(yz)$. For emphasis we point out that xy is always defined, and that there is only one identity, and only one inverse for each member of the group. If $S \subset X$, $e \in S$ and $xy^{-1} \in S$ for all $x, y \in S$, we call S a *subgroup* of X. A subgroup is itself a group. ⟦For example, $x \in S$ implies

$x^{-1} = ex^{-1} \in S$; and $x, y \in S$ imply $xy = x(y^{-1})^{-1} \in S$.] An *invariant* subgroup of X is a subgroup S such that $xSx^{-1} \subset S$ for all $x \in X$. A *commutative* group satisfies $xy = yx$ for all x, y.

NOTATION. By x^2 is meant xx, $x^3 = xxx$ and so on, $x^0 = e$, $x^{-n} = (x^{-1})^n$. For subsets A, B of a group X,

$$AB = \{ab: a \in A, b \in B\}, \qquad aB = \{ab: b \in B\},$$

$$A^{-1} = \{a^{-1}: a \in A\}.$$

$A^2 = AA$ and so on. For filterbases $\mathscr{F}_1$, $\mathscr{F}_2$, $\mathscr{F}$, the filterbases

$$\mathscr{F}_1\mathscr{F}_2 = \{AB: A \in \mathscr{F}_1, B \in \mathscr{F}_2\}, \qquad \mathscr{F}^{-1} = \{A^{-1}: A \in \mathscr{F}\},$$

$$\mathscr{F}^2 = \{A^2: A \in \mathscr{F}\}$$

and so on.

ADDITIVE NOTATION. Occasionally, and for commutative groups only, the group operation is written $(x, y) \to x + y$, the identity is written 0, and the inverse of x is written $-x$. Then $x + x$ is written $2x$ rather than x^2, and so on.

The left *cosets* of a subgroup S of X are the sets xS for $x \in X$. If S is an invariant subgroup, the multiplication $(xS)(yS) = (xy)S$ makes the collection of left cosets into a group called the *quotient group* by S and written X/S.

A topology T for a group X is called a *group topology* if it makes the group operations continuous. Specifically, the maps $(x, y) \to xy$ and $x \to x^{-1}$ which carry $X \times X$ to X, and X to X, respectively, are to be continuous; $X \times X$ has the product topology of course. A *topological group* is a pair (X, T) in which X is a group and T is a group topology. A *semimetric* group is a pair (X, d) in which d is a semimetric which induces a group topology.

★EXAMPLE 1. Suppose that a function $p: X \to \mathbf{R}$ is given, where X is a group. Let p have the properties: $p(x) \geq 0$ for all x, $p(e) = 0$, $p(x^{-1}) = p(x)$ for all x, $p(xy) \leq p(x) + p(y)$ for all x, y, and $p(x_n) \to 0$ implies $p(ax_na^{-1}) \to 0$ for all a. (The last condition is redundant for a commutative group.) Such a function p will be called an *absolute-value function* because of its resemblance to the ordinary absolute value. We shall show that *defining* $d(x, y) = p(x^{-1}y)$ *yields a semimetric* d *for* X *which induces a group topology.* It is clear that $d(x, y) \geq 0$ for all x, y, that $d(x, x) = 0$, and that $d(x, y) = d(y, x)$.

$$[\![d(x, y) = p(x^{-1}y), d(y, x) = p(y^{-1}x), y^{-1}x = (x^{-1}y)^{-1}.]\!]$$

Also

$$d(x, z) = p(x^{-1}z) = p(x^{-1}yy^{-1}z) \leq p(x^{-1}y) + p(y^{-1}z) = d(x, y) + d(y, z).$$

Thus d is a semimetric. It makes the map $x \to x^{-1}$ continuous. [Let $\{x_n\}$ be a sequence with $x_n \to x$. Then $d(x_n^{-1}, x^{-1}) = p(x_nx^{-1}) = p(xx^{-1}x_nx^{-1}) \to 0$

since $p(x^{-1}x_n) = d(x, x_n) \to 0.$⟧ Finally, the group operation is continuous. To see this, it is sufficient to check sequences since $X \times X$ is a semimetric space ⟦Theorem 6.4.2⟧. Let $x_n \to x$, $y_n \to y$. Then

$$d(x_n y_n, xy) = p(y_n^{-1} x_n^{-1} xy) = p(y_n^{-1} y y^{-1} x_n^{-1} xy)$$
$$\leq p(y_n^{-1} y) + p(y^{-1} x_n^{-1} xy) \to 0.$$
$$[\![p(x_n^{-1} x) = d(x_n, x) \to 0.]\!]$$

★EXAMPLE 2. Invariant semimetrics. A semimetric d for a group is called *left invariant* if $d(ax, ay) = d(x, y)$ for all a, x, y; *right invariant* if $d(xa, ya) = d(x, y)$; and *two-sided invariant* if it is both left and right invariant. The semimetric given in Example 1 is left invariant.

$$[\![d(ax, ay) = p(x^{-1}a^{-1}ay) = p(x^{-1}y) = d(x, y).]\!]$$

Conversely, *every left-invariant semimetric d which induces a group topology is derived from an absolute-value function p by means of the formula* $d(x, y) = p(x^{-1}y)$. Indeed if we set $p(x) = d(x, e)$, p will have all the cited properties. That $p(x) \geq 0$ and $p(e) = 0$ are trivial. Next,

$$p(x^{-1}) = d(x^{-1}, e) = d(xx^{-1}, xe) = d(e, x) = d(x, e) = p(x);$$
$$p(xy) = d(xy, e) = d(y, x^{-1}) \leq d(y, e) + d(e, x^{-1}) = p(x) + p(y).$$

The last condition on p is checked by means of the observation that $x_n \to e$ if and only if $p(x_n) \to 0$, ⟦$d(x_n, e) = p(x_n)$⟧, hence $p(x_n) \to 0$ implies $x_n \to e$; this implies that $ax_na^{-1} \to aea^{-1} = e$ ⟦the group operation is continuous⟧, and so, finally, $p(ax_na^{-1}) \to p(e) = 0$.

NOTE. We shall make immediate use of the results on right and two-sided invariant semimetrics given in Problems 7, 8, and 9.

★EXAMPLE 3. Let $I = [0, 1]$, $G = \{f : f$ is a homeomorphism of I onto itself$\}$. Let the group operation be composition, $fg = f \circ g$; thus $(fg)(x) = f(gx)$, $e(x) = x$ for all x, and f^{-1} is both the inverse of the function f and the inverse of $f \in G$. Define $p(f) = \max\{|f(x) - x| : x \in I\}$. Then $p(e) = 0$, $p(f) \geq 0$ for all f and, indeed, p is an absolute-value function in the sense of Example 1, for $p(f^{-1}) = p(f)$. ⟦Let $p(f) = |f(y) - y|$ and $x = f(y)$. Then $p(f^{-1}) \geq |f^{-1}(x) - x| = |y - f(y)| = p(f)$; also $p(f) = p[(f^{-1})^{-1}] \geq p(f^{-1})$⟧, also, $p(fg) \leq p(f) + p(g)$

$$[\![|f(gx) - x| \leq |f(gx) - g(x)| + |g(x) - x|$$
$$= |f(y) - y| + |g(x) - x| \leq p(f) + p(g),$$

where $y = g(x)$⟧, and, finally, if $p(f_n) \to 0$, then $p(gf_ng^{-1}) \to 0$ for all g as we now prove. Let $\varepsilon > 0$. Choose $\delta > 0$ such that $a \in I$, $b \in I$, $|a - b| < \delta$ imply

$|g(a) - g(b)| < \varepsilon$. ⟦$g$ is uniformly continuous on I, by Corollary 11.2.1.⟧ Choose N so that $n > N$ implies $p(f_n) < \delta$. Let $p(gf_ng^{-1}) = |(gf_ng^{-1})(x_n) - x_n|$ for $n = 1, 2, \ldots$. Then for $n > N$, $p(gf_ng^{-1}) = |(gf_n)(y_n) - g(y_n)| < \varepsilon$, where $y_n = g^{-1}(x_n)$; ⟦the expression is $|g(y_n + z_n) - g(y_n)|$ with $|z_n| = |f_n(y_n) - y_n| \le p(f_n) < \delta$⟧. Since, in addition to the properties listed, $p(f) > 0$ if $f \neq e$, p leads to a left-invariant metric as in Example 2, and a right-invariant metric as in Problem 7. A form of the right-invariant metric is $d(f, g) = \max\{|f(x) - g(x)| : x \in I\}$. ⟦$d(f, g) = p(fg^{-1}) = \max|fg^{-1}x - x| = \max|fg^{-1}x - gg^{-1}x| = \max\{|f(y) - g(y)| : y \in I\}$ since g^{-1} maps I onto itself.⟧ See also Problem 19.

★EXAMPLE 4. *Let Φ be a collection of group topologies for a group X. Then $\bigvee \Phi$ is a group topology.* Suppose that a net $x_\delta \to e$ in $\bigvee \Phi$. Then $x_\delta \to e$ in each $T \in \Phi$ ⟦Theorem 6.2.1⟧. Thus $x_\delta^{-1} \to e$ in each $T \in \Phi$ and so $x_\delta^{-1} \to e$ in $\bigvee \Phi$ ⟦Theorem 6.2.1⟧. Similarly $x_\delta \to x$, $y_\delta \to y$ implies $x_\delta y_\delta \to xy$. In contrast with the result of this Example, the inf of two group topologies need not be a group topology ⟦Sec. 12.2, Problem 125⟧.

★EXAMPLE 5. Let $f: X \to Y$, where X, Y are groups. We call f a *homomorphism* if $f(ab) = f(a)f(b)$. It follows that $f(e) = e$. ⟦$f(e) = f(ee) = f(e)f(e)$⟧, and $f(a^{-1}) = (fa)^{-1}$, ⟦$f(a^{-1})f(a) = f(a^{-1}a) = f(e) = e$⟧. If n is a positive integer

$$f(x^n) = f(xx \cdots x) = f(x)f(x) \cdots f(x) = (fx)^n.$$

If n is a negative integer,

$$f(x^n) = f[(x^{-1})^{-n}] = [f(x^{-1})]^{-n} = [(fx)^{-1}]^{-n} = (fx)^n.$$

Thus the equation $f(x^n) = (fx)^n$ holds for all integer n. An *isomorphism* is a one-to-one homomorphism, and a *topological isomorphism* is an isomorphism which is a homeomorphism. An *endomorphism* is a homomorphism of a group into itself, an *automorphism* is a one-to-one endomorphism (hence an isomorphism), and a *topological automorphism* is an automorphism which is a homeomorphism. *The weak topology on a group X by a homomorphism $f: X \to Y$, Y a topological group, is a group topology.* If $x_\delta \to e$ in $w(f)$ we have $f(x_\delta) \to e$ in Y, hence $(fx_\delta)^{-1} \to e$ in Y and so $f(x_\delta^{-1}) \to e$ in Y. Thus $x_\delta^{-1} \to e$ in $w(f)$ ⟦Lemma 6.3.1⟧. Similarly if $x_\delta \to x$ and $y_\delta \to y$ in $w(f)$ it follows that $x_\delta y_\delta \to xy$. ▌ It follows immediately that the weak topology on a group X by a collection of homomorphisms to topological groups is a group topology ⟦Example 4 and the definition of weak topology⟧.

★EXAMPLE 6. *Any product of topological groups is a topological group.* This is a special case of Example 5 when it is pointed out that every product of groups is a group under the definition $(xy)_\alpha = x_\alpha y_\alpha$ for $x, y \in \prod\{X_\alpha : \alpha \in A\}$.

Let X be a topological group and $a \in X$. Consider the function $L_a: X \to X$ given by $L_a(x) = ax$. This function is called *left translation by a*. Similarly *right translation by a* is defined by the equation $R_a(x) = xa$. We shall discuss only left translation; all such discussions will apply as well to right translation with the change of a few appropriate words. *Left translation by a is a homeomorphism of X onto itself.* It is continuous ⟦Sec. 6.6, Problem 4⟧, and has the continuous inverse $L_{a^{-1}}$. ⟦For example, $L_a[L_{a^{-1}}(x)] = a(a^{-1}x) = x$.⟧ From this it follows that *a set N is a neighborhood of a point x if and only if $x^{-1}N$ is a neighborhood of e.* ⟦A homeomorphism onto preserves neighborhoods, thus N is a neighborhood of x if and only if $L_{x^{-1}}[N]$ is a neighborhood of $L_{x^{-1}}(x)$.⟧ Because of this fact all discussions of a group topology may be localized at the identity element. Let $\mathscr{V}$ be the set of neighborhoods of e, then *N is a neighborhood of a point x if and only if there exists $U \in \mathscr{V}$ with $xU \subset N$.* ⟦If N is a neighborhood of x, let $U = x^{-1}N$. Conversely if $xU \subset N$ for some $U \in \mathscr{V}$, $x^{-1}N \supset U$ is a neighborhood of e, hence, as just proved, N is a neighborhood of x.⟧ As an example of the localization we have, for a net x_δ, *$x_\delta \to x$ if and only if $x^{-1}x_\delta \to e$.* ⟦If $x_\delta \to x$, $x^{-1}x_\delta \to x^{-1}x = e$ by continuity; and if $x^{-1}x_\delta \to e$, $x_\delta = x(x^{-1}x_\delta) \to xe = x$.⟧

The operation $x \to x^{-1}$ is also a homeomorphism of X onto itself ⟦it is its own inverse⟧. It follows that if U is a neighborhood of e, U^{-1} is a neighborhood of $e^{-1} = e$, and so $U \cap U^{-1}$ is a neighborhood of e. Since $U \cap U^{-1}$ is clearly symmetric (call a set S *symmetric* if $S = S^{-1}$), we have the result: *A topological group has a base of symmetric neighborhoods of e.*

We shall turn, in Theorems 12.1.5 and 12.1.6, to the problem of defining a group topology for a given group. As we have just seen, it is sufficient to specify the neighborhoods of e. The following result will help to decide which collections of sets are eligible to be the collection of neighborhoods of e.

LEMMA 12.1.1. *Let X be a group and T a topology for X. Then the map $x \to x^{-1}$ is continuous at e if and only if whenever U is a neighborhood of e, U^{-1} is also a neighborhood of e. The map $(x, y) \to xy$ is continuous at $(e, e) \in X \times X$ if and only if, for every neighborhood U of e, there exists a neighborhood V of e such that $VV \subset U$.*

If the map $x \to x^{-1}$ is continuous, it is a homeomorphism onto ⟦it is its own inverse⟧, hence carries a neighborhood of e onto a neighborhood of e. Conversely, suppose that U^{-1} is a neighborhood of e whenever U is; let $\mathscr{F}$ be a filterbase converging to e, and let U be a neighborhood of e. Then U^{-1} is a neighborhood of e hence includes some $A \in \mathscr{F}$. Then $A^{-1} \subset U$ and $A^{-1} \in \mathscr{F}^{-1}$, thus $\mathscr{F}^{-1} \to e$. Next let the map $(x, y) \to xy$ be continuous at (e, e) and let $\mathscr{F}$ be the filter of all neighborhoods of e. Then the filterbase $\mathscr{F}\mathscr{F} \to e$, in other words every neighborhood U of e includes a member

V_1V_2 of $\mathscr{F}\mathscr{F}$. Let $V = V_1 \cap V_2$. Then $VV \subset V_1V_2 \subset U$. Finally suppose that $\mathscr{F}_1$, $\mathscr{F}_2$ are filters converging to e, and let U be a neighborhood of e; if there exists a neighborhood V of e with $VV \subset U$ we have $V \in \mathscr{F}_1$ and $V \in \mathscr{F}_2$, hence $U \supset VV \in \mathscr{F}_1\mathscr{F}_2$ and so $U \in \mathscr{F}_1\mathscr{F}_2$. Thus $\mathscr{F}_1\mathscr{F}_2 \to e$. ▌

We now discuss procedures for introducing uniformities into groups. The end product will be three uniformities $\mathscr{L}$, $\mathscr{R}$ and $\mathscr{B}$ (left, right, and two-sided). We begin with $\mathscr{L}$. Each set U containing e leads to a connector $U_L = \{(x, y): x^{-1}y \in U\}$. ⟦$\Delta \subset U_L$ since $x^{-1}x = e \in U$ so that $(x, x) \in U_L$ for all x.⟧ Moreover, $U_L(x) = xU$. ⟦$y \in U_L(x)$ implies $(x, y) \in U_L$, $x^{-1}y \in U$, $y \in xU$ and conversely.⟧ We call U_L the *connector L-associated with U*. In particular every neighborhood of e leads to a connector and we now see that the set of all such connectors uniformizes the group topology. We shall make use of the formulas: $U_L \cap V_L = (U \cap V)_L$ ⟦if $(x, y) \in U_L \cap V_L$, $x^{-1}y \in U \cap V$, hence $(x, y) \in (U \cap V)_L$ and conversely⟧ $(U_L)^{-1} = (U^{-1})_L$ ⟦if $(x, y) \in (U_L)^{-1}$, then $(y, x) \in U_L$, $y^{-1}x \in U$, $(y^{-1}x)^{-1} = x^{-1}y \in U^{-1}$, $(x, y) \in (U^{-1})_L$, and conversely⟧ and $U_L \circ V_L = (VU)_L$ ⟦if $(x, y) \in U_L \circ V_L$, then $y \in (U_L \circ V_L)(x)$ so that there exists z with $y \in U_L(z)$ and $z \in V_L(x)$; that is, $z^{-1}y \in U$, $x^{-1}z \in V$, hence $x^{-1}zz^{-1}y = x^{-1}y \in VU$ and so $(x, y) \in (VU)_L$; the converse is similar⟧.

THEOREM 12.1.1. *Every topological group is a uniform space.*

Let $\mathscr{L}' = \{U_L: U \text{ a neighborhood of } e\}$, where U_L is the connector L-associated with U. We show that $\mathscr{L}'$ is a base for a uniformity. (See Sec. 11.1, Definition 2.) First, it is a filterbase: ⟦$\varnothing \notin \mathscr{L}'$, since each $U_L \supset \Delta$; $\mathscr{L}' \neq \varnothing$, since $X_L = X \times X \in \mathscr{L}'$, and $U_L \cap V_L = (U \cap V)_L$, as proved above⟧. That $(U^{-1})_L = (U_L)^{-1}$ was just proved. Finally, let $U_L \in \mathscr{L}'$. By Lemma 12.1.1, there exists a neighborhood V of e with $VV \subset U$. Then $V_L \circ V_L \subset U_L$ because $V_L \circ V_L = (VV)_L$ ⟦proved above⟧. The uniformity $\mathscr{L}$ generated by $\mathscr{L}'$ is called the *left uniformity* for the topological group. We still have to show that the uniform topology T_L coincides with the group topology T_G. Let N be a T_L neighborhood of x. Then there exists $U_L \in \mathscr{L}$ with $U_L(x) \subset N$; that is, $xU \subset N$ with U a T_G neighborhood of e. This makes N a T_G neighborhood of x. Conversely, if N is a T_G neighborhood of x, say $N \supset xU$ with U a T_G neighborhood of e; then $xU = U_L(x)$ and $N \supset U_L(x)$, making N a T_L neighborhood of x. ▌

The *right uniformity* $\mathscr{R}$ is defined similarly. (The notation is given in Problem 18.) The *two-sided uniformity* $\mathscr{B}$ is defined to be $\mathscr{R} \vee \mathscr{L}$. The three uniformities $\mathscr{R}$, $\mathscr{L}$, $\mathscr{B}$ are equivalent ⟦Problem 18; Theorem 11.4.1⟧, but may not be equal ⟦Sec. 12.2, Example 3⟧. They are all equal for a compact group, since a compact space has at most one uniformity ⟦Corollary 11.2.2⟧. For a neighborhood U of e we define the *connector B-associated with U* to be $U_B = U_L \cap U_R$.

THEOREM 12.1.2. *The two-sided uniformity $\mathscr{B}$ has as a base the set of all U_B such that U is a symmetric neighborhood of e.*

Clearly $U_B \in \mathscr{B}$ for each U. Conversely, let $C \in \mathscr{B}$. Then $C \supset V_L \cap W_B$ for some neighborhoods V, W of e. Let U be a symmetric neighborhood of e with $U \subset V \cap W$. Then $C \supset V_L \cap W_R \supset U_L \cap U_R = U_B$. ▮

We proceed to list a few properties which topological groups have by virtue of being uniform spaces.

THEOREM 12.1.3. *Let A be a set in a topological group. Then*

$$\bar{A} = \bigcap \{AU: U \text{ a neighborhood of } e\}.$$

Recall the formula of Theorem 11.1.3 which implies in particular that $\bar{A} = \bigcap \{U_L(A): U \text{ a neighborhood of } e\}$. [[$\{U_L: U$ a neighborhood of $e\}$ is a base for the left uniformity.]] But $U_L(A) = \bigcup \{U_L(a): a \in A\} = \bigcup \{aU: a \in A\} = AU$. ▮

It is left to the reader [[Problems 13 and 14]] to show that if b is a base for the the neighborhood system of e, then $\bar{A} = \bigcap \{AU: U \in b\}$.

COROLLARY 12.1.1. *For every set A and neighborhood U of e, $\bar{A} \subset AU$.*

See also Problem 103.

THEOREM 12.1.4. *Every topological group is completely regular. The following conditions on a topological group X are equivalent:*

(i) *X is a T_0 space.*
(ii) *X is a Tychonoff space.*
(iii) $\bigcap \{U: U$ *is a neighborhood of* $e\} = \{e\}$.

Every uniform space is completely regular [[Theorem 11.5.2]], hence Conditions (i) and (ii) are equivalent. Theorem 12.1.3 implies that $\bigcap \{U: U$ is a neighborhood of $e\}$ is $\overline{\{e\}}$. This is equal to $\{e\}$ if and only if $\{e\}$ is closed, hence if and only if all singletons are closed [[translation is a homeomorphism onto]], so, finally, if and only if the space is T_1. ▮

★EXAMPLE 7. *Let (X, d) be a semimetric group and suppose that d is left invariant. Then d induces the left uniformity $\mathscr{L}$.* Let $D^\varepsilon = \{(x, y): d(x, y) < \varepsilon\}$. Then each D^ε, $\varepsilon > 0$, is a connector in the d uniformity. But $D^\varepsilon \in \mathscr{L}$. [[Let $U = \{x: d(x, e) < \varepsilon\}$; then U is a neighborhood of e, and D^ε is the connector L-associated with U since $(x, y) \in D^\varepsilon$ if and only if $d(x, y) < \varepsilon$, equivalently, $d(e, x^{-1}y) < \varepsilon$; and this is true if and only if $x^{-1}y \in U$.]] Since every connector in the d uniformity includes such a D^ε, it follows that $d \subset \mathscr{L}$. Conversely, $\mathscr{L}$ has a base of connectors of the form U_L where U_L is the connector L-associated with a neighborhood U of e. Then U includes some cell N of radius ε, hence U_L includes the connector associated with N, namely D^ε.

Thus U_L belongs to the d uniformity, and so $\mathscr{L} \subset d$. ▌ Similarly, if d is right invariant, it induces the right uniformity $\mathscr{R}$. Any left-invariant semimetric immediately leads to a right-invariant one ⟦Problem 8⟧.

It follows from Example 7 that a two-sided-invariant semimetric induces all of $\mathscr{L}$, $\mathscr{R}$, $\mathscr{B}$ which must, then, be equal. We shall see in Sec. 12.2, Example 3, that these need not be equal in general.

Definition 1. *A B-semimetric is any semimetric D of the form $D(x, y) = d(x, y) + d(x^{-1}, y^{-1})$, where d is a left-invariant semimetric.*

The reason for this name is contained in Example 8. First note that, in Definition 1, D determines d; indeed let $p(x) = \frac{1}{2}D(x, e)$. Then $p(x) = \frac{1}{2}d(x, e) + \frac{1}{2}d(x^{-1}, e) = d(x, e)$ since d is left invariant. Thus p is the absolute-value function associated with d and so, by Example 2, $d(x, y) = p(x^{-1}y) = \frac{1}{2}D(x^{-1}y, e)$.

★EXAMPLE 8. *Let X be a topological group and suppose that its topology is induced by a B-semimetric D. Then D induces the two-sided uniformity $\mathscr{B}$.* (See Problem 101.) Let d be left invariant and D as in Definition 1. Let $p(x) = \frac{1}{2}D(x, e)$. Then, as just proved, p is the absolute-value function associated with d. In the following we shall use the formula $D(x, y) = p(x^{-1}y) + p(xy^{-1})$ ⟦the right-hand side is $d(x, y) + d(x^{-1}, y^{-1})$⟧. A basic connector for the D uniformity is a set of the form $U^{\varepsilon} = \{(x, y): D(x, y) < \varepsilon\} \supset \{(x, y): p(x^{-1}y) < \varepsilon/2\} \cap \{(x, y): p(xy^{-1}) < \varepsilon/2\} = V_L \cap V_R$, where $V = \{x: p(x) < \varepsilon/2\}$. Thus $\mathscr{B}$ includes the D uniformity. Conversely, a basic connector for $\mathscr{B}$ is a set of the form $U_B = U_L \cap U_R$ with U a neighborhood of e. Then, for some $\varepsilon > 0$, $U \supset \{x: D(x, e) < 2\varepsilon\} = \{x: p(x) < \varepsilon\}$, hence

$$U_B = U_L \cap U_R \supset \{(x, y): p(x^{-1}y) < \varepsilon\} \cap \{(x, y): p(xy^{-1}) < \varepsilon\}$$

$$\supset \{(x, y): p(x^{-1}y) + p(xy^{-1}) < \varepsilon\} = \{(x, y): D(x, y) < \varepsilon\}.$$

Thus $\mathscr{B}$ is included in the D uniformity.

We now turn to the problem of defining a group topology. We consider this in two steps. Given a topology for a group, check in some efficient way whether it is a group topology (Theorem 12.1.5); given a group, define a group topology for it (Theorem 12.1.6).

Theorem 12.1.5. *Let X be a group and T a topology for X. Then T is a group topology if and only if (omit* (d) *if X is commutative)*

(a) *every left translate of an open set is open;*

(b) *for every neighborhood U of e, U^{-1} is a neighborhood of e;*

(c) *for each neighborhood U of e, there exists a neighborhood V of e such that $VV \subset U$,*

(d) *for each neighborhood U of e and $a \in X$, there exists a neighborhood V of e with $aVa^{-1} \subset U$.*

Suppose first that T is a group topology. Property (a) is immediate since left translation is a homeomorphism of X onto itself. Properties (b), (c) follow from Lemma 12.1.1. Next let U be a neighborhood of e, and $a \in X$. The map $x \to axa^{-1}$ is continuous at e, hence there exists a neighborhood V of e such that this map carries it into U; that is $aVa^{-1} \subset U$. Conversely, suppose that T obeys the four conditions. We first observe that the group operation $(x, y) \to xy$ is continuous at $(e, e) \in X \times X$ ⟦Lemma 12.1.1⟧. To extend this result we shall prove two statements:

$$x_\delta \to e \quad \text{implies} \quad ax_\delta a^{-1} \to e \quad \text{for all} \quad a \tag{12.1.1}$$

and

$$\text{(i) } x_\delta \to x, \qquad \text{(ii) } x^{-1}x_\delta \to e, \qquad \text{(iii) } x_\delta x^{-1} \to e \text{ are equivalent} \tag{12.1.2}$$

⟦To prove statement (12.1.1), let U be a neighborhood of e, and $a \in X$. Choose V as in (d). Then $x_\delta \in V$ eventually, hence $ax_\delta a^{-1}$ belongs to aVa^{-1}, and hence to U eventually. To prove (12.1.2), assume that (i) holds. Let U be an open neighborhood of e. Then xU is a neighborhood of x and so $x_\delta \in xU$ eventually. Conversely, if (ii) holds, and N is an open neighborhood of x, $x^{-1}N$ is a neighborhood of e, so that $x^{-1}x_\delta \in x^{-1}N$ eventually; thus $x_\delta \in N$ eventually and so (i) holds. Assuming (ii) we obtain (iii) from statement (12.1.1) since $x_\delta x^{-1} = x(x^{-1}x_\delta)x^{-1}$; similarly (iii) implies (ii).⟧ Now we can prove that the group operation is continuous everywhere, for if $x_\delta \to x$, $y_\delta \to y$, we have $x_\delta y_\delta(xy)^{-1} = x(x^{-1}x_\delta)(y_\delta y^{-1})x^{-1} \to e$ by (12.1.1) and (12.1.2) and the fact that the operation is continuous at (e, e). Thus, again by statement (12.1.2), $x_\delta y_\delta \to xy$. Finally, the inverse operation is continuous. To see this, let $x_\delta \to x$; then $xx_\delta^{-1} = (x_\delta x^{-1})^{-1} \to e$. ⟦$x_\delta x^{-1} \to e$ by statement (12.1.2), and the result follows since the inverse operation is continuous at e, by Lemma 12.1.1.⟧ But $xx_\delta^{-1} = (x^{-1})^{-1}x_\delta^{-1}$ and so $x_\delta^{-1} \to x^{-1}$ by statement (12.1.2). ∎

It follows from Theorem 12.1.5 that the collection of symmetric neighborhoods of e satisfies the five conditions of the next theorem. We now see that these conditions are also sufficient for this purpose.

THEOREM 12.1.6. *Let X be a group and $\mathscr{F}$ a collection of subsets of X such that (omit* (d) *if X is commutative)*

(a) *$\mathscr{F}$ is a filterbase;*

(b) *every member of $\mathscr{F}$ is symmetric;*

(c) *for each $U \in \mathscr{F}$ there exists $V \in \mathscr{F}$ with $VV \subset U$;*

(d) *for each $U \in \mathscr{F}$ and $a \in X$, there exists $V \in \mathscr{F}$ with $aVa^{-1} \subset U$.*

Then there exists a unique group topology for X such that $\mathscr{F}$ is a local base for the neighborhoods of e.

For each $U \in \mathscr{F}$, let U_L be the connector L-associated with U. ⟦$e \in U$ for, with V chosen as in (c), V is not empty; let $a \in V$. Then $a^{-1} \in V$ and so $aa^{-1} \in U$.⟧ Let $\mathscr{L}' = \{U_L: U \in \mathscr{F}\}$. We shall show that $\mathscr{L}'$ is a base for a uniformity. ⟦That $\mathscr{L}'$ is a filterbase is clear since $\mathscr{F}$ is a filterbase. (Compare the proof of Theorem 12.1.1.) Each member of $\mathscr{L}'$ is symmetric, (see the formulas given before Theorem 12.1.1). Finally let $U_L \in \mathscr{L}'$. Choose $V \in \mathscr{F}$ with $VV \subset U$. Then $V_L V_L = (VV)_L \subset U_L$.⟧ Let $\mathscr{L}$ be the uniformity generated by $\mathscr{L}'$ and T the topology induced by this uniformity. A set N is a neighborhood of a point x if and only if there exists $U_L \in \mathscr{L}'$ with $U_L(x) \subset N$ ⟦Theorem 11.1.2⟧, hence if and only if there exists $U \in \mathscr{F}$ with $xU \subset N$ ⟦$xU = U_L(x)$⟧. In particular, each member of $\mathscr{F}$ is a neighborhood of e, and each neighborhood of e includes a member of $\mathscr{F}$. In other words, $\mathscr{F}$ is a base for the neighborhood system of e. We shall show that T is a group topology by checking the four conditions of Theorem 12.1.5. First, let G be open, $a \in X$. Then aG is open. ⟦Let $x \in aG$, then $a^{-1}x \in G$, so there exists a neighborhood U of e with $a^{-1}xU \subset G$. Then $xU \subset aG$ so that aG is a neighborhood of x. So aG is open.⟧ To check (b), let U be a neighborhood of e. Then U includes some $V \in \mathscr{F}$, hence $U^{-1} \supset V^{-1} = V$ and so U^{-1} is a neighborhood of e. To check Part (c), let U be a neighborhood of e, then U includes some $U_1 \in \mathscr{F}$. There exists $V \in \mathscr{F}$ with $VV \subset U_1 \subset U$. This sort of reasoning also yields Part (d). The fact that the topology is unique follows from the localization given above since the filter of neighborhoods of e is uniquely determined by $\mathscr{F}$. ∎

EXAMPLE 9. Let S_1 be the group of all complex numbers of absolute value 1 with ordinary multiplication. With the Euclidean topology this is a compact metric commutative group. Duality theory for groups consists of studying a group X by means of its homomorphisms into S_1. These are called *characters* of X. The set of all characters of X is made into a group by the definitions $(\chi_1\chi_2)(x) = \chi_1(x)\chi_2(x)$, $\chi^{-1}(x) = \chi(x^{-1}) = [\chi(x)]^{-1}$, $1(x) = 1$ for all x. Thus the constant character whose value is 1 is the identity for this group. The set of continuous characters is a subgroup of this group and is called the *character group of X* or *dual group of X*. In the case of noncommutative (even locally compact, metric) groups the character group may be trivial, (Example 10). The duality theory achieves its greatest success in the study of locally compact commutative groups. See [Rudin], [Hewitt and Ross].

EXAMPLE 10. Let X be the group (under ordinary matrix multiplication) of matrices $\left(\begin{smallmatrix} a & b \\ c & d \end{smallmatrix}\right)$ of real numbers with $ad - bc = 1$. The map $\left(\begin{smallmatrix} a & b \\ c & d \end{smallmatrix}\right) \to (a, b, c, d)$ is one-to-one from X onto a closed subspace of $\mathbf{R}^4$, thus X is made into a locally compact complete metric group. (The metric inherited from $\mathbf{R}^4$ has no variance properties.) Now let Y be any commutative group, and

$f: X \to Y$ a homomorphism. We shall show that $f(A) = e$ for all $A \in X$. First for any matrix A of the form $\left(\begin{smallmatrix} 1 & 0 \\ c & 1 \end{smallmatrix}\right)$, $f(A) = e$. ⟦Let

$$B = \begin{pmatrix} \sqrt{2}/2 & 0 \\ 0 & \sqrt{2} \end{pmatrix}.$$

Then $f(A) = f(BAB^{-1})$ (since Y is commutative) $= f(A^2) = (fA)^2$.⟧ If $A = \left(\begin{smallmatrix} 1 & b \\ 0 & 1 \end{smallmatrix}\right)$, $f(A) = e$. ⟦Let $B = \left(\begin{smallmatrix} 0 & 1 \\ -1 & 0 \end{smallmatrix}\right)$. Then $A = BCB^{-1}$, where $C = \left(\begin{smallmatrix} 1 & 0 \\ -b & 1 \end{smallmatrix}\right)$. By the earlier result $f(C) = e$.⟧ If $A = \left(\begin{smallmatrix} a & b \\ c & 1 \end{smallmatrix}\right)$, $f(A) = e$. ⟦$\left(\begin{smallmatrix} a & b \\ c & 1 \end{smallmatrix}\right) = \left(\begin{smallmatrix} 1 & b \\ 0 & 1 \end{smallmatrix}\right)\left(\begin{smallmatrix} 1 & 0 \\ c & 1 \end{smallmatrix}\right)$ since $a - bc = 1$.⟧ If $A = \left(\begin{smallmatrix} a & b \\ c & 0 \end{smallmatrix}\right)$, $f(A) = e$. ⟦$A = \left(\begin{smallmatrix} 1+c & 1 \\ c & 1 \end{smallmatrix}\right)\left(\begin{smallmatrix} a-c & b \\ c+c^2-ac & 1 \end{smallmatrix}\right)$ since $bc = -1$.⟧ Finally, if $A = \left(\begin{smallmatrix} a & b \\ c & d \end{smallmatrix}\right)$ with $d \neq 0$, $f(A) = e$.

$$⟦A = \begin{pmatrix} \dfrac{bd - b + 1}{d} & \dfrac{b - 1}{d} \\ d - 1 & 1 \end{pmatrix} \begin{pmatrix} \dfrac{c + 1}{d} & 1 \\ \dfrac{c + 1 - d}{d} & 1 \end{pmatrix}.⟧$$

Problems on Topological Groups

1. The semimetric of Example 1 is a metric if and only if $p(x) > 0$ for $x \neq e$.
2. A semimetric d is left invariant if and only if L_a is an isometry for each a.
★3. The *conjugation* or *inner automorphism* operator C_a, is defined for each a, by $C_a = L_a \circ R_{a^{-1}} = R_{a^{-1}} \circ L_a$. Thus $C_a(x) = axa^{-1}$. Show that each C_a is a topological automorphism. What is the operation of conjugation in a commutative group?
4. If a topology makes the operation $(x, y) \to xy^{-1}$ continuous, it is a group topology. ⟦Call this operation u, let $v(x, y) = (x, y^{-1})$, $w(x) = (e, x)$. Then $x \to x^{-1}$ is $u \circ w$, $(x, y) \to xy$ is $u \circ v$.⟧
★5. The discrete and indiscrete topologies for a group are group topologies. Show also that they can be given by two-sided invariant semimetrics. ⟦$p(x) = 1$ if $x \neq e$; $p = 0$, respectively.⟧
6. If G is an open set, GA and AG are open for every set A. ⟦$GA = \bigcup \{Ga: a \in A\}$, and each $Ga = R_a(G)$ is open.⟧
★7. Let p be an absolute-value function, and $d(x, y) = p(xy^{-1})$. Show that d is a right-invariant semimetric which induces a group topology. Conversely, a right-invariant semimetric which induces a group topology leads to an absolute-value function p by the formula $p(x) = d(x, e)$.
★8. Let $d'(x, y) = d(x^{-1}, y^{-1})$. Show that d' is a left-invariant semimetric if and only if d is a right-invariant semimetric; and, if they are invariant

thus, they lead to the same absolute-value function as in Example 2, that is $d'(x, e) = d(x, e)$.

★9. Let $p: X \to \mathbf{R}$ satisfy $p(x) \geq 0$ for all x, $p(e) = 0$, $p(x^{-1}) = p(x)$, $p(xy) \leq p(x) + p(y)$, $p(axa^{-1}) = p(x)$. Let $d(x, y) = p(x^{-1}y)$. Show that p is an absolute-value function and that d is a two-sided invariant semimetric. ⟦$d(xa, ya) = p(a^{-1}x^{-1}ya) = p(x^{-1}y) = d(x, y)$.⟧ Conversely, a two-sided invariant semimetric d leads to a function p with the above properties, by the formula $p(x) = d(x, e)$. (Note that the extra assumption of Problem 7 that d induces a group topology is not needed here. It is automatically satisfied. Compare Problem 205.)

10. Let d be a semimetric on a group, then the following are equivalent: (a) d is left invariant and right invariant, (b) d is left invariant, and *invariant under inversion* ($d(x^{-1}, y^{-1}) = d(x, y)$), (c) d is right invariant and invariant under inversion, (d) d is left invariant and invariant under inner automorphism ($d(x, y) = d(axa^{-1}, aya^{-1})$).

11. Let p be an absolute-value function on a group, and let $d(x, y) = p(x^{-1}y)$. The following are equivalent: the conditions of Problem 10, (e) $p(xy) = p(yx)$ (the group is "p-commutative"), (f) p is invariant under inner automorphism.

12. The conditions of Problems 10 and 11 are also equivalent to (g) $d(ax, by) \leq d(a, b) + d(x, y)$, (h) $p(abcd) \leq p(bc) + p(ad)$. ⟦If (h), $p(ab) = p(b^{-1}bab) \leq p(ba)$; hence (e). If (e), $p(abcd) = p(bcda)$, hence (h). If (h), $d(ax, by) = p(x^{-1}a^{-1}by)$, hence (g). If (g), $p(a^{-1}xa) = d(ea, xa) \leq p(x)$, hence (f).⟧

★13. Let b be a base for the neighborhood system of e. Let $b_L = \{U_L: U \in b\}$, where U_L is the connector L-associated with U. Show that b_L is a base for the left uniformity.

★14. With b as in Problem 13, $\bar{A} = \bigcap \{AU: U \in b\}$ for any set A ⟦Problem 13; Theorem 11.1.3⟧.

★15. Let T and T' be group topologies for a group X. Show that $T \supset T'$ if and only if every T' neighborhood of e is a T neighborhood of e. ⟦If $x \in G \in T'$, then $xU \subset G$ for some T' neighborhood U of e which is thus a T neighborhood of e, hence $G \in T$.⟧

16. $\overline{A^{-1}} = (\bar{A})^{-1}$. ⟦Inversion is a homeomorphism onto.⟧ In particular the closure of a symmetric set is symmetric.

17. A topological group has a local base of symmetric closed neighborhoods of e. ⟦If U is closed and $V \subset U$ is symmetric, then $\bar{V} \subset U$ is both symmetric and closed, by Problem 16.⟧

★18. For each neighborhood U of e, let $U_R = \{(x, y): xy^{-1} \in U\}$. Define $\mathscr{R}' = \{U_R: U$ is a neighborhood of $e\}$. Show that $\mathscr{R}'$ is a base for a uniformity; the uniformity $\mathscr{R}$ generated by $\mathscr{R}'$ is called the *right uniformity*. Show that $\mathscr{R}, \mathscr{L}$ are equivalent uniformities. ⟦As in Theorem 12.1.1, show $T_R = T_G$.⟧

19. The left-invariant metric induced by p in Example 3 is given by $d(f, g) = \max |f^{-1}(x) - g^{-1}(x)|$. (Compare Problem 8.)
20. If X is a compact group, $\mathscr{R} = \mathscr{L} = \mathscr{B}$ ⟦Corollary 11.2.2⟧.
21. For sets A, B, in X, $\bar{A}\bar{B} \subset \overline{AB}$. ⟦Let $y \in \bar{A}$, $z \in \bar{B}$ and let $\mathscr{F}$, $\mathscr{F}'$ be filterbases in A, B with $\mathscr{F} \to y$, $\mathscr{F}' \to z$. Then $\mathscr{F}\mathscr{F}'$ is a filterbase in AB and converges to yz. Thus $yz \in \overline{AB}$.⟧
22. Construct subsets A, B of **R** such that $\bar{A} + \bar{B} \neq \overline{A + B}$. Thus inequality may hold in Problem 21. ⟦Make A, B closed, $A + B$ not closed.⟧
23. The closure of a subgroup is a subgroup. ⟦$\bar{S}\bar{S} \subset \overline{SS} = \bar{S}$ by Problem 21. Similarly $(\bar{S})^{-1} = \overline{S^{-1}} = \bar{S}$.⟧ The closure of an invariant subgroup is invariant.

---

101. Suppose that d is a left-invariant semimetric which is not right invariant. Let $D(x, y) = d(x, y) + d(x^{-1}, y^{-1})$. Show that D is neither left nor right invariant.
102. $\bar{A} = \bigcap \{UA: U \text{ is a neighborhood of } e\}$. ⟦Use the right uniformity as in Theorem 12.1.3.⟧
103. For each set A, and neighborhood U of e, $\bar{A} \subset AU$ and $\bar{A} \subset UA$ ⟦Theorem 12.1.3 and Problem 102⟧.
104. If K is compact, G is open, and $K \subset G$, then there exists a neighborhood U of e with $KU \subset G$ ⟦Sec. 11.1, Problem 105⟧. If K is compact, F is closed and $K \pitchfork F$, there exists a neighborhood U of e such that $KU \pitchfork F$. ⟦This is the same result.⟧
105. If K is compact and F is closed, then KF and FK are closed. ⟦*Proof with nets:* Let $x_\alpha \in KF$, $x_\alpha \to x$. Then $x_\alpha = k_\alpha f_\alpha$; k_α has a convergent subnet $k_\beta \to k \in K$. Then $f_\beta = k_\beta^{-1} x_\beta \to k^{-1}x$, hence $k^{-1}x \in F$ so $x \in KF$. *Proof without nets:* Let $x \notin KF$. Then $K^{-1}x \pitchfork F$. By Problem 104, $K^{-1}xU \pitchfork F$ and so $xU \pitchfork KF$. Thus $x \notin \overline{KF}$.⟧
106. Let X be a group with a Hausdorff topology making the group operation continuous. Show that the operation $x \to x^{-1}$ has closed graph. ⟦Let $x_\delta \to x$, $x_\delta^{-1} \to y$. Then $xy = \lim x_\delta x_\delta^{-1} = e$.⟧
107. Show that the RHO topology is an example for Problem 106, and the inverse map is not continuous. (Compare Sec. 6.7, Example 3 which discusses the graph of this map.)
108. Deduce the fact that $[0, 1)$ is not RHO-compact from Problem 107 and Sec. 6.7, Problem 103. ⟦Its image under the inverse map is not closed.⟧
109. Let X be a group with a compact Hausdorff topology making the group operation continuous. Show that X is a topological group ⟦Problem 106; Sec. 7.1, Problem 108⟧.
110. Can the cofinite topology be a group topology? Must it be? ⟦Sec. 6.6, Problem 101.⟧

111. Let U be a neighborhood of e and U_L the connector L-associated with U. Show that $(\bar{U})_L = \bar{U}_L = vU_L$ (Sec. 11.1, Problem 113). ⟦Let $(x, y) \in \bar{U}_L$. Then $(x_\delta, y_\delta) \to (x, y)$ with $(x_\delta, y_\delta) \in U_L$, that is $x_\delta^{-1} y_\delta \in U$. Then $x x_\delta^{-1} y_\delta \to x x^{-1} y = y$ hence $y \in x\bar{U} = vU_L(x)$.⟧
112. **J** can be given a set of operations which makes it a topological group (with the Euclidean topology) ⟦Sec. 6.4, Problem 203⟧.
113. If a subgroup has nonempty interior, it is open and closed. ⟦If x is interior to S and $y \in S$, $x^{-1}S$ is a neighborhood of e so $yx^{-1}S$ is a neighborhood of y. But $S = yx^{-1}S$. Thus S is open. The complement of S is also open since it is a union of translates (cosets) of S.⟧
114. Let V be a symmetric neighborhood of e, then $S = \bigcup\{V^n : n = 1, 2, \ldots\}$ is an open and closed subgroup. ⟦If $a, b \in S$, $a \in V^m$, $b \in V^n$ so $ab^{-1} \in V^{m+n} \subset S$. S is open and closed by Problem 113.⟧
115. A *Baire group* is a topological group which is a Baire space. A topological group X is a Baire group if and only if it is of second category in itself. ⟦Let V be a symmetric neighborhood of e which is of first category in the space. Form S as in Problem 114. It is open and closed and of first category. Now X is a union of translates of S and the result follows from Sec. 9.3, Problem 123.⟧
116. Let S be a subgroup of a Baire group X. Then if S is a dense G_δ, $S = X$. ⟦S is residual by Sec. 9.3, Problem 10. If $S \neq X$, then S can be translated into $\tilde{S}$, hence S is of first category in X too. Thus X is of first category⟧.
117. Let S be a subgroup of a complete semimetric group. (The semimetric induces the group topology but no assumption is made concerning its invariance or the uniformity it induces.) Then if S is a G_δ, it is closed. ⟦We may assume S dense by Problem 23. The result follows from Problem 116.⟧
118. The box topology is a group topology for a product of topological groups.
119. If S is open or closed, $S\overline{\{e\}} = S$. ⟦S closed implies $S \subset S\overline{\{e\}} \subset \overline{Se} = S$; if S is open, $s \in S$, $s^{-1}S \supset \overline{\{e\}}$, $S \supset s\overline{\{e\}}$.⟧ The formula is false in general. ⟦$S = \{e\}$.⟧
120. A locally compact group is a disjoint union of subspaces, each of which is open, closed, σ-compact, locally compact, hemicompact, Lindelöf, normal. ⟦In Problem 114 with V compact, S has all these properties (Theorem 5.3.5; Sec. 8.1, Problems 10 and 125). Now consider the cosets of S.⟧
121. A locally compact group is normal and paracompact. If it is either separable or connected it is σ-compact, hemicompact, and Lindelöf. ⟦*First half:* Problem 120; Sec. 4.1, Problem 123; Sec. 10.2, Problems 122 and 123. *Second half:* There are only countably many (respectively, one) subspaces of the kind mentioned in Problem 120.⟧

122. A topological group need not be normal. ⟦Example 6; Sec. 6.7, Problem 203. This example is separated and separable. It shows also that a separable group need not be Lindelöf, by Theorem 5.3.5.⟧
123. A compact group need not be separable; *a fortiori*, a Lindelöf group need not be separable ⟦Sec. 7.4, Problem 106⟧.
124. There is no nonzero homomorphism of $\mathbf{Q}$ into $\mathbf{Z}$. ⟦If $f(q) > 0$, let $n > f(q)$. Then $nf(q/n) = f(q)$ and so n divides $f(q)$.⟧

---

201. Is the topology of the no-point compactification of $\mathbf{R}$ at 0 a group topology?
202. Let $\mathbf{2}$, $\mathbf{3}$ be groups with 2, 3 elements, respectively. Give them the discrete topology. Let $A = \mathbf{2}^\omega$, $B = \mathbf{3}^\omega$, each with the discrete topology. Let $X = A \times \mathbf{3}^\omega$, $Y = \mathbf{2}^\omega \times B$. Then X, Y are homeomorphic ⟦Sec. 6.4, Problem 202⟧ and isomorphic, but not topologically isomorphic. ⟦Define $a^n \in \mathbf{2}^\omega$ by $a_k^n = 0$ for $k < n$, 1 for $k \geq n$; $x^n = (a_n, 0) \in \mathbf{2}^\omega \times B$. Then $x^n \to 0$ and each x^n has order 2. This is impossible in $A \times \mathbf{3}^\omega$.⟧ This example is due to S. Kakutani.
203. $\mathbf{2}^\omega$ has a compact open proper subgroup.
204. A topological group is discrete if and only if it has a compact open subset K such that $xK \not\subset K$ for all $x \neq e$. ⟦Take $K = G$ in Problem 104.⟧
205. Must a left-invariant semimetric on a group always induce a group topology? ⟦It makes the operations continuous at e and $x \to ax$ continuous for each a.⟧
206. In Problem 117, can "complete semimetric" be replaced by "Baire"?
207. Give the additive group of $\mathbf{R}$ a compact, connected, group topology. ⟦See [Hewitt and Ross, p. 415].⟧
208. If a group is totally bounded in its left uniformity, must it be totally bounded in its right uniformity?

12.2 Group Concepts

In this section we shall study continuity and uniform continuity, metrization, completeness, and completion.

THEOREM 12.2.1. *A homomorphism which is continuous at some point is continuous everywhere.*

Suppose f is continuous at x. Let y_δ be a net with $y_\delta \to y$. Then $y_\delta y^{-1}x \to x$ and so $f(y_\delta y^{-1}x) \to f(x)$. Thus $f(y_\delta) = f(y_\delta y^{-1}x)\cdot f(x^{-1}y) \to f(x)\cdot f(x^{-1}y) = f(y)$ so that f is continuous at y. ∎

LEMMA 12.2.1. *Let $f\colon X \to Y$ be a homomorphism with the property that f*

maps each neighborhood of e in X onto a neighborhood of e in Y. Then f is an open map.

Let G be an open set in X and $y \in f[G]$, say $y = f(g)$. Then $g^{-1}G$ is a neighborhood of e in X, and so $f[g^{-1}G]$ is a neighborhood of e in Y. But then $yf[g^{-1}G]$ is a neighborhood of y, and this is equal to $f[G]$. ⟦It is $yf(g^{-1})f[G] = yy^{-1}f[G] = f[G]$.⟧ Thus $f[G]$, as a neighborhood of all its points, is open. ∎

In treating uniform continuity of a map f between groups we have to decide which uniformities to place on the domain and range. We shall restrict ourselves to the case in which both have their left uniformities. In the next result, it is not assumed that f is a homomorphism.

LEMMA 12.2.2. *A function $f: X \to Y$, with X, Y topological groups, is uniformly continuous when X, Y have their left uniformities if and only if for every neighborhood U of e in Y, there exists a neighborhood V of e in X with $f[xV] \subset f(x)U$ for all $x \in X$.*

Suppose first that f is uniformly continuous and U is a neighborhood of e in Y. Let U_L be the connector L-associated with U. There exists a connector V_L for X, L-associated with a neighborhood V of e, such that $(fa, fb) \in U_L$ whenever $(a, b) \in V_L$. This implies that $f[xV] \subset f(x)U$ for all x. ⟦Let $b \in xV = V_L(x)$. Then $(x, b) \in V_L$ and so $(fx, fb) \in U_L$; hence $f(b) \in U_L(fx) = f(x)U$.⟧ Conversely, let U_L be a connector for Y, L-associated with a neighborhood U of e. Choose V as in the statement of the theorem and let V_L be its L-associated connector. Then $(a, b) \in V_L$ implies $(fa, fb) \in U_L$. ⟦$b \in V_L(a) = aV$, hence $f(b) \in f(a)U = U_L(fa)$.⟧ ∎

THEOREM 12.2.2. *A continuous homomorphism $f: X \to Y$ is uniformly continuous when X, Y are both given their left uniformities.*

Let U be a neighborhood of e in Y. Choose a neighborhood V of e in X with $f[V] \subset U$. ⟦f is continuous at e.⟧ Then for every x, $f[xV] = f(x)f[V] \subset f(x)U$. By Lemma 12.2.2, the result follows. ∎

The result of Theorem 12.2.2 holds if X, Y are both given their right or two-sided uniformities. However the identity map from $(X, \mathscr{R})$ to $(X, \mathscr{L})$ is not uniformly continuous (although it is a continuous homomorphism) unless $\mathscr{R} \supset \mathscr{L}$ ⟦Theorem 11.2.1⟧.

EXAMPLE 1. **R**, the reals, is a commutative group with the operation $+$, identity 0, and inverse map $x \to -x$. With the Euclidean metric, it is a metric group, and its three uniformities are all equal, as is the case with every commutative group. Let f be an endomorphism of **R**. Then for all $x \in \mathbf{R}$ and all rational r we have $f(rx) = rf(x)$. ⟦For integer r this was given in Sec. 12.1,

Example 5. For $r = m/n$, $nf(rx) = f(nrx) = f(mx) = mf(x)$.⟧ Thus if f is continuous $f(ax) = af(x)$ for all a, x; in particular $f(a) = af(1)$. Thus *if f is a continuous endomorphism of* **R**, *there exists k with $f(x) = kx$ for all x.* ⟦$k = f(1)$.⟧

▲ EXAMPLE 2. Let H be a Hamel base for **R**, considering **R** as a linear space over **Q** ⟦Sec. 7.3, Problem 203⟧. Thus every real number is a unique finite linear combination of H with rational coefficients. Fix $a, b \in H$. Let $f: H \to \mathbf{R}$ be an arbitrary function save only that we require $bf(a) \neq af(b)$. Extend f to all of **R** by $f(\sum rh) = \sum rf(h)$, where $\sum rh$ denotes any finite sum of the form $\sum r_i h_i$, with $h_i \in H$. Then f is an endomorphism of **R**, and f is not continuous since if it were the result of Example 1 would give $bf(a) = af(b)$. *Note:* By making f one-to-one from H onto itself we can get a discontinuous automorphism of **R** onto itself. For an extended and elementary essay on endomorphisms of **R**. (See [May, pp. 97–124].)

We now turn to the metrization theorem for groups. Recall the list given at the beginning of Section 11.5: $CB \leftrightarrow M_u \to M_t \to FC$. We shall find that for topological groups, a little more is true, namely that these four conditions are equivalent. (The uniformity involved in CB and M_u may be $\mathscr{R}$, $\mathscr{L}$ or $\mathscr{B}$.) To see this, it is sufficient to prove $FC \to M_u$, as we now do. Because of Sec. 12.1, Example 8, it is sufficient to prove this for the left uniformity $\mathscr{L}$, and because of localization, it is sufficient to assume first countability at e. The semimetrics constructed have an additional very special character.

THEOREM 12.2.3. *If the topology of a topological group is first countable at e, it is given by a left-invariant semimetric, and by a B semimetric.*

It is sufficient to prove the first part only and use Sec. 12.1, Definition 1 and Example 8 for the B semimetric. It would be easy to prove that the left uniformity has a countable base, (see, for example, Sec. 12.1, Problem 13), then semimetrizability follows from Theorem 11.5.1. However, to construct a left-invariant semimetric will require a little more effort. Instead of using Theorem 11.5.1, we shall vary its proof to obtain an absolute-value function. In the following, the reader should keep in view the proof of Theorem 11.5.1, which we shall refer to as *the earlier proof*. Let X be a topological group with a countable base $\{U_n\}$ of neighborhoods of e. As in the earlier proof we construct a local base $\{V_n\}$ of symmetric neighborhoods of e with $V_n V_n V_n \subset V_{n-1}$ $n = 1, 2, \ldots$; $V_0 = X$. For each x, set $P(x) = 0$ if $x \in V_n$ for all n, and $P(x) = 2^{-n}$ if $x \in V_n \setminus V_{n+1}$. Then $P(x) \geq 0$, $P(x^{-1}) = P(x)$ since each V_n is symmetric, $P(e) = 0$. As in the earlier proof (Relation (11.5.3)), it is proved that $x_m \to e$ if and only if $P(x_m) \to 0$. Next, the analogue of (11.5.3) in the earlier proof is

$$P(abc) \leq 2 \max[P(a), P(b), P(c)].$$

Letting $2r$ denote the right-hand side, $r = 2^{-n}$. Then a, b, c all belong to V_n so that $abc \in V_{n-1}$. Hence $P(abc) \leq 2^{-(n-1)} = 2r$. The analogue of (11.5.4) is $P(\sum_{i=1}^n a_i) \leq 2 \sum_{i=1}^n P(a_i)$. The earlier proof can be modified by replacing $D(x_i, x_{i+1})$ by $P(a_i)$, and $D(x_1, x_{m+1})$ by $P(\sum_{i=1}^m a_i)$. We are now ready to define p. In place of Formula (11.5.1) we use

$$p(a) = \inf\left\{\sum_{i=1}^{k} P(a_i a_{i-1}^{-1}): a_0 = e, a_k = a\right\},$$

the inf taken over all finite collections $a_0, a_1, \ldots, a_k$ with $a_0 = e$, $a_k = a$. It follows that $p(ab) \leq p(a) + p(b)$ as in Lemma 11.5.1. ⟦Now choose $a_0, a_1, \ldots, a_k; b_0, b_1, \ldots, b_n$ with $a_0 = b_0 = e$, $a_k = a$, $b_n = b$,

$$\sum_{i=1}^{k} P(a_i a_{i-1}^{-1}) < p(a) + \varepsilon/2, \qquad \sum_{i=1}^{n} P(b_i b_{i-1}^{-1}) < p(b) + \varepsilon/2.$$

Let $c_0 = e$, $c_1 = b_1, \ldots, c_n = b$, $c_{n+1} = a_1 b$, $c_{n+2} = a_2 b$, $c_{n+k} = ab$. Then

$$p(ab) \leq \sum P(c_i c_{i-1}^{-1}) = \sum_{i=1}^{n} P(b_i b_{i-1}^{-1}) + \sum_{i=1}^{k} P(a_i a_{i-1}^{-1}) < p(b) + p(a) + \varepsilon.⟧$$

This function p is an absolute value as in Sec. 12.1, Example 1. To see this, note that the triangle inequality has just been proved. Clearly $p(x) \geq 0$ for all x, $p(e) = 0$, ⟦take $k = 1$ in the definition of p⟧, $p(x) = p(x^{-1})$ since $P(x) = P(x^{-1})$. Now, let $d(x, y) = p(x^{-1}y)$, then d is a left-invariant semimetric ⟦as in Sec. 12.1, Examples 1, 2⟧. We shall show that d induces the topology of the group. As in the earlier proof we show that $p \leq P \leq 2p$, (instead of $d \leq D \leq 2d$, as there). Then $x_m \to e$ in X if and only if $P(x_m) \to 0$ ⟦noted above⟧, and this is true if and only if $p(x_m) \to 0$, equivalently, $x_m \to e$ in the semimetric space (X, d). Since d is left invariant, $x_m \to x$ in (X, d) is equivalent to $x^{-1}x_m \to e$ in (X, d), this in turn is equivalent to $x^{-1}x_m \to e$ in X, that is, $x_m \to x$ in X. Thus d and the group topology coincide. ∎

It is clear that a right-invariant semimetric could have been constructed in Theorem 12.2.3. Indeed a right-invariant semimetric can be made directly out of a left-invariant one, call it d, by the formula $d'(x, y) = d(x^{-1}, y^{-1})$. However, there may be no two-sided-invariant semimetric ⟦Example 4⟧. Before the Examples, we need a few remarks on the nature of Cauchy filters. We shall use the notation given at the beginning of Section 12.1.

LEMMA 12.2.3. *A filter $\mathscr{F}$ is an $\mathscr{L}$-Cauchy filter if and only if $\mathscr{F}^{-1}$ is an $\mathscr{R}$-Cauchy filter. It is a $\mathscr{B}$-Cauchy filter if and only if both $\mathscr{F}$ and $\mathscr{F}^{-1}$ are $\mathscr{L}$-Cauchy filters. If $\mathscr{F}$ is a $\mathscr{B}$-Cauchy filter, $\mathscr{F}^{-1}$ is also a $\mathscr{B}$-Cauchy filter. The same results hold for nets.*

A filter $\mathscr{F}$ is an $\mathscr{L}$-Cauchy filter if and only if for each connector U_L L-associated with a neighborhood of U of e, $\mathscr{F}$ contains a set S with $S \times S \subset U_L$; that is, $S^{-1} S \subset U$. Similarly $\mathscr{F}$ is an $\mathscr{R}$-Cauchy filter if and only if it contains, for each U; a set S with $SS^{-1} \subset U$. Since $S^{-1}S = S^{-1}(S^{-1})^{-1}$, the first part is clear. The second part follows from the first and Theorem 11.4.2. If $\mathscr{F}$ is a $\mathscr{B}$-Cauchy filter, both $\mathscr{F}$ and $\mathscr{F}^{-1}$; that is, both $\mathscr{F}^{-1}$ and $(\mathscr{F}^{-1})^{-1}$, and $\mathscr{L}$-Cauchy filters; thus $\mathscr{F}^{-1}$ is a $\mathscr{B}$-Cauchy filter. The result for nets follows from consideration of the associated filters. ∎

★EXAMPLE 3. Let G be the metric group given in Sec. 12.1, Example 3. We shall show that $\mathscr{L} \neq \mathscr{R}$, where $\mathscr{L}$, $\mathscr{R}$ are the left and right uniformities. (Of course, $\mathscr{L}$ and $\mathscr{R}$ are equivalent by Sec. 12.1, Problem 18.) For this it is sufficient to show that $\mathscr{R} \neq \mathscr{B}$ ⟦$\mathscr{B} = \mathscr{L} \vee \mathscr{R}$, so if $\mathscr{L} = \mathscr{R}$, then $\mathscr{L} = \mathscr{R} = \mathscr{B}$.⟧ We shall accomplish this by showing that *$\mathscr{B}$ is complete and $\mathscr{R}$ is not*. Let $X = C(I)$, where

$$I = [0, 1], d(f, g) = \max\{|f(x) - g(x)| : x \in I\}.$$

Then (X, d) is a complete metric space ⟦Sec. 9.1, Example 4⟧, and d gives G its right uniformity ⟦Sec. 12.1, Example 3⟧. Thus $(G, \mathscr{R}) = (G, d)$ will be shown noncomplete when it is shown that G is a nonclosed subspace of X ⟦Sec. 11.3, Problem 8.⟧ This is very easy and is left to the reader. (See Problem 1.) Next we shall show that $(G, \mathscr{B})$ is complete. (See Problem 104 for an abstract setting of the proof.) Let $\{f_n\}$ be a $\mathscr{B}$-Cauchy sequence. Then both $\{f_n\}$ and $\{f_n^{-1}\}$ are $\mathscr{R}$-Cauchy sequences. ⟦Lemma 12.2.3, with $\mathscr{L}$, $\mathscr{R}$ interchanged.⟧ Thus they are Cauchy sequences in (X, d). Since X is complete there exist $f, g \in X$ with $f_n \to f, f_n^{-1} \to g$. But then $g = f^{-1}$. ⟦For all x,

$$|f(gx) - x| \leq |f(gx) - f(f_n^{-1}x)| + |f(f_n^{-1}x) - x| \to 0$$

as $n \to \infty$, since the second term is $|f(f_n^{-1}x) - f_n(f_n^{-1}x)| \leq d(f, f_n)$; and, considering the first term, $f_n^{-1}(x) \to g(x)$, and f is continuous. Similarly $g(fx) = x$.⟧ Thus $f \in G$ and $f_n \to f$ in G. ⟦$f_n \to f$ in X, and G is a topological subspace.⟧

★EXAMPLE 4. *The topology of G (Example 3) cannot be given by a two-sided-invariant metric.* If there were such a metric, Sec. 12.1, Example 7 would imply that $\mathscr{R} = \mathscr{L}$, in contradiction with Example 3.

EXAMPLE 5. Topological spaces of a certain kind have the property that, if first countable, they must be metrizable. In fact, if a first countable topological space can be given an operation which makes it a topological group, it must be semimetrizable ⟦Theorem 12.2.3⟧. As an example, let **2** stand for a discrete topological space with two points, and A any set. Then $\mathbf{2}^A$ can be made into a topological group, for we may write $\mathbf{2} = \{0, 1\}$ and define $0 + 0 = 1 + 1 = 0, 0 + 1 = 1 + 0 = 1$, and apply Sec. 12.1, Example 6. Thus $\mathbf{2}^A$ is metrizable if and only if it is first countable. The same is true for D^A, where D is a discrete space with countably many points ⟦take $D = \omega$⟧, or a continuum of points ⟦take $D = \mathbf{R}$ with the discrete topology⟧. These results were noted also in Sec. 6.4, Problem 6.⟧

Some easy results on completeness are listed in Problems 2 through 6. We shall make immediate use of these. We now consider the problem of completion. Of course every uniform space has a completion, and it is natural to expect a group completion. Our first result is negative. Example 6 shows a metric group which is not a topological subgroup of any group which is complete in its left uniformity. Thus with the left-invariant metric d, the metric completion cannot be made into a group by extending the operations. (But see Problem 203.) However, completions with the two-sided uniformity exist ⟦Theorem 12.2.4; Problems 202, 210⟧.

EXAMPLE 6. With G as in Example 3, $(G, \mathscr{L})$ has no group completion. This means that there is no topological group X which is complete in its left uniformity $\mathscr{L}_X$, and includes G as a topological subgroup. Suppose that such a group X exists. It follows that $\mathscr{R}_X$ is complete ⟦Problem 3⟧ and so $\mathscr{L}_X$, $\mathscr{R}_X$ are Cauchy equivalent. ⟦If $\mathscr{F}$ is $\mathscr{L}_X$-Cauchy, $\mathscr{F} \to x \in X$, hence $\mathscr{F}$ is $\mathscr{R}_X$-Cauchy.⟧ Now Corollary 11.5.2 shows that $\mathscr{L}$, $\mathscr{R}$ are the relative uniformities of $\mathscr{L}_X$, $\mathscr{R}_X$ on G, hence $\mathscr{L}$, $\mathscr{R}$ are Cauchy equivalent, contradicting Problem 6.

In the process of completion, techniques will be needed for extending the group operations and for showing that they are continuous. These techniques are given in the next two lemmas.

LEMMA 12.2.4. *The product of two $\mathscr{L}$-Cauchy filters is $\mathscr{L}$-Cauchy. The same is true for $\mathscr{R}$, $\mathscr{B}$.*

It will be sufficient to prove this for $\mathscr{L}$. The result for $\mathscr{R}$ follows by symmetry, and the result for $\mathscr{B}$ by Lemma 12.2.3 (or Theorem 11.4.2). Let $\mathscr{F}_1, \mathscr{F}_2$ be Cauchy filters in $(X, \mathscr{L})$ and let U be a neighborhood of e. Let V be a symmetric neighborhood of e with $VVV \subset U$. As in Lemma 12.2.3, $\mathscr{F}_2$ contains a set S_2 with $S_2^{-1}S_2 \subset V$. Let $b \in S_2$. There exists a neighborhood W of e such that $b^{-1}Wb \subset V$ ⟦Theorem 12.1.5⟧. Finally, $\mathscr{F}_1$ contains a set S_1 with $S_1^{-1}S_1 \subset W$. Then $S_1S_2 \in \mathscr{F}_1\mathscr{F}_2$ and $(S_1S_2)^{-1}S_1S_2 \subset U$. ⟦The left-hand side is $S_2^{-1}bb^{-1}S_1^{-1}S_1bb^{-1}S_2 \subset S_2^{-1}S_2b^{-1}WbS_2^{-1}S_2 \subset VVV \subset U$.⟧ Hence $\mathscr{F}_1\mathscr{F}_2$ is an $\mathscr{L}$-Cauchy filterbase. ∎

LEMMA 12.2.5. *Let D be a dense subspace of a topological space T, Y a* *regular space, and $f: T \to Y$ a function with the property that $f|D \cup \{t\}$ is continuous for each $t \in T$. Then f is continuous on T.*

Let $t \in T$, and let U be a closed neighborhood of $f(t)$ in Y. There is an open neighborhood of t in $D \cup \{t\}$ which maps into U, hence an open neighborhood G of t in T with $f[G \cap D] \subset U$. It follows that $f[G] \subset U$. ⟦Since U is closed, it is sufficient to prove that $f[G] \subset \overline{U}$. Let $g \in G$ and let V be a neighborhood of $f(g)$. Since f is continuous on $D \cup \{g\}$, there is, as before, an open neighborhood H of g with $f[H \cap D] \subset V$. Now $G \cap H \cap D$ is not empty, since D is dense, and $G \cap H$ is open and, since it contains g, nonempty; so we may choose $h \in G \cap H \cap D$. Then $f(h) \in V \cap f[G \cap D] \subset V \cap U$. Thus every neighborhood of $f(g)$ meets U in a nonempty set.⟧ It follows that f is continuous. ⟦Y has a local base of closed neighborhoods of $f(t)$.⟧ ∎

The completion theorem is now given for separated groups only. The semi-metric case is outlined in Problem 202, and the nonseparated commutative case in Problem 210.

THEOREM 12.2.4. *Let X be a separated topological group. Then there is a separated topological group Y which is complete in its two-sided uniformity and has X as a dense topological subgroup.*

Let $(Y, \mathscr{U})$ be the separated completion of $(X, \mathscr{B})$ ⟦Theorem 11.5.6⟧, where $\mathscr{B}$ is the two-sided uniformity of X. For $y_1, y_2 \in Y$, let $\mathscr{F}_i = \{U \cap X: U \in N_i\}$ for $i = 1, 2,$; where N_i is the neighborhood filter of y_i. Each $\mathscr{F}_i$ is a $\mathscr{B}$-Cauchy filter in X ⟦each N_i is convergent, hence Cauchy, in Y⟧, hence $\mathscr{F}_1\mathscr{F}_2$ is a $\mathscr{B}$-Cauchy filterbase in X ⟦Lemma 12.2.4⟧, hence is a Cauchy filterbase in Y. Let its limit be denoted by $y_1 y_2$. This agrees with the previous meaning of this symbol if $y_1, y_2 \in X$ since the group operation is continuous. Similarly define y^{-1} by applying Lemma 12.2.3. We first note that the operations are continuous. ⟦That the map $y \to y^{-1}$ is continuous follows from Lemma 12.2.5 with $D = X$, $T = Y$; and that $(y_1, y_2) \to y_1 y_2$ is continuous follows from Lemma 12.2.5 with $D = X \times X$, $T = Y \times Y$.⟧ It follows that (i) $ey = ye = y$ for all y ⟦let $f(y) = ey$, $i(y) = y$; then f, i are continuous and agree on the dense subspace X⟧; (ii) $yy^{-1} = e$ ⟦$y \to yy^{-1}$ is a continuous function which has the constant value e on the dense subspace X⟧; and (iii) $(y_1 y_2)y_3 = y_1(y_2 y_3)$ ⟦a similar argument involving continuous maps on $Y \times Y \times Y$ and its dense subspace $X \times X \times X$⟧. We now know that $(Y, \mathscr{U})$ is a complete topological group, and that $(X, \mathscr{B})$ is a dense uniform subgroup. Finally, $\mathscr{U}$ is the two-sided uniformity of Y ⟦Corollary 11.5.2⟧. ∎

Problems on Topological Groups

1. Let $f(x) = 2x$ for $0 \le x \le 1/2$, $f(x) = 1$ for $1/2 \le x \le 1$. Then $f \in \overline{G} \setminus G$ in Example 3. ⟦For example, let $f_n(x) = f(x) - x(1 - x)/n$.

(Note that f maps I onto itself.) Compare Sec. 7.1, Problems 118 to 120.⟧

★2. $\mathscr{L}$ and $\mathscr{R}$ are Cauchy equivalent if and only if, whenever $\mathscr{F}$ is an $\mathscr{L}$-Cauchy filter, $\mathscr{F}^{-1}$ is also an $\mathscr{L}$-Cauchy filter ⟦Lemma 12.2.3⟧.

★3. $(X, \mathscr{L})$ is complete if and only if $(X, \mathscr{R})$ is. ⟦Let $\mathscr{L}$ be complete and $\mathscr{F}$ $\mathscr{R}$-Cauchy. Then $\mathscr{F}^{-1}$ is $\mathscr{L}$-Cauchy, hence convergent, and so $\mathscr{F}$ is convergent.⟧

★4. If $(X, \mathscr{L})$ is complete, $(X, \mathscr{B})$ is also ⟦Corollary 11.3.1⟧. The converse is false ⟦Example 3⟧.

★5. X is $\mathscr{L}$-complete if and only if it is $\mathscr{B}$-complete and $\mathscr{L}$ and $\mathscr{R}$ are Cauchy equivalent ⟦Problems 3 and 4; Lemma 12.2.3; Sec. 11.3, Problem 17⟧.

★6. In Example 3, $\mathscr{L}$ and $\mathscr{R}$ are not Cauchy equivalent ⟦Problem 5⟧.

★7. The *kernel* of a homomorphism f between groups is $f^{-1}[\{e\}]$, sometimes written $f^{\perp}$. Show that f is one-to-one if and only if the kernel of f contains only the identity of its domain.

---

101. Show directly that the sequence given in Problem 1 is not a Cauchy sequence in the left uniformity.

102. For $f \in X$ (Example 3) let $u(f) = f(0)$ and $v(f) = f(1) - 1$. Then u, v are continuous on X. ⟦$|u(f) - u(g)| \le d(f, g)$,⟧ hence $u^{\perp}, v^{\perp}$ are closed. Thus $\bar{G}_0 \subset u^{\perp} \cap v^{\perp}$. Show that $\bar{G}_0$ is the set of monotonely increasing functions in $u^{\perp} \cap v^{\perp}$ and G_0 contains exactly those which are strictly increasing. ($G_0 = \{$increasing members of $G\}$.)

103. With the notation of Example 3, define $u: G \to X$ by $u(f) = f^{-1}$. Since u is continuous, it has closed graph. Show that the graph of u is a closed subset of $X \times X$. ⟦If $f_n \to f$, $u(f_n) \to g$, show that $g = f^{-1}$ as in Example 3.⟧

104. Deduce the completeness of $(G, \mathscr{B})$ in Example 3 from Problem 103 and Sec. 11.4, Problem 101.

105. The map $x \to x^{-1}$ is $\mathscr{R}\mathscr{L}$-uniformly continuous (that is, a uniformly continuous map from $(X, \mathscr{R})$ to $(X, \mathscr{L})$.) It is also $\mathscr{L}\mathscr{R}$ and $\mathscr{B}\mathscr{B}$ uniformly continuous. It is $\mathscr{L}\mathscr{L}$-uniformly continuous if and only if $\mathscr{L} = \mathscr{R}$.

106. Let (X, d) be a metric group with d left invariant. Form a B-metric D from d. Let $(Y, \bar{D})$ be the completion (Theorem 12.2.4). Form $\bar{d}$ from $\bar{D}$, ($\bar{d}(x, e) = \frac{1}{2}\bar{D}(x, e)$, it is left invariant. Why is $(Y, \bar{d})$ not a metric completion of (X, d)?

107. The graph G of a discontinuous endomorphism of $\mathbf{R}$ is dense in $\mathbf{R}^2$. ⟦We may assume $f(1) = 0$, (consider $f(x) - xf(1)$,) and $f(t) = 1$ for some t. Then $(at + b, a) \in G$ for all rational a, b.⟧

108. An endomorphism of $\mathbf{R}$ is continuous if its graph is a G_δ ⟦Problem 107;

Sec. 12.1, Problem 117⟧. Thus, such a graph could never be homeomorphic with **J** ⟦Sec. 9.1, Problem 205⟧.

109. There is a natural isomorphism between the group E of endomorphisms of **R**, and $\mathbf{R}^H$. (See Example 2.) The group operation on E is $(f, g) \to f + g$, where $(f + g)(x) = f(x) + g(x)$. ⟦Each member of $\mathbf{R}^H$ leads to a member of E as in Example 2.⟧
110. Let X be a group and Y a separated topological group. Let $H(X, Y)$ be the set of all homomorphisms from X to Y. Show that $H(X, Y)$ is a closed subspace of Y^X.

$$⟦f(xy) = \lim f_\delta(xy) = \lim f_\delta(x)f_\delta(y) = f(x)f(y).⟧$$

111. Let $f: X \to Y$ be a homomorphism. Show that the kernel of f is dense if and only if $f[U] = f[X]$ for every neighborhood U of e.
112. Let $f: X \to Y$ be a homomorphism with the property that $f[U]$ is somewhere dense for each neighborhood U of e. Show that f is almost open. ⟦Let V be a symmetric neighborhood of e with $VV \subset U$. Let y be interior to $\overline{f[V]}$. Then $\overline{f[U]} \supset \overline{f[V]}\,\overline{f[V]} \supset \overline{f[V]}y^{-1}$ which is a neighborhood of e.⟧
113. Let f be a homomorphism with the property that the image of every set with nonempty interior has nonempty interior. Show that f is an open map ⟦as Problem 112, without closures⟧. Also, replace "image" by "preimage," "open" by "continuous." (These results are false if "homomorphism" is omitted; see Sec. 6.2, Problem 113.)
114. Let X be separable, Y a Baire group, and $f: X \to Y$ a homomorphism onto. Then f is almost open. ⟦Let U be a neighborhood of e in X and D a dense countable set, $DU = X$ by Sec. 12.1, Problem 103, hence $Y = f[D]f[U]$. Since $f[D]$ is countable, $f[U]$ is somewhere dense. See Problem 112.⟧
115. In Problem 114, replace "separable" by "Lindelöf." ⟦Same proof except that D may not be dense.⟧
116. Let X be locally compact, Y separated, and f a continuous, almost open homomorphism of X onto Y. Then f is an open map. ⟦The image of a compact neighborhood is closed by Theorems 5.4.4 and 5.4.5. Now see Problem 113.⟧
117. Let X be locally compact and separable, and let Y be a separated Baire group. Then every continuous homomorphism of X onto Y is an open map. In particular every continuous isomorphism of X onto Y is a homeomorphism ⟦Problems 114 and 116⟧.
118. Two comparable separated separable group topologies are equal if the larger is locally compact and the smaller is of second category ⟦Problem 117⟧.
119. In Problem 117 we cannot omit "separable" ⟦in Problem 118, take

the larger topology discrete⟧, "separated" ⟦smaller topology indiscrete⟧ or "Baire." ⟦$Y = \mathbf{Q}$, $X = \mathbf{Q}$ with discrete topology.⟧

120. In Problem 117 "locally compact" cannot be omitted. ⟦Let f be a discontinuous onto automorphism of $\mathbf{R}$ (Example 2), and let X be the graph of f. Then X is a separable metric group, $\mathbf{R}$ is a complete metric group, and $P_1 : X \to \mathbf{R}$ is not an open map since $f = P_2 \circ P_1^{-1}$ is not continuous. (Another way to see that X is not locally compact is given in Problems 128 and 122.)⟧
121. Two comparable separated separable locally compact group topologies must be equal. In particular $\mathbf{R}$ has no strictly smaller locally compact separated group topology ⟦Problem 118; Theorem 9.3.6⟧. This solves Sec. 12.1, Problem 201. (Compare Sec. 8.1, Problem 121.)
122. Let X, Y be locally compact separable metric groups, and $f: X \to Y$ a homomorphism with closed graph. Then f is continuous. ⟦$f = P_2 \circ P_1^{-1}$, where P_1, P_2 are the projections from the graph G; and $P_1 : G \to X$ is open by Problem 117.⟧
123. Let X be a separable complete metric group. Let Y be a separated Baire group. Let f be a continuous isomorphism of X onto Y. Then f is a homeomorphism. ⟦Problems 113 and 114; Sec. 9.1, Problem 118.⟧ Formulate analogues of Problems 118, 121, and 122.
124. In Problem 123 we cannot omit "separable", "separated," "Baire," or "complete" ⟦Problems 119 and 120⟧. (See Sec. 12.3, Problem 101.)
125. Let X be a group and T, T' two different group metrics which are separated, separable, and locally compact. Show that $T \cap T'$ is not a group topology. ⟦Because it is T_1 but not T_2. It is T_1 since T, T' are. If it were T_2, the identity map from (X, T) to (X, T') would have closed graph by Sec. 6.7, Problem 112. The result follows from Problem 122.⟧
126. The intersection of two group topologies need not be a group topology. ⟦Take $X = \mathbf{R}$ in Problem 125, $T =$ Euclidean, and $T' = w(f)$ with f as in the note in Example 2, or as in Problem 206. A commutative group and complete metrics!⟧
127. A locally compact group is $\mathscr{L}$-complete. ⟦If $S^{-1}S \subset U$ then $S \subset xU$ for some $x \in S$. Thus, with U a compact neighborhood of e, every $\mathscr{L}$-Cauchy filter must contain a compact subset, namely xU.⟧
128. A locally compact subgroup of a separated group must be closed ⟦Problem 127⟧.
129. Let f be a discontinuous endomorphism of $\mathbf{R}$ with $f(1) = 1$, and $p(x) = |x| + |f(x)|$. Then p is an absolute-value function. Let $X = (\mathbf{R}, d)$ with $d(x, y) = p(x - y)$. Show that the closure in X of $\mathbf{Q}$ is the set of fixed points of f. (*Note*: X is topologically isomorphic with the graph of f; it is separable, but $\mathbf{Q}$ is not dense.)
130. Let X be a commutative topological group and V a symmetric neighborhood of e. Show that there exists a continuous absolute-value

function p, and $\varepsilon > 0$, such that $(p < \varepsilon) \subset V$. ⟦Imitate Lemma 11.5.2.⟧

131. A commutative topological group is topologically isomorphic into a product of commutative semimetric groups. ⟦Imitate Theorem 11.5.3, using Problem 130.⟧

---

201. Let (X, D) be a semimetric group with D a B-semimetric. Let $\{u_n\}$ be a D-Cauchy sequence. Then if $x_n \to e$, it follows that $u_n x_n u_n^{-1} \to e$. ⟦With

$$p(x) = \tfrac{1}{2}D(x, e), p(u_n x_n u_n^{-1}) \leq p(u_n u_k^{-1}) + p(u_k x_n u_k^{-1}) + p(u_k u_n^{-1}).$$

Fix k large.⟧

202. Let X be a semimetric group. There is a complete semimetric group (Y, D), with D a B-semimetric such that X is a dense topological subgroup. ⟦Give X its B-semimetric. Let (Y, D) be the semimetric completion of Theorem 9.2.2. For $u = \{u_n\}$, $v = \{v_n\}$ in Y let $uv = \{u_n v_n\}$, $u^{-1} = \{u_n^{-1}\}$ (Lemma 12.2.3). To show D is a group semimetric let $q(u) = \frac{1}{2}D(u, e)$. Prove that q is an absolute-value function using $q(u) = \lim p(u_n)$, where $p = q \mid X$. The only nontrivial part of this follows from Problem 201.⟧
203. Let (X, d) be a semimetric group with d left invariant. Show that X has a semimetric semigroup completion; that is, the group multiplication can be extended to an associative operation on the semimetric completion ⟦the proof of Problem 202⟧.
204. A left-invariant-semimetric group has a left completion if and only if it satisfies the condition analogous to that given in Problem 201.
205. Find the completion in the sense of Problem 203, of G, in Examples 3 and 4. ⟦Its closure in $C(I)$.⟧
206. $\mathbf{R}$ and $\mathbf{R}^2$ are isomorphic ⟦their Hamel bases must have equal cardinality, c⟧, but not topologically isomorphic ⟦Sec. 5.2, Problem 8⟧.
207. Two group topologies for a group X with the same dense sets must be equal. ⟦Take complements in Problem 113.⟧ Can "dense" be replaced by "fundamental"? (A set is called *fundamental* if it is not included in any closed proper subgroup.)
208. If a topological group is topologically complete, it is complete in its two-sided uniformity $\mathscr{B}$. ⟦Let Y be the completion of $(X, \mathscr{B})$ as in Problem 202. Then $Y = X$ by Sec. 9.1, Problem 206; and Sec. 12.1, Problem 117.⟧ (This result is due to Victor Klee.) In particular the matrix group of Sec. 12.1, Example 10 is $\mathscr{B}$-complete.
209. Problem 208 becomes false if "two-sided" is replaced by "left."
210. Every commutative topological group has a completion. ⟦Imitate Theorem 11.5.4, using Problems 131 and 202.⟧

12.3 Quotients

THEOREM 12.3.1. *The quotient topology by a homomorphism f from a topological group X onto a group Y is a group topology. Furthermore it has a local base of neighborhoods of e consisting of $\mathscr{F} = \{U: U$ is symmetric and $f^{-1}[U]$ is a neighborhood of e in $X\}$.*

We shall first check that the collection $\mathscr{F}$ satisfies the conditions of Theorem 12.1.6. Consider, for example, Condition (d). Let $U \in \mathscr{F}$ and $b \in Y$ then $b = f(a)$ for some $a \in X$, and there exists a symmetric neighborhood W of e in X with $aWa^{-1} \subset f^{-1}[U]$. Let $V = f[W]$. Then $V \in \mathscr{F}$ and $bVb^{-1} = f[aWa^{-1}] \subset U$. The other conditions are easily checked. We show that the group topology T which has $\mathscr{F}$ as its base of neighborhoods of e is precisely the quotient topology T_Q by f. Let $G \in T$ and $x \in f^{-1}[G]$. Then $(fx)^{-1}G$ is a T neighborhood of e hence includes some $U \in \mathscr{F}$. Then $xf^{-1}[U]$ is a neighborhood of x and is included in $f^{-1}[G]$. [[If $a \in f^{-1}[U]$, $f(xa) = f(x)f(a) \in f(x)U \subset G$.]] This proves that $f^{-1}[G]$ is open, hence $G \in T_Q$. Conversely, let $G \in T_Q$ and $y \in G$; say $y = f(x)$. Then $x^{-1}.f^{-1}[G]$ is a neighborhood of e hence includes a symmetric neighborhood W of e. Let $U = f[W]$. Then $U \in \mathscr{F}$ and $yU \subset G$. [[If $u \in U$, say $u = f(w)$, then $w \in x^{-1}f^{-1}[G]$ so that $yu = f(xw) \in G$.]] Thus $G \in T$. ∎

Now let S be an invariant subgroup of a topological group X, and let $q: X \to X/S$ be defined by $q(x) = xS$. (See the definition of coset and quotient group in Section 12.1.) Then q is a homomorphism onto X/S; the quotient topology by q is called the quotient topology of X/S. In the notation of Section 6.5, this is X/ρ where $x \rho y$ means $x^{-1}y \in S$, or, equivalently, since S is an invariant subgroup, $yx^{-1} \in S$. [[Since $x^{-1}y \in S$ if and only if $q(x^{-1}y) = e$ (the identity in X/S, thus $e = S$), and this holds if and only if $q(x) = q(y)$. See Sec. 6.5, Example 4.]] The equivalence classes determined by ρ are the left cosets of S (they are also the right cosets).

Conversely if $f: X \to Y$ is a homomorphism onto, the kernel S of f is an invariant subgroup of X; and Y, with the quotient topology is topologically isomorphic with X/S under the map $xS \to f(x)$. [[This map is well defined since $xS = x'S$ implies $f(x) = f(x')$. It is an isomorphism and the topological results are given in Theorem 6.5.5.]]

THEOREM 12.3.2. *The quotient topology by a homomorphism $f: X \to Y$ is separated if and only if the kernel S of f is closed. The quotient of a topological group X by an invariant subgroup S is separated if and only if S is closed.*

(There is no assumption on the separation of X.) The equivalence of the two conditions holds since Y is topologically isomorphic with X/S as just pointed out. Now S is closed if and only if all of its translates are closed. These are cosets of S, hence members of the quotient group X/S. The result follows from Theorem 6.7.5. ∎

NOTE. The reader is warned of an inconsistency between topological and algebraic notation. If S is a subset of a topological space X, topologists find it very useful to have a name for X/ρ where the equivalence classes of ρ are S and all singletons in $X \setminus S$. This is usually denoted by X/S. No ambiguity can arise in a case where "subgroup" has no meaning, so that X/S could not be a quotient group in the algebraic sense. We shall reserve the use of the notation X/S for the quotient group.

By a *quotient homomorphism* is meant a homomorphism between topological groups which is also a quotient map.

THEOREM 12.3.3. *A quotient homomorphism $f: X \to Y$ is an open map. Hence a continuous homomorphism is a quotient map if and only if it is open and onto.*

The last part follows from Theorem 6.5.1. Let V be a symmetric neighborhood of e in X. Then $f[V]$ is a neighborhood of e in Y. ⟦It is symmetric, and its inverse image under f includes V, hence is a neighborhood of e in X; that is, $f[V] \in \mathscr{F}$ in the notation of Theorem 12.3.1.⟧ The result now follows from Lemma 12.2.1. ∎

This result may be contrasted with the fact that a quotient map in general need not be open, and a quotient homomorphism need not be closed ⟦Sec. 6.5, Problems 7 and 8⟧.

★EXAMPLE 1. Let $X = \mathbf{R} \setminus \{0\}$ with ordinary multiplication as the group operation. The identity is 1, the inverse map is $x \to 1/x$, and the Euclidean topology is a group topology. The subgroup $S = \{x: x > 0\}$ is open and closed (compare Sec. 12.1, Problem 113), thus X is not connected. S has exactly two cosets, S and $-S$; X/S is the discrete group with two members. (See Problem 2.) The inclusion map $i: S \to X$ is an open homomorphism, but not a quotient map since it is not onto. Like $\mathbf{R}$, X is commutative and its three uniformities are equal. A basic connector is $\{(x, y): x \neq 0, y \neq 0, |x/y - 1| < \varepsilon\}$. This uniformity is metrizable of course. (See Problem 8.)

LEMMA 12.3.1. *Let X be a semimetric group with absolute-value function p, and $f: X \to Y$ a homomorphism onto a group Y. Then the quotient topology on Y is semimetrizable, indeed $h(y) = \inf\{p(x): f(x) = y\}$ defines an absolute-value function h for the quotient topology of Y.*

(See Problem 13.) To show the quotient topology semimetrizable, it would be easy to check that Y is first countable, and apply Theorem 12.2.3. We shall instead, check the given formula. That $h(e) = 0$, $h(y) \geq 0$ for all y are trivial. Next let $y, z \in Y, \varepsilon > 0$. Choose $a, b \in X$ with $f(a) = y, f(b) = z$,

$$p(a) < h(y) + \varepsilon, \qquad p(b) < h(z) + \varepsilon.$$

Then $f(ab) = yz$ so that

$$h(yz) \leq p(ab) \leq p(a) + p(b) < h(y) + h(z) + 2\varepsilon.$$

The triangle inequality follows. That $h(y^{-1}) = h(y)$ for all y is clear. Finally, let $h(y_n) \to 0$ and $b \in Y$; say $b = f(a)$. For $n = 1, 2, \ldots$, let $f(x_n) = y_n$, $p(x_n) < h(y_n) + 1/n$. Then $p(x_n) \to 0$, hence $p(ax_na^{-1}) \to 0$. Now $h(by_nb^{-1}) \leq p(ax_na^{-1}) \to 0$ since $f(ax_na^{-1}) = by_nb^{-1}$. Having checked that h is an absolute-value function, observe that $f: (X, p) \to (Y, h)$ is continuous. ⟦$h(fx) \leq p(x)$ so that f is continuous at e, hence everywhere by Theorem 12.2.1.⟧ Thus h is smaller than the quotient topology. Finally, let U be a basic quotient neighborhood of e; that is, $U \in \mathscr{F}$ in Theorem 12.3.1. Then $f^{-1}[U] \supset (p < \varepsilon)$ for some $\varepsilon > 0$. But then $U \supset (h < \varepsilon)$. ⟦Let $h(y) < \varepsilon$. Choose x with $f(x) = y$, $p(x) < \varepsilon$. Then $x \in f^{-1}[U]$ and so $y = f(x) \in U$.⟧ Since U is an h neighborhood of e, h is larger than the quotient topology. ∎

THEOREM 12.3.4. *Let (X, p) be a semimetric group with absolute-value function p, and S an invariant subgroup. Then X/S is a semimetric group with absolute-value function h given by $h(xS) = \inf\{p(xs): s \in S\}$. It is a metric group if and only if S is closed.*

This is Lemma 12.3.1 with $Y = X/S$, $f =$ quotient map. See also Theorem 12.3.2. ∎

LEMMA 12.3.2 *In Lemma 12.3.1, if X is complete in its left uniformity $\mathscr{L}$, Y is also.*

Let $\{y_n\}$ be an $\mathscr{L}$-Cauchy sequence. It has a subsequence $\{z_n\}$ satisfying $h(z_n^{-1}z_{n+1}) < 2^{-n}$ ⟦Sec. 9.1, Problem 11⟧. Choose x_n, $n = 0, 1, 2, \ldots$, with $f(x_0) = z_1$, and for $n > 0$, $f(x_n) = z_n^{-1}z_{n+1}$, $p(x_n) < 2^{-n}$, and let $a_n = x_0x_1x_2\cdots x_n$. Then $\{a_n\}$ is an $\mathscr{L}$-Cauchy sequence ⟦for $m > n$,

$$p(a_n^{-1}a_m) = p(a_{n+1}a_{n+2}\cdots a_m) \leq p(a_{n+1}) + \cdots + p(a_m)$$
$$\leq 2^{-(n+1)} + \cdots + 2^{-m} < 2^{-n}⟧,$$

hence convergent, say $a_n \to a$. Then $z_n \to f(a)$. ⟦$z_n = f(x_0x_1x_2\cdots x_{n-1}) = f(a_{n-1}) \to f(a)$.⟧ It follows that $y_n \to f(a)$ ⟦Sec. 9.1, Problem 3⟧. ∎

THEOREM 12.3.5. *In Theorem 12.3.4, if X is complete in its left uniformity, X/S is also.* (*But see Sec.* 13.2, *Problem* 110.)

This is Lemma 12.3.2 with $Y = X/S$, $f =$ quotient map. ∎

THEOREM 12.3.6. *Let X, Y be topological groups and $f: X \to Y$ a continuous homomorphism. Then there exists a continuous isomorphism $g: X/S \to Y$, with S the kernel of f, and $f = g \circ q$.*

Define g by $g(xS) = f(x)$; g is well defined since if $xS = aS$, $a^{-1}x \in S$ so that $f(a^{-1}x) = e$ and $f(x) = f(a)$. Also, g is one-to-one ⟦$g(xS) = e$ implies $f(x) = e$ so that $x \in S$ and $xS = S$, the identity of X/S⟧, a homomorphism. ⟦$g(aSbS) = g(abS) = f(ab) = f(a)f(b) = g(aS)g(bS)$.⟧ Finally g is continuous ⟦$g \circ q = f$ and the result follows from Theorem 6.6.1⟧. ∎

Problems on Topological Groups

In this list, X, Y are topological groups.

1. The quotient topology by a homomorphism f is the largest group topology which makes f continuous.
2. Let S be a subgroup of X. Then X/S is indiscrete if and only if S is dense; X/S is discrete if and only if S has nonempty interior. (See Example 1.)
3. The component S of e is an invariant subgroup of X. ⟦$SS^{-1} = \bigcup \{Sx: x \in S^{-1}\}$ is the union of continuous images of S, hence connected sets, all containing e. Thus SS^{-1} is connected and so $SS^{-1} \subset S$. Similarly $aSa^{-1} \subset S$ for each a.⟧
4. The *center* ($= \{a: xa = ax$ for all $x\}$) of a separated group is a closed subgroup. ⟦It is the kernel of the map $x \to xax^{-1}a^{-1}$.⟧
5. In Problem 4, "separated" cannot be omitted. ⟦Indiscrete.⟧
6. In Theorem 12.3.6, $g(Sx) = f(x)$.
7. For any homomorphism f with kernel S, $f^{-1}[fA] \subset AS$. ⟦If $f(x) \in f[A]$, $f(xa^{-1}) = e$ for some $a \in A$.⟧
8. Show that p is an absolute-value function for $\mathbf{R}\backslash\{0\}$ (Example 1), where $p(x) = |\log x|$ if $x > 0$, $p(x) = 1$ if $x < 0$, and that it induces the Euclidean topology.
9. In Problem 8, S and $-S$ are isometric and unit distance apart.
10. Find all absolute-value functions p such that $p(xy) = p(x)p(y)$ for all x, y. ⟦There is only one.⟧
11. Every quotient of a locally compact group is locally compact ⟦Sec. 5.4, Problem 15⟧.
12. For $A \subset X$, $q^{-1}q[A] = AS = SA$, where S is an invariant subgroup of X, and q is the quotient map onto X/S.
13. In Lemma 12.3.1, $h(y) = d(e, C)$, where C is a certain coset of S, the kernel of f; namely $C = xS$ for any x with $f(x) = y$.

---

101. Let X be a separable semimetric group which is complete in its left uniformity, and let Y be a separated Baire group. Let f be a continuous homomorphism of X onto Y. Then f is an open map. ⟦Let S be the kernel of f, $g: X/S \to Y$ the map induced by f as in Theorem 12.3.6. By Theorems 12.3.4 and 12.3.5, and Sec. 12.2, Problem 123, g is an open map. Hence $f = g \circ q$ is open, by Theorem 12.3.3.⟧

102. **R** has a proper subgroup which is of second category in **R**. ⟦Let a Hamel base be $A \cup \{b_n\}$. Let S_n be the subgroup generated by $A \cup (b_1, b_2, \ldots, b_n)$. Then $\mathbf{R} = \bigcup S_n$ so not all S_n are of first category.⟧
103. The subgroup in Problem 102 is a Baire metric space but is not topologically complete. ⟦Otherwise it is closed by Sec. 12.1, Problem 117, and Sec. 9.1, Problem 206, hence has interior since it is not nowhere dense, hence is open by Sec. 12.1, Problem 113. But **R** is connected.⟧
104. Let $f: X \to Y$ be a quotient homomorphism in which the kernel of f is $\overline{\{e\}}$. Then a filterbase in Y is $\mathscr{L}$-Cauchy if and only if it is the image of an $\mathscr{L}$-Cauchy filterbase in X. ⟦Let $f[\mathscr{F}]$ be Cauchy, and U a neighborhood of e in X. Choose $A \in \mathscr{F}$ with $f[A^{-1}A] \subset f[U]$. Then $A^{-1}A \subset f^{-1}fU \subset U\overline{\{e\}}$ (by Problem 7) $\subset UU$. Thus $\mathscr{F}$ is Cauchy.⟧ (The condition on the kernel of f cannot be omitted; see Sec. 6.5, Problem 203.)
105. Let $S = \overline{\{e\}}$. Then S is an invariant subgroup of X and X/S is separated. (Compare Sec. 6.7, Examples 4 and 5. The present construction is a special case.) Show that X/S is $\mathscr{L}$-complete if and only if X is ⟦Problem 104⟧.
106. Show that the following two uniformities of $\mathbf{R} \setminus \{0\}$ are noncomparable: The relative uniformity of $(\mathbf{R}, +)$, left = right = two-sided; and the uniformity of Example 1.
107. A locally compact G_δ group is semimetrizable ⟦Sec. 5.4, Problem 201⟧. (A G_δ group is a topological group which is a G_δ space.) It is sufficient to assume that $\{e\}$ is a G_δ and the group is locally compact.
108. The *order* of x is the least positive integer k with $x^k = e$, or $+\infty$ if no such k exists. Let X be a compact commutative group of which every member has finite order. Show that there is a member of maximum order. ⟦Let $S_n = \{x\colon \text{order of } x \text{ divides } n\}$; $X = \bigcup S_n$ so some S_n has interior (Baire category theorem). X/S_n is discrete hence finite.⟧
109. If X is connected, so is X/S, but not conversely. ⟦$S = X$; Theorem 5.2.2.⟧ However $X/\overline{\{e\}}$ is connected if and only if X is. ⟦If S is open and closed, $q[S]$ is open and closed by Sec. 6.7, Problem 108. If S is proper, $q[S]$ is also since $q^{-1}q[S] = S\overline{\{e\}}$ by Problem 12; this is S by Sec. 12.1, Problem 119.⟧

---

201. Let T, T' be group topologies for a group Y. Let $f\colon (Y, T) \times (Y, T') \to Y$ be given by $f(x, y) = xy$. Show that the quotient topology is $T \wedge T'$, that is, the largest group topology included in $T \cap T'$. (Compare Sec. 12.2, Problem 126.
202. With T, T' as in Problem 102, $T \wedge T'$ is separated if and only if the

identity map from (Y, T) to (Y, T') has closed graph ⟦Problem 201; Theorem 12.3.2⟧. (Compare Sec. 6.7, Problem 113.)

203. The conditions of Problem 202 ($T \wedge T'$ separated) are not sufficient to imply that $T \cap T' = T \wedge T'$; that is, that $T \cap T'$ be a group topology.
204. Does Theorem 12.3.5 hold for the two-sided uniformity?
205. Interpret the equation $C^*(X)/C_0(X) = C(\beta X \setminus X)$.

12.4 Topological Vector Spaces

In this section we present some of the basic tools of functional analysis, a major branch of today's pure and applied mathematics. Most of our presentation will be very special forms of important results. For more general forms of these results, and their applications, see [Wilansky (a)], [Kelley and Namioka], and [Dunford and Schwartz], where many more references will be found.

A vector space is a set X with two operations. First, it is a commutative group with an operation called *addition* written $(x, y) \to x + y$, with identity 0, and inverse $x \to -x$. Second, a function on $\mathbf{R} \times X$, called *scalar multiplication*, is assumed, written $(\alpha, x) \to \alpha x$ satisfying $1x = x$, $(\alpha\beta)x = \alpha(\beta x)$, $\alpha(x + y) = \alpha x + \alpha y$, $(\alpha + \beta)x = \alpha x + \beta x$. The usual conventions of arithmetic are used; for example, $\alpha x + \beta y$ means $(\alpha x) + (\beta y)$. We shall use such elementary facts as $0x = 0$ (these are $0 \in \mathbf{R}$, $0 \in X$, respectively), and $\alpha(x - y) = \alpha x - \alpha y$. (See [Wilansky (a), Section 2.1].) In this context, real numbers are called *scalars*, members of X are called *vectors*. Examples of vector spaces are $\mathbf{R}^n$, $C(X)$, $C^*(X)$ with the usual operations.

NOTATION. The additive notation is used. (See the beginning of Section 12.1.) Also $\alpha S = \{\alpha x : x \in S\}$ for scalar α.

A vector topology for a vector space X is a group topology for $(X, +)$ making the map $(\alpha, x) \to \alpha x$ continuous. Since $(X, +)$ is commutative, its right, left and two-sided uniformities are equal, and we shall speak of the uniformity of X; a basic connector being $\{(x, y) : x - y \in U\}$, where U is a neighborhood of 0. Also, formulas of the form aba^{-1} which arose, for example, in Theorem 12.1.5, no longer need be considered since $a + b - a = b$. A *topological vector space* is a pair (X, T) in which X is a vector space, and T is a vector topology. A *semimetric vector space* is a topological vector space whose topology is given by a semimetric. All the results of topological groups carry over; some of them are listed as problems.

★EXAMPLE 1. *The discrete topology is not a vector topology* (except in the trivial case $X = \{0\}$). This fact, contrasted with Sec. 12.1, Problem 5, already shows the influence of the scalars. The proof consists of observing that if $x \neq 0$, $x/n \not\to 0$ in the discrete topology, as $n \to \infty$; hence scalar multiplication is not even separately sequentially continuous. The indiscrete topology is always a vector topology ⟦Problem 2⟧.

★EXAMPLE 2. A *paranorm* on a vector space X is an absolute-value function p which satisfies the additional condition: If $\alpha_n \to \alpha$, $p(x_n - x) \to 0$, then $p(\alpha_n x_n - \alpha x) \to 0$. Here $\{\alpha_n\}$, $\{x_n\}$ are sequences of scalars and vectors, respectively. The induced invariant semimetric d is given by $d(x, y) = p(x - y)$. It induces a vector topology. ⟦Sec. 12.1, Example 1; continuity of scalar multiplication is precisely the extra assumption on p.⟧ Every first countable topological vector space is semimetrizable, indeed its topology can be given an equivalent invariant semimetric d ⟦Theorem 12.2.3⟧, which is then given by an absolute-value function p; $p(x) = d(x, 0)$. But then p must be a paranorm since the topology it induces is a vector topology. A *Fréchet space* is a complete metric vector space.

★EXAMPLE 3. A *seminorm* on a vector space is an absolute-value function p which satisfies the additional condition $p(\alpha x) = |\alpha| p(x)$ for scalar α, vector x. Every seminorm is a paranorm,

$$[\![p(\alpha_n x_n - \alpha x) = p[\alpha_n(x_n - x) + (\alpha_n - \alpha)x] \leq |\alpha_n| p(x_n - x) + |\alpha_n - \alpha| p(x)]\!],$$

but not conversely. ⟦$p(x) = |x|/(1 + |x|)$ for $x \in \mathbf{R}$.⟧ A *seminormed space* is a pair (X, p) in which X is a vector space, p is a seminorm, and X is given the (vector) topology induced by p via d, where $d(x, y) = p(x - y)$. If a seminorm p satisfies $p(x) > 0$ for all $x \neq 0$ it is called a *norm*; finally, a complete normed space is called a *Banach space*, in honor of S. Banach. The function defined on $\mathbf{R}^n$ in Section 1.2 is a norm, as is the one defined on $L[S]$ in Section 10.3.

A function $f: X \to Y$, with X, Y vector spaces, is called *linear* if $f(\alpha a + \beta b) = \alpha f(a) + \beta f(b)$ for all scalars α, β and vectors a, b. A *linear functional* is a linear map $f: X \to \mathbf{R}$, and the *dual space* X' of a topological vector space X is the space of all continuous linear functionals on X. The weak topology $w(X')$ by the collection X' (see Problem 9) is referred to, simply, as *the weak topology* of X. The *weak topology is smaller than the original topology of* X. ⟦It is the smallest vector topology which makes all the members of X' continuous.⟧ There is also a natural topology that may be placed on X'. Each $x \in X$ defines a linear functional $\hat{x}$ on X' by the formula $\hat{x}(f) = f(x)$ for all $f \in X'$. For any nonempty subset S of X, let $\hat{S} = \{\hat{x}: x \in S\}$, then $\hat{S}$ is a set of linear functionals on X' and the weak topology by $\hat{S}$, $w(\hat{S})$, for X' is called the *weak-star* (or *weak**) *topology* by S. The weak* topology by X is called simply, the *weak* topology*.

LEMMA 12.4.1. *A linear functional on a topological vector space is continuous if and only if it is bounded on some neighborhood of* 0.

If f is continuous, $f^{-1}[(-1, 1)]$ is a neighborhood of 0 and $|f|$ is less than 1 on it. Conversely, if f is bounded on a neighborhood U of 0, say $|f| < M$ on U, then $f^{-1}[(-\varepsilon, \varepsilon)] \supset (\varepsilon/M)U$ and so f is continuous at 0. By Theorem 12.2.1, f is continuous everywhere. ∎

★EXAMPLE 4. Let X be a seminormed space and denote the value of the seminorm at x by $\|x\|$. For a real-valued function f on X, let $\|f\|$ be $\sup\{|f(x)|: \|x\| \leq 1\}$. Then Lemma 12.4.1 implies that a linear functional f is continuous if and only if $\|f\| < \infty$. ⟦If $\|f\| < \infty$, $|f|$ is bounded on the unit disc, while if f is continuous, it is bounded, say $|f(x)| < M$, for x in some disc, say for $\|x\| \leq \varepsilon$. Then if $\|x\| \leq 1$, $\|\varepsilon x\| \leq \varepsilon$ and so $|f(\varepsilon x)| < M$. Thus $\|f\| \leq M/\varepsilon$.⟧ If $\|f\| < \infty$, we have the useful inequality $|f(x)| \leq \|f\| \cdot \|x\|$ for all x. ⟦Fix x, f; let $t > \|x\|$ so that $t \neq 0$. Then $\|x/t\| < 1$ and so $|f(x/t)| \leq \|f\|$. Thus $|f(x)| \leq t\|f\|$. But t is arbitrary.⟧

EXAMPLE 5. Let c_0 be the Banach space of real null sequences with $\|x\| = \sup|x_n|$, and let a be a real sequence with $\sum |a_n| < \infty$. The equation $f(x) = \sum a_n x_n$ defines a linear function on c_0. It is continuous as $|f(x)| \leq \|x\| \cdot \sum |a_n|$ implies $\|f\| \leq \sum |a_n|$. Indeed $\|f\| = \sum |a_n|$. ⟦Fix a positive integer m and let $x_i = \operatorname{sgn} a_i$ for $i = 1, 2, \ldots, m$; $x_i = 0$ for $i > m$. Then $\|x\| \leq 1$ and $f(x) = \sum_{i=1}^{m} |a_i|$. Thus $\|f\| \geq \sum_{i=1}^{m} |a_i|$ for all m.⟧

The following extension theorem, given by H. Hahn in 1927 and S. Banach in 1929 is one of the foundation stones of functional analysis. For our purposes it yields a sufficient supply of linear functionals for a successful duality theory. (See, for example, Corollaries 12.4.2 and 12.4.3.)

THEOREM 12.4.1 (THE HAHN–BANACH THEOREM). *Let X be a seminormed space, S a vector subspace, and $f \in S'$. Then f can be extended to $F \in X'$ with $\|F\| = \|f\|$.*

We may assume $\|f\| = 1$. ⟦If $\|f\| = 0$ take $F = 0$; otherwise consider $f/\|f\|$.⟧ It is sufficient to construct F with $\|F\| \leq 1$ since $\|F\| \geq 1$ is a trivial consequence of the fact that F extends f. Let $P = \{g: g \in A'$ for some vector subspace $A = A_g$ of X with $A \supset S$, $\|g\| = 1$, $g = f$ on $S\}$. It is not empty since $f \in P$; P is made into a poset by the ordering $g_2 \geq g_1$ means g_2 is an extension of g_1. Let C be a maximal chain in P and $D = \bigcup \{A_g: g \in C.\}$ D is a vector subspace of X, ⟦let α, β be scalars and $x, y \in D$. Then $x, y \in A_g$ for some $g \in C$, hence $\alpha x + \beta y \in A_g \subset D$⟧. Define $F: D \to \mathbf{R}$ by $F(x) = g(x)$ whenever $x \in A_g$ with $g \in C$; F is well defined since C is a chain; F is linear by the same argument which showed D to be a vector subspace, and $\|F\| \leq 1$.

⟦For any x, $x \in A_g$ for some g, hence $|F(x)| = |g(x)| \leq \|x\|$.⟧ Thus the proof is concluded by showing that $D = X$. If possible, let $y \in X \setminus D$. Let $u = \sup\{-F(x) - \|x + y\| : x \in D\}$, $v = \inf\{-F(x) + \|x + y\| : x \in D\}$. Then $u \leq v$. ⟦For any a, $b \in D$, $F(b) - F(a) = F(b - a) \leq \|b - a\| \leq \|b + y\| + \|a + y\|$ and so $-F(a) - \|a + y\| \leq -F(b) + \|b + y\|$.⟧ Now let D_1 be the smallest vector subspace of X which includes D and y ⟦Problem 4⟧, and define F_1 on D_1 by $F_1(d + \alpha y) = F(d) + \alpha u$. We shall show that $F_1 \in P$. First F_1 extends f ⟦it extends F⟧, F_1 is linear ⟦Problem 5⟧, and $\|F_1\| \leq 1$. ⟦Let $x \in D_1$, $x = d + \alpha y$. If $\alpha = 0$, $|F_1(x)| = |F(d)| \leq \|d\| = \|x\|$; if $\alpha \neq 0$, $-F(d/\alpha) - \|y + d/\alpha\| \leq u \leq v \leq -F(d/\alpha) + \|y + d/\alpha\|$, hence $-F(d) - \|\alpha y + d\| \leq \alpha u \leq -F(d) + \|\alpha y + d\|$ (treat $\alpha > 0$ and $\alpha < 0$ separately; the result is exactly the same), thus $-\|x\| \leq F(d) + \alpha u \leq \|x\|$ and so $|F_1(x)| \leq \|x\|$.⟧ The existence of F_1 contradicts the definition of C, for $C \cup \{F_1\}$ would be strictly larger. ∎

COROLLARY 12.4.1. *Let X be a seminormed space and $x \in X$ with $\|x\| \neq 0$. Then there exists $f \in X'$ with $\|f\| = 1, f(x) = \|x\|$.*

Let $S = \{\alpha x : \alpha \in \mathbf{R}\}$. Then S is a vector subspace of X. Define $f: S \to \mathbf{R}$ by $f(\alpha x) = \alpha\|x\|$. Then $f(x) = \|x\|$, f is linear and $\|f\| = 1$. ⟦$|f(\alpha x)| = |\alpha| \cdot \|x\| = \|\alpha x\|$.⟧ The Hahn–Banach theorem yields the result. ∎

COROLLARY 12.4.2. *The weak topology of a normed space X is separated.*

In view of Theorem 6.3.2 it is sufficient to prove that X' is separating. Let $x \neq y$. By Corollary 12.4.1, there exists $f \in X'$ with $f(x - y) \neq 0$. This implies $f(x) \neq f(y)$. ∎

COROLLARY 12.4.3. *Let X be an infinite-dimensional normed space, and C its unit circumference, $C = \{x : \|x\| = 1\}$. Then 0 is in the closure of C in the weak topology.*

Let U be a neighborhood of 0. Then there exist $f_1, f_2, \ldots, f_n \in X'$ and $\varepsilon > 0$ with $U \supset \bigcap (|f_i| < \varepsilon)$. Let $x_1, x_2, \ldots, x_{n+1}$ be linearly independent and consider the n equations in $n + 1$ unknowns $\alpha_1, \alpha_2, \ldots, \alpha_{n+1}$: $\sum_{j=1}^{n+1} f_i(x_j)\alpha_j = 0$. This has a solution in which not all α_j are 0. Let $x = \sum \alpha_j x_j$, $y = x/\|x\|$. Then $y \in C \cap U$. ⟦For each i, $|f_i(x)| = |\sum f_i(x_j)\alpha_j| = 0 < \varepsilon$.⟧ ∎

COROLLARY 12.4.4. *For an infinite-dimensional normed space, the norm and weak topologies are different.*

This follows from Corollary 12.4.3, since C is norm closed. ∎

The next Example shows how a topological restriction leads to an algebraic conclusion. It also hints at the value of having sufficiently many continuous linear functionals.

EXAMPLE 6. *A locally compact normed space is finite dimensional.* The unit disc is compact since it is closed and included in some multiple of a compact neighborhood of 0. ⟦Say some compact neighborhood K of 0 includes $D(0, \varepsilon)$; then $D(0, 1) = (1/\varepsilon)D(0, \varepsilon) \subset K/\varepsilon$. See Problem 6.⟧ Then the unit circumference $C = \{x: \|x\| = 1\}$ is compact since it is a closed subset of $D(0, 1)$. Thus C is compact in the weak topology since it is smaller than the norm topology ⟦Theorem 5.4.4⟧, thus C is closed in the weak topology ⟦Corollary 12.4.2; Theorem 5.4.5⟧. The result follows from Corollary 12.4.3. ∎

The following basic result was given for separable spaces by S. Banach, before 1932. It was extended in 1940 by L. Alaoglu.

THEOREM 12.4.2 (THE BANACH–ALAOGLU THEOREM). *Let X be a seminormed space and $H = \{f \in X': \|f\| \leq 1\}$. With the weak* topology H is a compact Hausdorff space.*

Let $Y = \prod \{[-\|x\|, \|x\|]: x \in X\}$; Y is the product of a family of closed intervals of real numbers. It is, of course, compact, by Tychonoff's theorem, and a Hausdorff space; the proof is concluded by showing that H is a closed subspace of Y. First, $H \subset Y$ ⟦$f \in H$ implies that $f: X \to \mathbf{R}$ and $|f(x)| \leq \|x\|$ for each x by Example 4⟧, H is a topological subspace of Y since they both have the weak topology by all the maps $f \to f(x)$ as x ranges over X. (These are the projections from Y onto the factor spaces $[-\|x\|, \|x\|]$.) Finally H is a closed subspace of Y. ⟦Let $f \in \bar{H}$. Then f, as a member of Y is a real-valued function on X. It remains to show f linear. Fix scalars α, β, vectors x, y. Let f_δ be a net in H converging to f. Then

$$\begin{aligned} f(\alpha x + \beta y) &= P_{\alpha x + \beta y}(f) = \lim P_{\alpha x + \beta y}(f_\delta) = \lim f_\delta(\alpha x + \beta y) \\ &= \lim \alpha f_\delta(x) + \beta f_\delta(y) = \alpha f(x) + \beta f(y). \end{aligned}$$

Thus f is linear. Moreover since $f \in Y$ we have $f(x) \in [-\|x\|, \|x\|]$ and so $|f(x)| \leq \|x\|$ which implies $\|f\| \leq 1$ and $f \in H$.⟧ ∎

The Banach–Alaoglu theorem has many applications. One of these is given in Problems 119, 120. Another is a representation theorem (Corollary 12.4.5) which is a key step in the study of Banach algebras and topological groups. (See [Wilansky (a), Chapter 14]; [Rudin].) A *congruence* is a linear isometry; and X *congruent into* Y means there exists a congruence of X with a subspace of Y.

In the next result $C(H)$ is taken to be a normed space with

$$\|x\| = \sup\{|x(t)|: t \in H\}.$$

It is easily seen to be a Banach space by an application of Theorem 4.2.10.

COROLLARY 12.4.5. *Every normed space X is congruent into $C(H)$ where H is some compact Hausdorff space.*

The definition of H is given in the statement of the Banach–Alaoglu theorem. The congruence is $x \to \hat{x} \mid H$. This is clearly a linear map, moreover it is an isometry. ⟦For any $f \in H$, $|\hat{x}(f)| = |f(x)| \leq \|x\|$, thus $\|\hat{x}|H\| \leq \|x\|$. Conversely, use Corollary 12.4.1 to choose $f \in H$ with $f(x) = \|x\|$. Then $\|\hat{x}|H\| \geq |\hat{x}(f)| = \|x\|$.⟧ ∎

The procedure for defining a vector topology follows its group analogue with some adjustment because of the extra operation (scalar multiplication.) A set S is called *balanced* if $\alpha S \subset S$ for $|\alpha| \leq 1$, and *absorbing* if for every $x \in X$, there exists a number k such that $|\alpha| > k$ implies $x \in \alpha S$.

THEOREM 12.4.3. *Let X be a vector space and $\mathscr{F}$ a collection of subsets of X such that*

(a) *$\mathscr{F}$ is a filterbase,*

(b) *for each $U \in \mathscr{F}$ there exists $V \in \mathscr{F}$ with $V + V \subset U$,*

(c) *every member of $\mathscr{F}$ is balanced and absorbing.*

Then there exists a unique vector topology for X such that $\mathscr{F}$ is a local base for the neighborhoods of 0.

The conditions of Theorem 12.1.6 are satisfied as they apply to the additive group of X. ⟦A balanced set is symmetric, and Condition (d) is automatic in a commutative group.⟧ Thus, that theorem yields a unique topology with continuous addition. It remains to show that scalar multiplication is continuous. Fix a vector y and a scalar β. Let $U \in \mathscr{F}$. The various choices now to be made are guided by the identity

$$\alpha x - \beta y = \alpha(x - y) + (\alpha - \beta)y. \tag{12.4.1}$$

Let m be a positive integer such that $2^m > |\beta| + 1$, and choose $V \in \mathscr{F}$ with $2^m V \subset U$. ⟦Choose $V_1 \in \mathscr{F}$ with $V_1 + V_1 \subset U$. Then $2V_1 \subset V_1 + V_1 \subset U$. Choose $V_2 \in \mathscr{F}$ with $V_2 + V_2 \subset V_1$. Then $2^2 V_2 \subset 2(V_2 + V_2) \subset 2V_1 \subset U$. Similarly $V_3 + V_3 \subset V_2$ implies $2^3 V_3 \subset U$ and so on.⟧ Choose $W \in \mathscr{F}$ with $W + W \subset V$. Since W is absorbing $y \in kW$ for some number $k > 1$. The proof of continuity is concluded by showing that if $0 < \varepsilon < 1/k$, then $|\alpha - \beta| < \varepsilon$, $x - y \in W$ imply $\alpha x - \beta y \in U$. This follows from Equation (12.4.1), because $\alpha(x - y) \in \alpha W \subset 2^m W$, ⟦$|\alpha| < |\beta| + 1 < 2^m$, hence $\alpha 2^{-m} W \subset W$⟧, $(\alpha - \beta)y \in \varepsilon k W \subset W \subset 2^m W$, and so by (12.4.1), $\alpha x - \beta y \in 2^m W + 2^m W \subset 2^m V \subset U$. ∎

We now present a form of the famous Uniform Boundedness Principle and two applications, Example 7 and Problem 121. A development of this principle and its place in functional analysis may be found in such texts as [Wilansky (a)] and [Kelley and Namioka].

THEOREM 12.4.4. *Let $\{f_n\}$ be a sequence of continuous linear functionals on a Banach space X. If $\{f_n(x)\}$ is bounded for each $x \in X$, then $\{\|f_n\|\}$ is bounded.*

For each $n = 1, 2, \ldots$, let $B_n = \{x \in X: |f_m(x)| \leq n \text{ for all } m\}$. Each B_n is closed ⟦it is an intersection of closed sets⟧, and $X = \bigcup B_n$. By Theorem 9.3.5, there exists n such that B_n has interior, say $D(y, \varepsilon) \subset B_n$. For any x with $\|x\| \leq 1$, let $z = y + \varepsilon x$. Then $z \in D(y, \varepsilon)$ and so, for every m, $|f_m(z)| \leq n$. Then

$$|f_m(x)| = (1/\varepsilon)|f_m(z) - f_m(y)| \leq (n/\varepsilon) + (1/\varepsilon) \sup_r |f_r(y)| = M,$$

say. Thus $\|f_m\| \leq M$ for all m. ∎

The following application, which also has an easy direct proof, is given to illustrate the use of the uniform boundedness principle.

EXAMPLE 7. A classical resonance theorem. *Suppose that $\{a_n\}$ is a sequence of real numbers such that $\sum a_n x_n$ is convergent whenever x is a null sequence. Then $\sum |a_n| < \infty$.* (It follows that $x \to \sum a_n x_n$ is continuous; see Example 5.) The reason for the term "resonance" is that if $\sum |a_n| = \infty$, there is "resonance" at some point x; that is $\sum a_n x_n$ diverges. To prove the result, for $m = 1, 2, \ldots$, let $f_m(x) = \sum_{n=1}^{m} a_n x_n$. As in Example 5, $\|f_m\| = \sum_{n=1}^{m} |a_n|$ and it follows from Theorem 12.4.4 that $\{\|f_m\|\}$ is bounded. Hence $\sum |a_n| < \infty$.

Problems

In this list X is a topological vector space.

1. A topological vector space is completely regular, and is $T_{3\frac{1}{2}}$ if and only if $\{0\}$ is closed ⟦Theorem 12.1.4⟧.
2. The indiscrete topology is always a vector topology. ⟦Take $F = \{X\}$ in Theorem 12.4.3.⟧
3. In an attempt to obtain the discrete topology, take $\mathscr{F} = \{0\}$ in Theorem 12.4.3. Which parts of the hypotheses fail?
4. Let S be a vector subspace of a vector space L, and let $x \in L \setminus S$. Show that $\{s + \alpha x: s \in S, \alpha \in \mathbf{R}\}$ is the smallest vector subspace of L which includes S and x. It is called the *span of S and x*. Show also that the representation is unique in the sense that if $s + \alpha x = s' + \alpha' x$, then $s = s'$, $\alpha = \alpha'$. ⟦If $\alpha \neq \alpha'$, $x = (s - s')/(\alpha' - \alpha) \in S$.⟧
5. With x, L, S as in Problem 4, let f be a linear functional on S, and $u \in \mathbf{R}$. Define F on the span of S and x by $F(s + \alpha x) = f(s) + \alpha u$. Show that F is linear.
6. Multiplication by a nonzero scalar is a homeomorphism of X onto itself. ⟦$x \to \alpha x$ has inverse $x \to x/\alpha$.⟧

7. For $\alpha \in \mathbf{R}$, $A \subset X$, $\alpha\bar{A} \subset \overline{\alpha A}$. ⟦For $\alpha \neq 0$, they are equal, by Problem 6.⟧
8. The closure of a vector subspace of X is a vector subspace. ⟦Imitate Sec. 12.1, Problem 23; using Problem 7.⟧
★9. The sup of vector topologies is a vector topology. ⟦Imitate Sec. 12.1, Example 4.⟧ State and prove the same for weak topologies by linear maps and product topologies. ⟦Sec. 12.1, Examples 5 and 6.⟧
10. Let $f: X \to Y$ be linear and onto. Show that the quotient topology for Y is a vector topology and has a local base of neighborhoods of 0 consisting of $\{U: U$ is balanced and $f^{-1}[U]$ is a neighborhood of 0 in $X\}$. Obtain analogues of Theorems 12.3.1, 12.3.2, and 12.3.3.
11. Let (X, p) be a semimetric vector space, where p is a paranorm, and S a vector subspace. Show that X/S is semimetrizable with paranorm h, given by $h(x + S) = d(0, x + S)$ ⟦Sec. 12.3, Problem 13⟧; and it is complete if X is ⟦Theorem 12.3.5⟧.
★12. Every neighborhood of 0 is absorbing. ⟦$x/n \to 0$ for each x.⟧
13. Let X be a seminormed space, S a closed vector subspace and $x \notin S$. Show that there exists $f \in X'$ with $f = 0$ on S, $f(x) = 1$. ⟦Define $f(s + \alpha x) = \alpha$. For $s \in S$, $\alpha \neq 0$, we have

$$d(x, S) \leq \|x - (-s/\alpha)\| = \|s + \alpha x\|/|f(s + \alpha x)|.$$

Hence $\|f\| \leq 1/d(x, S)$. By the Hahn–Banach theorem, f may be extended to all of X.⟧
14. The weak* topology for X' by S is separated if S is dense in X ⟦Theorem 6.3.2⟧. For a seminormed space X, the weak* topology by S is separated if and only if the span of S is dense.
15. If X is a seminormed space, the seminorm and weak topologies have the same closed vector subspaces. ⟦Let S be a seminorm-closed vector subspace and $x \notin S$. Choose f as in Problem 13. Then $(f \leq \frac{1}{2})$ is weakly closed and $x \notin (f \leq \frac{1}{2}) \supset S$. Thus x is not in the weak closure of S and so S is weakly closed.⟧ Compare Corollary 12.4.4 which shows that they may not have the same closed subsets.
16. A set S is called *bounded* if for every neighborhood U of 0, $S \subset tU$ for some scalar t. Show that S is bounded if and only if whenever $\{s_n\}$ is a sequence in S and $\{t_n\}$ is a null sequence of scalars, $t_n s_n \to 0$.

---

101. A topological vector space need not be normal. ⟦Problem 9; Sec. 12.1, Problem 122.⟧
102. Let X be a seminormed space. Then X', with the weak* topology is σ-compact, Lindelöf, normal. ⟦It is $\bigcup nH$ with H as in the Banach–Alaoglu theorem. Also Sec. 8.1, Problem 10; Theorem 5.3.5.⟧
103. Every Tychonoff space X is homeomorphic into the dual of some Banach space with its weak* topology. Compare Sec. 13.4, Problem

109. ⟦Let $Y = C^*(X)$. The map $x \to \hat{x}$ where $\hat{x}(f) = f(x)$ for all $f \in Y$ carries X to Y' since

$$\|\hat{x}\| = \sup\{|\hat{x}(f)| : f \in Y, \|f\| \leq 1\} = \sup\{|f(x)| : f \in Y, |f(x)| \leq 1\}.$$

This map is a homeomorphism when Y' has the weak* topology, by a slight modification of Theorem 8.2.1.⟧

104. Let f be a linear functional on X. Then f is continuous if and only if $f^{\perp}$ is closed. ⟦If f is not continuous and $x \in X$, let U be any balanced neighborhood of 0; $f[U] = \mathbf{R}$ by Lemma 12.4.1. The result follows by Sec. 12.2, Problem 111.⟧
105. Problem 104 is false for group characters. ⟦Sec. 12.2, Example 2; consider $\exp(i\pi f)$ with f an automorphism of $\mathbf{R}$.⟧
106. A one-dimensional separated topological vector space X is *linearly homeomorphic* with $\mathbf{R}$. (That is, there is a linear homeomorphism from X onto $\mathbf{R}$.) ⟦Fix $x \neq 0$. Let $f: X \to \mathbf{R}$ be $f(\alpha x) = \alpha$. Then f^{-1} is continuous by definition of topological vector space; f is continuous by Problem 104.⟧
107. $\mathbf{R}$ has only two possible vector topologies, Euclidean and indiscrete. ⟦The dimension of $\overline{\{0\}}$ is 0 or 1; see Problems 8 and 106.⟧
108. Deduce from Problem 106 that the discrete topology is not a vector topology. ⟦A one-dimensional subspace would not be $\mathbf{R}$.⟧
109. Every topological vector space is connected. ⟦Every point x lies in the set $S_x = \{\alpha x: 0 \leq \alpha \leq 1\}$ which is connected since it is a continuous image of $[0, 1]$ under the map $\alpha \to \alpha x$ (Theorem 5.2.2). The result follows from Theorem 5.2.1.⟧
110. A proper vector subspace S cannot be absorbing.

$$⟦S = \bigcup \{nS: n = 1, 2, \ldots\}.⟧$$

(That it cannot have interior follows from Problems 12 and 109; and Sec. 12.1, Problem 113. This is false for groups ⟦Sec. 12.3, Example 1⟧.)

111. A vector subspace of a vector space L is called *maximal* if it is a proper subspace and maximal among proper subspaces. These are equivalent: S is a maximal subspace of L; there exists $x \notin S$ with $L =$ span of S and x; $L =$ span of S and x for every $x \notin S$; L/S is one-dimensional; $S = f^{\perp}$ for some linear functional f on L. Show also that $f^{\perp}$ is a maximal subspace for each linear functional $f \neq 0$.
112. A maximal subspace of X is either closed or dense ⟦Problem 8⟧. (Compare the hint for Problem 104.)
113. Let S be a vector subspace of a seminormed space X such that $S + N(0, r) \supset N(0, 1)$ for some $r < 1$. Show that S is dense in X. ⟦By an easy induction, $N(0, 1) \subset S + r^n N(0, r)$, using $N(0, r) = rN(0, 1)$. By Theorem 12.1.3, $N(0, 1) \subset \bar{S}$. By Problems 8 and 110, $\bar{S} = X$.⟧

114. If a seminormed space has a totally bounded neighborhood of 0, it has a dense finite-dimensional subspace. ⟦Say $N(0, 1)$ is totally bounded. Then $N(0, 1) \subset F + N(0, \frac{1}{2})$ for some finite F. Apply Problem 113 with $S = \text{span } F$.⟧ It need not be finite dimensional. ⟦Indiscrete.⟧ Compare Example 6.
115. In Corollary 12.4.3, the weak closure of C is $\{x: \|x\| \leq 1\}$.
116. Let X be a seminormed space, H the unit disc in X', (see the Banach–Alaoglu theorem), and S a dense subset of X. Show that on H the weak* topology is equal to the weak* topology by S ⟦Problem 14; Theorem 5.4.10⟧. In particular, if X is separable, $(H, \text{weak*})$ is a compact metric space ⟦Theorem 6.3.4⟧.
117. A separable normed space is congruent into $C(H)$, where H is some compact metric space ⟦Corollary 12.4.5; Problem 116⟧. (Actually we can make $H = [0, 1]$. See [Banach, p. 185, Theorem 9].)
118. With X, S as in Problem 116 let $\{f_n\}$ be a sequence of linear functionals on X with $\{\|f_n\|\}$ bounded. Then if $f_n(s) \to 0$ for all $s \in S$ it follows that $f_n(x) \to 0$ for all $x \in X$. ⟦Assume $\|f_n\| \leq 1$ for all n and apply Problem 116.⟧
119. If $\{x_n\}$ is a sequence of real numbers with $x_n \to b$, it follows that $(1/n) \sum_{k=1}^{n} x_k \to b$. ⟦Apply Problem 118. Take X = all real sequences $x = \{x_n\}$ with $x_n \to 0$, $S = \{x \in X: x_n = 0 \text{ eventually}\}$, $\|x\| = \sup |x_n|$, $f_n(x) = (1/n) \sum_{k=1}^{n} x_k$. This yields the result for $b = 0$. In general consider $\{x_n - b\}$.⟧
120. Let $A = (a_{nk})$ be a real infinite matrix satisfying the famous *Silverman–Toeplitz conditions*: $\lim_{n\to\infty} a_{nk} = a_k$ exists for each k, $\lim_{n\to\infty} \sum_{k=1}^{\infty} a_{nk} = t$ exists, and $\sup_n \sum_{k=1}^{\infty} |a_{nk}| < \infty$. Show that if $x = \{x_n\}$ is a sequence of real numbers with $x_n \to b$ it follows that $\lim_n \sum_{k=1}^{\infty} a_{nk}x_k = tb + \sum a_k x_k$. ⟦Apply Problem 118; with the same X, S as in Problem 119; let $f_n(x) = \sum a_{nk}x_k$.⟧
121. If $\sum a_k x_k$ is convergent for all x such that $\sum |x_k| < \infty$, show by the method of Example 7 that x is bounded.
122. A linear map from any topological vector space onto a Banach space is almost open. ⟦$\bigcup \{nf[N]: n = 1, 2, \ldots\}$ is the whole range space for every neighborhood N of 0. Thus some $nf[N]$, hence $f[N]$ itself, is somewhere dense, by Theorem 9.3.2.⟧
123. A continuous linear one-to-one map from any Banach space onto another is a homeomorphism ⟦Problem 122; Sec. 9.1, Problem 118⟧.
124. Two comparable complete norms must be equivalent ⟦Problem 123⟧.
125. "One-to-one" may be dropped in Problem 123 if "a homeomorphism" is replaced by "an open map." ⟦Write $f: X \to Y$ as $g \circ h$, where $h: X \to X/S$, $g: X/S \to Y$, $S = f^{\perp}$. Use Theorems 12.3.4 and 12.3.5.⟧
126. A linear map with closed graph from one Banach space to another is

continuous. ⟦The graph G is a Banach space; $f = P_2 \circ P_1^{-1}$, where P_1, P_2 are projections from G; P_1 is a homeomorphism by Problem 123.⟧

---

201. The dual of a Banach space with its weak* topology is hemicompact. (Compare Problem 102), and "Banach" cannot be replaced by "normed". ⟦See [Wilansky (a), Sec. 13.5, Problems 12, 13, and 14].⟧
202. If X has a local base of convex neighborhoods of 0, the two definitions of boundedness given in Problem 16 and Sec. 11.3, Problem 203 coincide. They do not coincide in general. ⟦See [Wilansky (a), Sec. 10.3, Problem 21].⟧
203. Let X be an infinite-dimensional second category space. Then X has a maximal subspace which is of second category in X. ⟦Let $H \cup \{x_n\}$ be a Hamel base and S_n the span of $H \cup \{x_1, x_2, \ldots, x_n\}$. Then $X = \bigcup S_n$ so some $S_n \in$ cat II.⟧
204. On every infinite-dimensional second category space, there is defined a discontinuous linear functional ⟦Problem 104 and 203⟧.
205. Let X be an infinite-dimensional vector space and T the largest vector topology, namely $T = \bigvee \Phi$, where Φ is the family of all vector topologies for X. (See Problem 9.) Show that (X, T) is of first category. ⟦By Problem 204, since the weak topology by any linear functional belongs to Φ.⟧

Function Spaces

13

13.1 The Compact Open Topology

One of the most important early applications of abstract analysis to classical analysis was a key step in the proof of the Riemann mapping theorem in which the existence of a function solving a certain minimum problem is asserted. A real-valued function h is defined on a set S of functions and a function $f \in S$ is sought with $h(f) = \min\{h(g): g \in S\}$. One way to do this is to topologize S in such a way that it is compact and h is continuous. The existence of $f \in S$ minimizing h is then assured by Theorem 5.4.4. ⟦$h[S]$ is a compact subset of $\mathbf{R}$.⟧ It is also sufficient that h be lower semicontinuous ⟦Sec. 5.4, Problem 107⟧. For a discussion of this point see [Courant, pp. 23, 24].

Topologies are placed on function spaces for several purposes. As just mentioned, if a compact topology can be introduced, certain problems of analysis can be solved. In Section 13.2 topologies will be discussed whose convergence is of some required form. In this section we shall consider topologies on function spaces which make $f(x)$ a continuous function of both f and x (jointly). The main result is Theorem 13.1.1.

Let Y be a topological space, X a set and $F \subset Y^X$. Thus F is a certain family of functions from X to Y. Let Φ be a collection of subsets of X. We assume that F and Φ are not empty. For $S \in \Phi$ and $G \subset Y$, let

$$\begin{aligned}[S, G] &= \{f \in F: f[S] \subset G\} = \bigcap \{\{f \in F: f(s) \in G\}: s \in S\} \\ &= \bigcap \{F \cap P_s^{-1}[G]: s \in S\},\end{aligned}$$

where $P_s: Y^X \to Y$ is the projection $P_s(f) = f(s)$. Since $\bigcup \{[S, G]: S \in \Phi, G$ an open set in $Y\} = F$, ⟦$[S, Y] = F$⟧, the collection of all $[S, G]$ with $S \in \Phi$, G an open set in Y, is a subbase for a unique topology for F which will be called *the Φ open topology*.

★EXAMPLE 1. Let Φ be the set of all singletons in X, (equivalently, by Problem 1, the set of all finite subsets of X). Then the Φ open topology is the product topology. ⟦Let f_δ be a net in F, and $f \in F$. By Theorem 6.4.1, it will suffice to show that $f_\delta \to f$ in the Φ open topology if and only if $f_\delta(x) \to f(x)$ for each $x \in X$. Suppose first that $f_\delta \to f$, $x \in X$, and that G is an open neighborhood of $f(x)$. Then $f_\delta \in [\{x\}, G]$ eventually since the latter is a Φ open set containing f. Thus $f_\delta(x) \in G$ eventually and so $f_\delta(x) \to f(x)$. Conversely, if $f_\delta(x) \to f(x)$ for all $x \in X$, let U be a neighborhood of f. Then, by definition of the Φ open topology and subbase, U includes a finite intersection of sets, each of which is of the form $[\{x\}, G]$ and each of which contains f; for each such set $f_\delta(x) \in G$ eventually; that is, $f_\delta \in [\{x\}, G]$ eventually. Hence $f_\delta \in U$ eventually by Lemma 3.4.1.⟧

The special case of the Φ open topology in which X is a topological space and Φ is the collection of compact subsets of X is called the *compact open topology*. It was introduced by R. H. Fox in 1945 and R. F. Arens in 1946 as an aid in the study of continuous convergence and joint continuity. (See the definitions following Example 2.)

EXAMPLE 2. Suppose that X is discrete. Then the compact open topology is equal to the product topology. ⟦A set is compact if and only if it is finite. The result follows by Example 1.⟧ As pointed out in Problem 4, the compact open topology is always larger than the product topology.

Let X be a topological space and $S \subset X$; a net f_δ of functions: $X \to Y$ is said to *converge continuously on S* to a function f if, for each net x_δ in S with $x_\delta \to x \in S$, it follows that $f_\delta(x_\delta) \to f(x)$. (Note that both nets are defined on the same directed set.) A topology T for $F \subset Y^X$ is called *jointly continuous on a subset S of X* if, whenever $f_\delta, f \in F$ and $f_\delta \to f$ in T, it follows that $f_\delta \to f$ continuously on S. Thus *T is jointly continuous on S if and only if the map $(f, x) \to f(x)$ from $F \times S \to Y$ is continuous when F has the topology T.* ⟦If T is jointly continuous, let u_δ be a net in $F \times S$ with $u_\delta \to u$; say $u_\delta = (f_\delta, x_\delta)$, $u = (f, x)$. Then $f_\delta \to f$ and $x_\delta \to x$ since the projections of $F \times S$ on F, S are continuous. Hence $f_\delta(x_\delta) \to f(x)$. Conversely, if the map $(f, x) \to f(x)$ is continuous, let f_δ, x_δ be nets in F, S, with $f_\delta \to f$, $x_\delta \to x$. Then $f_\delta(x_\delta) \to f(x)$ by hypothesis; in other words, $f_\delta \to f$ continuously.⟧ To avoid excessive symbolism we shall designate this map as the *joint map*; thus the joint map from $F \times S$ to Y is the map $(f, x) \to f(x)$, and a topology T for F is jointly continuous on S if and only if the joint map from $F \times S$ to Y is continuous when F has the topology T.

LEMMA 13.1.1. *Let F be a set of continuous maps: $X \to Y$. Then the compact open topology for F is jointly continuous on all locally compact sets in X.*

Let S be a locally compact subspace of X. Let $(g, s) \in F \times S$ and let G be an open neighborhood of $g(s)$ in Y. We must find a neighborhood of (g, s) which the joint map carries into G. Since g is continuous and S is locally compact, s has a compact neighborhood $N \subset S$ with $g[N] \subset G$. (*Note:* N is a neighborhood of s in the relative topology of S.) Let $U = [N, G] \times N$. Then U is an open set in $F \times S$ when F has the compact open topology; it contains (g, s); and U is mapped into G by the joint map. ⟦Let $(f, x) \in U$. Then $f(x) \in G$ since $f \in [N, G]$, and $x \in N$.⟧ ∎

LEMMA 13.1.2. *Let F be a set of continuous maps: $X \to Y$. Then any topology for F which is jointly continuous on all compact sets in X is larger than the compact open topology.*

(A proof with nets is outlined in Problem 101.) Let T be a topology for F which is jointly continuous on all compact sets in X, and let $\mathscr{F}$ be a filter in F which converges in the topology T to $f \in F$. Let $U = [K, G]$ be a subbasic open neighborhood of f in the compact open topology of F. The proof is concluded by showing that $U \in \mathscr{F}$. ⟦This will imply that $\mathscr{F} \to f$ in the compact open topology.⟧ We first show that for each $x \in K$, there exists a neighborhood N_x of x in K with $[N_x, G] \in \mathscr{F}$. ⟦Let N be the neighborhood filter of x in K. Since $\mathscr{F} \to f$ in (F, T), and $N \to x$ in K, and since T is jointly continuous on K, we have $\mathscr{F}(N) \to f(x)$ so that G includes some member of $\mathscr{F}(N)$, say $A(N_x) = \{u(t): u \in A,\ t \in N_x\}$, $A \in \mathscr{F}$. Since $A(N_x) \subset G$ this means $A \subset [N_x, G]$.⟧ Reduce the cover $\{N_x: x \in K\}$ to a finite cover $(N_1, N_2, \ldots, N_n)$ of K. Each $[N_i, G] \in \mathscr{F}$, hence

$$U = [K, G] = \bigcap \{[N_i, G]: i = 1, 2, \ldots, n\} \in \mathscr{F}. \quad ∎$$

Lemmas 13.1.1 and 13.1.2 may be put together in any topological space all of whose compact subsets are locally compact. All regular spaces and all Hausdorff spaces have this property ⟦Theorem 5.4.11⟧.

THEOREM 13.1.1. *Let X be a topological space of the type mentioned in the preceding lines. Let F be a set of continuous functions from X to a topological space Y. The compact open topology for F is the smallest topology for F which is jointly continuous on all compact sets in X, and it is also the smallest topology for F which is jointly continuous on all locally compact sets in X.*

Any topology which is jointly continuous on all locally compact sets is also jointly continuous on all compact sets ⟦they are locally compact⟧; thus

Lemma 13.1.2 may be applied with Lemma 13.1.1, to get the second statement. The first statement follows similarly, applying Lemma 13.1.1. ∎

COROLLARY 13.1.1. *If X is a locally compact regular space, and F is a set of continuous functions from X to a topological space Y, the compact open topology is the smallest topology for F which is jointly continuous on X.*

The situation without local compactness is described in Problem 112.

We now give a result concerning the metrizability of the compact open topology. It appears in final form, with some discussion, as Theorem 13.2.4.

LEMMA 13.1.3. *Let X be a Tychonoff space such that the compact open topology for $C(X)$ is metrizable. Then X is hemicompact.*

Let d be a metric for the topology. Let $U_n = \{f \in C(X): d(f, 0) < 1/n\}$ for $n = 1, 2, \ldots$. Here 0 stands for the identically 0 function. For each n, there exists a compact set K_n and $\varepsilon_n > 0$ such that $U_n \supset [K_n, N_n]$, where $N_n = (-\varepsilon_n, \varepsilon_n) \subset \mathbf{R}$. ⟦$U_n$ is a neighborhood of 0, hence, by definition of the compact open topology, it includes $\bigcap \{[A_i, G_i]: i = 1, 2, \ldots, k\}$ for certain compact sets A_i, and open sets G_i in $\mathbf{R}$ such that $0 \in \bigcap [A_i, G_i]$; that is $0 \in G_i$ for each i. Let $K_n = \bigcup \{A_i: i = 1, 2, \ldots, k\}$, and $(-\varepsilon_n, \varepsilon_n) \subset \bigcap \{G_i: i = 1, 2, \ldots, k\}$.⟧ The proof is concluded by showing that $\{K_n\}$ is a cobase for the compact sets. Let K be a compact subset of X. Then $[K, (-1, 1)]$ is a neighborhood of 0, hence includes U_n for some n. Then $K \subset K_n$. ⟦If not, let $f \in C(X)$ with $f = 0$ on K_n, $f(k) = 1$ for some $k \in K$. Then $f \in U_n$ but $f \notin [K, (-1, 1)]$.⟧ ∎

Problems

In this list Y is a topological space, X a set, $F \subset Y^X$, and Φ is a collection of subsets of X, all nonempty. When the compact open topology is mentioned, X is a topological space.

1. Let Φ_1 be the collection of all finite unions of members of Φ. Then the Φ_1 open topology is equal to the Φ open topology. ⟦Half is trivial. Conversely if $[S, G]$ is a Φ_1 open set, $[S, G] = \bigcap [S_i, G]$ where $S = \bigcup S_i$, so that $[S, G]$ is Φ open.⟧
2. Lemma 13.1.1 holds with "locally compact set" replaced by "set in which each point has a compact (relative) neighborhood." (See Sec. 8.1, Problem 126.)
3. Every topology is a Φ open topology. ⟦Take X = singleton. Then $F = Y$.⟧
★4. If Φ contains every singleton, the Φ open topology is larger than the product topology, hence is T_1 or T_2 if Y is.
5. If Φ contains every singleton, $[S, H]$ is closed for every $S \in \Phi$ and

closed $H \subset Y$. ⟦It is $\bigcap \{[\{s\}, H]: s \in S\}$. Each $[\{s\}, H] = P_s^{-1}[H]$ is closed in the product topology; now see Problem 4.⟧

★6. Any topology larger than a jointly continuous topology is itself jointly continuous.

---

101. Write out this proof of Lemma 13.1.2. Let $f_\delta \to f$ in T, where T is jointly continuous. If $f_\delta \not\to f$ in the compact open topology, there exists compact K in X and open G in Y with $f \in [K, G]$ and $f_\delta \notin [K, G]$ frequently. Let f_α be a subnet with $f_\alpha \notin [K, G]$. So there exists x_α in K with $f_\alpha(x_\alpha) \notin G$. Let x_β be a subnet with $x_\beta \to x$ in K ⟦Theorem 7.1.4 and Sec. 7.1, Problem 207⟧. Then $f_\beta(x_\beta) \to f(x)$ contradicting $f(x) \in G$.
102. Let $\Phi = \{X\}$. Show that Δ is dense in the Φ open topology, where Δ is the set of constant functions, $\Delta = \{f: f(x) = f(x') \text{ for all } x, x' \in X\}$.
103. Let Φ_1 be the set of all subsets of members of Φ. Show that the Φ_1 open topology is larger than the Φ open topology, and may be strictly larger. ⟦With Φ as in Problem 102, Δ is not dense in the Φ_1 open topology since the latter is larger than the product topology.⟧
104. Give an example in which Φ_1 refines Φ and the Φ_1 and Φ open topologies are not comparable. ⟦$\Phi = \{X\}$, Φ_1 = all singletons. See Problem 103.⟧
105. The compact open topology need not be normal if Y is. ⟦Example 2; Sec. 6.7, Problem 203. Y may be **R** or even a discrete space. Or X may be a discrete space with two points; see Sec. 6.7, Example 3.⟧
106. The compact open topology need not be first or second countable if Y is. ⟦See the hint for Problem 105.⟧
107. If X is pseudofinite, the compact open topology is equal to the product topology.
108. Suppose that Φ contains every singleton, and that for every $f \in F$, $S \in \Phi$, and open G in Y such that $f[S] \subset G$, there exists a closed neighborhood H of $f[S]$ with $H \subset G$. Show that the Φ open topology is regular. ⟦If $f \in [S, G]$, then $f \in [S, H] \subset [S, G]$; apply Problem 5.⟧
109. Suppose that Y is regular and that all the members of F are continuous. Show that the compact open topology is regular ⟦Problem 108; Theorem 5.4.6⟧. See Sec. 13.2, Problem 112.
110. For $f \in F$, $G \subset Y$, let $E(f, G) = [f^{-1}[G], G]$. Show that the topology T with subbase $\{E(f, G): f \in F, G \text{ open in } Y\}$ is jointly continuous on X. ⟦Clearly $E(f, G)$ maps into G under the joint map.⟧
111. Let X be regular and let 0 be an open cover of X. Let T be the topology with subbase $\{[S, G]: S$ is a closed set in X which is included in one of the sets of 0; G open in $Y.\}$ Show that T is jointly continuous on X. ⟦For $x \in X$, $f \in F$, say $x \in A \in 0$. Let G be open in Y. Choose a closed neighborhood S of x such that $S \subset f^{-1}[G] \cap A$. Then $[S, G] \times S$ is a neighborhood of (f, x) which maps into G under the joint map.⟧

112. Let F be the set of all continuous maps from a non-locally-compact Tychonoff space X into $[0, 1]$. Then F has no smallest topology which is jointly continuous on X. ⟦Let x have no compact neighborhood. Let T' be a jointly continuous topology. Let U be a T' neighborhood of 0 in F, and V a neighborhood of x in X such that $U \times V$ maps into $[0, \frac{1}{2})$ under the joint map. Since $\overline{V}$ is not compact, X has an open cover $\mathcal{O}$ which cannot be reduced to a finite cover of $\overline{V}$. Form T as in Problem 111. Then $T \not\supset T'$. To see this, let $W = \bigcap \{[S_i, G_i]: i = 1, 2, \ldots, n\}$ be a basic T neighborhood of 0. Then $\overline{V} \not\subset \bigcup S_i$ and so $V \not\subset \bigcup S_i$. So there exists $f \in F$ with $f = 0$ on each S_i and $f(v) = 1$ for some $v \in V$. Then $f \in W \setminus V$; for $0 \in G_i$ for each i since $0 \in W$; and $(f, v) \not\to [0, \frac{1}{2})$. Hence $V \not\supset W$ and so $V \notin T$. This result is due to Richard Arens.⟧

13.2 Topologies of Uniform Convergence

Various forms of convergence of sequences of functions arise naturally in classical analysis; examples are pointwise, uniform, and mean convergence, as well as convergence in measure and uniform convergence on the members of certain families of sets. Topological methods are applied by considering function spaces and placing topologies on them whose associated convergence is of the desired form. In this section we consider the classical property: uniform convergence on compact sets; and show that the topology associated with it is the compact open topology. It will be introduced by specialization of topologies of uniform convergence on certain families of sets.

Note the close resemblance between the following notations and those given at the beginning of Section 13.1. Let $(Y, \mathscr{U})$ be a uniform space, X a set, $F \subset Y^X$, and Φ a collection of subsets of X which is directed by containment (that is, for $S', S'' \in \Phi$ there exists $S \in \Phi$ with $S \supset S' \cup S''$). We assume that F and Φ are not empty. For each $S \in \Phi$ and connector U, let $(S, U) = \{(f, g) \in F \times F: (fs, gs) \in U \text{ for all } s \in S\}$. It is easy to check Sec. 11.1, Definition 2, to see that $\mathscr{B} = \{(S, U): S \in \Phi, U \in \mathscr{U}\}$ is a base for a uniformity on F. ⟦For example $\Delta \subset (S, U)$ since $(fs, fs) \in U$ for all s; and each (S, U) is symmetric if U is. Some other computations are $(S', U') \cap (S'', U'') \supset (S, U)$ if $S \supset S' \cup S''$ and $U \subset U' \cap U''$. (This helps check that $\mathscr{B}$ is a filterbase.) Also $(S, U) \circ (S, U) \subset (S, U \circ U) \subset (S, V)$ if $U \circ U \subset V$.⟧ The uniformity generated by $\mathscr{B}$ is called the *uniformity of Φ convergence*, and the corresponding topology is called the *topology of Φ convergence*. They are also called *the uniformity (the topology) of uniform convergence on the members of Φ*. The reason for this name is contained in the following theorem.

THEOREM 13.2.1. *A net $f_\delta \to f$ in the topology of Φ convergence if and only if $f_\delta \to f$ uniformly on each $S \in \Phi$.*

Let $f_\delta \to f$ in the topology of Φ convergence, $S \in \Phi$, and let U be a connector in Y. Then $(f_\delta, f) \in (S, U)$ eventually. This implies the conclusion. Conversely, if $f_\delta \to f$ uniformly on each $S \in \Phi$, let G be a neighborhood of f. There exists a basic connector (S, U) with $(S, U)(f) = \{g \in F: (f, g) \in (S, U)\} \subset G$. Since $f_\delta \to f$ uniformly on S, it follows that $(f_\delta, f) \in (S, U)$ eventually, and so, eventually, $f_\delta \in (S, U)(f) \subset G$. Hence $f_\delta \to f$. ▮

It is natural to ask of any given topology for a space of functions if it is jointly continuous in the sense of Section 13.1.

THEOREM 13.2.2. *Assume that X is a topological space, that $\bigcup \Phi = X$ and that all members of F are continuous. Then the topology of Φ convergence is jointly continuous on each $S \in \Phi$.*

Fix $g \in F$, $S \in \Phi$, and $s \in S$. Let G be a neighborhood of $g(s)$; then $G \supset U(gs)$ for some connector U. Let V be a symmetric connector with $V \circ V \subset U$. Let N_s be a relative neighborhood of s in S such that $g[N_s] \subset V(gs)$. ⟦g is continuous.⟧ Let

$$N_g = (S, V)(g) = \{h \in F: (g, h) \in (S, V)\}.$$

Then N_g is a neighborhood of g since (S, V) is a connector in the uniformity of Φ convergence. The proof is concluded by showing that $N_g \times N_s$ maps into G under the joint map. ⟦Let $(f, x) \in N_g \times N_s$. Then $(g, f) \in (S, V)$ so that $(gx, fx) \in V$; also $(gx, gs) \in V$ by definition of N_s. Putting these together yields $(fx, gs) \in V \circ V \subset U$, hence $f(x) \in U(gs) \subset G$.⟧ ▮

THEOREM 13.2.3. *Let X be a topological space. Let F be a set of continuous functions from X to a uniform space Y. Then, on F, the topology T_u of uniform convergence on compact sets is equal to the compact open topology T_k.*

First $T_u \supset T_k$ by Theorem 13.2.2, and Lemma 13.1.2. To prove the converse, let $f \in F$ and let N be a T_u neighborhood of f. Then $N \supset (K, U)(f) = \{g \in F: (f, g) \in (K, U)\}$ for some compact set K, and connector U in Y. Let V be a symmetric closed connector with $V \circ V \circ V \subset U$ ⟦Theorem 11.1.6⟧. Now $\{V(fx)^i: x \in K\}$ is an open cover of $f[K]$, which is compact. ⟦f is continuous.⟧ It may be reduced to a finite cover $\{V(fx_j)^i: j = 1, 2, \ldots, n\}$. For $j = 1, 2, \ldots, n$, let $G_j = [V \circ V(fx_j)]^i$ and let $K_j = K \cap f^{-1}[V(fx_j)]$ for $j = 1, 2, \ldots, n$. Each K_j is compact since f is continuous, and V is a closed connector. Thus for each j, $[K_j, G_j]$ is a T_k neighborhood of f, and the proof is concluded by showing that $N \supset \bigcap \{[K_j, G_j]: j = 1, 2, \ldots, n\}$. Let $g \in [K_j, G_j]$ for all j, and let $k \in K$. Then $k \in K_j$ for some j ⟦$f(k) \in V(fx_j)$ for some j⟧, and so $f(k) \in V(fx_j)$ by definition of K_j. Thus $f(x_j) \in V(fk)$. But also $g(k) \in G_j \subset V \circ V(fx_j)$ and so $g(k) \in V \circ V \circ V(fk) \subset U(fk)$. This says that $(fk, gk) \in U$ for all $k \in K$; in other words, $(f, g) \in [K, U]$. Hence $g \in (K, U)(f) \subset N$. ▮

The following Example shows that the topology of Y is not sufficient to determine the topology of Φ convergence. It is entirely expected that the uniformity of Φ convergence will depend on the uniformity of Y; less expected that equivalent uniformities for Y can induce nonequivalent uniformities of Φ convergence; that is, different topologies of Φ convergence. Since the Φ open topology depends only on the topology of Y, this shows easily that the Φ open topology and the topology of Φ convergence can be different. (See Problem 2.) These facts lend more interest to Theorems 13.2.2 and 13.2.3, which show that under certain conditions, the uniformity chosen for the topology of Y is irrelevant.

EXAMPLE 1. Let $Y = \mathbf{R}$, $X = [0, \infty)$, $F = Y^X$. Let

$$d_1(y, z) = |y - z|, \qquad d_2(y, z) = |y/(1 + |y|) - z/(1 + |z|)|.$$

Thus d_1 induces the Euclidean uniformity, d_2 induces an equivalent uniformity ⟦Sec. 11.1, Example 2⟧. Let $f(x) = x$, $f_n(x) = x(1 + 1/n)$. Then $f_n \to f$ uniformly on X when Y has d_2; ⟦$d_2(f_n x, fx) = x/(1 + x)(n + nx + x) < 1/n$ for all x⟧; but not when Y has d_1; ⟦$d_1(f_n x, fx) = x/n$⟧. Hence with $\Phi = \{X\}$, the topology of Φ convergence is different, depending on whether Y has d_1 or d_2, even though these are equivalent metrics.

It is interesting to compare properties of $C(X)$ with those of X. See for example Sec. 5.2, Problem 112; Sec. 8.6, Problem 108. The treatise [Gillman and Jerison] is devoted to $C(X)$ as a ring. A study of $C(X)$ as a topological vector space, and the pairing of properties of X with those of $C(X)$ is given in [Warner]. These results are included in the Tables in the Appendix.

LEMMA 13.2.1. *Let X be hemicompact. Then $C(X)$ with the compact open topology is metrizable.*

By Theorem 13.2.3, and Theorem 11.5.1, it is sufficient to show that the uniformity of compact convergence has a countable base. It is automatically separated ⟦Sec. 13.1, Problem 4⟧. Let $\{K_n\}$ be a cobase for the compact sets of X. Let $\mathscr{B} = \{(K_n, N_m): n, m = 1, 2, \ldots\}$, where $N_m = \{(x, y): |x - y| < 1/m\}$. Then $\mathscr{B}$ is countable, consists of connectors for the uniformity of compact convergence, and is a base for that uniformity ⟦$(K, V) \supset (K_n, N_m)$ as soon as $K \supset K_n$, $V \supset N_m$⟧. ∎

The converse of Lemma 13.2.1 should not be expected to hold since it requires information about X from knowledge of $C(X)$ without any concessions from X sufficient to ensure that $C(X)$ is large enough to yield such information. An explicit counterexample is given in Problem 111.

THEOREM 13.2.4. *Let X be a Tychonoff space. Then the compact open topology for $C(X)$ is metrizable if and only if X is hemicompact.*

This is Lemma 13.2.1 and Lemma 13.1.3. ▮

EXAMPLE 2. The difference between pointwise and uniform convergence is intimately related to a property which filters fail to have, namely, closure under intersection. Consider the uniformity $\mathscr{U}$ of uniform convergence on X, and P the product uniformity, for Y^X. A subbasic P connector is of the form (x, U) with $x \in X$ and U a connector for Y. Since $(x, U) \supset (X, U)$ we have $P \subset \mathscr{U}$. But the converse fails; that is, (X, U) may not be a P connector even though $(X, U) = \bigcap \{(x, U): x \in X\}$ just because the filter P may not be closed under arbitrary intersection. If X is finite we have the trivial result that $\mathscr{U} = P$.

Problems

In this list, X, Y, F, Φ have the meanings given at the beginning of the section, except that X is a topological space when appropriate.

1. In Theorem 13.2.2, it is sufficient to assume about F that for each $f \in F$, and $S \in \Phi$, $f | S$ is continuous. ⟦The same proof.⟧
2. In Example 1, the Φ open topology is equal to Φ convergence when Y has the Euclidean metric d_1, but not when Y has d_2.
3. Suppose Φ refines Φ_1. Show that the uniformity of Φ_1 convergence is larger than that of Φ convergence. ⟦$(S, U) \supset (S_1, U)$ if $S \subset S_1$.⟧ Compare Sec. 13.1, Problem 104.
4. If Φ_1 is a cobase for Φ, the uniformities of Φ and Φ_1 convergence are equal ⟦Problem 3⟧.
5. If $\bigcup \Phi = X$ and Y is separated, then the uniformity of Φ convergence is separated. ⟦If $f \neq g$, $f(x) \neq g(x)$ for some x, and $(f, g) \notin (S, U)$ if $x \in S$ and $(fx, gx) \notin U$.⟧
6. If the members of F are continuous, it is sufficient in Problem 5 to assume $\bigcup \Phi$ dense in X, and Y separated. ⟦Same proof.⟧
7. In Theorem 13.2.3, if X is compact, the compact open topology is equal to the topology of uniform convergence on X.

---

101. Let f, K, U have the meanings given in the proof of Theorem 13.2.3, and let $M = [K, U(f[K])]$. Show that $g \in M$, $k \in K$ imply $g(k) \in U(f[K])$. Why is this not sufficient to imply that $M \subset N$ and hence $T_u \subset T_k$?
102. Suppose that for each $f \in F$, and $S \in \Phi$, every neighborhood of $f[S]$ is a uniform neighborhood. Then the topology of Φ convergence is larger than the Φ open topology. ⟦If $[S, G]$ is a subbasic neighborhood of f in the Φ open topology, $U(f[S]) \subset G$ for some connector U. Hence $[S, G] \supset (S, U)(f)$.⟧
103. Deduce the first half of Theorem 13.2.3 ($T_u \supset T_k$) from Problem 102 and Sec. 11.1, Problem 105.

104. Let Y be a commutative topological group, and F a subgroup of Y^X. Show that the topology of Φ convergence is a group topology. ⟦In Theorem 12.1.6, take $\mathscr{F}$ to be $\{[S, U]: S \in \Phi, U$ a symmetric neighborhood of e in Y. The uniformity of the topological group F has as base, all sets of the form $\{(f, g): f^{-1}g \in [S, U]\} = (S, U_L)$, where U_L is the connector L-associated with U. This is Φ convergence.⟧
105. Let Y be a topological vector space and F a vector subspace of Y^X such that for every $f \in F$ and $S \in \Phi$, $f[S]$ is a bounded set in Y. Show that the topology of Φ convergence is a vector topology. ⟦Apply Theorem 12.4.3 as in Problem 104. The assumption implies that each member of $\mathscr{F}$ is absorbing.⟧
106. The topology of uniform convergence on $\mathbf{R}$ is not a vector topology for $\mathbf{R}^{\mathbf{R}}$. ⟦x/n does not converge uniformly to 0, hence multiplication is not continuous.⟧
107. Let $S \subset X$ and $g: C(X) \to C(S)$ be given by $g(f) = f \mid S$. Show that g is continuous when both $C(X)$, $C(S)$ have the compact open topology. ⟦It is easy to apply Theorem 13.2.1.⟧
108. Let S be a closed subset of a Tychonoff space X and define g as in Problem 107. Then g is an open map onto its range $E = g[C(X)]$. ⟦Let $U = [K, I]$ for compact $K \subset X$ and $I = (-1, 1) \subset \mathbf{R}$. If $K \not\pitchfork S$ and $h \in E$, $h = g(f)$, let $u = 0$ on K, $u = 1$ on S, $v = u \circ f$. Then $v \in U$ and $g(v) = h$. So $g[U] = E$. If K meets S, let $H = K \cap S$. For $h \in [H, I] \cap E$, $h = g(f)$, let $u = 0$ on $(|f| \geq 1) \cap K$, $u = 1$ on S, $0 \leq u(x) \leq 1$ for all $x \in X$ (using Sec. 8.3, Problem 5), $v = u \circ f$. Then $v \in U$ and $g(v) = h$. So $g[U] \supset [H, I] \cap E$, a neighborhood of 0. By Lemma 12.2.1, g is open.⟧
109. In Problem 108, E is a dense subspace of $C(S)$. ⟦With $h \in C(S)$, $K \subset S$ compact, extend $h \mid K$ to $v \in C(X)$ using Corollary 8.5.1. Then $g(v) = h$ on K.⟧
110. A quotient of a complete topological vector space by a closed vector subspace need not be complete. ⟦Let X be a nonnormal, Tychonoff k space; for example, the one in Sec. 6.7, Example 3; and S a closed subspace which is not C-embedded (Sec. 8.5, Problem 102). The map of Problems 106 and 107 is a quotient map by Theorem 6.5.1. Its range is not complete by Problem 109. This result is due to G. Köthe and this example is due to V. Ptak.⟧
111. Suppose that X has no continuous real functions except constants. (For example, X indiscrete; or see Sec. 5.2, Example 7.) Show that the compact open topology, the topology of uniform convergence on X, the pointwise topology are all the same, and are given by the metric $d(f, g) = |f(x) - g(x)|$ for any $x \in X$. ⟦If a net of constants converges pointwise, it must converge uniformly on X.⟧
112. Suppose that Y is a completely regular topological space and that all

the members of F are continuous. Show that the compact open topology is completely regular. Compare Sec. 13.1, Problem 109 ⟦Theorems 13.2.3; 11.4.5, and 11.5.2⟧.

113. Let X be the complex plane and F the set of entire functions. The compact open topology is metrizable by Theorem 13.2.4. Show that it is given by the paranorm

$$p(f) = \sum \frac{2^{-n}\|f\|_n}{1 + \|f\|_n},$$

where $\|f\|_n = \max\{|f(z)|: |z| \leq n\}$ ⟦Theorem 3.5.2⟧.

---

201. Let Y be a complete uniform space, and X a k space. Show that the set of continuous functions from X to Y is complete in the uniformity of uniform convergence on compact sets ⟦Sec. 8.1, Problem 120⟧.
202. Discuss uniform joint continuity.

13.3 Equicontinuity

As we have seen, compact sets in topological spaces possess many desirable properties. In function spaces, it is possible to discuss a property, related to compactness, which is called equicontinuity. If the reader thinks of continuity in terms like: "f is continuous if for each ε, there exists a δ (depending on ε and f), such that . . ." then he will think of equicontinuity of a family F of functions in the same terms, except that the δ chosen will depend on ε and the family F; that is, one δ will do for all the members of F. Here is the definition. Let F be a set of functions $f: X \to Y$, where X is a topological space and Y is a uniform space. Then F is called *equicontinuous at* $x \in X$ if for each connector U for Y, there exists a neighborhood N of x such that $f[N] \subset U(fx)$ for each $f \in F$. Also, F is called *equicontinuous* if it is equicontinuous at each point of X. Theorem 13.3.2 below shows that an equicontinuous set of functions is sufficiently "small" that certain topologies must coincide. The discussion immediately following Theorem 13.3.1 indicates the kind of conclusion derivable from the assumption of equicontinuity.

EXAMPLE 1. For $n = 1, 2, \ldots,$ define $f_n: \mathbf{R} \to \mathbf{R}$ by $f_n(x) = nx$ and $g_n: \mathbf{R} \to \mathbf{R}$ by $g_n(x) = x/n$. Then $\{g_n\}$ is equicontinuous ⟦take $\delta = \varepsilon$; then $|y - x| < \delta$ implies $|y/n - x/n| < \varepsilon$⟧, but $\{f_n\}$ is not. ⟦For any $\delta > 0$ and real x, $|ny - nx| = \varepsilon$ for $y = x + \varepsilon/n$; and certainly $|y - x| < \delta$ if $n > \varepsilon/\delta$.⟧

In the following computational lemma, we use the $[S, G]$ notation introduced at the beginning of Section 13.1.

LEMMA 13.3.1. *Let Y be a uniform space, X a set, $x \in X$, and $F \subset Y^X$. Let U be a connector for Y, and set $M = M_x = \bigcap \{f^{-1}[V(fx)]: f \in F\}$, where V is* *a connector such that $V \circ V \subset U$. Then for $f \in F$, $[\{x\}, V(fx)] \times M$ maps into $U(fx)$ under the joint map (from $F \times X$ to Y).*

Let $g \in [\{x\}, V(fx)]$, $m \in M$. Then $g(m) \in V(gx)$ by definition of M; also $g(x) \in V(fx)$ by definition of g. Thus $g(m) \in V \circ V(fx) \subset U(fx)$. ∎

Let us interpret Lemma 13.3.1 in terms of joint continuity. First, $[\{x\}, V(fx)]$ is a neighborhood of f in the pointwise ($=$ product) topology of F; thus this topology is jointly continuous on a topological space X providing that M_x (in Lemma 13.3.1) is a neighborhood of x for each $x \in X$. Now M_x is a neighborhood of x if and only if F is equicontinuous at x; hence the following result.

THEOREM 13.3.1. *Let F be an equicontinuous set of functions from a topological space X to a uniform space Y, and let Φ be a collection of subsets of X which contains every singleton. Then the Φ open topology is jointly continuous on X.*

The preceding discussion shows that the pointwise topology is jointly continuous on X; *a fortiori* the Φ open topology is jointly continuous on X ⟦Sec. 13.1, Problems 4 and 6⟧. ∎

(Theorem 13.3.1 should be compared with Theorem 13.1.1 which asserts in particular, that the compact open topology is jointly continuous on each compact subset of X. The conclusion of Theorem 13.3.1 is better, and is typical of the improved results obtainable in the presence of equicontinuity.)

EXAMPLE 2. Define $f_n: \mathbf{R} \to \mathbf{R}$ for $n = 1, 2, \ldots$, by $f_n(x) = 0$ for $x \leq n$, 1 for $x \geq n + 1$, and a "straight line," for $n \leq x \leq n + 1$. Then $\{f_n\}$ is equicontinuous. Now $f_n \to 0$ pointwise, and it follows from Theorem 13.3.1 that $f_n \to 0$ continuously; that is, if $\{x_n\}$ is a convergent sequence in $\mathbf{R}$, $f_n(x_n) \to 0$.

EXAMPLE 3. Define $f_n: \mathbf{R} \to \mathbf{R}$ by $f_n(x) = 0$ for $x \leq 0$ and for $x \geq 2/n$; $f_n(1/n) = 1$, and f_n a "straight line" between 0 and $1/n$, and between $1/n$ and $2/n$. Each f_n is continuous; $f_n \to 0$ pointwise but not continuously because $f_n(1/n) \not\to 0$. It follows from Theorem 13.3.1 that $\{f_n\}$ is not equicontinuous.

THEOREM 13.3.2. *Let F be an equicontinuous set of functions from a topological space X to a uniform space Y. On F the product topology of Y^X, the compact open topology, and the topology of uniform convergence on compact sets are all equal.*

This follows immediately from Theorem 13.3.1 (Φ = singletons), Lemma 13.1.2 and Theorem 13.2.3; taking account of the fact that the compact open topology must be larger than the product topology ⟦Sec. 13.1, Problem 4⟧. ∎

EXAMPLE 4. Define $f_n: \mathbf{R} \to \mathbf{R}$ by $f_n(x) = x/n$ for $n = 1, 2, \ldots$. Then $\{f_n\}$ is equicontinuous and $f_n \to 0$ pointwise. It follows from Theorem 13.3.2 that $f_n \to 0$ uniformly on every compact set in $\mathbf{R}$. Of course it is false that $f_n \to 0$ uniformly on $\mathbf{R}$.

THEOREM 13.3.3. *Let F be an equicontinuous set of functions from a topological space X to a uniform space Y. The closure of F in Y^X is also equicontinuous.*

Let V be a closed connector for Y and fix $x \in X$. There exists a neighborhood N of x such that $f[N] \subset V(fx)$ for every $f \in F$ ⟦by definition of equicontinuity⟧. But then $f[N] \subset V(fx)$ for every $f \in \bar{F}$. ⟦For $n \in N$, $f \in \bar{F}$, $f(n) = P_n(f) \in \overline{P_n[F]}$ by Theorem 4.2.4, and the definition of the product topology. Since $P_n[F] \subset V(fx)$ and $V(fx)$ is closed, the result follows.⟧ Hence $\bar{F}$ is equicontinuous. ∎

COROLLARY. *Let f_δ be an equicontinuous net of functions from a topological space X to a uniform space Y. Then if $f_\delta \to f$ pointwise, it follows that f is continuous.*

This follows from Theorem 13.3.3 and Problem 1. ∎

This Corollary should be compared with Theorem 4.2.10, which is actually a special case in view of Problem 4.

In view of the preceding results, it is natural to compare equicontinuity with compactness. Theorem 13.3.4 is a descendant and generalization of a classical theorem due to C. Arzela in 1889, and G. Ascoli in 1883. Its most important content is the conclusion of compactness from the assumption of equicontinuity. Some applications are shown in [Goffman and Pedrick, pp. 30–31], [Buck, pp. 152–158], and [Banach, p. 97].

In the following result C is the set of all continuous functions from X to Y and it has the compact open topology.

THEOREM 13.3.4. *Let X be a locally compact topological space, Y a uniform space, F a closed subspace of C. Then F is compact if and only if it is equicontinuous and $\{f(x): f \in F\}$ is a compact subset of Y for each $x \in X$.*

Suppose first that F is compact and $x \in X$. Then $P_x: F \to Y$ (the projection, $P_x(f) = f(x)$), is continuous. ⟦It is continuous when F has the (smaller) product topology.⟧ Thus $P_x[F] = \{f(x): f \in F\}$ is compact. To prove that F is equicontinuous at x, let U be a connector for Y, and V a symmetric connector with $V \circ V \subset U$. For each $f \in F$, there exists a neighborhood N_f of f in the compact open topology, and a neighborhood M_f of x with $N_f \times M_f$ mapping into $V(fx)$ under the joint map. ⟦The compact open topology is jointly continuous on X by Lemma 13.1.1.⟧ Reduce the cover $\{N_f: f \in F\}$ of F to a finite cover $(N_1, N_2, \ldots, N_n)$, where N_i is N_{f_i}, and let $M_1, M_2, \ldots, M_n$

be the corresponding choices of M_f. With $M = \bigcap M_i$, equicontinuity will follow when we prove that $f[M] \subset U(fx)$ for all $f \in F$. ⟦To prove this, let $f \in F$ and $m \in M$. Suppose $f \in N_i$. Then $m \in M_i$, and so $f(m) \in V(f_i x)$. We also have $f(x) \in V(f_i x)$ since $x \in M_i$. From these two relations we have $(fx, fm) \in V \circ V \subset U$, hence $f(m) \in U(fx)$ as required.⟧ Next, assume that F is equicontinuous and that $A_x = \{f(x) : f \in F\}$ is compact for each $x \in X$. (For this half of the theorem, local compactness of X is not used.) Let F_1 be the closure of F in Y^X. Then F_1 is a compact subspace of Y^X. ⟦$F \subset \prod \{A_x : x \in X\} \subset Y^X$ and the middle term is compact by Tychonoff's product theorem.⟧ Hence F_1 is compact in the compact open topology ⟦Theorems 13.3.2 and 13.3.3⟧ and F, as a closed subset of F_1 in this topology, is also compact. ∎

Problems

In this list, X is a topological space, Y is a uniform space, and C is the set of all continuous functions from X to Y.

★1. Every member of an equicontinuous set of functions is continuous.
2. With F as in Theorem 13.3.2, suppose that Φ is a family of compact subsets of X which contains all singletons. Then the Φ open and compact open topologies are equal.
3. Give C any jointly continuous topology larger than the product topology of Y^X. Then a compact subset of C must be equicontinuous. ⟦The proof of Theorem 13.3.4.⟧
4. A uniformly convergent sequence in C must be equicontinuous. ⟦Problem 3. Apply Theorem 13.2.2 with $\Phi = \{X\}$.⟧
5. A subset of C is equicontinuous if it is compact in the topology of Φ convergence, and $\bigcup \Phi = X$ ⟦Problem 3 and Theorem 13.2.2⟧.

---

101. Write out a direct proof for the result of Problem 4.
102. In Theorem 13.3.4, the last condition cannot be omitted; that is, a closed equicontinuous set need not be compact. ⟦Let $F = \{f_n\}$ with $f_n(x) = n$ for all x.⟧
103. Let X be a compact T_2 or regular space and F a closed subset of $C(X)$ in the norm topology, $\|f\| = \max\{|f(x)| : x \in X\}$. (This is the compact open topology! See Sec. 13.2, Problem 7.) Then F is compact if and only if it is equicontinuous and pointwise bounded ⟦Theorem 13.3.4⟧.

---

201. Let X be a normed space and $F \subset X'$. Show that F is equicontinuous if and only if it is norm bounded.
202. In Theorem 13.3.4 replace "locally compact" by "k," assuming both X, Y to be Hausdorff.

13.4 Weak Compactness

Many standard theorems of classical analysis assert the existence of convergent subsequences of a given sequence, in other words sequential compactness. Partly for these reasons, it is useful to explore relations among the three main kinds of compactness. Although useful and instructive, the equivalence theorem for semimetric space, Theorem 7.2.1, is too special for many applications since the most frequently occurring function space topologies are usually not semimetrizable. The Eberlein–Smulian theorem gives the same equivalence in a very general setting. The theorem developed from results given by W. F. Eberlein in 1947, and V. L. Smulian in 1940 and 1943. Our treatment is based on the exposition in [Kelley and Namioka].

During the course of the following demonstration, certain properties related to first countability will be used.

DEFINITION 1. *A topological space is called closure sequential (respectively, closure countable) if for every set A and $x \in \bar{A}$, A contains a sequence converging to x; (respectively, a countable subset A_0 with $x \in \bar{A}_0$).*

Then first countable implies closure sequential, which implies closure countable; but neither implication can be reversed. ⟦Sec. 3.1, Problem 203; Sec. 3.1, Problem 201 or Sec. 8.3, Problem 103. See also Problem 116.⟧

The setting for the main theorem will be any space C of the form given in the following definition.

DEFINITION 2. *Let X be a compact topological space, Y a compact metric space. Let C be the set of all continuous functions from X to Y. Throughout this section, C will have the product topology of Y^X (that is, the pointwise topology).*

The first step in the development (Theorem 13.4.1) will be to show that C is closure countable; the resemblance of this property to first countability will help bring together the various forms of compactness. A related, but more refined and difficult result (Theorem 13.4.2) is that every compact subspace of C is closure sequential.

DEFINITION 3. *For $f \in C, x \in X, \varepsilon > 0$; $U(f, x, \varepsilon) = \{h \in C: d[h(x), f(x)] < \varepsilon\}$.*

LEMMA 13.4.1. *Let $A \subset C$, $f \in C$, $f \in \bar{A}$. Let n be a positive integer and $\varepsilon > 0$. Then A contains a finite set $F = F_{n, \varepsilon}$ which meets*

$$\bigcap \{U(f, x_i, \varepsilon): i = 1, 2, \ldots, n\}$$

for every set $(x_1, x_2, \ldots, x_n)$ of n points of X.

For each point $z \in X^n$, say $z = (x_1, x_2, \ldots, x_n)$, choose $f_z \in A$ with

$d[f_z(x_i), f(x_i)] < \varepsilon$ for all i. ⟦Possible since $f \in \bar{A}$, and by definition of the product topology.⟧ For each z, let $V_z = \{x \in X: d[f_z(x), f(x)] < \varepsilon\}$. Each V_z is open ⟦since f_z and f are continuous⟧, and nonempty ⟦$x_i \in V_z$ for all i⟧, hence $W_z = V_z \times V_z \times \cdots \times V_z$ (n factors) is an open set in X^n and contains z. Thus $\{W_z: z \in X^n\}$ is an open cover of the compact space X^n and can be reduced to a finite cover $(W_{z_1}, W_{z_2}, \ldots, W_{z_m})$. Let $F = (f_{z_1}, f_{z_2}, \ldots, f_{z_m})$. To see that F satisfies the requirements of the statement, fix $z = (x_1, x_2, \ldots, x_n)$. Say $z \in W_{z_k}$. Then, for each i, $f_{z_k} \in U(f, x_i, \varepsilon)$. ⟦$d[f_{z_k}(x_i), f(x_i)] < \varepsilon$ since $z \in W_{z_k}$ implies $x_i \in V_{z_k}$.⟧ ∎

THEOREM 13.4.1. *C is closure countable.*

Let $A \subset C, f \in C, f \in \bar{A}$. With $F_{n,\varepsilon}$ as in Lemma 13.4.1, let

$$F = \bigcup \{F_{n,1/m}: n, m = 1, 2, \ldots\}.$$

Then F is countable, and we shall show that $f \in \bar{F}$. Let U be a neighborhood of f. Then there exist $x_1, x_2, \ldots, x_n \in X$ and $\varepsilon > 0$ with $U \supset \bigcap \{U(f, x_i, \varepsilon): i = 1, 2, \ldots, n\}$. Let $m > 1/\varepsilon$. By definition of $F_{n, 1/m}$, it meets $\bigcap U(f, x_i, 1/m) \subset \bigcap U(f, x_i, \varepsilon) \subset U$. Hence F meets U. ∎

LEMMA 13.4.2. *Let S be a compact subset of C, $f \in S$, and $\{f_n\}$ a sequence of members of S. Then if $f_n(x) \to f(x)$ for each x in some subset B of X, it follows that $f_n(x) \to f(x)$ for each $x \in \bar{B}$.*

Let $b \in \bar{B}$ and suppose that $f_n(b) \not\to f(b)$. This means that for some neighborhood U of $f(b)$, $f_n(b) \notin U$ for infinitely many n, say for $n = n_k$, $k = 1, 2, \ldots$. Let $g \in S$ be a cluster point of $\{f_n\}$ ⟦Theorem 7.1.4⟧, then for each $x \in X$, $g(x)$ is a cluster point of $\{f_{n_k}(x)\}$. ⟦The projection $P_x: C \to Y$ is continuous and carries g to $g(x)$, etc.⟧ For $x \in B$ this sequence has only one cluster point, $f(x)$, thus $g(x) = f(x)$ for $x \in B$, hence also for $x = b$. This makes $f(b)$ a cluster point of $\{f_{n_k}(b)\}$, since $g(b)$ is, contradicting $f_{n_k}(b) \notin U$ for all k. ∎

As a first step toward Theorem 13.4.2, we prove a special case of it.

LEMMA 13.4.3. *In Definition 2, assume that X is semimetrizable. Let K be a compact subset of C. Then K is closure sequential.*

Let $A \subset K$, $u \in \bar{A}$ (the closure is taken in K), and let B be a countable dense subset of X. ⟦X is separable, by Theorem 5.3.4.⟧ Let $u_B = u \mid B$ and $A_B = \{g \mid B: g \in A\}$, K_B is defined similarly. Then $u \in \bar{A}_B$ in the space $K_B \subset Y^B$. Now Y^B is a metric space ⟦Theorem 6.4.2⟧, hence there exists a sequence $\{v_n\}$ in A_B with $v_n \to u_B$. To say $v_n \in A_B$ means $v_n = f_n \mid B$ for some $f_n \in A$. Then $f_n(x) \to u(x)$ for all $x \in B$, hence, by Lemma 13.4.2, for all $x \in \bar{B} = X$. Thus finally $f_n \to u$. ∎

We now outline, temporarily omitting justification of the details, the extension of Lemma 13.4.3 by omission of the assumption that X is semimetrizable; the final result being Theorem 13.4.2. First we shall use Theorem 13.4.1 to assume that A (in the proof of Lemma 13.4.3) is countable. Then we shall replace the topology of X by $w(B)$, the weak topology by B, where $B = A \cup \{u\}$. This will be seen to be compact and semimetrizable so that Lemma 13.4.3 can be applied as soon as we show that B is contained in a compact set in $C_1 = \{f \in Y^X : f \text{ is continuous when } X \text{ has the topology } w(B)\}$. (This is not trivial since C_1 may be expected to be strictly smaller than C.) Since the closure $\bar{B}$ of B in Y^X is compact, this will be accomplished by showing that $\bar{B} \subset C_1$ (Lemma 13.4.4). It is a crucial fact for the final success of the theory that this can be done under the assumption that B is included in a countably compact set.

LEMMA 13.4.4. *Let S be a countably compact subset of C. Let $B \subset S$ and let $h \in \bar{B}$, the closure taken in Y^X. Then h is continuous on X when X is given $w(B)$, the weak topology by B.*

Assume on the contrary that h is not continuous in this topology. Let x_0 be a point in X where h is not continuous. There exists an open neighborhood U of $h(x_0)$ such that $h^{-1}[U]$ is not a $w(B)$ neighborhood of x_0. Choose $f_1 \in B$ with $d[f_1(x_0), h(x_0)] < 1$. ⟦Possible since $h \in \bar{B}$ so that, by definition of the product topology, we may approximate h by a member of B on any finite subset of X.⟧ Choose $x_1 \in X \setminus h^{-1}[U]$ such that $d[f_1(x_1), f_1(x_0)] \le 1$. ⟦Possible since f_1 is $w(B)$ continuous.⟧ Next choose $f_2 \in B$ with

$$d[f_2(x_1), h(x_1)] \le \tfrac{1}{2} \qquad \text{and} \qquad d[f_2(x_0), h(x_0)] \le \tfrac{1}{2}.$$

⟦Same justification as for the choice of f_1.⟧ Choose $x_2 \in X \setminus h^{-1}[U]$ with $d[f_1(x_2), f_1(x_0)] \le \tfrac{1}{2}$ and $d[f_2(x_2), f_2(x_0)] \le \tfrac{1}{2}$. ⟦Possible since both f_1 and f_2 are $w(B)$ continuous.⟧ Continuing in this way we get sequences $\{x_n\}$, $\{f_n\}$, with

$$d[f_m(x_n), h(x_n)] \le \frac{1}{m} \qquad \text{for } n = 0, 1, \ldots, m-1; m = 1, 2, \ldots; \tag{13.4.1}$$

and

$$d[f_m(x_n), f_m(x_0)] \le \frac{1}{n} \qquad \text{for } m = 1, 2, \ldots, n; n = 1, 2, \ldots; \tag{13.4.2}$$

$$h(x_n) \notin U \text{ for } n = 1, 2, \ldots. \tag{13.4.3}$$

Now, $Y \setminus U$ is a compact metric space and so $\{h(x_n)\}$ has a convergent subsequence ⟦Theorem 7.2.1⟧. Let $\{t_n\}$ be a subsequence of $\{x_n\}$ such that $h(t_n) \to b \in Y \setminus U$. Summarizing the construction to date we have

$$\lim_n \lim_m f_m(t_n) = b, \tag{13.4.4}$$

$$\lim_m \lim_n f_m(t_n) = h(x_0). \tag{13.4.5}$$

⟦First, $b = \lim_n h(t_n)$, and $h(t_n) = \lim_m f_m(t_n)$ by Formula (13.4.1). This proves Formula (13.4.4). Also $h(x_0) = \lim f_m(x_0)$ by Formula (13.4.1), and $f_m(x_0) = \lim_n f_m(x_n)$ [by Formula (13.4.2)], $= \lim_n f_m(t_n)$ since $\{t_n\}$ is a subsequence of $\{x_n\}$. This proves Formula (13.4.5).⟧ Note that the right-hand sides of Formulas (13.4.4) and (13.4.5) cannot be equal since $b \notin U$; we shall deduce that this inequality is contradicted by the fact that S is countably compact. (Up till now neither this hypothesis nor any property of the original topology of X has been used.) We are going to replace x_0 and h by $x \in X$ and $f_0 \in S$ with the properties that x is a good deal like x_0 [see Equation (13.4.7)], but is a cluster point of $\{t_n\}$; and f_0 is a good deal like h [see Equation (13.4.6)], but is continuous. Precisely, $\{t_n\}$ has a cluster point x in (X, T), where T is the original topology of X ⟦Theorem 7.1.4⟧, and $\{f_n\}$ has a cluster point f_0 in S ⟦Theorem 7.1.2⟧. We have, taking $t_0 = x_0$,

$$f_0(t_n) = h(t_n) \quad \text{for} \quad n = 0, 1, 2, \ldots. \tag{13.4.6}$$

⟦Fix n and let $t = t_n$. Then $\lim_m f_m(t) = h(t)$ by Formula (13.4.1). Since Y is a Hausdorff space, the sequence $\{f_m(t)\}$ has no cluster point other than $h(t)$. But $f_0(t)$ must be a cluster point for f_0 is a cluster point of $\{f_m\}$ in the product topology, and the projection P_t which carries f_m to $f_m(t)$ is a continuous map onto Y.⟧ Also

$$f_m(x) = f_m(x_0) \quad \text{for } m = 0, 1, 2, \ldots. \tag{13.4.7}$$

⟦Fix m and let $f = f_m$. Then $\lim_n f(x_n) = f(x_0)$ by Formula (13.4.2). The case $m = 0$ is allowed in Formula (13.4.2) as well, since $f_0(x_n)$ is a cluster point of $\{f_m(x_n)\}$ for each n by the argument used in proving Equation (13.4.6); similarly $f_0(x_0)$ is a cluster point of $\{f_m(x_0)\}$; and so $d[f_0(x_n), f_0(x_0)]$ is a cluster point of $\{d[f_m(x_n), f_m(x_0)]\}$, hence is less than or equal to $1/n$ by Formula (13.4.2). Since Y is a Hausdorff space, the sequence $\{f(x_n)\}$ has no cluster point other than $f(x_0)$. But $f(x)$ must be a cluster point, for x is a T-cluster point of $\{x_n\}$, and f is T-continuous.⟧ We are now ready to obtain the contradiction announced after Equation (13.4.5). From Equation (13.4.4), $\lim_m f_m(t_n)$ exists for each n. Its value can only be $f_0(t_n)$ since $f_0(t_n)$ is a cluster point of the convergent sequence $\{f_m(t_n)\}$. Hence $b = \lim_n f_0(t_n)$. Again, since $\lim_n f_0(t_n)$ exists, its value can only be $f_0(x)$ since $f_0(x)$ is a cluster point of the convergent sequence $\{f_0(t_n)\}$. ⟦f_0 is T-continuous, and x is a T-cluster point of $\{t_n\}$.⟧ Thus finally $b = f_0(x)$. From Equation (13.4.7) $b = f_0(x_0)$, and from Equation (13.4.6), $b = h(x_0)$. This contradicts the facts that $b \notin U$ and U is a neighborhood of $h(x_0)$. ∎

THEOREM 13.4.2. *Let S be a compact subspace of C. Then S is closure sequential.*

Let $A \subset S$. We shall show that every point u in the closure of A (in S) is a sequential limit point of A. By Theorem 13.4.1, we may assume that A is countable. Let T be the topology of X, and w the weak topology by $B = A \cup \{u\}$. Then $w \subset T$ ⟦T makes all the members of B continuous⟧, and (X, w) is a compact semimetric space ⟦Sec. 5.4, Problem 4; Theorem 6.3.4⟧. Let K be the closure in Y^X of B. Then $K \subset C_1 = \{f \in Y^X : f \text{ is } w\text{-continuous}\}$ by Lemma 13.4.4, and K is compact. ⟦It is a closed subset of Y^X and we may apply Tychonoff's theorem (Theorem 7.4.1) and Theorem 5.4.2.⟧ By Lemma 13.4.3, K is closure sequential and so u is the limit of a sequence of members of A since both are contained in K. ∎

The program of relating the various forms of compactness can now be successfully concluded. *A relatively (countably, sequentially) compact set in a topological space is one whose closure is (countably, sequentially) compact.*

LEMMA 13.4.5. *Every countably compact subset S of C is relatively compact.*

Let K be the closure in Y^X of S. Applying Lemma 13.4.4 with $B = S$ we see that every member of K is continuous when X is given the topology $w(S)$; *a fortiori*, every member of K is continuous when X is given its original topology T, which is larger than $w(S)$. ⟦Every member of S is T-continuous.⟧ This proves that $K \subset C$. Now K is a closed subset of Y^X which is compact, by Tychonoff's theorem ⟦Theorem 7.4.1⟧, hence K is compact ⟦Theorem 5.4.2⟧. Since $S \subset K \subset C$, S is relatively compact in C. ∎

THEOREM 13.4.3 (THE EBERLEIN–SMULIAN THEOREM).

(a) *Let X be a compact topological space, Y a compact metric space, S a subspace of Y^X consisting entirely of continuous functions. Then S is compact if and only if it is countably compact, and if and only if it is sequentially compact.*

(b) *Let $C = \{f \in Y^X : f \text{ is continuous}\}$. Then a subset of C is relatively compact if and only if it is relatively countably compact and if and only if it is relatively sequentially compact.*

If S is sequentially compact, it is certainly countably compact ⟦Sec. 7.1, Problem 4⟧. Let S be countably compact. Then the closure K of S in C is compact ⟦Lemma 13.4.5⟧ and so to prove S compact it suffices to show that S is closed in C. ⟦Let $h \in K$. By Theorem 13.4.2, h is the limit of a sequence $\{f_n\}$ of points of S. This sequence can have no cluster point other than h, and it must have a cluster point in S since S is countably compact. Hence $h \in S$ and S is closed.⟧ Finally (for part (a)) assume that S is compact and let A be an infinite subset of S. Then A has an accumulation point in S ⟦Theorem 7.1.2, along with the trivial fact that a compact space is countably compact⟧. By Theorem 13.4.2, A has a sequence converging to a point of S. Thus S is sequentially compact ⟦Sec. 7.1, Problem 12⟧. Part (b) is now trivial from Part (a) by considering the closure of S in C. ∎

EXAMPLE 1. Since sequential compactness has, classically, been the most used form of compactness, it is worthwhile to look for results in which it is the conclusion. One such is the following. Let X be a compact T_2 space, Y a compact metric space, and F an equicontinuous closed subset of C (Definition 2), such that $\{f(x): f \in F\}$ is a compact subset of Y for each $x \in X$. Then F is sequentially compact in the product topology, the compact open topology, and the topology of uniform convergence on compact sets. (They are all equal by Theorem 13.3.2.) ⟦This follows from Theorem 13.4.3, and Theorem 13.3.4. Note that F is closed in the compact open topology since this is larger than the product topology.⟧

EXAMPLE 2. *In the weak topology of a normed space N we have the equivalences expressed in Theorem 13.4.3.* If S has any one of the compactness properties involved, it is norm bounded, hence it suffices to prove the result for (D, w), where D is the unit disc, and w is the weak topology. Let X be the unit disc of N' with the weak* topology, a compact Hausdorff space by the Banach–Alaoglu theorem. The map $u \to \hat{u}$, defined by $\hat{u}(f) = f(u)$ for all $f \in X$, carries D into $C(X, Y)$, the set of all continuous functions from X to Y where $Y = [-1, 1]$. ⟦$\hat{u}$ is continuous since if $f_\delta \to f$, $\hat{u}(f_\delta) = f_\delta(u) \to f(u) = \hat{u}(f)$, and $\hat{u}$ carries X into Y since $|\hat{u}(f)| = |f(u)| \leq \|f\| \cdot \|u\| \leq 1$.⟧ The map $u \to \hat{u}$ is a homeomorphism when $C(X, Y)$ has the product topology; that is, the relative topology of Y^X. ⟦If $u_\delta \to u$ and $f \in X$, $\hat{u}_\delta(f) = f(u_\delta) \to f(u) = \hat{u}(f)$ thus $\hat{u}_\delta \to \hat{u}$. Conversely, if $\hat{u}_\delta \to \hat{u}$, for every $g \in N', f = g/\alpha \in X$ for some real $\alpha \neq 0$, then $g(u_\delta) = \alpha f(u_\delta) = \alpha \hat{u}_\delta(f) \to \alpha \hat{u}(f) = g(u)$. Since g is arbitrary, $u_\delta \to u$ weakly.⟧ We have now proved that (D, w) is homeomorphic with a space of the type given in Theorem 13.4.3. ∎

Problems

1. Show that the equivalence of the various forms of compactness in semimetric space (Theorem 7.2.1) is a special case of the Eberlein–Smulian theorem. ⟦Take X to be a singleton.⟧
2. In Definition 2, C is dense in Y^X if X is Hausdorff and $Y = [0, 1]$. ⟦A product neighborhood of f involves a finite subset of X. Use Corollary 8.5.1 or Sec. 8.5, Example 1; and Theorem 5.4.7.⟧
3. A closure countable topology larger than the cocountable topology must be discrete. Hence give an example of a T_2 space which is not closure countable. ⟦Sec. 6.2, Problem 110.⟧

---

101. The result of Problem 2 is false for $Y = \{0, 1\}$. ⟦Let $X = [0, 1]$, $f(x) = 0$ for $0 \leq x < 1, f(1) = 1$. No continuous map approximates f at 0 and 1.⟧

102. The results of Lemma 13.4.1 and Theorem 13.4.1 hold without assuming Y compact.
103. The result of Lemma 13.4.2 holds with no assumption on X, with Y an arbitrary Hausdorff space, and with S countably compact.
104. The result of Lemma 13.4.3 holds with X compact and separable, with Y compact metric, and with K countably compact.
105. The result of Lemma 13.4.4 holds if X is countably compact instead of compact.
106. In Definition 2 take X to be Hausdorff and $Y = [0, 1]$. Show that C is countably compact if and only if X is finite. ⟦By Problem 2 and Lemma 13.4.4, with $B = S = C$, every member of Y^X is continuous.⟧
107. In Lemma 13.4.4, "countably compact" cannot be omitted ⟦Problem 106⟧.
108. The result of Example 2 is false for the weak* topology of a dual Banach space ⟦Sec. 12.4, Problem 103⟧.
109. Show that Sec. 12.4, Problem 103 becomes false if "the dual of" is deleted, and "weak*" is replaced by "weak" ⟦Example 2⟧.
110. Theorem 13.4.3 holds when Y^X is given the compact open topology, if we assume S equicontinuous ⟦Theorems 13.3.2 and 13.3.3⟧.
111. Say that $F \subset C$ (Definition 2) obeys the *iterated limit condition* if whenever the left-hand sides of Equations (13.4.4) and (13.4.5) (selections made from X, F) both exist, they are equal. Show that if F obeys this condition and the assumptions of Definition 2 hold, then $\bar{F} \subset C$. ⟦The proof of Lemma 13.4.4.⟧
112. On the assumption of Problem 111 prove also that F is relatively compact in C.
113. In Problems 111 and 112, the assumption that Y is compact cannot be dropped. ⟦Let $X = Y = \mathbf{R}$; $f_n(x) = nx$ for $|x| \le n^{-1/2}$, $1/x$ for $|x| \ge n^{-1/2}$.⟧
114. If F is relatively compact in C (Definition 2) it must obey the iterated limit condition. ⟦The last part of the proof of Lemma 13.4.4.⟧
115. In Problem 114, the assumption that X is compact cannot be dropped. ⟦Let $f_n = 0$ on $(-\infty, n-1]$, 1 on $[n, \infty)$. Then $\{0, f_1 f_2, \ldots\}$ is compact.⟧
116. In contrast with Theorem 13.4.1, C need not be first countable. ⟦Take $X = Y = [0, 1]$. The argument of Sec. 6.4, Problem 6 works, applying Sec. 8.5, Example 1 to $B_n \cup \{\alpha\}$.⟧

---

201. In the hint to Problem 116, must C be closure sequential?
202. Is every metric space homeomorphic into the dual of some normed space with the weak topology? (This is suggested by Example 2, and Theorem 7.2.1.)

Miscellaneous Topics

14

14.1 Extremally Disconnected Spaces

Extremally disconnected spaces arise naturally in the study of certain categories of topological spaces (see Section 14.3), and in the study of the lattice structure of $C(X)$ (see Example 3). An *extremally disconnected space* is a topological space in which the closure of every open set is open. A discrete space is extremally disconnected; so are an indiscrete space, a cofinite space, and a cocountable space. Other examples are given below. (See Example 1.) We shall see that these spaces have interesting special properties which, up till now, the reader might not have seen in any spaces other than the trivial ones just mentioned. The first result is an example of this, in that it shows that there are no "contiguous" open sets (such as (0, 1) and (1, 2) in **R**).

THEOREM 14.1.1. *A space is extremally disconnected if and only if every two disjoint open sets have disjoint closures.*

If X is extremally disconnected and A, B are disjoint open sets, then $\bar{A} \not\pitchfork B$ ⟦Sec. 2.5, Problem 5⟧. For the same reason, $\bar{A} \not\pitchfork \bar{B}$. ⟦Since $\bar{A}$ is open.⟧ Conversely, if disjoint open sets have disjoint closures, let G be open, and $F = \bar{G}$. Then $G, \tilde{F}$ are disjoint open sets so that $\bar{G}$ does not meet $\bar{\tilde{F}}$; that is F does not meet $\bar{\tilde{F}} = F^{i\sim}$. Hence $F \subset F^i$ and so F is open. ∎

THEOREM 14.1.2. *A space X is extremally disconnected if and only if whenever F_1, F_2 are closed sets with $F_1 \cup F_2 = X$, then also $F_1^i \cup F_2^i = X$.*

Since $F_1 \cup F_2 = X$ if and only if $\tilde{F}_1 \cap \tilde{F}_2 = \varnothing$, and $F_1^{i\sim} = \bar{\tilde{F}}_1$ this follows from Theorem 14.1.1. ▮

LEMMA 14.1.1. *Let X be an extremally disconnected Tychonoff space. Then βX is extremally disconnected.*

Let G, H be disjoint open sets in βX, and set $A = G \cap X$, $B = H \cap X$. Then A, B are disjoint open subsets of X. Let $f: X \to \mathbf{R}$ be the characteristic function of $\bar{A}$. (That is, $f = 1$ on $\bar{A}$, 0 on $X \setminus \bar{A}$; $\bar{A}$ is the closure of A in X.) Then f is continuous on X ⟦Sec. 4.2, Problem 14⟧, and so f has a continuous extension $F: \beta X \to \mathbf{R}$. Now $F^{\perp}$ is closed, and includes H. ⟦$B \not\pitchfork \bar{A}$ since B is open in X, hence $F = f = 0$ on B. But B is dense in H since H is open and X is dense, hence $F = 0$ on H.⟧ Similarly $(F = 1)$ is closed and includes G. Thus G, H are included in disjoint closed sets and the result follows by Theorem 14.1.1. ▮

Lemma 14.4.1 yields a rich supply of nondiscrete extremally disconnected spaces.

★EXAMPLE 1. *Let D be a discrete space. Then βD is extremally disconnected.* This follows from Lemma 14.1.1. ▮ Of course βD is not discrete unless D is finite. ⟦It is compact and infinite.⟧

LEMMA 14.1.2. *Let X be extremally disconnected and $S \subset X$. Suppose that every two disjoint (relatively) open subsets A, B of S are included in disjoint open subsets G, H, of X. Then S is extremally disconnected.*

Let A, B be disjoint open subsets of S. Then, with G, H as in the statement of the Lemma, we have $\bar{A} \cap \bar{B} \subset \bar{G} \cap \bar{H} = \varnothing$ by Theorem 14.1.1. The result follows, again by Theorem 14.1.1. ▮

The criterion of Lemma 14.1.2 enables us to show that extremal disconnectedness is inherited by three kinds of subspace.

THEOREM 14.1.3. *Let S be a subspace of an extremally disconnected space X. Then any one of the following three assumptions implies that S is extremally disconnected:* (i) *S is open,* (ii) *S is dense,* (iii) *S is a retract of X.*

Let A, B be disjoint open sets in S. If S is open, A, B are open in X and Lemma 14.1.2 applies with $G = A$, $H = B$. If S is dense, $A = G \cap S$, $B = H \cap S$ for certain open G, H in X; G, H must be disjoint. ⟦If $G \cap H$ is not empty it must meet S since S is dense; say $x \in G \cap H \cap S$. But then $x \in (G \cap S) \cap (H \cap S) = A \cap B$ which is impossible.⟧ Thus again Lemma 14.1.2 applies. Finally let $r: X \to S$ be a retraction, and let $G = r^{-1}[A]$, $H = r^{-1}[B]$. Then G, H are open, ⟦r is continuous⟧, and $A \subset G$, $B \subset H$. The result follows from Lemma 14.1.2. ▮

Some earlier results which may be compared with Theorem 14.1.4 are: βX is connected if and only if X is ⟦Sec. 8.3, Problem 104⟧, and, $\mathbf{Q}^+$ is connected ⟦Sec. 8.1, Problem 127⟧.

THEOREM 14.1.4. *Let X be a Tychonoff space. Then βX is extremally disconnected if and only if X is.*

Half of this is Lemma 14.1.1, and the other half follows from Theorem 14.1.3 since X is dense in βX. ∎

Extremally disconnected Hausdorff spaces have the interesting and unusual property that sequential convergence must be trivial. (See Theorem 14.1.5.)

LEMMA 14.1.3. *Let X be a Hausdorff space, $\{x_n\}$ a sequence in X with $x_n \to x$, and all $x, x_1, x_2, \ldots$ unequal. Then there exists a disjoint sequence $\{G_n\}$ of open sets with $x_n \in G_n$ for each n.*

For each n, let $K_n = \{x\} \cup \{x_i : i \neq n\}$. Then each K_n is compact since it is a convergent sequence together with its limit; thus x_n, K_n can be separated by open sets ⟦Theorem 5.4.6⟧, say $x_n \in A_n$, $K_n \subset B_n$ with A_n, B_n disjoint open sets. For each n, let $G_n = B_1 \cap B_2 \cap \cdots \cap B_{n-1} \cap A_n$; then each G_n is open, ⟦a finite intersection of open sets⟧, $x_n \in G_n$ for each n ⟦$x_n \in K_i$ if $i \neq n$⟧ and any two of the G_n are disjoint. ⟦If $m > n$, then $G_m \cap G_n \subset B_n \cap A_n = \varnothing$.⟧ ∎

THEOREM 14.1.5. *In an extremally disconnected Hausdorff space, a convergent sequence must be eventually constant. Hence a first countable (in particular a metrizable) extremally disconnected Hausdorff space must be discrete.*

Suppose that X is a Hausdorff space which contains a convergent sequence which is not eventually constant. Then X surely contains a sequence $\{x_n\}$ with $x_n \to x$, and all $x, x_1, x_2, \ldots$ unequal. ⟦A subsequence of the first mentioned sequence.⟧ Let $\{G_n\}$ be the disjoint sequence of open sets given by Lemma 14.1.3, and let $G = \bigcup \{G_{2n} : n = 1, 2, \ldots\}$. Then G is open, but $\bar{G}$ is not since $x \in \bar{G}$ ⟦for all n, $x_{2n} \in G_{2n} \subset G$; and $x_{2n} \to x$⟧; but x is not interior to $\bar{G}$. ⟦For all n, $x_{2n+1} \in G_{2n+1}$, hence $x_{2n+1} \notin \bar{G}$ since $\bar{G} \not\pitchfork G_{2n+1}$. ($G$ and G_{2n+1} are disjoint open sets.) But $x_{2n+1} \to x$.⟧ It follows that X is not extremally disconnected. The second statement of Theorem 14.1.5 follows since if X is first countable and not discrete, it has a point x which is an accumulation point of $X \setminus \{x\}$, hence a sequential limit point ⟦Theorem 3.1.1⟧. ∎

EXAMPLE 2. Let X be an extremally disconnected compact Hausdorff space ⟦Example 1⟧ and let x be a nonisolated point. *Then $X \setminus \{x\}$ is not com-*

pact ⟦it is not closed, see Theorem 5.4.5⟧ *but it is countably compact.* ⟦Suppose that it is not. Then $X \setminus \{x\}$ contains a countably infinite set S with no accumulation point. Since X is compact, S has an accumulation point in X and this can only be x. It follows by Sec. 7.1, Problem 13, that S is a sequence converging to x, an impossibility by Theorem 14.1.5.⟧ (It appears that a space is countably compact when each infinite set has a limit point, and compact when each infinite set has enough limit points.)

EXAMPLE 3. A partially ordered set L is called a *lattice* if to each a, b correspond a unique least upper bound $a \vee b$ and greatest lower bound $a \wedge b$. (For example, $a \vee b$ is, by definition, a member c of L such that $c \geq a, c \geq b$, and if, for some d, $d \geq a$ and $d \geq b$, then also $d \geq c$.) A lattice is called *complete* if every nonempty subset S which is bounded above has a least upper bound, written $\bigvee S$. It follows that every nonempty subset S which is bounded below has a greatest lower bound, written $\bigwedge S$. ⟦Namely $\bigwedge S = \bigvee \{x: x \text{ is a lower bound of } S\}$.⟧ The connection of these ideas with the topological concepts of this section is contained in the following result: *Let X be a zero-dimensional topological space, and L the collection of all simultaneously open and closed subsets of X. Then L is a lattice, moreover L is a complete lattice if and only if X is extremally disconnected.* Suppose first that X is extremally disconnected; and let $S \subset L$. Then $\overline{\bigcup \{A: A \in S\}}$ is closed and open ⟦it is the closure of an open set⟧ and is clearly the smallest such set which includes every $A \in S$. Conversely, if L is complete, let G be an open subset of X and $C = \{S: S \in L, S \subset G\}$; thus C is the collection of all open and closed subsets of G. By hypothesis, C has a least upper bound which we shall denote by H; thus $H = \bigvee C$. Then H is open and closed, and the proof is concluded by showing that $H = \bar{G}$. ⟦This implies that $\bar{G}$ is open.⟧ First, $H \supset \bar{G}$. ⟦Since X is zero-dimensional, $G = \bigcup C \subset H$ since every member of C is included in H. Hence $\bar{G} \subset \bar{H} = H$.⟧ Next $H \subset \bar{G}$. ⟦$H \setminus \bar{G}$ is open, hence includes an open and closed set S. Then $H \setminus S$ is open and closed, and includes every set in C since it includes G. This makes $H \setminus S$ an upper bound for C and forces S to be empty since H is the least upper bound. So $H \setminus \bar{G}$ is empty also, otherwise S could have been chosen nonempty.⟧ ∎

A famous theorem of M. H. Stone says that every Boolean algebra is the algebra of all open and closed subsets of a zero-dimensional compact Hausdorff space X. (See [Simmons, p. 344]. Note that a totally disconnected compact T_2 space is zero-dimensional.) The result of this example shows that the algebra is complete if and only if X is extremally disconnected.

Problems

In this list e.d. stands for extremally disconnected.

★1. A space is e.d. if and only if the interior of each closed set is closed.

2. In Lemma 14.1.3 and Theorem 14.1.5, "Hausdorff" cannot be replaced by "T_1," ⟦cofinite⟧, or "regular" ⟦indiscrete⟧.
3. Every subset of an e.d. T_2 space is sequentially closed ⟦Theorem 14.1.5⟧.
4. Every subset of $\beta\omega$ which includes ω is e.d. ⟦Theorem 14.1.3⟧.

---

101. Describe all topologies such that every two disjoint sets have disjoint closures.
102. If any finite subset of $\beta\omega \setminus \omega$ is removed from $\beta\omega$, the resulting space is countably compact ⟦Example 2⟧.
103. Let S be a subspace of an e.d. T_2 space which, in its relative topology, is closure sequential (for example, metrizable). Then S is discrete.
104. An e.d. space is zero-dimensional if and only if it is regular. However an e.d. Hausdorff space need not be regular or zero-dimensional. ⟦Sec. 6.2, Example 2 and Problem 112.⟧
105. A countable product of discrete spaces (each with more than one point) cannot be e.d. ⟦Theorem 14.1.5; Theorem 6.4.2⟧. (Hence e.d. is not productive; see the next problem.)
106. The product of two e.d. spaces need not be e.d.; indeed $\beta\omega \times \beta\omega$ is not e.d. ⟦Consider $\{(n, n) : n \in \omega\}$.⟧ Thus $\beta\omega \times \beta\omega$ is not even homeomorphic with $\beta(\omega \times \omega)$. Compare Sec. 8.5, Problem 105.
107. In an e.d. Hausdorff space, every compact set is finite or noncountable ⟦Sec. 10.1, Problem 109⟧.
108. Let X be either a regular or a Hausdorff space. Then sequential convergence is trivial in X if and only if every compact subset of X is finite or noncountable ⟦Sec. 10.1, Problem 109; conversely, a convergent sequence leads to a compact countable set⟧.
109. A sequentially compact e.d. T_2 space must be finite ⟦Theorem 14.1.5⟧.
110. For $t \in \beta\omega$, the space $\omega \cup \{t\}$ is not pseudocompact. ⟦It has an open–closed discrete subspace by Theorem 14.1.1. (Or use Sec. 8.6, Problem 1; Theorem 8.5.3.)⟧
111. A countably compact space may have two different uniformities. Compare Theorem 11.4.6; Sec. 11.4, Example 4. ⟦Let X be $\beta\omega$ with two points removed; see Problem 102. Consider X^+ and βX. The uniformities are different since they have different completions.⟧
112. Lemma 14.1.3 holds for the space of Problem 110 with $x_n = n$. ⟦Take $G_n = \{n\}$.⟧ Why does the proof of Theorem 14.1.5 fail to yield a contradiction?
113. Call a space completely T_2 if any two points can be completely separated. Show that a T_2 space is completely T_2 if it is either zero-dimensional or e.d.
114. Disjoint open sets in an e.d. space can be completely separated, but not necessarily disjoint closed sets ⟦Problem 104⟧.

115. Let $f \in C(X)$, where X is e.d. compact T_2, and let x be a nonisolated point of X. Then $f(y) = f(x)$ for some $y \neq x$. (For $X = \beta\omega$ and $f(x) = 0$, this yields a result similar to that of Sec. 8.5, Problem 9.) ⟦Let $f(x) = t$, $g = (f - t)^{-1}$. Then g is bounded on $X \setminus \{x\}$ by Example 2, and Sec. 7.1, Problem 114.⟧

---

201. Let A, B be open sets in an e.d. space. Does the inclusion $\bar{A} \cap \bar{B} \subset A \cup B$ necessarily hold?
202. No complete Boolean algebra can be countably infinite. ⟦Its Stone space would be second countable by Problem 104, hence metrizable, hence discrete, hence finite.⟧
203. $\beta\omega \setminus \omega$ is not e.d. ⟦See [Gillman and Jerison, 6R and 6W].⟧ Hence e.d. is not F-hereditary; also $\beta\omega \setminus \omega$ is not a retract of $\beta\omega$ ⟦Theorem 14.1.3⟧. What result follows from this with Example 3 and Sec. 12.3, Problem 205?
204. Give an example of an e.d. compact T_2 space without isolated points ⟦See [Bourbaki (b), II.4 #12b]. It is also possible to use techniques similar to those of Sec. 7.3, Problems 216 and 217.⟧

14.2 The Gleason Map

Let a function $u: X \to Y$ be called *minimal* if it is continuous, and no closed proper subset of X is carried onto $u[X]$ by u. For example, a one-to-one continuous map would be minimal; but a minimal map need not be one-to-one; for example, every continuous map from an infinite cofinite space onto an infinite space is minimal ⟦the closed proper subsets are finite⟧, and there are such maps which are not one-to-one. ⟦A finite-to-one self-map of a cofinite space will do, since the inverse image of any closed set is closed.⟧ See also Problem 2. Important sufficient conditions that a minimal map be a homeomorphism are given in Theorems 14.2.3, 14.2.4, and 14.2.5. Our first result shows that every continuous map between suitable spaces can be cut down to a minimal map without reducing its range.

THEOREM 14.2.1 *Let X be a compact and Y a T_1 space, and let $u: X \to Y$ be a continuous map. Then there exists a closed subspace F of X such that $u: F \to Y$ is minimal and $u[F] = u[X]$.*

(Of course the second u should have been written $u \mid F$.) Let P be the collection of all closed subsets S of X such that $u[S] = u[X]$; P is not empty since $X \in P$. Order P by inclusion, let C be a maximal chain, and let $F = \bigcap \{S: S \in C\}$. F is closed, ⟦it is an intersection of closed subsets⟧. In order to show that $u[F] = u[X]$, let $y \in u[X]$. For each $S \in C$, let $A_S = u^{-1}[\{y\}] \cap S$. Each A_S is closed, and $\{A_S: S \in C\}$ has the finite intersection property. ⟦Any finite collection of A_S has a smallest member (that is, one included in the others) and the smallest member is nonempty since $y \in u[S]$ for all $S \in P$.⟧ Thus since X is compact, $\bigcap \{A_S : S \in C\} \neq \varnothing$ ⟦Theorem 5.4.3⟧. Let $x \in \bigcap \{A_S : S \in C\}$. Then $x \in F$ ⟦$\bigcap A_S \subset F$⟧, and $u(x) = y$. ⟦For any $S \in C$, $x \in A_S$, hence $x \in u^{-1}[\{y\}]$.⟧ This shows that $u[F] = u[X]$, and we note, finally, that $u \mid F$ is minimal. ⟦If not, there is a closed proper subset F' of F with $u[F'] = u[X]$, and $C \cup \{F'\}$ is a chain in P which is strictly larger than C.⟧ ▮

We now describe the interesting and important *Gleason map* which was given by Andrew Gleason in 1958. Let X be a compact Hausdorff space and let D be X, but with the discrete topology. (D is merely a discrete space with the same cardinality as X.) The identity map $i: D \to X$ is continuous, and so it has a continuous extension $f: \beta D \to X$ ⟦Theorem 8.3.1⟧. By Theorem 14.2.1, there is a closed (hence compact) subspace F of βD such that $u = f \mid F$ is a minimal map of F onto X. We pause to state the principal theorem of this section, which contains this and some further information.

THEOREM 14.2.2. *Let X be a compact Hausdorff space. Then there exists an extremally disconnected compact Hausdorff space F and a minimal map of F onto X. (This is the Gleason map.)*

Let F be the space defined in the preceding discussion. It is a compact Hausdorff space, and it remains to show that it is extremally disconnected. By Theorem 14.1.3, it is sufficient to show that F is a retract of βD. Define $h: X \to F$ to be any right inverse for u ⟦possible because f is onto, Sec. 1.1, Problem 11⟧. Now we may consider $h: D \to F$ since $D = X$ (except with a different topology) and then h is continuous ⟦D is discrete⟧, and so h has a continuous extension $r: \beta D \to F$ ⟦Theorem 8.3.1⟧. We shall derive a contradiction from the assumption that r is not a retraction, and this will conclude the proof. If r is not a retraction, there exists $t \in F$ with $r(t) \neq t$. Let U, V be disjoint open neighborhoods in F, of $r(t)$, t, respectively, let $A = V \cap r^{-1}[U]$, and let $B = F \setminus A$. Then B is a closed subset of F; moreover, it is a proper subset of F, ⟦$t \notin B$⟧. A contradiction will follow from the fact that u is minimal, and that $u[B] = X$, the latter of which we now prove. Let $x \in X$.

Case I. Suppose $h(x) \in B$, then $x = u(hx) \in u[B]$.

Case II. Suppose $h(x) \in A$. Then $r(hx) \in B$.

⟦Since $h(x) \in A \subset r^{-1}[U]$, it follows that $r(hx) \in U$, hence $r(hx) \notin V$ and so $r(hx) \notin A$.⟧ Also $u[r(hx)] = x$. ⟦$u \circ r = f$ since this is true on the dense subspace D where it takes the form $u \circ h = i$; thus $u[r(hx)] = f(hx) = u(hx) = x$.⟧ Thus $x = u[r(hx)] \in u[B]$.

Since Cases I, II are exhaustive, we have proved that $u[B] = X$. ▮

We continue with a derivation of some properties of minimal maps, beginning with a lemma showing that a minimal map u is "almost one-to-one." (If u is one-to-one and A, B are disjoint sets, then $u[A] \not\pitchfork u[B]$. The result is of this form.)

LEMMA 14.2.1. *Let $u: X \to Y$ be minimal. Let F, G be disjoint subsets of X with F closed, G open. Then $u[G] \not\pitchfork \{u[F]\}^i$.*

Let $y \in u[G]$, and let N be an open neighborhood of y. We are required to show that $N \not\subset u[F]$. Let $A = F \cup \{u^{-1}[N]\}^{\sim}$. Then A is a closed proper subset of X ⟦let $y = u(x)$, $x \in G$; then $x \notin F$, and $u(x) \in N$ so that $x \notin A$⟧, and so, since u is minimal, $u[A] \neq u[X]$. Let $z \in u[X] \setminus u[A]$. Then $z \in N$. ⟦$z = u(x)$ for some $x \notin A$; in particular $x \in u^{-1}[N]$.⟧ Also $z \notin u[F]$. ⟦$F \subset A$, hence $u[F] \subset u[A]$; but $z \notin u[A]$.⟧ Thus $N \not\subset u[F]$ and so y is not an interior point of $u[F]$. ▮

THEOREM 14.2.3. *An open minimal map u of a Hausdorff space X must be a homeomorphism (into).*

Let $x_1, x_2 \in X$ with $x_1 \neq x_2$. Let G_1, G_2 be disjoint open neighborhoods of x_1, x_2. Then $G_1 \not\pitchfork \bar{G}_2$, and Lemma 14.2.1 shows that $u[G_1] \not\pitchfork \{u[\bar{G}_2]\}^i$. But $u(x_1) \in u[G_1]$, and $u(x_2) \in \{u[\bar{G}_2]\}^i$. ⟦$u(x_2) \in u[G_2] \subset u[\bar{G}_2]$, and $u[G_2]$ is open.⟧ Thus $u(x_1) \neq u(x_2)$, and so u is one-to-one. Since it is also continuous and open, it is a homeomorphism. ▮

Theorem 14.2.3 fails for closed maps ⟦Problem 101⟧ but has an important analogue, Theorem 14.2.4.

LEMMA 14.2.2. *Let u be a closed minimal function from a topological space X onto an extremally disconnected space Y, and let G be an open subset of X. Then $u[\bar{G}]$ is open.*

Let $F = \tilde{G}$. Then $u[F]$, $u[\bar{G}]$ are closed, and $u[F] \cup u[\bar{G}] = Y$. ⟦$u[F] \cup u[\bar{G}] = u[F \cup \bar{G}] = u[X] = Y$.⟧ Hence $\{u[F]\}^i \cup \{u[\bar{G}]\}^i = Y$ ⟦Theorem 14.1.2⟧. Thus $u[G] \subset \{u[\bar{G}]\}^i$ ⟦Lemma 14.2.1⟧. The latter set is closed ⟦Sec. 14.1, Problem 1⟧, and so $\overline{u[G]} \subset \{u[\bar{G}]\}^i$. Hence $u[\bar{G}] \subset \{u[\bar{G}]\}^i$ ⟦Theorem 4.2.4⟧, and so $u[\bar{G}]$ is open. ▮

LEMMA 14.2.3. *A closed minimal function u from a regular space X onto an extremally disconnected space Y is open.*

Let G be an open subset of X, and $x \in G$. Let N be an open neighborhood of x with $\bar{N} \subset G$. Then $u[\bar{N}]$ is open ⟦Lemma 14.2.2⟧ and $u(x) \in u[\bar{N}] \subset u[G]$. Thus $u[G]$ is a neighborhood of $u(x)$. Since x is an arbitrary member of G, $u[G]$ is open. ▮

THEOREM 14.2.4. *A closed minimal function from a T_3 space onto an extremally disconnected space is a homeomorphism.*

This follows from Lemma 14.2.3 and Theorem 14.2.3. ▮

THEOREM 14.2.5. *A minimal function from a compact Hausdorff space onto an extremally disconnected Hausdorff space is a homeomorphism.*

This follows from Theorem 14.2.4, with Theorems 5.4.8 and 5.4.7. ▮

COROLLARY. *The Gleason map to an extremally disconnected compact Hausdorff space is a homeomorphism.*

Problems

1. A minimal map from **R** to **R** must be one-to-one.
2. A minimal map from **R** to $\mathbf{R}^2$ need not be one-to-one. ⟦Draw a curve in $\mathbf{R}^2$ which intersects itself once.⟧
3. Let f: $[0, 2\pi] \to [-1, 1]$ be the trigonometric sine. Find $F \subset [0, 2\pi]$ such that $f \mid F$ is minimal.
4. Let u be a minimal map from X onto Y, and let F, G be disjoint sets in X with F closed, G open. Let $A = Y \setminus u[F]$. Show that $u[G] \subset \bar{A}$.

---

101. A closed minimal map of **R** need not be a homeomorphism ⟦Problem 2⟧.
102. Let B be an e.d. compact T_2 space. Then B can be embedded in βD for some discrete space D. ⟦Let D be B with the discrete topology. Extend i: $D \to B$ to u: $\beta D \to B$ and apply Theorems 14.2.1 and 14.2.5⟧.
103. In Problem 102, D can be chosen to have the cardinality of any dense subspace of B. ⟦The map u in Problem 102 is still onto.⟧
104. Every e.d. Tychonoff space X can be embedded in βD for some discrete D, and D can be chosen to have the cardinality of any dense subspace of X. In particular, a separable e.d. Tychonoff space is a subspace of $\beta\omega$. ⟦In Problem 103, take $B = \beta X$. This result is due to M. Henriksen and J. Isbell.⟧
105. In Problem 102 we cannot improve the result to conclude that B is homeomorphic with βD ⟦Sec. 14.1, Problem 204; Sec. 8.3, Problem 107⟧.
106. In Theorem 14.2.1, compact cannot be omitted. ⟦Give **R** the topology in which a nonempty set is open if and only if it is of the form (a, b) with $-\infty \le a < 0 < b \le +\infty$. Take $Y = \{0\}$ and $u(x) = 0$ for all x.⟧

---

201. Must a minimal function from a compact T_2 space onto an e.d. T_2 space be open?
202. Let $X = \beta\omega \setminus \{t\}$, $t \notin \omega$. Is it possible to have a net f_δ of homeomorphisms of X onto itself such that f_δ tends uniformly to a constant function? (Compare Sec. 7.1, Problems 118 to 120. Use the unique uniformity of X.)

14.3 Categorical Algebra

A valuable tool, which developed since the middle of the nineteenth century, consists of considering collections of objects and possible structures of such collections, rather than individual objects. Such collections are called spaces. Thus one considers a compact topological space whose members are functions, and deduces the existence of a function with certain properties. The middle of the twentieth century has witnessed a further step in this direction, namely, that of considering collections of spaces. These collections are called categories. In discussing sets whose members are sets, the famous classical paradoxes suggest that an explicit set-theoretical foundation be laid for the subject. We shall not take the space for this. The reader's intuition will be a sufficient guide for our discussion. Further study might require reference to such a text as [Halmos], or the Appendix of [Kelley].

We now define the concept *category*. We begin with a class Γ, whose members are called *objects*. For example, we might consider the class of all topological spaces. The objects of Γ are sets in most cases (all, in the following discussion), but the theory ignores the possibility that the objects can be sets and refrains from any mention of "points" (that is, members of objects), except in discussions of examples. Next, with each ordered pair (X, Y) of objects of Γ, we associate a set called $\hom(X, Y)$. In most examples (all, in the following discussion) the members of $\hom(X, Y)$ are maps from X to Y. The members of $\hom(X, Y)$ are called *morphisms* from X to Y, and if $\alpha \in \hom(X, Y)$, X is called the *domain* of α and Y the *codomain*. The notation $\alpha: X \to Y$ will sometimes be used instead of $\alpha \in \hom(X, Y)$. It is assumed that $\hom(X, Y)$ and $\hom(Z, W)$ are disjoint unless $X = Z$ and $Y = W$.

★EXAMPLE 1. Let Γ be the class of all sets, and let $\hom(X, Y)$ be the set of all maps from X to Y. Now suppose that $X \subset Y$ and $i: X \to Y$ is the inclusion map. Then $i \in \hom(X, Y)$. Further, let $i_X: X \to X$ be the identity map. Now, assuming $X \neq Y$, $\hom(X, X)$ and $\hom(X, Y)$ are disjoint, in particular $i_X \notin \hom(X, Y)$, $i \notin \hom(X, X)$, and so, indeed $i \neq i_X$. The distinction between these two functions is a nicety not usually observed, since, according to the usual definition of function as a set of ordered pairs, they are equal. However, it is important for many purposes to distinguish between them. See, for example, Problem 121.

Finally, as part of the definition of category, we make one further assumption: to every $\alpha \in \hom(X, Y)$ and $\beta \in \hom(Y, Z)$ there is associated a morphism, called $\beta\alpha \in \hom(X, Z)$ with the properties:

$$\text{For each object } X \text{ there exists a morphism } 1_X \in \hom(X, X) \text{ such that for any } Y \text{ and } \alpha \in \hom(X, Y),\ \beta \in \hom(Y, X) \text{ we have } \alpha 1_X = \alpha,\ 1_X\beta = \beta, \tag{14.3.1}$$

and

$$\text{whenever } \alpha \in \hom(X, Y),\ \beta \in \hom(Y, Z),\ \gamma \in \hom(Z, W), \text{ then } \gamma(\beta\alpha) = (\gamma\beta)\alpha. \tag{14.3.2}$$

The morphism $\beta\alpha$ is called the *composition* of β with α.

This concludes the definition. A *category*, then is a class of objects, which has a set of morphisms associated with each ordered pair of objects, a morphism $\beta\alpha$ associated with each ordered pair α, β of morphisms such that domain β = codomain α, and morphism 1_X associated with each object X such that conditions (14.3.1) and (14.3.2) hold.

Note that if 1_X and $1'_X$ both satisfy (14.3.1), then $1_X = 1_X 1'_X = 1'_X$ so that 1_X is unique.

★EXAMPLE 2. We may consider the category Γ of all sets and maps with ordinary composition and identity. Another example is the category T of Hausdorff spaces and continuous maps with ordinary composition and identity.

The early stages of the theory consist of attempts to show that morphisms behave like maps, and to extend to categories definitions which make sense for maps. We begin with an attempt to generalize the concept of onto map; a morphism α is called *epic*, if, for every β, γ for which $\beta\alpha = \gamma\alpha$, it follows that $\beta = \gamma$. Suppose Γ is a category of sets and maps. If $\alpha: X \to Y$ is onto, then α is an epic morphism (called *epimorphism*, for short). ⟦Say $\beta, \gamma \in \hom(Y, Z)$. Let $y \in Y$. Then $y = \alpha x$ for some $x \in X$ so that $\beta y = \beta\alpha x = \gamma\alpha x = \gamma y$. Thus $\beta = \gamma$.⟧ However, an epimorphism need not be onto ⟦Example 3⟧, nor need it have dense range ⟦Problem 12⟧.

★EXAMPLE 3. Let T be as in Example 2. Let X be a dense proper subspace of Y; then the inclusion map $i: X \to Y$ is not onto, but is an epimorphism. ⟦If $\beta, \gamma \in \hom(Y, Z)$ and $\beta i = \gamma i$, then for $x \in X$, $\beta x = \gamma x$, hence $\beta = \gamma$ since they agree on a dense subspace of Y.⟧

★EXAMPLE 4. Let K be the category of compact Hausdorff spaces and continuous maps. A morphism is epic if and only if it is onto. ⟦If $\alpha \in \hom(X, Y)$ is not onto, $\alpha[X]$ is a compact proper subspace of Y, hence, with $I = [0, 1]$, there exists $\beta \in \hom(Y, I)$, $\gamma \in \hom(Y, I)$ with $\beta(y) = \gamma(y)$

for $y \in \alpha[X]$, but $\beta \neq \gamma$; for example, we may take $\beta = 0$, $\gamma(y) = 0$ for $y \in \alpha[X]$, $\gamma(y) = 1$ for some $y \in Y$, using Theorem 5.4.7. Then $\beta\alpha = \gamma\alpha$ but $\beta \neq \gamma$ so that α is not epic.⟧

A morphism α is called *monic* if for every β, γ for which $\alpha\beta = \alpha\gamma$, it follows that $\beta = \gamma$. If $\alpha \in \hom(X, Y)$ and α is one-to-one, then α is a monic morphism (called *monomorphism*, for short). However, a monomorphism need not be one-to-one ⟦Problem 4⟧. A discussion of interpretations of the terms "monic" and "epic" may be found in MR *33* #161, and [Burgess].

Suppose X, Y are objects and $\alpha: X \to Y$ is monic. We say that X is a *subobject of Y by means of* α. In the category T (Example 3) X is a subobject of Y by means of α if and only if α is a one-to-one continuous map of X into Y.

A morphism is called a *retraction* if it has a right inverse; thus $\alpha \in \hom(X, Y)$ is a retraction if and only if there exists $\beta \in \hom(Y, X)$ with $\alpha\beta = 1_Y$. Under these circumstances Y is called a *retract* of X.

EXAMPLE 5. In T (Example 2) this definition does not specialize to the definition of retraction given earlier. Indeed the map $x \to x + 1$ from **R** to itself has the right inverse $x \to x - 1$ but is not a retraction in the earlier sense since it is not the identity on its range. However suppose $\alpha: X \to Y$ is a retraction in the sense of category, so that there exists $\beta: Y \to X$ with $\alpha\beta = 1$. Then $\beta: Y \to \beta[Y]$ is a homeomorphism ⟦it has the continuous inverse $\alpha \mid \beta[Y]$⟧, and $\beta\alpha: X \to \beta[Y]$ is a retraction in the topological sense. ⟦For $x \in \beta[Y]$, $\beta\alpha(x) = \beta\alpha[\beta(y)] = \beta[\alpha\beta(y)] = \beta y = x$.⟧ Thus the codomain of a (categorical) retraction in T is homeomorphic with a (topological) retract of its domain. This proves that *Y is a retract of X in the category T if and only if Y is homeomorphic with a retract of X in the topological sense.*

Note that *a retract of an object is a subobject.* ⟦If $\alpha: X \to Y$ is a retraction, let $\alpha\beta = 1_Y$. Then β is a monomorphism, since if $\beta\gamma = \beta\delta$ then $\gamma = 1_Y\gamma = \alpha\beta\gamma = \alpha\beta\delta = 1_Y\delta = \delta$; thus Y is a subobject of X.⟧

Let Γ be a category and X an object. We call X *projective* if for all objects Y, Z, every $\alpha \in \hom(X, Z)$, and every epimorphism $\beta \in \hom(Y, Z)$, there exists $\gamma \in \hom(X, Y)$ such that $\alpha = \beta\gamma$. This definition is illustrated in the diagram:

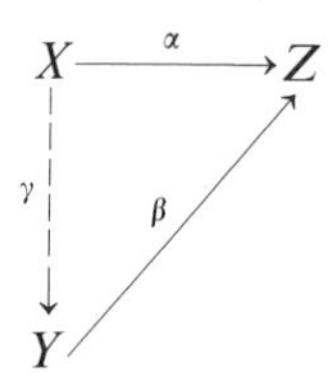

The dotted line indicates the morphism whose existence is postulated. The diagram is commutative ($\beta\gamma = \alpha$).

Every retraction is an epimorphism ⟦suppose that α is a retraction and that $\beta\alpha = \gamma\alpha$; let $\alpha\delta = 1$, then $\beta = \beta 1 = \beta\alpha\delta = \gamma\alpha\delta = \gamma 1 = \gamma$⟧, but not conversely ⟦Problem 3⟧. However a partial converse holds.

THEOREM 14.3.1. *If Z is a projective object, every epimorphism with codomain Z is a retraction.*

Let $\beta \in \hom(Y, Z)$ be epic. Applying the definition of projective object with $X = Z$, and $\alpha =$ identity map, we obtain the existence of $\gamma \in \hom(Z, Y)$ with $\beta\gamma = \alpha$; that is, β has a right inverse, hence is a retraction. ∎

★EXAMPLE 6. In K (Example 4) every discrete space is projective. ⟦For $x \in X$ (see the diagram, above), define γx to be any member of $\beta^{-1}[\alpha x]$.⟧ Also, if D is discrete and $X = \beta D$, the Stone–Cech compactification of D, then X is projective. ⟦Define $\gamma_0 \colon D \to Y$ by choosing for $\gamma_0 d$ any member of $\beta^{-1}[\alpha d]$. Extend γ_0 to $\gamma \colon X \to Y$. Then $\beta \circ \gamma = \alpha$ on D, hence on X, since D is dense.⟧

THEOREM 14.3.2. *A retract of a projective object is projective.*

Let $\delta \colon X \to S$ be a retraction with right inverse $\varepsilon \colon S \to X$, where X is projective. Let $\alpha \colon S \to Z$, and let $\beta \colon Y \to Z$ be epic. By hypothesis, there exists $\gamma \colon X \to Y$ with $\beta\gamma = \alpha\delta$. ⟦$\alpha\delta \colon X \to Z$.⟧ Then $\beta(\gamma\varepsilon) = \alpha$. ∎

EXAMPLE 7. *In the category K (Example 4) an object is projective if and only if it is a retract of βD for some discrete topological space D.* First βD is projective; ⟦as proved in Example 6⟧; hence any retract of βD is projective ⟦Theorem 14.3.2⟧. Conversely, let X be a projective object in K, and let D be X with the discrete topology. The identity $i \colon D \to X$ is continuous, and so it has a continuous extension $\alpha \colon \beta D \to X$ ⟦Theorem 8.3.1⟧. Clearly α is onto, hence an epimorphism, hence a retraction, by Theorem 14.3.1. ∎

EXAMPLE 8. *In the category K (Example 4) an object is projective if and only if it is extremally disconnected.*

If X is projective, it is extremally disconnected, by Example 7 and Theorem 14.1.3. (There is no conflict in the two definitions of retract since extremal disconnectedness is preserved by homeomorphism; see Example 5.) Conversely, if X is extremally disconnected, let $\alpha \colon F \to X$ be the Gleason map (Theorem 14.2.1). By the Corollary to Theorem 14.2.5, α is a homeomorphism, hence it suffices to prove that F is projective. But F was proved, in the course of proving Theorem 14.2.2, to be a retract of βD for some discrete D; hence, by Example 7, F is projective. ∎

Due to limitations of space we must pass very briefly over some of the most important and beautiful concepts of the theory. Consider the possibility of "dualizing" all the preceding work by interchanging domain and codomain, monic and epic, and so on. For example let us call an object X *injective* (dualizing the definition of "projective") if for all objects Y, Z, every $\alpha \in \hom(Z, X)$, and every monomorphism $\beta \in \hom(Z, Y)$, there exists $\gamma \in \hom(Y, X)$ such that $\alpha = \gamma\beta$. The dual of Theorem 14.3.1 is that if Z is injective, every monomorphism with domain Z is a coretraction (Problem 1). There is a concept, which we shall not describe, called *dual category*, and a dual result due to J. L. Kelley, to that of Example 8, in which the injective Banach spaces are shown to be precisely those of the form $C(X)$, with X extremally disconnected, compact, Hausdorff. See [Cohen].

We next turn to the important concept known as "*functor*." If A, B are categories a functor $F: A \to B$ is a map which carries each object X of A to an object $F(X)$ of B, and each morphism $\alpha \in \hom(X, Y)$ of A to a morphism $F(\alpha) \in \hom(F(X), F(Y))$ of B such that

$$F(1_X) = 1_{F(X)}, \tag{14.3.3}$$

and

$$F(\alpha\beta) = F(\alpha)F(\beta), \tag{14.3.4}$$

whenever $\alpha\beta$ is defined. Examples are the so-called *forgetful functors* which carry objects onto objects with less structure. For example, with Γ, T as in Example 2, define $F: T \to \Gamma$ by $F(X) = X$, $F(\alpha) = \alpha$. The most significant functor occurring in this book is the Stone–Cech compactification functor. With T as the category of Tychonoff spaces and continuous maps, and K as in Example 4, let $B: T \to K$ have $B(X) = \beta X$, and, for $\alpha \in \hom(X, Y)$, let $B(\alpha): \beta X \to \beta Y$ be the unique extension of the continuous map α. On the other hand, there is no natural one-point compactification functor F since we have no natural way of defining $F(\alpha): X^+ \to Y^+$. ⟦For example, with $X = \omega$, $Y = Y^+ = \{-1, 1\}$, $\alpha(n) = (-1)^n$, α cannot be extended to ω^+.⟧ See also Problem 111.

A *contravariant functor* F is a map carrying objects X of A to objects $F(X)$ of B and $\alpha \in \hom(X, Y)$ of A to $F(\alpha) \in \hom(F(Y), F(X))$ of B such that Equation (14.3.3) holds and $F(\alpha\beta) = F(\beta)F(\alpha)$ whenever $\alpha\beta$ is defined.

EXAMPLE 9. Let A be a category, and let Γ be as in Example 2. Fix an object $0 \in A$ and define a contravariant functor $*: A \to \Gamma$ by (write $*(X)$ and $*(\alpha)$ as X^*, α^*, respectively), for an object X of A, $X^* = \hom(X, 0)$, for a morphism $\alpha: X \to Y$, $\alpha^*: Y^* \to X^*$ is given by $\alpha^*(\lambda) = \lambda\alpha$ for all $\lambda \in Y^*$. This is the famous *adjoint functor*. See Problems 106, 107, and 201.

We conclude with the definition of *adjoint of a functor*. (Not every functor has an adjoint ⟦Problems 117 and 118⟧.) Suppose that $F: A \to B$ is a functor and that there is a map G carrying the objects of B to the objects of A with the following property: for each object Z of B there exists $\lambda_Z: Z \to F[G(Z)]$

such that for every $\alpha: Z \to F(X)$, there is a unique $u_\alpha: G(Z) \to X$ satisfying $F(u_\alpha)\lambda_Z = \alpha$. Defining, for $\delta \in \hom(W, Z)$, $G(\delta) = u_{\lambda_Z\delta}$ yields a functor G, called a functor adjoint to F. Problems 108 and 109 give examples.

Problems

1. In Γ (Example 2), the following are equivalent for a morphism α: α is epic, α is onto, α is a retraction. Also equivalent are: α is monic, α is one-to-one, α is a *coretraction* (this is, α has a left inverse). Show also that every object is projective and injective.
2. Let X be $\mathbf{R}$ with the discrete topology; show that there exists no monomorphism α such that $\mathbf{R}$ is a subobject of X by means of α. ⟦The inverse image of a singleton by a one-to-one continuous map: $\mathbf{R} \to X$ would have to be open.⟧
3. With X as in Problem 2, the identity map $i: X \to \mathbf{R}$ is an epimorphism in T (Example 2) but not a retraction ⟦Problem 2⟧.
4. Consider the system which contains as objects exactly two sets X, Y with $X = (a, b)$, $Y = (u, v, w)$; let $\hom(X, Y)$ contain the two functions β, γ such that $\beta(a) = \beta(b) = u$, $\gamma(a) = \gamma(b) = v$; let $\hom(Y, X)$ contain one function α such that $\alpha(u) = a$, $\alpha(v) = \alpha(w) = b$; let $\hom(X, X)$ contain $\alpha \circ \beta$, $\alpha \circ \gamma$ and 1_X; and let $\hom(Y, Y)$ contain $\beta \circ \alpha$, $\gamma \circ \alpha$ and 1_Y. Show that α is monic but not one-to-one.
5. Show that $(0, 1)$ is not a retract of $\mathbf{R}$ in the topological sense ⟦Sec. 4.2, Problem 27⟧, but is a retract of $\mathbf{R}$ in the category T (Example 2) ⟦Example 5; $(0, 1)$ is homeomorphic with $\mathbf{R}$, a retract of $\mathbf{R}$⟧.
6. In T, (Example 2), a monomorphism must be one-to-one.
7. Show that every coretraction is a monomorphism, but the converse fails. ⟦As in Problem 2, consider the identity map from X to $\mathbf{R}$.⟧
8. An *isomorphism* is a morphism with inverse; that is $\alpha: X \to Y$ is an isomorphism if there exists $\beta: Y \to X$ such that $\alpha\beta = 1_Y$, $\beta\alpha = 1_X$. Show that a morphism is an isomorphism if and only if it is both a retraction and a coretraction.
9. In the category T, (Example 2), the identity map from $\mathbf{R}$ with the discrete topology to $\mathbf{R}$ is epic and monic, but not an isomorphism.
10. In the category K (Example 4) a monic and epic morphism must be an isomorphism ⟦Theorem 5.4.9⟧.
11. X is called a *proper subobject* of Y by means of α if $\alpha: X \to Y$ is monic but not an isomorphism. In the category T (Example 2), $\mathbf{R}$ with the discrete topology is a proper subobject of $\mathbf{R}$ by means of the identity map.
12. Let X be $\mathbf{R}$ with the simple extension of the Euclidean topology by $\mathbf{Q}$. Let C be the category with two objects, X and $\mathbf{J}$, and all continuous

maps. Show that the inclusion map from **J** to X is epic, but not onto; indeed its range is not dense.

---

101. A category is called *balanced* if no object is a proper subobject of another object by means of an epimorphism. Show that K (Example 4) is balanced ⟦Theorem 5.4.9⟧; so is Γ (Example 2); but T (Example 2) is not balanced ⟦Problems 9 and 11⟧.
102. An object is called *reversible* if it is not a proper subobject of itself by means of an epimorphism. Show that in the category T (Example 2), this definition agrees with that of Sec. 5.1, Problem 207.
103. In the category T (Example 2), **R** is a proper subobject of itself by means of some morphism α, yet **R** is reversible.
104. Call Y a *quotient object* of X by means of α if $\alpha: X \to Y$ is an epimorphism. Show that in the category K (Example 4), every epimorphism is a quotient map in the sense of Section 6.5.
105. Carry out the dual program described in the text, obtaining dual forms of Theorems 14.3.1 and 14.3.2.
106. In Example 9, let $A = \Gamma$, and assume that 0 has at least two points. Show that $\alpha^* = \beta^*$ implies $\alpha = \beta$.
107. In Problem 106, α^* is epic (monic) if and only if α is monic (epic). ⟦For example, if $\alpha^*: Y^* \to X^*$ is epic and $\alpha\beta = \alpha\gamma$, then $\beta^*\alpha^* = \gamma^*\alpha^*$ implies $\beta^* = \gamma^*$, hence, by Problem 106, $\beta = \gamma$. If α is epic it is a retraction, hence α^* is a coretraction and is monic.⟧
108. Let Γ, T be as in Example 2. Let $F: T \to \Gamma$ be the forgetful functor, and $D: \Gamma \to T$ be defined by $D(X) = X$ with the discrete topology, $D(\alpha) = \alpha$. Show that D is adjoint to F. ⟦$\lambda_Z =$ identity, $u_\alpha = \alpha$.⟧
109. Let K be as in Example 4, T the category of Tychonoff spaces and continuous maps. Let $F: K \to T$ be the forgetful functor, and $B: T \to K$ the Stone–Cech compactification functor. Show that B is adjoint to F. ⟦Take λ_Z to be inclusion, u_α to be the extension of α.⟧
110. Let C be a category whose objects are sets and morphisms are maps. If X, Y are objects and $\alpha: X \to Y$ is inclusion, show that α^* is the restriction to X of maps defined on Y.
111. Describe a completion functor on the category of uniform spaces and uniformly continuous maps ⟦Theorems 11.5.4 and 11.3.4⟧.
112. Say that a collection $\{X_\delta\}$ of objects has a *product* $Y = \prod X_\delta$ if Y is an object, and there exist morphisms $p_\delta: Y \to X_\delta$, called *projections*, such such that if Z is any object and $\alpha_\delta: Z \to X_\delta$ for each δ, there exists a unique $\beta: Z \to Y$ such that $p_\delta\beta = \alpha_\delta$. Show that the product of Section 6.4 is a product for the categories Γ, T, K of Examples 2 and 4 ⟦Theorem 7.4.1⟧.
113. Let C be the category of all nonempty subsets of **R**, and inclusions.

(hom(X, Y) = $\varnothing$ unless $X \subset Y$). Show that C has no product; indeed that a collection of two nonintersecting sets has no product.

114. Let Y be a product of $\{X_\delta\}$ (Problem 112.) Suppose that $\alpha\colon Z \to Y$, $\beta\colon Z \to Y$ satisfy $p_\delta\alpha = p_\delta\beta$ for all δ. Show that $\alpha = \beta$. ⟦See the word "unique" in the definition.⟧
115. If Y, Z are both products of $\{X_\delta\}$, they are isomorphic. ⟦With $p_\delta\colon Y \to X_\delta$, $q_\delta\colon Z \to X_\delta$. Let $\alpha\colon Y \to Z$ have $q_\delta\alpha = p_\delta$, $\beta\colon Z \to Y$ have $p_\delta\beta = q_\delta$. Then $p_\delta = q_\delta\alpha = p_\delta\beta\alpha$ so that $\beta\alpha = 1_Y$ by Problem 114.⟧
116. If a functor F has an adjoint, it preserves products; that is, $F(\prod X_\delta) = \prod F(X_\delta)$. ⟦Let $Y = F(\prod X_\delta)$. The projections are $q_\delta = F(p_\delta)$. For $\alpha_\delta\colon Z \to F(X_\delta)$, choose $\gamma\colon G(Z) \to \prod X_\delta$ such that $p_\delta\gamma = u_{\alpha_\delta}$. Finally, take $\beta = F(\gamma)\lambda_Z$. Then $q_\delta\beta = F(p_\delta\gamma)\lambda_Z = \alpha_\delta$.⟧
117. The functor D of Problem 108 has no adjoint ⟦Problem 116; Sec. 6.4, Problem 101⟧.
118. The Stone–Cech compactification functor has no adjoint ⟦Problem 116; Sec. 14.1, Problem 106⟧.
119. Suppose that a collection $\{X_\delta\}$ of objects has a product and satisfies hom(X_δ, $X_{\delta'}$) $\neq \varnothing$ for all δ, δ'. Show that each p_δ is a retraction.
120. In K (Example 4), the object [0, 1] is injective ⟦Theorem 8.5.3⟧.
121. Let X, Y, Z be objects of Γ (Example 2) with $Y \subset Z$, and let $\alpha\colon X \to Y$. Define $\beta\colon X \to Z$ by $\beta(x) = \alpha(x)$ for all x, so that α, β are the "same" map. (See Example 1.) Show that it is possible that $\alpha^* \neq \beta^*$ (Example 9); indeed α^* may be monic but β^* usually is not ⟦Problem 106⟧.

---

201. Let B be the category of Banach spaces and linear maps α with $\|\alpha\| \leq 1$ and define the adjoint functor * taking 0 = **R**. Show that * is one-to-one, that α^* is monic (epic) if and only if α is epic (monic) and that α^* is an isomorphism if and only if α is. ⟦See [Wilansky (a), Theorem 7.2.2 and Problem 8].⟧
▲202. The category of Problem 201 is balanced. ⟦See [Wilansky (a), Corollary 11.2.1].⟧
203. Let k be the category of k spaces and continuous maps. Take the category product to be the k *extension* of the ordinary product (the largest topology which agrees with the given topology on compact sets). Show that this is a product in the sense of Problem 112. ⟦See [Steenrod, Theorem 4.2].⟧
204. There is a category concept which specializes to "onto" in the category T of T_2 spaces and continuous maps, namely "α is onto in T if and only if whenever $\alpha = \beta\gamma$ with β monic, γ must be epic." ⟦See [Burgess, Theorem 10].⟧

14.4 Paracompact Spaces

The reader should, at this stage, be familiar with the material of Sec. 10.2, Problems 122 through 125, Sec. 11.1, Problems 107 through 110.

DEFINITION 1. *Let X be a topological space; $\mathcal{N}$ is the set of all neighborhoods of Δ, the diagonal, in $X \times X$; T_N is the topology generated by $\mathcal{N}$ as in Sec. 11.1, Problem 108.*

An infinite intersection of members of $\mathcal{N}$ need not belong to $\mathcal{N}$; the following is a useful sufficient condition.

LEMMA 14.4.1. *Let F be a locally finite family of closed sets in X. For each $S \in F$, let G_S be an open neighborhood of S, and $H_S = (G_S \times G_S) \cup (\tilde{S} \times \tilde{S})$. Then $I = \bigcap \{H_S: S \in F\} \in \mathcal{N}$.*

For $x \in X$, choose a neighborhood U of x which meets only finitely many members of F, perhaps none. Then $I = I_1 \cap I_2$ where $I_1 = \bigcap \{H_S: S \not\pitchfork U\}$, $I_2 = \{H_S: S \text{ meets } U\}$. Now $I_1 \supset U \times U$. ⟦For $y, z \in U$, $(y, z) \in \tilde{S} \times \tilde{S} \subset H_S$ for all S involved in I_1.⟧ Hence I_1 is a neighborhood of (x, x). Also I_2 is a finite intersection of sets, each of which is open and contains (x, x). ⟦If $x \in S$, $(x, x) \in G_S \times G_S \subset H_S$; if $x \notin S$, $(x, x) \in \tilde{S} \times \tilde{S} \subset H_S$.⟧ So $I = I_1 \cap I_2$ is also a neighborhood of (x, x). ∎

Recall the definition of uniform cover, Section 11.2. Namely C is a uniform cover of the uniform space X if there exists a connector U such that $\{U(x): x \in X\}$ refines C. Now if $\mathcal{N}$, Definition 1, is a uniformity, we may speak of a uniform cover of $(X, \mathcal{N})$. But the definition makes sense even if $\mathcal{N}$ is not a uniformity, although the phrase "uniform cover" seems inappropriate in that case. Instead, "even cover" is used: thus a cover C of a topological space X is called *even* if there exists $U \in \mathcal{N}$ such that $\{U(x): x \in X\}$ refines C. We shall use the result of Problem 1. Keep in mind that if $\mathcal{N}$ is a uniformity, a cover of X is a uniform cover of $(X, \mathcal{N})$ if and only if it is an even cover of (X, T_N). We are going to see that if (X, T) is paracompact, $\mathcal{N}$ is indeed a uniformity, and $T = T_N$, so that the distinction between even and uniform covers will vanish.

LEMMA 14.4.2. *Every open cover C of a paracompact space X is even.*

Let F be a locally finite cover of X, consisting of closed sets, which refines C ⟦Sec. 10.2, Problem 125⟧. For each $S \in F$, choose $G_S \in C$ with $S \subset G_S$ and form H_S and I as in Lemma 14.4.1. Then $I \in \mathcal{N}$ by Lemma 14.4.1, and it remains to prove that $\{I(x): x \in X\}$ refines C. Let $x \in X$ so that $x \in S$ for some $S \in F$. But then $y \in I(x)$ implies $(x, y) \in I$, hence $(x, y) \in G_S \times G_S$. ⟦Certainly $(x, y) \in H_S = (G_S \times G_S) \cup (\tilde{S} \times \tilde{S})$ and $(x, y) \notin \tilde{S} \times \tilde{S}$ just because $x \notin \tilde{S}$.⟧ Thus $y \in G_S$. This proves that $I(x) \subset G_S$. ∎

THEOREM 14.4.1. *Let X be a paracompact space. Then $\mathscr{N}$ is a uniformity and the topology of X is T_N.*

Consulting Sec. 11.1, Problems 107 and 110, we see that we have only to prove that for each $U \in \mathscr{N}$ there exists $V \in \mathscr{N}$ with $V \circ V \subset U$. For each $x \in X$, U is a neighborhood of (x, x) so $U \supset G_x \times G_x$ for some open neighborhood G_x of x. By Lemma 14.4.2, $C = \{G_x : x \in X\}$ is an even cover of X; that is, there exists symmetric $V \in \mathscr{N}$ such that $\{V(x) : x \in X\}$ refines C. The proof is concluded by showing that $V \circ V \subset U$. Let $(x, y) \in V \circ V$. Then, since V is symmetric, there exists z with $x, y \in V(z)$. Now $V(z) \subset G_w$ for some w, thus $(x, y) \in G_w \times G_w \subset U$. ∎

COROLLARY 14.4.1. *A paracompact space is a normal uniform space.*

Normality follows from Theorem 14.4.1, and Sec. 11.1, Problem 107. ∎

For paracompact S, we shall refer to $\mathscr{N}$ as its *largest uniformity*. (See Corollary 14.4.4.)

We can now extend several compactness properties to paracompact spaces. We have seen that a compact regular space is a normal uniform space ⟦Theorem 5.4.7; Sec. 4.3, Problem 3; Theorem 11.4.5⟧. The extension of this result is Corollary 14.4.1. (Of course, the proof of Corollary 14.4.1 supplies an independent proof of the compactness result.)

Next recall (Theorem 11.2.2), that every open cover of a compact regular space is uniform (in the unique uniformity of the space).

COROLLARY 14.4.2. *Let a paracompact space have its largest uniformity. Then every open cover is a uniform cover.*

This is Lemma 14.4.2. ∎

The next result extends Corollary 11.2.1.

COROLLARY 14.4.3. *Let a paracompact space X have its largest uniformity and let Y be a uniform space. Then every continuous function from X to Y is uniformly continuous.*

This is proved in exactly the same way as Corollary 11.2.1. ∎

COROLLARY 14.4.4. *The largest uniformity of a paracompact space is larger than any other uniformity for the space.*

This follows from Corollary 14.4.3 when Theorem 11.2.1 is applied to the identity map. It is also trivial from Sec. 11.1, Problem 106. ∎

Certain uniform spaces have the property that their topology can be given by a complete uniformity. Let us call such a space *u-complete*. A complete uniform space is u-complete, so also are any realcompact space ⟦Sec. 11.4, Example 5⟧, and any metric space ⟦Sec. 11.4, Example 1⟧. If $\mathscr{N}$ is a uniformity for X, X is u-complete if and only if $\mathcal{N}$ is complete ⟦Corollary 11.3.1⟧, although X may be u-complete without having $\mathcal{N}$ as one of its uniformities. ⟦Let X be realcompact and not normal; for example, Sec. 8.6, Problem 8.⟧ We now see that every paracompact space is u-complete.

THEOREM 14.4.2. *A paracompact space X is complete in its largest uniformity.*

Let $\mathcal{F}$ be a Cauchy filter which is not convergent. For each $x \in X$, there is an open neighborhood G_x of x with $\tilde{G}_x \in \mathcal{F}$. ⟦By Lemma 11.3.1, x is not a cluster point of $\mathcal{F}$, hence for some $A \in \mathcal{F}$, $x \notin \bar{A}$. Then with $G_x = \tilde{\bar{A}}$ we have $\tilde{G}_x \supset A$.⟧ By Lemma 14.4.2, there exists symmetric $V \in \mathcal{N}$ such that $\{V(x): x \in X\}$ refines $\{G_x: x \in X\}$. Choose $A \in \mathcal{F}$ small of order V. Then, for any $x \in A$, $A \subset V(x) \subset G_y$ for some y, so that $G_y \in \mathcal{F}$. But this contradicts $\tilde{G}_y \in \mathcal{F}$. ∎

A deep theorem of A. H. Stone states that every semimetric space is paracompact. (See [M. E. Rudin].) Also K. Morita has proved that every regular Lindelöf space is paracompact. (See [Dugundji, Theorem 8.6.5].) A beautiful example of H. H. Corson shows that it is possible for $\mathcal{N}$ to be a complete uniformity for X without X being paracompact (see table in Appendix). Thus, even though $\mathcal{N}$ is a maximal uniformity it need not have the property that every open cover is uniform, since this implies paracompactness. (See [Kelley, Theorem 5.28].) (I am indebted to J. W. Taylor for this remark.) See also Sec. 14.5, Problem 201.

Paracompactness seems destined to play a role in analysis. As examples we cite a theorem on semicontinuity [Kelley, 5X], subordinate partitions of unity [Dugundji, 8.4.2], and an application of the latter to the construction of a Riemann metric [Auslander and McKenzie, p. 102].

Problems

1. In the definition of even cover, U may be chosen symmetric.
2. If every open cover of a uniform space is uniform, the space is complete. ⟦The proof of Theorem 14.4.2.⟧
3. A pseudocompact paracompact space is compact ⟦Theorem 14.4.2; Sec. 11.3, Problem 22⟧.

---

101. A topological group need not be paracompact ⟦Corollary 14.4.1; Sec. 12.1, Problem 122⟧.
102. Let X be a uniform space without isolated points such that $\mathcal{N}$ is a metrizable uniformity. Show that X is compact. ⟦If not, let $\{x_n\}$ be closed and $d(x_m, x_n) \geq \varepsilon$ for all m, n. If $\{U_n\}$ is any sequence in $\mathcal{N}$, choose a neighborhood V_n of x_n such that $V_n \not\supset U_n(x_n)$.⟧
103. The assumption on isolated points cannot be omitted from Problem 102. ⟦Discrete.⟧

104. Accepting the fact that **R** is paracompact, its largest uniformity is complete and not metrizable ⟦Problem 102; Theorem 14.4.2⟧.
105. Let a uniformity $\mathscr{U}$ be called a *u uniformity* (*u* uniformity*) for X if every $f \in C(X)$ (every $f \in C^*(X)$) is uniformly continuous. (Compare Sec. 4.3, Problem 203). Show that a u^* uniformity need not be a u uniformity. (Compare Sec. 8.5, Problem 116 which gives an implication for semimetrics. The same implication holds for topological groups; see [Comfort and Ross, Theorem 2.8].) ⟦Sec. 11.4, Problem 111.⟧
106. υ is a u uniformity; β is a u^* uniformity; and if X is paracompact, $\mathscr{N}$ is a u uniformity.
107. A u uniformity need not be complete, ⟦Sec. 11.4, Problem 102⟧, hence it need not have the property that every open cover is a uniform cover ⟦Problem 2⟧.
108. Let X, Y, F, Φ be as in Section 13.2. Let Y be paracompact and have its largest uniformity. Let F and all members of Φ be closed. Show that the topology of Φ convergence is larger than the Φ open topology ⟦Sec. 13.2, Problem 102⟧.

---

201. Is it possible to have inequality in Problem 108?

14.5 Ordinal Spaces

Suppose given an uncountable well-ordered set X with a last member; let 0 be its first member, and let Ω be the first member of X which has the property that $[0, \Omega] = \{x : x \leq \Omega\}$ is uncountable. We shall write $W = [0, \Omega) = \{x : x < \Omega\}$, and $W^+ = [0, \Omega]$. We make W^+ into a topological space by means of the *interval topology* which has as base the set of all intervals $(a, b) = \{x : a < x < b\}$, $[0, x)$, and $(x, \Omega]$. For $x < \Omega$ we write $x + 1$ for the first member of $(x, \Omega]$.

We first note that *W^+ is a compact Hausdorff space.* ⟦Let $x < y$; then $[0, x + 1)$ and $(x, \Omega]$ are disjoint neighborhoods of x, y. Next, let C be an open cover of W^+. Imitating Sec. 5.4, Example 1, let $S = \{x: [0, x]$ is covered by a finite subset of $C\}$. Assume that $S \neq W^+$ and let t be the first member of W^+ which is not in S. Then t belongs to some $G \in C$. Since G is open, $G \supset (a, b]$ with $a < t \leq b$. Now $a \in S$ by definition of t, and so also $t \in S$ since we may adjoin G to the subset of C which covers $[0, a]$. This is a contradiction. Thus, actually, $S = W^+$, and since $\Omega \in S$, compactness is proved.⟧

Thus, also *W is a locally compact Hausdorff space*; *moreover, it is not compact*. To prove this it is sufficient to show that Ω is not an isolated point of W^+ since this implies that W is not closed. ⟦If Ω were an isolated point there would be x such that $(x, \Omega] = \{\Omega\}$. Since $x < \Omega$, $[0, x]$ is countable, and $[0, \Omega] = [0, x] \cup \{\Omega\}$ is also countable.⟧

It was no accident that the notation W^+ was chosen, indeed W^+ is the one-point compactification of the locally compact T_2 space W.

The following result is basic in discussions of W and W^+.

LEMMA 14.5.1. *Every sequence in W has a least upper bound in W.*

If $\{x_n\}$ has an upper bound; that is, a point t such that $t > x_n$ for all n; it surely has a least upper bound; namely, the first member of the set of upper bounds; so we merely have to show the existence of an upper bound. But if no such upper bound exists, we have $W = \bigcup \{[0, x_n]: n = 1, 2, \ldots\}$, a countable union of countable sets. ⟦Each $[0, x_n]$ is countable, by definition of Ω.⟧ Thus W is countable; but this is false. ∎

We now show that W *is countably compact*. Let S be a countably infinite set in W. Let t be the first member of W satisfying $t > s$ for infinitely many $s \in S$. ⟦The set of upper bounds of S is not empty by Lemma 14.5.1.⟧ Then t is an accumulation point of S. ⟦If not, there exist a, b, with $a < t < b$ such that $(a, b) \not\pitchfork S$. But then $t > s \in S$ implies $a > s$ (with possibly one exception) so that $a > s$ for infinitely many $s \in S$. This contradicts the definition of t.⟧

Since W is not compact, it follows that W and W^+ *are not metrizable* ⟦Theorem 7.2.1⟧ and W *is not realcompact* ⟦Sec. 8.6, Problem 1⟧.

However W *is normal.* ⟦Let A, B be disjoint closed sets. The result will follow if we can show either A or B compact, since we may cite Theorem 5.4.6 and Sec. 4.1, Problem 7. So suppose that neither A nor B is compact. It follows that they are both cofinal, since for example, if $A \subset [0, x]$, A is a closed subset of a closed subset of W^+, hence compact. Thus we may choose $x_1 \in A$, $x_2 \in B$ with $x_2 > x_1$, $x_3 \in A$ with $x_3 > x_2$, and in general, $x_{2n-1} \in A$ $x_{2n} \in B$ with $x_{2n+1} > x_{2n} > x_{2n-1}$. Let t be the least upper bound of $\{x_n\}$ ⟦Lemma 14.5.1⟧. As in the proof of countable compactness, t is a cluster point of $\{x_n\}$ hence an accumulation point of both A, B. Since A, B are closed, they both contain t, which is impossible.⟧

Let us now examine $f \in C(W)$. For $x \in W$, $K_x = f[[x, \Omega)]$ is a compact subset of $\mathbf{R}$. ⟦Exactly as before, $[x, \Omega)$ is countably compact, and a countably compact set in $\mathbf{R}$ is compact.⟧ Since $\{K_x\}$ has the finite intersection property, its intersection is not empty, so we may choose $y \in \bigcap K_x$. For each positive integer n, let $A_n = \{x: |y - f(x)| \geq 1/n\}$. Now $A_n, f^{-1}[\{y\}]$ are disjoint closed sets of which the latter is not compact. ⟦It is not closed in W^+.⟧ Hence A_n is compact. ⟦See the proof that W is normal.⟧ Thus $A_n \subset [0, b_n]$ for some b_n. Our conclusion about f is that *there exists $a \in W$ such that $f(x) = y$ for all $x \in [a, \Omega)$.* ⟦Let a be the least upper bound of $\{b_n\}$ (Lemma 14.5.1). If $x \geq a$, we have, since $x > b_n$, $|y - f(x)| < 1/n$ for all n.⟧ It follows immediately that W is C-embedded in W^+ (this is the same as C^*-embedded since W is pseudocompact) and so, that $W^+ = \beta W$.

Finally, we note that *W is not paracompact.* This is immediate from the fact that W is countably compact and not compact; taking account of Sec. 14.4, Problem 3.

REMARK. The uncountable well-ordered set X mentioned at the beginning of this section plays no role. If some other set X' is chosen and W' formed in the same way as W, then W, W' can be put in one-to-one order-preserving correspondence. To prove this let $S = \{x \in W: \exists x' \in W' \text{ with } [0, x], [0, x']$ able to be placed in one-to-one order-preserving correspondence $u\}$. Then $S = W$. ⟦If not we could choose the first member y of $W \setminus S$, then $S = [0, y)$, $u[S] \neq W$ since $u[S]$ is countable and we may take y' to be the first member of $W' \setminus u[S]$. Defining $u(y) = y'$ yields $y \in S$.⟧ Now $u(W) = W'$. ⟦$u(W)$ is uncountable, thus for each $x' \in W'$, there exists $a \in W$ with $u(a) > x'$. Then $x' \in [0', u(a)] = u[0, a]$ so that $x' \in u[W]$.⟧ Thus also W, W' are homeomorphic. The (unique) space W is referred to as the *first uncountable ordinal number* and is sometimes designated by the symbol Ω. More generally, the interval $[0, a) \subset W$ is designated by a. Thus $a + 1$ is identified with $[0, a] = [0, a + 1)$.

Problems

★1. How do we know that there exists an uncountable well-ordered set? ⟦Theorem 10.2.1.⟧
2. For every $x \in W$, $x + 1$ is an isolated point.

★3. For every $x \in W$, (x, Ω) is not empty.
4. W, W^+ are not second countable ⟦Theorem 10.1.1⟧.
5. W^+ is not first countable ⟦Theorem 8.3.2⟧.
6. W is first countable. ⟦For $x \in W$, $[0, x) = \{a_n\}$ consider
$$\{(a_n, x]: n = 1, 2, \ldots\}.⟧$$
7. ω^+ is (homeomorphic with) a subspace of W. ⟦Consider $\{0, 0 + 1, 0 + 1 + 1, \ldots\}$, and its least upper bound.⟧
8. Sequential convergence in W and W^+ is not trivial; indeed every increasing sequence is convergent ⟦Lemma 14.5.1⟧.
9. W, W^+ are not extremally disconnected ⟦Theorem 14.1.5⟧.
10. W has a unique uniformity. ⟦See Sec. 11.5, Problem 108.⟧
11. W is sequentially compact ⟦Theorem 7.1.3⟧ and so is W^+. ⟦Use Lemma 14.5.1.⟧ Thus the Stone–Cech compactification of a normal space can be sequentially compact.

101. Let $f: W \to W^+$. We may consider f to be a net $(f_w: W)$. Suppose that $f_w \to \Omega$. Show that $f_w \geq w$ for some w.
102. Fix $a \in W^+$. Let b be the first member of W^+ satisfying $b \geq x$ for all $x < a$. Show that either $b = a$ or $b + 1 = a$. ⟦If $b + 1 < a$, we would have $b \geq b + 1$.⟧
103. Show that a is an isolated point of W^+ if and only if $a = b + 1$ for some b ⟦Problem 102⟧.
104. Any nonisolated point of W^+ is called a *limit ordinal*. Show that the set of limit ordinals in W is cofinal. ⟦For any x, consider the least upper bound of $\{x, x + 1, x + 1 + 1, \ldots\}$. Another proof stems from Problems 101 and 103. If the result is false, the net $(w - 1: W)$ converges to Ω, contradicting Problem 101.⟧
105. Every compact set in W is included in some $[0, a]$. ⟦Otherwise it would have Ω as an accumulation point in W^+.⟧
106. W is not σ-compact. ⟦Let $S = \bigcup K_n$. Each $K_n \subset [0, a_n]$ by Problem 105. Thus $S \subset [0, a]$ by Lemma 14.5.1.⟧
107. W is not Lindelöf ⟦Sec. 8.1, Problem 124⟧.

---

201. For W, $\mathcal{N}$ (Sec. 14.4, Definition 1) is a uniformity.
202. For W, the uniformity $\mathcal{N}$ is not complete ⟦Sec. 11.3, Problem 21⟧.

14.6 The Tychonoff Plank

Let the space $\omega^+ \times W^+$ be denoted by the symbol P^+, and let $u = (\infty, \Omega)$, a point of P^+. The space $P = P^+ \setminus \{u\}$ is called the *Tychonoff plank*, in honor of A. Tychonoff. It is clear that P^+ is a compact Hausdorff space ⟦Theorem 7.4.1⟧ and that P is a dense open subspace. Thus P is locally compact, and P^+ is its one-point compactification. Unlike W, *P is not countably compact*, indeed P contains a sequence converging to u. ⟦Namely $\{(n, \Omega): n = 1, 2, \ldots\}$. Refer to Theorem 6.4.1.⟧

We are now going to show that *P is C-embedded in P^+*. Let $f \in C(P)$. In particular, f is continuous on $\{n\} \times W$ for each $n \in \omega$, and so by the discussion in Section 14.5, there exists $a_n \in W$ such that f is constant on $\{n\} \times [a_n, \Omega)$, hence, also, on $\{n\} \times [a_n, \Omega]$. Repeating this argument for $\{\infty\} \times W$ yields $a_0 \in W$ such that f is constant, say, $f = \alpha$, on $\{\infty\} \times [a_0, \Omega)$. Now define $f(u) = \alpha$, and f is extended to P^+. ⟦Choose $a \in W$ with $a > a_n$ for $n = 0, 1, 2, \ldots$, by Lemma 14.5.1. Let (x_δ, y_δ) be a net converging to u. We may assume $y_\delta > a$ for all δ since $y_\delta > a$ eventually. Now if $x_\delta \neq \infty$, we have $f(x_\delta, y_\delta) = f(x_\delta, a) \to f(\infty, a) = \alpha$, while if $x_\delta = \infty$, we have $f(x_\delta, y_\delta) = \alpha$. Thus $f(x_\delta, y_\delta) \to \alpha = f(u)$.⟧ Since P is C-embedded in a compact space, *P is pseudocompact, and $P^+ = \beta P$.*

The chief value of P is as a source of counterexamples; the relevant properties are given in the problems. Note that it supplies us with the only example in this book of a locally compact Hausdorff space which is not normal; equivalently, a normal space (P^+) with a nonnormal open subspace ⟦Problem 4⟧.

Problems on the Tychonoff Plank P

★1. u is a sequential limit point of P. ⟦Namely, $(n, \Omega) \to u$.⟧

2. $\omega \times \{\Omega\}$ is a closed discrete subspace of P. ⟦It is a sequence converging to u.⟧

3. The subspace of Problem 2 is not C^*-embedded in P. ⟦If it were, it would be C^*-embedded in $P^+ = \beta P$, hence in $\omega^+ \times \{\Omega\}$. This would make ω C^*-embedded in ω^+.⟧

4. The plank is not normal. ⟦Many proofs are available: (i) Sec. 8.5, Problem 103; (ii) Problem 1 and Theorem 8.3.2; (iii) Problems 2, 3, and Theorem 8.5.3; (iv) Problem 201.⟧

5. The plank is not first countable. ⟦$\{1\} \times W$ has $(1, \Omega)$ as an accumulation point, but not a sequential limit point since $\{1\} \times W^+$ is homeomorphic with $W^+ = \beta W$. Apply Theorem 8.3.2.⟧ (The same proof shows that P is not closure-sequential.)

---

101. The plank is neither realcompact ⟦Sec. 8.6, Problem 1⟧, nor paracompact ⟦Corollary 14.4.1⟧.

102. The plank has a unique uniformity. ⟦See Sec. 11.5, Problem 108.⟧

---

201. Show that $\{\infty\} \times W$ and $\omega \times \{\Omega\}$ are disjoint closed sets in P which cannot be separated by open sets.

14.7 Completely Regular and Normal Spaces

Complete regularity is a very tractable condition with convenient properties. For example, it is hereditary and productive, and a T_1 space is completely regular if and only if it is a subspace of a cube ⟦Theorem 8.2.2⟧. It is a sufficient condition for the existence of a uniformity ⟦Theorem 11.4.5⟧, and of the Stone–Cech compactification. Very important classes of spaces must be completely regular, namely the uniform spaces, with topological groups and topological vector spaces as notable special cases. In contrast, normality and T_4 are difficult to treat. They are neither productive ⟦Sec. 6.7, Example 3⟧, nor hereditary ⟦Sec. 14.6, Problem 4; or any nonnormal Tychonoff space⟧. It is difficult to tell whether a given space is normal; for example, it is unknown whether $X \times [0, 1]$ is normal if X is a T_4 space. Topological groups and vector spaces of a very special kind need not be normal ⟦Sec. 6.7, Problem 203; Sec. 12.1, Problem 122⟧, even the weak topology of a Banach space! ⟦See [Corson (a), p. 12].⟧

Of course a T_4 space will have all the properties of a $T_{3\frac{1}{2}}$ space, and more; but this fact is cold comfort if we wish to apply a theorem to a space not known to be T_4. What is helpful is that there is a body of techniques and results designed to yield conclusions from the $T_{3\frac{1}{2}}$ assumption which are analogous to, and as useful as, those obtainable from the assumption T_4. Sometimes, of course, a more refined argument actually yields the *same* conclusion. (See Example 3.)

EXAMPLE 1. If X is T_4, no point of $\beta X \setminus X$ is a sequential limit point of X ⟦Theorem 8.3.2⟧. This is false for $T_{3\frac{1}{2}}$ ⟦Sec. 14.6, Problem 1⟧, but we have the consolation result that βX cannot be first countable at any point $\beta X \setminus X$ ⟦Sec. 10.2, Problem 112⟧.

EXAMPLE 2. Every closed subspace of a normal space is C-embedded ⟦Theorem 8.5.3⟧, and this fails for Tychonoff spaces ⟦Sec. 8.5, Problem 102⟧. However, certain types of subspaces are C-embedded; for example, compact ones ⟦Corollary 8.5.1⟧, certain sequences ⟦Theorem 10.2.4⟧. In a normal space, disjoint closed sets are completely separated ⟦Theorem 4.2.11⟧; in a Tychonoff space, disjoint closed sets are completely separated if one of them is compact. ⟦Completely separate the compact one from the closure in βX of the other one.⟧

EXAMPLE 3. Suppose that X has an infinite locally finite disjoint family of open subsets. If X is normal we see immediately that X is not pseudocompact. ⟦There is an infinite discrete closed subspace, to which we may apply Theorem 8.5.3.⟧ We can obtain exactly the same result for completely regular spaces, but it requires a special argument ⟦Theorem 10.2.5⟧.

EXAMPLE 4. Let A, B be disjoint closed sets in a $T_{3\frac{1}{2}}$ space X and let $\bar{A}$, $\bar{B}$ be their closures in βX. If X is normal, $\bar{A} \not\cap \bar{B}$. ⟦Let $f = 0$ on A, 1 on B; such f exists by Theorem 8.5.3. Then $f = 0$ on $\bar{A}$, 1 on $\bar{B}$.⟧ However, these closures may meet if X is not normal. ⟦If they do not meet, they are completely separated in βX, and so A, B are completely separated in X.⟧ A result, true for Tychonoff spaces, which serves the same purpose as the above, is this: If A, B are disjoint zero-sets, $\bar{A} \not\cap \bar{B}$. ⟦Say $A = f^{\perp}$, $B = g^{\perp}$. Let $u = |f|/(|f| + |g|)$. Then $u = 0$ on A, 1 on B. The extension of u to βX is 0 on $\bar{A}$, 1 on $\bar{B}$.⟧

A systematic technique for replacing T_4 by $T_{3\frac{1}{2}}$ is to discuss zero-sets instead of closed sets. In a normal space, disjoint closed sets are completely separated ⟦Theorem 4.2.11⟧; in an arbitrary space, disjoint zero-sets are completely separated, ⟦if $f, g \in C(X)$ with $f^{\perp} \not\cap g^{\perp}$, then $|f|/(|f| + |g|)$ separates $f^{\perp}$, $g^{\perp}$⟧, and a completely regular space is well endowed with zero-sets; indeed every closed set is an intersection of zero-sets. ⟦Let F be closed and $x \notin F$. Let f be a real continuous function with $f = 0$ on F, $f(x) = 1$. Then $F \subset f^{\perp}$, $x \notin f^{\perp}$.⟧ This program requires the replacement of filters by z filters; these are appropriate collections of zero-sets; an example is given in Sec. 8.6, Definition 1 and Problem 117. We refer to [Gillman and Jerison], in which an important branch of mathematics is exposed, and in which the setting is that of Tychonoff spaces.

Note: The existence of a T_4 space X for which $X \times [0, 1]$ is not normal is consistent with set theory. See MR *35* #6564 (Jech); MR *17*, p. 391 (M. E. Rudin); and MR *13*, p. 264 (C. H. Dowker).

Appendix

Tables of Theorems and Counterexamples

Introduction

Each table is headed by a property which is the conclusion of all theorems. One uses the tables as follows: suppose it is required to know whether every compact Hausdorff space is normal. The conclusion is "normal," hence one looks at Table 22, NORMAL. (The tables are in alphabetical order.) This table has two lists; in the list headed *Implied by*, a short search reveals the entry "K.T_2 θ5.4.7." Recognizing the standard abbreviation "K" for "compact" (see LIST OF EQUIVALENCES, below), we conclude that every compact Hausdorff space is normal, the proof given in Theorem 5.4.7. As another example of the use of the tables, suppose it is required to know whether every first countable, compact Hausdorff space is metrizable. The conclusion is "metrizable" hence one looks at Table 21, METRIZABLE, then, as suggested there, Table 30, SEMIMETRIZABLE. In the second column of this table, under *Not implied by*, is given the entry FC.K.T_2.Σ.Γ [Kelley, 5J, 5M]. This indicates that a first countable, compact, Hausdorff, separable, connected space need not be metrizable; *a fortiori*, a first countable, compact Hausdorff space need not be metrizable. (The relevant counterexample is given in the cited reference.) A third use of the tables is to discover whether a specific space has a property; for example, to see if **Q** (the rationals) is hemicompact, note that Table 13, HEMICOMPACT, has a list under *Spaces*; in this list, **Q** appears, followed by "no 8.1 #112." This means that **Q** is not hemicompact as proved in Sec. 8.1, Problem 112. (A hint is given.) This entry "yes" after **R** means that **R** is hemicompact.

The entry "G $\subset$ B" in the first column of Table 1, BAIRE, means "an open subset of a Baire space is itself a Baire space"; the entry "r(B)" means "a retract of a Baire space need not be a Baire space." The entry $(\text{B}.T_{3\frac{1}{2}})^2$ in the second column means "$X \times X$ need not be a Baire space if X is a $T_{3\frac{1}{2}}$ Baire space."

The entry "<B" in the second column of Table 1 means: "if a topological space can be given a larger Baire topology, it need not, itself, be a Baire space."

References such as "Note 1" refer to the Notes at the end of the Tables.

EXAMPLE. Must a separable T_4 space be Lindelöf? Table 18 says no, with the remark that L.T_4 implies PK, so it is sufficient that Σ.T_4 does not imply PK. Table 23, the PK table, verifies this with an appropriate reference, concluding the investigation. It follows that a separable T_3 or $T_{3\frac{1}{2}}$ space need not be Lindelöf; however these entries are given explicitly because the appropriate references are easier.

LIST OF EQUIVALENCES

aa	almost all	H	homogeneous
AC	absolutely closed	Hed	hereditarily extremally disconnected
AMM	American Mathematical Monthly	HK	hemicompact
AMS	American Mathematical Society	HL	hereditarily Lindelöf
ASC	absolutely sequentially closed	HRK	hereditarily real compact
$\aleph_0$	countably infinite	HΣ	hereditarily separable
B	Baire	I	indiscrete
b	bounded	i	invertible
β	Stone–Cech compactification	id	identity map
bγ	boundedly complete	ID	infinite dimensional
C	continuous; Corollary	il	inverse limit of
CC	convex closure	ip	isolated points
CI, CII	of first (second) category	**J**	the irrationals
Cγ	Cech complete	k	*k* space
cfi	continuous closed image of	k′	*k′* space (see Note 11)
cfgi	continuous closed open image of	K	compact
cg	closed graph	KC	*KC* space
cgi	continuous open image of	KO	compact open topology
ch	characteristic	L	Lindelöf; lemma
CH	assuming the continuum hypothesis	LC	locally convex topological vector space
ci	continuous image of	LCK	locally countably compact
CK	countably compact	Lf	locally finite
cl	closure	LK	locally compact
closed	F	Lγ	locally complete
coc	cocountable	LΓ	locally connected
cof	cofinite	LM	locally metrizable
compact	K	LO	linearly ordered
complete	γ	loc	locally
connected	Γ	LpK	locally peripherally compact

cor	corollary
countably infinite	$\aleph_0$
cpi	continuous perfect image
CPK	countably paracompact
CR	completely regular
CS	closure sequential
c*uc	every ϕ in $C^*(X)$ is uc
cuc	every $\phi \in C(X)$ is uc
$C(X)$, $C^*(X)$	see pp. 54 and 59
$C_0(X)$	continuous functions vanishing at ∞
d	dense
D	discrete
δ	dispersion character
Δ	diagonal
dim	dimension
D-Σ	D-separable
E	Example
ed	extremally disconnected
EP	extreme points
ϕ	a function
f	finite
F	closed, or closed subspace
FC	first countable
FD	finite dimensional
G	open
G_δ	every point is a G_δ
Γ	connected
γ	complete
$\gamma(X)$	the completion of X
h	homeomorphic image of
LΣ	locally separable
LT_2	locally T_2
L2C	locally second countable
lsc	lower semicontinuous
LSK	locally sequentially compact
M	metrizable
MA	Mathematische Annalen
MI	maximal ideal in $C(X)$
MK	metacompact
MR	Mathematical Reviews
MS	Moore space
min	minimal
N	normal
n	normed space
nhd	neighborhood
nmc	nonmeasurable cardinal
Nγ	von Neumann complete
nip	nonisolated point
NS	*NS* space
open	G
OC	every disjoint family of open sets is countable
Π	product (of)
PAMS	Proceedings of the American Mathematical Society
PK	paracompact
PPK	pointwise paracompact
ψ	pseudo
ψB	pseudobounded
ψK	pseudocompact

ψF	pseudofinite
$\Pi(P)$	product of spaces each of which has property P
Q	the rationals
q	quotient of
R	the reals
R	regular
r	retract of
Rx	reflexive
res	residual
RHO	see Sec. 2.6, Example 3
RK	realcompact
Σ	separable
S	sequential
Sγ	sequentially complete
se	simple extension of
seq	sequential(ly)
SF	sequentially closed
SG	subgroup
SK	sequentially compact
second countable	2C
SM	semimetric
Sn	seminormed
SR	semiregular
σK	σ-compact
σLf	σ-locally finite
sym	symmetric
2C	second countable
θ	theorem
TB	totally bounded
TD	totally disconnected
TG	topological group
TVS	topological vector space
$T_{2\frac{1}{2}}$	2 points have disjoint closed neighborhoods
U	Uniform (space) or uniformity
u	uncountable
uc	uniformly continuous
uf	ultrafilter
US	*US* space
usc	upper semicontinuous
von Neumann complete	Nγ
w	weight
w^*	weak* topology
$w(A)$	weak topology by A
Z	zero-set
ZD	zero-dimensional
$< P$, $(> P)$	having a topology smaller (larger) than some topology with property P
+	one-point compactification

TABLE 1. BAIRE

Implied by	*Not implied by*
γ.SM θ9.3.2	$(\mathrm{B}.T_{3\frac{1}{2}})^2$ [Oxtoby]
CII.TG 12.1 #115	CII $\subset$ B ⟦**Q** $\cup$ [0, 1]⟧
Cγ 9.3 #120	ci(B) ⟦consider D⟧
cof.u 9.3 #112	K.KC 9.3 #113
K.T_2 θ9.3.6	d $\subset$ B ⟦**Q** $\subset$ **R**⟧
K.R θ9.3.6	F $\subset$ B 9.3 #13
C(HK.k) θ9.3.2	$\mathrm{G}_\delta \subset$ B 9.3 #118
CK.$T_{3\frac{1}{2}}$ 7.1 #114	LK.K.T_1 9.3 #113
(d.G_δ) $\subset$ B 9.3 #118	$<$ B ⟦consider D⟧
f 9.3 #14	$>$ B 6.2 E3
factor in B.Π	TVS. $>$ B 12.4 #205
G $\subset$ B 9.3 #13	r(B) 9.3 #13
$\mathrm{G}_\delta \subset$ (γ.SM) 9.3 #118	
$\mathrm{G}_\delta \subset$ K.T_2	*Spaces*
LCK.R 9.3 #3	**J** yes 9.1 #205
LK.T_2 θ9.3.6	**Q** no 9.3 E1
LK.R θ9.3.6	**R** yes θ9.3.2
LK.SM 9.3 #4	(**R**, RHO) yes
Lγ.SM 9.2 #102	C(**R**) yes θ13.2.4 and 13.2 #201
Π(γ.M) [Bourbaki(a), Chapter 3, p. 4]	C(**Q**) no
Π(2C.B) [Oxtoby]	LF space no [Kelley and Namioka, 22C]
ψK.$T_{3\frac{1}{2}}$ [Bourbaki(b), IX, 5 #10]	
res $\subset$ B 9.3 #117	
each G $\in$ CII θ9.3.3	
υX is B [Comfort(b), p. 115]	
υX if X is B [Comfort(b), p. 115]	
$T_{3\frac{1}{2}}.\Pi(\psi\mathrm{K})$ MR*31* #6209	

TABLE 2. CATEGORY TWO

Implied by†	*Not implied by*
B θ9.3.4	K.KC ⟦$\mathbf{Q}^+$⟧
$\aleph_0$.min T_2 MR*29* #579	γ.U 114 #118
$\aleph_0$. > (T_1.CII) 9.3 #204	$C(X)$ with $X\gamma$.M ⟦X = **J**⟧
cof.u	ci(CII) ⟦consider D⟧
min T_3 MR*29* #579	F ⊂ CII 9.3 #13
	G ⊂ CII 9.3 #103
Spaces	min T_2 MR*32* #6392
J yes 9.1 #205	n. barreled [Wilansky(c), pp. 45, 53]
Q no	<(CII) ⟦consider D⟧
R yes θ9.3.2	>(CII) Note 1
RHO yes	TVS. > CII 12.4 #205
C(**R**) yes θ13.2.4 and 13.2 #201	CII in itself 9.3 #6
C(**Q**) no	
	Denied by
	$\aleph_0.\Gamma.T_1$ 9.3 #104
	$\aleph_0.T_1$. ($\delta > 1$)
	cof. $\aleph_0$
	σK.ID.TVS Note 8
	T_2.TVS of $\aleph_0$ dim [Wilansky(a), 10.6, Corollary 3]
	$C(T_{3\frac{1}{2}}.\psi$K.non-K) Note 36

† See Table 1.

TABLE 3. COMPACT

Implied by	*Not implied by*†
AC.$T_{3\frac{1}{2}}$ 8.3 #116	AC.CK.T_2 8.3 #207
AC.T_3 [Alexandroff and Urysohn, p. 47]	CK.LK.Σ.$T_{3\frac{1}{2}}$ 14.1 E2
∀ F.F is AC [M. H. Stone]	CK.SK.LK.FC.T_4 ⟦W⟧
CC(K) ⊂ bγ.LC [Kelley and Namioka, 13.4]	CK ⊂ γ.TVS 12.4 #103
CK.$\aleph_0$ 5.4 #2	CK.TG [Kister]
CK ⊂ C(K, K.M) θ13.2.4	cuc.M 4.2 #113 and #115
CK.L θ5.4.1	HK.RK.σK.$\aleph_0$.M ⟦$\boldsymbol{\omega}$⟧
CK.γ 11.3 #21	K ∩ K in US 5.4 #115; X ed
CK.LK.TG 12.2 #127	K ∩ K in R.N. Note 2
CK.PK 14.4 #3	$\overline{\text{K}}$ in T_1 5.4 #114
CK.RK 8.6 #1	il(K) [Bourbaki(b), Part 1, p. 142]
CK.σK θ5.4.1	L ⟦**R**⟧
CK.SM θ7.2.1; 11.4 #124	min T_0
ci(K) θ5.4.4	min T_3 MR*27* #2949
⊂ cof 5.4 #14	min T_2 5.4 E4
$C^*(X)$ Σ.$T_{3\frac{1}{2}}$ 12.4 #117; [Wilansky(a), p. 269]	ψK.LK.Σ.$T_{3\frac{1}{2}}$ 14.1 E2

† See Table 7.

TABLE 3. COMPACT *(continued)*

Implied by	*Not implied by*†
cuc.LK.TG. not D [Kister]	ψK.σK.2C.T_2 5.2 E7, and #113
F ⊂ K θ5.4.2	ψK.TG.T_2 [Comfort and Ross, p. 485]
factor in a KΠ 6.7 #102	ψK.T_4 14.5
ϕ^{-1}[K], ϕ perfect [Dugundji, 11.5.3]	Π(K) with box 7.4 #104
γ ∀ U.$T_{3\frac{1}{2}}$ 11.4 E2	se(K) θ5.4.10
$\gamma(X)$, XTB 9.1 #2, 11.5 #1	K ⋁ K 6.2 #107
i.∃ (K.$\overline{\mathrm{G}}$) [Ryeburn]	TG.non-D.CK.cuc [Kister]
⋂ K in T_2 or KC 5.4 #115	every G cover is U C14.4.2
$\overline{\mathrm{K}}$ ⊂ R 5.4 #9	∃ unique U.LK.CK.T_4 14.5 #10
LK.R.i [Ryeburn]	$>$ K 6.2 #107
M. all Mγ or b AMM *58*, p. 389	
min P,$P = T_1, T_{3\frac{1}{2}}, T_4$ or LK.T_2 MR*27* #4204	*Spaces*
min $T_2.T_{2\frac{1}{2}}$ [Bourbaki(b), Part 1, p. 146]	$[a, b]$ yes 5.4 E1
min $T_3.T_{3\frac{1}{2}}$ MR*27* #2949	long line no [Hocking and Young, p. 56]
MK.CK.T_2 [Dugundji, θ11.3.3]	W no 14.5
Nγ.TB ⟦trivial⟧	W^+ yes 14.5
ψK.$\aleph_0.T_3$ θ8.5.3	
ψK.γ 11.3 #22	
ψK.HK.LK.T_2 10.2 #118	
ψK.HK.LK.KC 8.1 #117	
ψK.HK.FC.T_2 8.1 #117	
ψK.L.T_3 8.6 #1	
ψK.LK.TG 12.2 #127	
ψK.NS	
ψK.PK 14.4 #3	
ψK.RK 8.6 #1	
ψK.σK.LK.T_2 10.2 #118	
ψK.σK.T_3 8.1 #10; 8.5 #103; θ5.3.5	
ψK.SM 8.5 #103	
Π(K) θ7.4.1	
q(**R**, +) by F.SG [Kelley, 1J]	
r(K) θ5.4.4	
SK ⊂ C(K,K.M) θ13.2.4	
SK.SM θ7.2.1	
se(K) by S with $\tilde{\mathrm{S}}$K [Levine]	
TB.γ θ11.3.7	
TG.LK.cuc.non-D MR*24* #A3226	
K ∪ K 5.4 #10	
$<$ K 5.4 #4	
b.F ⊂ convex.γ.LK.M MR*36* #4520	
$T_{3\frac{1}{2}}.C^*(X)\,\Sigma$ Note 15	
every subbasic cover reducible to f [Kelley, 5.6]	
$\beta X = \beta\omega \setminus \omega$ (CH) [Fine and Gillman(a), θ4.6]	
$\beta X = \beta\mathbf{R} \setminus \mathbf{R}$ (CH) [Fine and Gillman(a), θ4.6]	
U. no ip. {nhds of Δ}M 14.4 #102	

TABLE 4. COMPLETE

Implied by	*Not implied by*
c*uc.SM 9.1 #114; 11.4 #102	bγ.TVS Note 3
Cγ.SM 9.1 #208	CK.$T_{3\frac{1}{2}}$ 11.3 #21
$\bar{\gamma}$ 11.3 #106	cuc.U 11.4 #102
γ + K in TVS [Wilansky(a), 10.2 Fact (i)]	D.M 9.1 E2
K.U θ11.3.7	LK.L.B.M ⟦consider (0, 1)⟧
$C^*(X)$ 9.1 E4	Nγ.TVS Note 3
$C(k)$ 13.2 #201	n.CII [Wilansky(a), 7.5 #18]
f.U θ11.3.7	ψK.$T_{3\frac{1}{2}}$ 11.3 #21
F ⊂ γ 11.3 #8	q(γ.TVS) by F 13.2 #110
factor in γ.Π 11.4 #119	Rx.TVS [Schaefer, p. 148]
LF space [Kelley and Namioka, 22C]	Sγ.FC 11.5 #111
Nγ.TB ⟦⇒K⟧	Sγ.2C MR*37* #3512
Nγ.SM 11.5 #110	C(RK) Note 4
Πγ 11.4 #7	γ + γ in **R** [Wilansky(a), 10.2 Fact (i)]
PK.nhds of Δ θ14.4.2	γ ∨ γ [Wilansky(a), 7.5 E8]
some q 9.1 E5; 12.3 #105; θ12.3.5	> γ [Wilansky(a), 7.5 E8]
r(γ) 11.4 #119	< γ ⟦D⟧
Rx.n [Wilansky(a), 7.2 Fact (ii)]	
Sγ.SM θ9.1.3	*Spaces*
TG.LK 12.2 #127	**R** yes 9.1 E3
υ if X is RK 11.4 E5	
uc.$h^{-1}(\gamma)$ θ11.3.3	
> equiv.γ.U C11.3.1	
every G cover is U 14.4 #2	

TABLE 5. COMPLETELY REGULAR
(See also Table 37)

Implied by	*Not implied by*
$\aleph_0$.R θ5.3.5	T_3 [Dugundji, 7.7 E3]
factor in (CR.Π) 6.7 #102	N 4.3 #3
K.R or K.T_2 θ5.4.7	se CR 6.2 E2
KO to CR 13.2 #112	>CR 6.2 E2
LK.R 8.1	< CR ⟦D⟧
Π(CR) θ6.7.3	Γ.LΓ.γ.Moore MR*20* #277
PK C14.4.1	*Denied by*
R.L. θ5.3.5	$\aleph_0$.Γ.T_2 5.2 E7
R.N 4.3 #3	
R.2C θ5.3.2	
se CR by F [Ryeburn]	
SM θ4.3.3	
$T_{3\frac{1}{2}}$	
TG θ12.1.4	
TVS 12.4 #1	
U θ11.5.2	
$w[C(X)]$ θ6.7.4	
w(CR) θ6.7.2	
ZD 4.3 #103	
⋁(CR) θ6.7.1 or 6.7 #115	
⊂ (CR) 4.3 #2	
certain ⋃ Note 5	
every lsc is ⋁C [Dugundji, p. 159]	

TABLE 6. CONNECTED

(See also Table 8)

Implied by	*Not implied by*
$\beta(\Gamma)$ θ5.2.3	$(X/S)\Gamma$ 12.3 #109
$\beta X\,\Gamma$ 8.3 #104	$\Gamma \cup \Gamma$ θ5.2.1
ci(Γ) θ5.2.2	$\Gamma \vee \Gamma$ 6.7 #118
cof ∞	$>\Gamma$ [[D]]
factor in ($\Gamma.\Pi$) 6.7 #102	
Γ^+ θ5.2.3	*Spaces*
$\bar{\Gamma}$ θ5.2.3	$\mathbf{R}^n$ yes 5.2 E4; 12.4 #109
I	$\mathbf{Q}$ no 5.2 #2
K-embedded $\subset \Gamma$ 8.5 #13	$\mathbf{Q}^+$ yes 8.1 #127
$\Pi\Gamma$ θ6.6.3	$\mathbf{R}^2\backslash\aleph_0$ yes
qΓ θ5.2.2	conv. $\subset$ R^2 yes
se(Γ) by d.Γ [Levine, θ9]	
TVS 12.4 #109	
$G \subset \beta Y.(G \cap Y)\Gamma$ MR*20* #2688	
$<\Gamma.$ 5.2 Cl	
$\phi'[\Gamma]$, $\phi: \mathbf{R} \to \mathbf{R}$ AMM *75*, p. 887	

TABLE 7. COUNTABLY COMPACT
(See also Note 7, and θ10.2.5)

Implied by	*Not implied by*
ASC.T_3 [Alexandroff and Urysohn]	$\beta X \setminus \{t\}$, $t \in \beta X \setminus X$ ⟦Plank⟧
$\overline{\text{CK}} \subset$ N Note 35	CK $\times$ CK 7.4, Remark
CK $\times$ (CK.FC.T_2) [Franklin, p. 111]	$\overline{\text{CK}} \vee$ CK AMM *73*, p. 358
CK $\times$ SK [Franklin, p. 111]	$\overline{\text{CK}} \subset$ TG Note 34
CK $\times$ (CK.seq.T_2) [Franklin, p. 111]	$\overline{\text{CK}} \subset T_{3\frac{1}{2}}$ Note 34
ci(CK) θ5.4.4	L ⟦**Q**⟧
factor in (CKΠ) 6.7 #102	ψK.T_2 7.1 #115
φ^{-1} [CK], φ perfect [Dugundji, 11.5.3]	ψK.FC.$T_{3\frac{1}{2}}$ [Gillman and Jerison, 5I]
F $\subset$ CK θ5.4.2	ψK.LK.$T_{3\frac{1}{2}}$ ⟦Plank⟧
i. $\exists$ (CK.$\overline{\text{G}}$) [Ryeburn]	ψK.TG Note 34
K	se CK [Levine, θ6]
K $\times$ CK [Gaal, p. 147]	$\overline{\text{SF}} \subset$ K θ8.3.2
(K.ed.T_2)$\setminus$1 point 14.1 E2	$\overline{\text{SK}}$ [Grothendieck]
ψK.N 8.5 #103	TB.TG ⟦**R**, β⟧
ψK.CPK.$T_{3\frac{1}{2}}$ MR*20* #1964	$\exists$ 1U [Gillman and Jerison, 15R]
ψK.$T_{3\frac{1}{2}}$ + separation MR*20* #1964	>CK ⟦D⟧
ψK.(F.ψB $\Rightarrow$ ψK) 8.6 #113	
($\aleph_0$Π)(CK.FC) [Dugundji, 11.3.6]	*Spaces*
($\leq \aleph_1$)Π(SK) [Scarborough and Stone, p. 144]	Plank no 14.6
q(CK) θ5.4.4	∞.D no 5.4 #13
SK 7.1 #4	long line yes [Hocking and Young, p. 56]
se CK by S with $\tilde{\text{S}}$CK [Levine]	W yes 14.5
$\aleph_0$ sum(K) PAMS *13* (1962), 37	$\beta\omega \setminus$ nip yes 14.1 E2
TG.N.cuc.loc b.non-D [Kister]	$\omega \cup \{t\}$, $t \in \beta\omega$ no θ5.4.1 or 14.1 #110
<CK θ5.4.4	β**Q** $\setminus$ **Q** no 8.3 #110; 5.4 #109
every $\aleph_0$.F is CK 7.1 #104	β**Q** $\setminus \{x\}$, $x \in \beta$**Q** $\setminus$ **Q** yes Note 33
(f ⋃)(F.CK) [Ryeburn]	β**Q** $\setminus \{x\}$, $x \in$ **Q** no 5.4 #109
$\beta X \setminus \{x\}$, $x \in \beta X \setminus X$, with $X\sigma$K and βX TD Note 33	
$X = \Pi Y_\alpha$, XN, each Y_αSK MR*37* #3511	

TABLE 8. DISCONNECTED
(All spaces have more than one point)

Implied by	*Not implied by*
$\aleph_0$.LK.T_2 5.4 #101	$\aleph_0$.T_2.2C
$\aleph_0$.M θ4.3.3	ed. T_1 5.2 #111
$\aleph_0$.T_3 5.3 #102	f.R.N ⟦I⟧
D	(TD.M)$^+$ 8.1 #127
ed.T_2 5.2 #111	ZD.R.N 5.2 #111
f.T_1 4.1 #3	
TD	
ZD.T_0 5.2 #111	
$C^*(X)$ has three idempotents 5.2 #112	

TABLE 9. DISCRETE

Implied by	*Not implied by*
cuc.LK.non-K.TG [Kister]	$\aleph_0$.ed.T_4.ψF 8.3 #103
ed. CS.T_2 θ14.1.5	cuc.CK.non-K.TG [Kister]
ed.FC.T_2 θ14.1.5	ed.K.T_2 ⟦$\beta\omega$⟧
ed.$\aleph_0$.LK.T_2 10.1 #109	f.T_0
ed.LK.TG [Rajagopalan, θ1]	Π(D) 6.4 #101
ed.SK.T_2 14.1 #109	q(TG) by component 6.5 #105
f.T_1 4.1 #3	TD.M.Σ ⟦**Q**⟧
K.T_2.Hed [Gillman and Jerison, 6R4]	
ψF.k.T_1 8.1 #115	
ψF.FC.T_2 8.1 #119	
ψF.LK.T_2 8.1 #117	
q(TG) by G subgroup 12.3 #2	
T_1.$\exists$ Lf base 10.2 #109	
$>$D	
$\Delta \subset \Pi$(D) 6.7 #121	
$XT_{3\frac{1}{2}}$, $C(X)$Rx [Warner, θ10]	

TABLE 10. EXTREMALLY DISCONNECTED

Implied by	*Not implied by*
β (ed) θ14.1.4	$(\text{ed.K.}T_2)^2$ 14.1 #106
βX ed θ14.1.4	$\overline{\text{ed}}$ $[\![\omega^+ = \bar{\omega}]\!]$
coc	F $\subset$ ed 14.1 #203
cof	$(\aleph_0\Pi)$(ed) 14.1 #105 or #106
d $\subset$ ed θ14.1.3	TD.ZD.M. $[\![\mathbf{Q}]\!]$
G $\subset$ ed θ14.1.3	$\exists$ ed.d.G subset $[\![\omega^+]\!]$
I	$>$ed 6.2 E3
loc ed.T_2 [Bourbaki(b), I.11 #22b]	$<$ed $[\![\text{D}]\!]$
r(ed) θ14.1.3	
se(ed) by d 6.2 #112	*Spaces*
semi D 6.7 #111	$(\text{ed.LK.}T_2)^+$ no MR*30* #3352
Z $\subset X \subset \beta$Z, Z ed θ14.1.3 and θ14.1.4	$\beta\omega \setminus \omega$ no 14.1 #203
$T_{3\frac{1}{2}}$. every G is C^*-embedded MR*21* #3824	$\beta\omega$ yes θ14.1.4
$T_{3\frac{1}{2}}$. every d is C^*-embedded MR*21* #3824	Plank no θ14.1.4 and θ14.1.5
	RHO no θ14.1.5
	$\omega \cup \{t\}$, $t \in \beta\omega$ yes θ14.1.3
	W, W^+ no 14.5 #9

TABLE 11. FINITE

Implied by	*Not implied by*
$\aleph_0$.K.T_2.ed 14.1 #107	$\aleph_0$.T_4.ed $[\![\omega]\!]$
K.T_2.division ring 7.3 #212	K.M.TD.ZD.Σ $[\![\omega^+]\!]$
K.T_2.Hed [Gillman and Jerison, 6R4]	K.M.TD.ZD.Σ.TG $[\![2^{\aleph_0}]\!]$
ψF.K	
ψK.P-space [Gillman and Jerison, 14K2]	
SK.ed.T_2 14.1 #109	
$T_{3\frac{1}{2}}$.dim $C^*(X) < \infty$	
$T_{3\frac{1}{2}}$.$C^*(X)$Rx [Warner, θE]	

TABLE 12. FIRST COUNTABLE

Implied by	*Not implied by*
$\aleph_0$.cof	$\aleph_0$.K $[\![\mathbf{Q}^+]\!]$
$\aleph_0$.LK.R 5.4 #202	$\aleph_0.T_4$ 8.3 #103
βX at x if X FC at x. 8.3 #117	$\aleph_0.T_4.G_\delta.\psi$F 8.3 #103
cgi(FC)	$\aleph_0.T_4$. every subset is a G_δ 8.3 #103
f	$\aleph_0$.TG [Hewitt and Ross, 4.22]
(F $\Rightarrow$ G_δ).CK.T_3 MA*92*, p. 267	cof, coc 3.1 #4
factor in (FCΠ) 6.7 #102	CS.$\aleph_0$.K.KC 8.1 #130
G_δ.LK.R 5.4 #201	CS.$T_2.\aleph_0$ 3.1 #203
G_δ.LCK.R [Alexandroff and Urysohn, p. 66]	CS.T_2.K 8.1 #132
(HK.R)$^+$ (at ∞) 8.1 #113	cfi(FC) [Kelley, 3R]
i.$\exists$FC.G.subspace [Ryeburn]	$G_\delta.T_2$ 6.2 #110
LK.TG. every K subgroup FC	G_δ.K.KC.$\aleph_0$ 8.1 #112, #113, #203, and #205
($\aleph_0\Pi$)(FC) 6.2 #6	KO to FC 13.1 #106
q(FC.TG) θ12.3.3	Π(FC) 6.4 #6
se(FC)	q(FC) [Kelley, 3R, 5N]
SM	SK $[\![W^+]\!]$
2C	Σ.K.T_2 θ8.3.2
($\aleph_0 \bigvee$)(FC) 6.2 #6	$\bigvee$ (FC) 6.4 #6
(f $\bigcup$)(G.FC subspaces) [Ryeburn]	$<$FC $[\![$D$]\!]$
$\subset$FC	$>$M 12.4 #205
	Spaces
	RHO yes 5.3 E1
	$\mathbf{Q}^+$ no 8.1 #113
	W^+ no 14.5 #5
	W yes 14.5 #6
	$\omega \cup \{t\}$, $t \in \beta\omega$ no 8.3 #103
	Plank no 14.6 #5

TABLE 13. HEMICOMPACT

Implied by	*Not implied by*†
$\aleph_0$.D	$\aleph_0$.M 8.1 #112
F ⊂ HK	ci(HK) Note 17
K	G ⊂ HK ⟦D ⊂ D^+⟧
LK.Γ.M [Hocking and Young, pp. 79–80]	G_δ ⊂ HK ⟦**J** ⊂ **R**⟧
LK.Γ.TG 12.2 #121	LK.Σ.T_2 8.3 #113
LK.L.R 8.1 #125	Π(HK) 9.3 #110, 6.4 #203
LK.Σ.M 8.3 #113	σK.M 8.1 #112
LK.Σ.TG 12.1 #121	w* dual of n 12.4 #201
ψF.$\aleph_0$	>HK 6.2 E3
σK.LK.R 8.1 #125	<HK Note 17
w* dual of Banach space 12.4 #201	*Denied by*
X^+FC at ∞ 8.1 #113	$\aleph_0$.Γ.FC.T_2 Note 9
$T_{3\frac{1}{2}}$.$C(X)$M θ13.2.4	TVS.M.γ.ID Note 8
	D.u
	Spaces
	R yes
	ω yes
	Q no 8.1 #112
	J no 8.1 #112
	$\beta X \setminus \{t\}$, $t \notin X$ no θ8.3.2
	$\omega \cup \{t\}$, $t \in \beta\omega$ yes 8.3 #103(b)
	RHO no 9.3 #115
	W no 8.1 #10

† See also Table 28.

TABLE 14. HEREDITARILY SEPARABLE

Implied by	*Not implied by*
$\aleph_0$	K.Σ.U.Π(K.M) 6.7 #201
2C	Σ.K.T_2 [Kelley, 3N]
Σ.SM	Σ.TG 6.7 #201
Σ.ordered space PAMS *15*, p. 867	
	Spaces
	$\mathbf{R}^{\mathbf{R}}$ no 6.7 #201
	RHO × RHO no 6.7 E3
	$\beta\omega$ no ⟦$\beta\omega \setminus \omega$⟧

TABLE 15. HOMOGENEOUS

H	*non*-H
any TG ⟦translation⟧	$\beta X \setminus X$ if X is not ψK(CH) MR*20* #1965
Q	$\beta\omega \setminus \omega$ [Frolik]
J 12.1 #112	$\beta\omega$ θ8.3.2
$[0, 1]^\omega$ 6.4 #204	W, W^+
Π(H) Note 10	
Cantor discontinuum ⟦**2** is H⟧	

A $\aleph_0$.Γ.T_2 may be homogeneous or not MR*23* #A2181. The closure of a H set in TVS need not be H Note 6

TABLE 16. *k* SPACE
(See also Note 41)

Implied by	*Not implied by*
Cγ	$\aleph_0.T_4$.L.RK 8.3 #103
CS.T_2 8.1 #116	d $\subset$ k 8.3 #103
FC 8.1 #119	k$'.T_2$ [Comfort(b), pp. 109, 116]
F $\subset$ k Note 14	(k.M) $\times$ (k.T_2) MR*33* #693
G $\subset$ k.T_2 Note 14	Π(M.k) $[\![\mathbf{R}^{\mathbf{R}}]\!]$
K 8.1 #118	υ(LK.nmc) MR*36* #5896
KC.X^+ is KC 8.1 #203	$<$k $[\![$D$]\!]$
LK 8.1 #118	$>$k 6.2 E3
q(k)	
q(k.T_2) [Dugundji, p. 248]	*Denied by*
S.T_2 10.1 #111	υ(D.mc) [Comfort(b), p. 115]
(k.FC.T_2) $\times$ (k.FC.T_2) 8.1 #119	ψF. not D 8.3 #103
(k.T_2) $\times$ (LK.T_2) [Dugundji, p. 249]	
each point has a K. nhd 8.1 #117	*Spaces*
$\exists$ Lf cover by F.K. sets 10.2 #115	W yes 8.1 #118
	$\mathbf{R}^{\mathbf{R}}$ no [Kelley, p. 240]
	$\omega \cup \{t\}$, $t \in \beta\omega \setminus \omega$ no 8.3 #103
	RHO yes 8.1 #119
	$\beta X \setminus \{x\}$ yes Note 14
	Plank yes 8.1 #118

TABLE 17. *KC* SPACE

Implied by	*Not implied by*
CS.US [Franklin]	US.LK [Wilansky(b), p. 1241]
coc 5.4 #106	$<$KC D
(k.KC)$^+$ 8.1 #203	(KC)^2 Note 40
(k.T_2)$^+$ 8.1 #203	
maximal K 5.4 E4	*Spaces*
ψF.T_1 5.4 #106	cof no 5.4 #14
T_2 θ5.4.5	$\mathbf{Q}^+$ yes 8.1 #203
$>$KC	

TABLE 18. LINDELÖF

(See also Table 29; for M *or* SM *spaces see Table 31)*

Implied by †	*Not implied by*
$\aleph_0$	AC.T_2 [Bourbaki(b), Part 1, p. 147]
ci(L) θ5.4.4	D ⟦u⟧
cof ⟦it is K⟧	G $\subset$ L, G $\subset$ K.T_2 ⟦D $\subset$ D$^+$⟧
F $\subset$ L θ5.3.3	HL $\times$ HL 6.7 E3; [Kelley, 1K]
$F_\sigma \subset$ L	$\bar{L} \subset T_4$ Note 30
factor in LΠ 6.7 #102	L $\vee$ L AMM *73*, p. 358
HK 8.1 #10	(L.T_4) $\times$ (L.T_4) 6.7 E3
i. $\exists$ (L.G) [Ryeburn]	LK.TG ⟦D⟧
K	LK.T_4.CK.SK.FC.LM 14.5
LK.Γ.TG 12.1 #121	M ⟦D⟧
LK.Σ.TG 12.1 #121	Π (L.TG) 12.1 #122
L $\times$ K [Gaal, p. 145]	RK 6.7 E3
PK.Σ [Dugundji, p. 176]	Σ.T_3 5.3 #201
PK. weak Banach space [Corson(b)]	Σ.$T_{3\frac{1}{2}}$.TG 12.1 #122
PK. every M. continuous image is Σ [Corson(a)]	Σ.T_4 ⟦L.$T_4 \Rightarrow$ PK⟧
q(L) θ5.4.4	se L [Levine]
Σ.SM θ5.3.4	$\exists\ \sigma$ Lf base.T_4 10.3 #2
2C θ5.3.2	ZD.T_4 ⟦D⟧
σK 8.1 #10	$>$L 6.2 E3
se L by set S with $\tilde{S}$ L [Levine]	every subbasic cover reducible to $\aleph_0$ [Kelley, 1J(e)]
T_4.C_0(LK.TG) with weak topology [Corson(b), θ2]	
C_0(LK.M) with weak topology [Corson(b), θ1]	*Spaces*
$C(X, Y)$ with X, Y, Σ.M [Klee and Rudin]	$\beta X \setminus \{x\}$, $x \in \beta X \setminus X$ 8.3 #112; 8.1 #124
w^* dual of SN 12.4 #102	coc yes
$<$L θ5.4.4	cof yes
ϕ^{-1}[L], ϕ perfect [Dugundji, 11.5.3]	RHO yes 5.3 E3
(f ⋃)(G.L)	RHO is HL [Kelley, 1K]
	RHO $\times$ RHO no 6.7 E3
	$\omega \cup \{t\}$, $t \in \beta\omega$ yes ⟦$\aleph_0$⟧
	W no 14.5 #107
	Plank no θ5.3.5

† See also Table 28.

TABLE 19. LOCALLY COMPACT

Implied by	*Not implied by*
cgi(LK) 5.4 #15	cfi(**R**) [Kelley, 5N]
factor in LK.Π 6.7 #102	CK.$T_{3\frac{1}{2}}$ Note 12
F ⊂ LK 5.4 #3 and #103	HK.T_4.$\aleph_0$ 8.3 #103
G ⊂ LK θ5.4.12	K.T_1 8.1 #6
G ⊂ K.R C5.4.1	K.KC 8.1 #6
G ⊂ K.T_2 θ5.4.7	LK ∨ K AMM *73*, p. 358
F ∩ G ⊂ LK 5.4 #103	($\aleph_0$Π)(LK) 7.4 #107
HK.FC.R 8.1 #129	r(K) ⟦r = identity⟧
HK.FC.T_2 8.1 #129	Σ.B.TG 12.4 E6
HK.FC.US 4.1 #117	T_1 each point has a K nhd 8.1 #6
i. ∃ (LK.G) [Ryeburn]	υ(LK.σ-ψK) [Comfort(b), p. 116]
K.R or K.T_2 θ5.4.11	>LK 6.2 E3
K.TG θ12.1.4	<LK ⟦D⟧
Lγ in every M AMM *58*, p. 391	⊂LK ⟦**Q** ⊂ **R**⟧
Π (LK) if aa factors K 7.4 #108	
(f Π)(LK)	*Denied by*
q(LK.TG) 12.3 #11	$\aleph_0$.T_2.Γ 5.3 #102
q(LK) by usc K decomposition [Kelley, θ5.20]	(∞Π)(non-K) 7.4 #107
r(LK) 5.4 #16	*Spaces*
X^+R 8.1 #123	cof yes 5.4 #12
X^+T_2 θ8.1.2	**Q** no θ5.4.13
T_2. each point has a K nhd 8.1 #126	$\mathbf{Q}^+$ no 8.1 #6
$T_{3\frac{1}{2}}$.∃ larger topology: $\lvert \beta X \setminus X \rvert < \infty$ MR*19* #1069	RHO no 5.4 #205
	$\omega \cup \{t\}$, $t \in \beta\omega \setminus \omega$ no 8.3 #103
	W yes 14.5
	Plank yes 14.6

TABLE 20. LOCALLY CONNECTED

(See Note 13)

Implied by	*Not implied by*
β(LΓ, ψ K.$T_{3\frac{1}{2}}$) *MR20* #2688	β(LΓ) ⟦β**R**⟧
cgi(LΓ) 6.5 #109	Γ 5.2 #109
D	all components G and F 5.2 #115
G ⊂ LΓ 5.2 (Remark)	⊂LΓ ⟦**Q** ⊂ **R**⟧
(fΠ)(LΓ) [Simmons, p. 152 #5]	<LΓ ⟦D⟧
i. ∃ (LΓ.G) [Ryeburn]	>LΓ 6.2 E3
Π(Γ.LΓ) [Simmons, p. 152 #7]	
r(LΓ) [Hu, p. 27]	*Denied by*
q(LΓ) 6.5 #109	β (non-ψK) MR*20* #2688
(f⋃)(G.Γ subspaces)	
βXLΓ MR*20* #2688	*Spaces*
	Q, **J** no
	Q$^+$ no
	β**R** no MR*20* #2688

TABLE 21. METRIZABLE

(See also Table 30)

Implied by	Not implied by†
$\aleph_0$.LK.T_2 10.1 #109	*Spaces*
bidual (LC.M) [Robertsons, 6.3.16]	βX no C8.3.1
cfgi(M) [A. H. Stone(b)]	(X, β) no 11.4 E3
ci(KM).T_2 [Bourbaki(b), 9.2.10.17]	RHO no 5.3 E1
cpi(M) [Dugundji, 11.5.2]	long line no [Hocking and Young, p. 56]
factor in MΠ 6.7 #102	W, W^+ no θ7.2.1
i. ∃ G.M. subspace [Ryeburn]	$\omega \cup \{t\}$, $t \in \beta\omega \setminus \omega$ no 8.3 #103a
K.T_2. ⋃ (two M subspaces)	LF space no [Kelley and Namioka, 22C]
LK.T_2. components G and M 10.1 #103	Plank no θ7.2.1
K.T_2.($\aleph_0$⋃)(ΣM) MR*18*, p. 813	
LCK.($\aleph_0$⋃)(2C)T_2 MR*18*, p. 813	
(Lf⋃)(F.M) 10.3 #202	
LM.K.T_2 [Alexandroff and Urysohn, p. 82]	
PK.LM [Hocking and Young, θ2.68]	
PK.MS Note 43	
PK.T_1.cgi(LΣ.M) [Hanai, θ4]	
PK.T_1.ZD.cgi(γ.M) [Michael], [Hanai]	
S.TG.T_2.Cγ [Arhangelskij]	
σ(K.M).LK [Warner]	
σ(ΣM).K.T_2 θ10.1.1	
PPK.LΣ.MS MR*29* #1622	
se M by F subspace [Ryeburn]	
2C.LCK.T_2 MR*18*, p. 813	
2C.LK.T_2 θ10.1.1	
2C.T_3 θ10.1.1	
TG.LK.$C_0(X)$ weakly L [Corson(b), θ2]	
T_2.q(cuc.M) [Corson(b), θ1.2]	
$T_{3\frac{1}{2}}$.$C^*(X)$Σ 12.4 #103 and #116	
(Σ.LK.M)$^+$ 10.1 #112	
(σK.LK.M)$^+$ 10.1 #112	
(LK.Γ.M)$^+$ [Hocking and Young, pp. 79–80]	

† See Table 30.

TABLE 22. NORMAL

(See also Table 38)

Implied by	*Not implied by*
$\aleph_0$.ZD 4.1 #103	G ⊂ N ⟦Plank⟧
cfi(N) [Dugundji, 7.3.3]	KO to N 13.1 #105
F ⊂ N 4.1 #6	N ∨ N AMM *73*, p. 358
factor in N.Π 6.7 #102	q(N) [Kelley, 4H]
i.∃ N.G. subspace [Ryeburn]	se N by F [Levine, θ5]
L.C_0(LK.TG) [Corson(b), θ2]	TG.Σ 12.1 #122
LK.TG 12.1 #121; θ12.1.4	TG.cuc [Kister]
PK C14.4.1	TVS.Σ 6.7 #203
r(N) 4.3 #7	U.Σ 11.4 #121; 12.1 #122
R.$\aleph_0$ 5.3 E4	w(f), f into N 8.3 #101
R.f 5.3 E4; θ5.4.7	every F.ψB set is ψK 8.6 #113
R.HK 8.1 #112	⊂N 8.3 #101
R.K θ5.4.7	$>$N 6.2 E3 or E2
R.L θ5.3.5	$<$N ⟦D⟧
R.2C θ5.3.2	
R.σ-Lf base L10.2.2	*Denied by*
R.σK 8.1 #10	uΠ(T_1.non-CK) 6.7 #203
σK.TG 8.1 #10	free group of non-N [Hewitt and Ross, p. 74]
SM θ4.3.3	Σ.∃(u.F.D subset) 5.3 E3
se N by F with F̃N [Ryeburn]	
T_4	
(f∪)(F.N) [Ryeburn]	
∪(N.F.G), disjoint 4.1 #123	
associated k topology of HK [Warner, p. 267]	
{nhds of Δ} is a U 11.1 #110	
w(f), f onto N	
every F is *C*-embedded 8.5 #102	
*w** dual of *Sn* 12.4 #102	

TABLE 23. PARACOMPACT

Implied by	*Not implied by*
$\aleph_0$.R ⟦It is L⟧	$\aleph_0.T_2$ 5.2 E7
cfi(PK) [Dugundji, 8.2.6]	LK.$T_{3\frac{1}{2}}$ ⟦Plank⟧
D 10.2 #123	LK.T_4.FC.LM.CK.SK 14.5
F $\subset$ PK 10.2 #124	$(\mathrm{PK.L.}\Sigma)^2$ 6.7 E3
factor in PKΠ [Dugundji, 8.2.4]	RK 6.7 E3
factor in certain $T_4\Pi$ Note 16	RK.T_4 Note 18
$\mathrm{F}_\sigma \subset$ PK [Dugundji, 8.2.5]	$\Sigma.T_4$ MR*18*, p. 496
i. $\exists$ (PK.$\overline{\mathrm{G}}$) [Ryeburn]	TG.Σ 14.4 #101
K.R 10.2 #122	$\subset$PK ⟦$X \subset \beta X$⟧
L.R [Dugundji, 8.6.5]	$<$PK ⟦D⟧
LK.TG 12.1 #121	$>$PK 6.2 E3
Π(M) with $\leq\aleph_0$ non-K [A. H. Stone(a), θ4]	{nhds of Δ} is a γ.U Note 18
M $\times$ (K.T_2) [A. H. Stone(a)]	$\exists$ unique U.LK.CK.T_4 14.5 #10
PK $\times$ (K.T_2) MR*24* #A2365	*Denied by*
PK $\times$ [PK.σ(LK.F)] MR*36* #5894	$\aleph_0.\Gamma.T_2$ 5.2 E7
q(PK.TG) θ12.3.3	*Spaces*
se PK by F with $\tilde{\mathrm{F}}$ PK [Ryeburn]	RHO yes 5.3 E2
σK.R 8.1 #10	Plank no C14.4.1
SM [M. E. Rudin]	*W* no 14.5
uniformly LK [Kelley, 6T]	$\omega \cup \{t\}$, $t \in \beta\omega$ yes ⟦It is L⟧
$\bigcup$ (F.G.PK), disjoint 10.2 #123	
(Lf $\bigcup$)(F.PK) MR*20* #2678	
C(X, Y), X, Y are Σ.M [Klee and Rudin, p. 470]	
φ^{-1} [PK], φ perfect [Dugundji, 11.5.3]	
{nhds of Δ} a U. every G cover even [Kelley, θ5.28]	

TABLE 24. PSEUDOCOMPACT

Implied by	*Not implied by*
$\aleph_0$.Γ 5.2 E7	d ⊂ ψK ⟦ω ⊂ βω⟧
Ascoli θ [Glicksberg, θ2]	F ⊂ ψK 7.1 #116
ci(ψK) 4.2 #104	(F.C^*-embedded) ⊂ ψK.$T_{3\frac{1}{2}}$ [Gillman and Jerison, 6P4]
CK 7.1 #114	G ⊂ ψK
C-embedded ⊂ ψK	(ψK.CK.$T_{3\frac{1}{2}}$)^2 [Gillman and Jerison, 9.15]
Dini Lemma 7.2 #109	$T_{3\frac{1}{2}}$.non-RK [Gillman and Jerison, 9L]
F ⊂ (ψK.N). 8.5 #103	TB.TG ⟦(**R**, β)⟧
$\overline{\text{G}}$ if G ⊂ ψK 10.2 #108	TG.cuc.non-D MR*34* #7699
i. ∃ (ψK.G) [Ryeburn]	>ψK 6.2 #7
K θ5.4.4; 5.4 E2	$X \cup S$, S d in $\beta X \setminus X$ [Fine and Gillman(b)]
Π(ψK.TG) [Comfort and Ross, p. 487]	$X \times Y \times Z$, where $X \times Y$, $Y \times Z$, $Z \times X$ are ψK MR*35* #966
Π(SK) [Stephenson, p. 444]	ΠX_n, where $\Pi^m X_n$ is ψK ∀ m MR*35* #966
ψB.F ⊂ T_4 8.6 #112	Z ⊂ ψK.$T_{3\frac{1}{2}}$ MR*21* #3821
ψK × K [Gillman and Jerison, 9.14]	
ψK × SK [Stephenson, p. 444]	*Spaces*
(ψK.k) × (ψK.$T_{3\frac{1}{2}}$) [Stephenson, p. 446]	W yes 7.1 #114
ψK × (ψK.S) [Stephenson, p. 446]	Plank yes 14.6
SK 7.1 #114	$\beta\omega \setminus$ nip yes 14.1 E2
<ψK 4.2 #104	$\omega \cup \{t\}$ no 14.1 #110
∃ 1 U 11.4 E4	
(f ⋃)(F.ψK) [Ryeburn]	
(f ⋃)(G.ψK) [Ryeburn]	
$\upsilon = \beta$ 8.6 #1; 11.4 E4	
βXLΓ MR*20* #2688	
cl_XG if $\text{cl}_{\upsilon X}$G is K [Comfort(a), p. 97]	
$X \cup S$ XLK, S d in $\beta X \setminus X$ [Fine and Gillman(b)]	
$\varphi[X]$K ∀ $\varphi \in C^*(X)$ 8.3 #114 and #115	
$\varphi \in C^*(X)$, $\varphi \neq 0$ on $X \Rightarrow \varphi \neq 0$ on βX 8.3 #115	

TABLE 24. PSEUDOCOMPACT *(continued)*

Implied by	*Not implied by*
$\beta X \setminus X$H MR*38* #1656	
$\beta X \setminus X$ has $<2^{c}$ points [Hewitt(a), p. 69]	
$\beta X \setminus X$ has no FG_δ [Hewitt(a), p. 68]	
$T_{3\frac{1}{2}}$. every 2 non-ψKZ's intersect [Gillman and Jerison, 1G4]	
$T_{3\frac{1}{2}}$. every MI is real [Gillman and Jerison, 5.8]	
$X \times Y$ if $\beta(X \times Y) \simeq \beta X \times \beta Y$ 8.5 #105	
$X \times X$ if $X\psi$K and $\upsilon(X \times X) \simeq \upsilon X \times \upsilon X$ 11.4 #123	
ΠX_α with every $\aleph_0$ subproduct ψK MR*21* #4405	

TABLE 25. PSEUDOFINITE

Implied by	*Not implied by*
coc 5.4 #106	cof 5.4 #14
D	*Spaces*
W-max, $\delta > 1$	$\omega \cup \{t\}$, $t \in \beta\omega$ yes 8.3 #103
	W no 14.5 #7
	Plank no 14.6 #5

TABLE 26. REALCOMPACT

Implied by	*Not implied by*
$\aleph_0.T_3$ θ8.6.3	ci(RK) ⟦D⟧
B ⊂ RK MR*36* #3314	cuc ⟦υ⟧
$C(X)$ bornological [Nachbin], [Shirota]	FC.CK.LK.k.T_4 ⟦W⟧
D.(cardinal ≤ c) 8.6 #111	G ⊂ RK ⟦$X \subset \beta X$⟧
D.nmc [Gillman and Jerison, 15.24]	LK.CK.Σ.$T_{3\frac{1}{2}}$ ⟦$\beta\omega \setminus$ nip⟧
F ⊂ RK 8.6 #110	NS [Nachbin], [Shirota]
factor in RKΠ 6.7 #102	q(RK).T_4 [Gillman and Jerison, 8I]
HK.T_3 θ8.6.3	$T_{3\frac{1}{2}}$.(RK.D) ∪ (RK.D.$\aleph_0$) [Gillman and Jerison, 8H6]
⋂ (RK) 8.6 #109	$T_{3\frac{1}{2}}$.(RK.F) ∪ (RK.F) MR*21* #1572
K 8.6	TG.T_2 Note 20
L.T_3 θ8.6.3	<RK ⟦D⟧
M.nmc [Gillman and Jerison, 15.24]	>RK 6.2 E3
Π(RK) θ8.6.5	$\overline{\text{RK}}$ ⟦$\overline{\omega} \subset \beta\omega \setminus$ nip⟧
PK.nmc θ14.4.2; [Gillman and Jerison, 15.20]	
RK ∪ K [Gillman and Jerison, 8.16]	*Spaces*
r(RK) L8.6.3	$\beta\omega \setminus$ nip no 8.6 #1
σK.T_3 θ8.6.3	**R** yes 8.6 #2 or θ8.6.3
⊂(Σ.M) C8.6.2	RHO yes 8.6 #7
$T_{3\frac{1}{2}}$. > HRK [Gillman and Jerison, 8.17]	*W* no 8.6 #1
U.γ.T_2.nmc [Gillman and Jerison, 15.20]	Plank no 14.6 #101
υX θ8.6.1	$\omega \cup \{t\}$, $t \in \beta\omega$ yes θ8.6.3
υ is γ 11.4 E5	
⊂(RK.G_δ) 8.6 #4	
cozero set ⊂RK [Gillman and Jerison, 8.14]	
T_4.($\aleph_0$ ⋃)(F.RK) MR*21* #1572	
φ^{-1} [RK] ⊂ RK [Gillman and Jerison, 8.13]	

TABLE 27. REGULAR

(See Table 36)

Implied by	*Not implied by*
cpi(R) [Dugundji, 11.5.2]	f.N 4.1 #8
CR θ4.3.1	N.K 4.1 #8
ed. semi-R	M ∧ M 12.2 #126
f.sym 4.1 #113	R ∧ R Note 19; 12.2 #126
factor in RΠ 6.7 #102	R^+ ⟦$\mathbf{Q}^+$⟧
i. ∃ (G.R) [Ryeburn]	<R ⟦D⟧
KO to R 13.1 #109	>R 6.2 E2
N.sym 4.1 #12	>M 6.2 E2
Π(R) θ6.7.3	{nhds of Δ} is a uniformity 11.1 #110
q(R) by usc K decomposition [Kelley, 5.20]	
$(R.LK)^+$ θ8.1.2	*Denied by*
se R by F [Levine]	$\aleph_0.\Gamma.T_2$ 5.2 E7
SM θ4.3.3	se by d 6.2 E2
T_3	
TG θ12.1.4	*Spaces*
TVS θ12.1.4	$\mathbf{Q}^+$ no θ8.1.2
U θ11.1.4	
(f ⋃)(F.R) [Ryeburn]	
w(R) θ6.7.2	
ZD 4.1 #103	
⋁ (R) θ6.7.1	
each point has a F.R nhd [Bourbaki(b), I.8.4.13]	

TABLE 28. σ-COMPACT

Implied by†	*Not implied by*‡
$\aleph_0$	D ⟦u⟧
ci(σK) θ5.4.4	FC.LK.CK.T_4 ⟦W⟧
cof	G $\subset$ σK ⟦D $\subset$ βD⟧
F $\subset$ σK	LK.M.TG ⟦D⟧
$F_\sigma \subset \sigma$K	LK.Γ.T_2 [Hocking and Young, p. 79]
factor in σKΠ	LK.Σ.$T_{3\frac{1}{2}}$ 8.3 #113
HK	Π(HK) 9.3 #110; 6.4 #203
i. ∃ (σK.G) [Ryeburn]	Π(σK) 9.3 #110; 6.4 #203
K	Σ.M.γ.TG 6.4 #203
LK.L 8.1 #124	TVS.T_4 Note 8
LK.Σ.M 8.3 #113	uniformly LK ⟦D⟧
LK.2C θ5.3.2	>σK 6.2 E3
(fΠ)(σK) [Hewitt and Ross, θ3.9]	(∞Π)(non-K) [Hewitt and Ross, θ3.9]
se(σK) by F with $\tilde{\text{F}}$ σK [Ryeburn]	ID.TVS.γ.M Note 8
<σK θ5.4.4	
(SN)$'$.w* 12.4 #102	*Denied by*
(f⋃)(G.σK) [Ryeburn]	$\beta X \setminus \{x\}$, $x \in \beta X \setminus X$ 8.3 #112
uniformly LK.Γ [Kelley, 6T]	
	Spaces
	Q, $\mathbf{R}^n$ yes
	J no 9.3 #110
	RHO no 9.3 #115
	D.u no
	$\beta\omega \setminus$ nip no 8.3 #113
	W no 14.5 #106

† See also Table 13.
‡ See also Table 18.

TABLE 29. SECOND COUNTABLE
(For M *or* SM *spaces, see Table 31)*

Implied by	*Not implied by*
$\aleph_0$.FC 5.3 E4	ci(2C) 5.3 #8
$\aleph_0$.LK.R 6.7 #123	cof 5.3 #3
cgi(2C) [Dugundji, 8.6.2]	FC ⟦D⟧
cpi(2C) [Dugundji, 11.5.2]	FC.Γ.K.T_2 [Kelley, 5J]
factor in (2C)Π 6.7 #102	FC.HΣ.HL ⟦RHO⟧
i.(∃ G.2C) [Ryeburn]	HΣ.HL.T_4.ZD.ed.$\aleph_0$ 8.3 #103
K.(Δ is a G_δ) 6.7 #209	KO to 2C 13.1 #106
KO from K.M to 2C [Hocking and Young, p. 35]	L.(Δ is a G_δ) 6.7 #210
($\aleph_0$Π)(2C) [Dugundji, 8.6.2]	Π(2C) ⟦$\mathbf{R}^{\mathbf{R}}$⟧
q(2C) by usc K decomposition [Kelley, 5.20]	Σ.FC.OC.K.T_2 [Kelley, 5M]
Σ.SM θ5.3.1	σ.(2C.K.M) ⟦$\aleph_0$⟧
se(2C) [Levine]	U with $\aleph_0$ base ⟦M⟧
q(2C.TG) θ12.3.3	$>$(2C) ⟦I⟧
⊂(2C) 5.3 #1	T_4.[$>$(2C.T_4)] ⟦RHO⟧
($\aleph_0$⋁)(2C) AMM*73*, p. 358	$<$2C 5.3 #8
(f⋃)(G.2C) [Ryeburn]	*Spaces*
	$\aleph_0$.cof yes 5.3 #3
	RHO no 5.3 E1; 10.1 #102
	$\omega \cup t$, $t \in \beta\omega$ no 8.3 #103
	$\mathbf{R}^{\mathbf{R}}$ no θ10.1.1
	W, W^+ no θ10.1.1
	Plank no θ10.1.1

TABLE 30. SEMIMETRIZABLE
(See also Table 21)

Implied by	*Not implied by*
$\aleph_0$.FC.R 10.1 #108	$\aleph_0.T_2$.2C.Γ.σ(K.M) 10.1 #101
$\aleph_0$.K.R 10.1 #109	$\aleph_0.T_1$.LK.K ⟦cof⟧
$\aleph_0$.LK.R 10.1 #109	$\aleph_0.T_4$ 10.1 #107
C(HK) L13.2.1	$\aleph_0$.TG.T_2 [Hewitt and Ross, 4.22]
f.R 10.1 #2	bornological [Kelley and Namioka, 20G]
f.sym 4.1 #113	cgi(M).T_2 [A. H. Stone(b)]
FC.TG θ12.1.3	FC.K.T_2.Σ.Γ [Kelley, 5J, 5M]
factor in SMΠ 6.7 #102	K.Σ.T_2 C.8.3.1
free ∪ (SM) 10.1 #103	K.T_1.(Δ is a G_δ) 10.1 #116
K.R.(Δ is a G_δ) 10.1 #115	LC.(dim = $\aleph_0$) [Kelley and Namioka, 6I]
LK.G_δ.TG 12.3 #107	LC.B.T_2.γ [Bourbaki(a), Ch. 3, p. 4]
($\aleph_0$Π)(SM) θ6.4.2	LM.CK.T_4.FC.LK.SK 14.5
q(TG.SM) θ12.3.4	q(M).T_2 [Rainwater, θ1, 2]
2C.R θ10.1.1	σ(K.M).T_4 8.3 #103
TVS. ∃ simply ordered base [Wilansky(a), 10.4 #9]	2C.N 10.1 #101
U.$\aleph_0$ base θ11.5.1	TG. ∃ simply ordered base MR*17*, p. 508
w({φ_n}), each φ_n to SM θ6.3.4	U.FC θ11.4.5; 5.3 E1 and E2
($\aleph_0$⋁)(SM) θ6.2.2	LK.$C_0(X)$ weakly L [Corson(b), p. 5]
∃ σ-LF base.R θ10.3.1	(disjoint)(F.M) ∪ (G.M) 10.1 #106
	<M ⟦D⟧
	>M 6.2 E2
	T_4.(>M) ⟦RHO⟧

TABLE 31. SEPARABLE

Implied by

$\aleph_0$
$\exists\ \aleph_0\psi$-base
OC.SM [Sikorski]
ci(Σ) 5.3 #7
cof 5.3 #3
C(K.M) [Dunford and Schwartz, p. 437]
factor in $\Sigma\Pi$ 5.3 #7
i.($\exists$ G.Σ) [Ryeburn]
G $\subset \Sigma$ 2.5 #12
γ(Σ.M)
K.SM θ5.3.4
LK.Γ.M [Alexandroff and Urysohn, p. 85]
L.SM θ5.3.4
L2C.Γ.M [Alexandroff and Urysohn, p. 92]
$\Pi\Sigma$ with $\leq c$ factors [Ross and Stone, p. 398]
q(Σ) 5.3 #7
se(Σ) by Σ [Levine]
se(HΣ) [Levine]
σ(K.M) θ5.3.4
2C
TB.SM L7.2.2
$<\Sigma$ 5.3 #7
$\subset\Sigma$.M
(Σ.M) $\vee$ (Σ.M) 6.7 #114
X^+M 10.1 #112
LC. weak Σ [Wilansky(a), θ12.2.4]

Not implied by

(βX)Σ(TG) MR*28* #4503
OC [Ross and Stone, θ1 and θ2]
D-Σ ⟦K⟧
F $\subset \Sigma$ 5.3 #4
F $\subset \Sigma.T_{3\frac{1}{2}}$ 6.7 E3
F $\subset \Sigma$.K.T_2 ⟦$\beta\omega \setminus \omega$⟧
Γ.LO.card $\leq$ c [Gaal, p. 125]
HL.T_2 [Miscenko]
K.T_2.TG 12.1 #123
L.LK.PK.TG 12.1 #123
L.σK.T_4.TG.TVS Note 22
LK.M ⟦D⟧
se Σ 6.2 E3
$>\Sigma$ ⟦D⟧
subgroup of Σ.TG ⟦see next entry⟧
TVS $\subset \Sigma$. barreled.TVS MR*30* #429
TB.U 12.1 #123
TVS. every b is Σ(CH) PAMS *6*, p. 729
(Banach)$'.w^*\Sigma$ 12.4 #102
T_2. every D is $\aleph_0.\psi$F.G_δ Note 24
(Σ.L.T_4) $\vee$ (Σ.L.T_4) 6.7 #122
(Σ.LC) $\vee$ (Σ.LC) [Klee]

Denied by

coc.u ⟦$\aleph_0 \Rightarrow$ F⟧
D.u
ΠX_α, $>c$ non-I factors [Ross and Stone]

Spaces

C(0, 1) yes 5.3 #202
RHO yes 5.3 E1
$\mathbf{R}^{\mathbf{R}}$ yes 6.7 #201
W, W^+ no L14.5.1
Plank no Note 37
$\beta\omega \setminus \omega$ no [Gillman and Jerison, 6Q2]

TABLE 32. SEQUENTIALLY COMPACT

Implied by	*Not implied by*†
CK ⊂ C(K, K.M) θ13.2.4	CK.T_2 7.4 E2
CK.CS	K.T_2 7.4 E2
CK.FC θ7.1.3	K.$T_2\Sigma$.ed 8.3 #3
CK.FC at non P-points [Scarborough and Stone, p. 143]	Π(SK.K.M) 7.4 E2
CK.seq.T_2 [Franklin]	*Spaces*
CK.$T_{3\frac{1}{2}}$. weak Σ MR*20* #2681	$\omega \cup \{t\}$, $t \in \beta\omega$ no 14.1 #110
CK.SM θ7.2.1	$2^{\mathbf{R}}$ no 7.4 #102
i. ∃ (SK.$\overline{\text{G}}$) [Ryeburn]	$\beta\omega$ no 8.3 #3
($\aleph_0$Π)SK θ7.4.2	$\mathbf{Q}^+$ yes 8.1 #130
se(SK) by F with $\tilde{\text{F}}$ SK [Ryeburn]	W, W^+ yes 14.5 #11
(f⋃)(F.SK) [Ryeburn]	Plank no 7.1 #4

† See also Table 7.

TABLE 33. T_0

Implied by	*Not implied by*
factor in $T_0\Pi$ 6.7 #102	cgi(M.TG) 12.3 #2
i. ∃ (G.T_0) [Ryeburn]	q(M.TG) 12.3 #2
KO to T_0 13.1 #4	TG.TVS ⟦I⟧
maximal K	ZD ⟦I⟧
ΠT_0	
T_1 θ4.3.1	
(f⋃)(G.T_0) [Ryeburn]	
$>T_0$ 4.1 #4	
$\subset T_0$	

TABLE 34. T_1

Implied by	Not implied by†
factor in $T_1\Pi$ 6.7 #102	
i. $\exists$ (G.T_1) [Ryeburn]	
KC 5.4 #106	
KO to T_1 13.1 #4	
ΠT_1	
q with F equivalence classes θ6.7.5	
sym T_0 4.1 #10	
T_2 θ4.3.1	
T_1^+ 8.1 #4	
$T_1 \wedge T_1$ 6.2 #105	
$>T_1$ 4.1 #4	
US 4.1 #117	
(f$\bigcup$)(G.T_1) [Ryeburn]	

† See Table 33.

TABLE 35. T_2

Implied by	Not implied by†
cpi(T_2) [Dugundji, 11.5.2]	cfi(T_2) [Kelley, 4G]
factor in $T_2\Pi$ 6.7 #102	KC.K.$\aleph_0$.CS 8.1 #130 and #205
i. $\exists$ (G.T_2) [Ryeburn]	LT_4 3.2 #110
KC.FC 4.1 #117	(M.TG) $\wedge$ (M.TG) 12.2 #126
KO to T_2 13.1 #4	maximal K 5.4 E4
ΠT_2 θ6.7.3	$T_2 \wedge T_2$ 6.2 #106 and #107
q(T_2) by usc.K decomposition [Kelley, 5.20]	TD 3.2 #110
$>T_2$	US 4.1 #117
T_3 θ4.3.1	US.LK [Wilansky(b), p. 1241]
US.FC 4.1 #117	US.K.$\aleph_0$.CS 8.1 #130 and #205
US.seq.(LCK or LSK) [Franklin]	US.K.$\aleph_0$.seq [Franklin]
w (separating family) θ6.3.2	$\bigwedge$ decreasing sequence of T_2 [Bourbaki(b), I.1.8 #26]
(LK.T_2)$^+$ θ8.1.2	*Denied by*
every point has F.T_2 nhd	(K.T_2) $\wedge$ (K.T_2)($\neq$) 6.2 #107
Δ is F 6.7 E1	(γ.n) $\wedge$ (γ.n)($\neq$) [Wilansky(a), C11.3.1]
id has cg 6.7 E2	

† See also Table 33.

TABLE 36. T_3

Implied by	Not implied by†
cpi(T_3) [Dugundji, 11.5.2]	$\aleph_0.T_2$.2C.Γ 5.2 E7
i. ∃ (T_3.G) [Ryeburn]	KC.K 8.1 #205
KC.LK 8.1 #7	q($T_{3\frac{1}{2}}$).T_2 [Gillman and Jerison, 3J]
R.T_0 4.1 #11	T_1.K.LK ⟦cof⟧
T_2.LCK.seq Note 39	T_2 6.2 E2
T_2.LK 5.4 #101	T_2.CK [Alexandroff and Urysohn, p. 26]
T_2.CK.FC [Dugundji, 11.3.5]	T_2.ed 14.1 #104
T_2.CPK.FC [Aull(a)]	T_2.k.FC Note 31
T_2.LpK [Hocking and Young, p. 104]	T_2.ψK 5.3 #103
$T_{3\frac{1}{2}}$ θ4.3.1	T_2. semi-R [Bourbaki(b), I.8 #24]
T_4 θ4.3.1	$>T_3$ 6.2 E2
$\subset T_3$	every 2 points have disjoint F nhds [Dugundji, 7.2 #3]
certain q 6.7 #109	
every point has F.T_3 nhd [Bourbaki(b), I.8.4.13]	*Spaces*‡
K. each point = ⋂ (F.G) 5.4 #207	

† See also Table 35.
‡ See Tables 38 and 37.

TABLE 37. $T_{3\frac{1}{2}}$
(See also Table 5)

Implied by	Not implied by†
$\aleph_0.T_3$ θ5.3.5	min T_3 MR *27* #2949
cfgi(T_4) [Dugundji, p. 157]	T_2.CK [Alexandroff and Urysohn, p. 66]
CR.T_0 4.1 #11	T_2.q($T_{3\frac{1}{2}}$) [Gillman and Jerison, 3J]
TG.T_0 θ12.1.4	$>T_{3\frac{1}{2}}$ 6.2 E2
T_2.K θ5.4.7	
T_2.LK 8.1	*Spaces*‡
T_4 θ4.3.1	$\beta\omega \setminus$ nip yes 4.3 #2
	Plank yes
	$\mathbf{R}^{\mathbf{R}}$ yes θ6.7.3

† See Table 36.
‡ See Table 38.

TABLE 38. T_4
(See also Table 22)

Implied by	*Not implied by*†
cfi(T_4) [Whyburn]	$\aleph_0.T_2.\Gamma.L.\sigma K$ 5.2 E7
cpi(T_4)	$G \subset T_4$; $G \subset K.T_2$ ⟦Plank⟧
i. $\exists$ (T_4.G) [Ryeburn]	HRK [Gillman and Jerison, 8.18]
K.T_2 θ5.4.7	N.T_0 4.1 #11
N.T_0.sym 4.1 #10	RK.[($\aleph_0$ ⋃)(F.PK)] MR*21* #1573
N.T_1	T_2.2C 10.1 #101
q (some) 6.7 #109	$T_3.\Sigma$ 5.3 #201
M × (K.T_2) [A. H. Stone(a)]	$T_{3\frac{1}{2}}$.CK [Gillman and Jerison, 8L]
PK × (K.T_2) MR*24* #A2365	$T_{3\frac{1}{2}}$.Hed [Gillman and Jerison, 6R(2)]
(CPK.T_4) × (M.σ-LK) MR*29* #4034	$T_{3\frac{1}{2}}$.LK.ψK ⟦Plank⟧
(CK.T_4) × (PK.FC.T_2) MR*21* #2218	$T_{3\frac{1}{2}}.\Sigma$.FC [Gillman and Jerison, 3K, 5I]
Spaces	$T_{3\frac{1}{2}}$.RK.Σ.ZD 8.6 #8
$\mathbf{R}^{\mathbf{R}}$ no 6.7 #203	$T_{3\frac{1}{2}}$. completely separate $\aleph_0$. F from F MR*20* #1964
RHO yes 5.3 E2	$T_{3\frac{1}{2}}$.($\aleph_0$.F) are C^*-embedded MR*20* #1964
W yes 14.5	$T_{3\frac{1}{2}}$.(F $\Rightarrow$ G_δ) [Katetov, p. 74]
Plank no 14.6 #4	weak Banach space [Corson(b), p. 12]
$\omega \cup \{t\}$, $t \in \beta\omega$ yes 8.3 #102	(T_4 × T_4).Σ 6.7 E3
$\beta\omega \setminus$ nip no (CH) Note 32	(LK.T_4) × (K.T_2) [Kelley, 5K]
$\beta\omega \setminus \omega \setminus \{t\}$, no (CH) Note 42	$> T_4$ 6.2 E2
$\beta\mathbf{R} \setminus \{x\}$, $x \in \beta\mathbf{R} \setminus \mathbf{R}$ no Note 32	$< T_4$ ⟦D⟧
$\beta\omega \setminus \omega$ yes θ5.4.7	
$W \times W^+$ no [Kelley, 4E]	

† See also Table 37.

TABLE 39. TOTALLY BOUNDED

(Uniform space only)

Implied by	*Not implied by*
β 11.4 E2	b $\subset$ M 9.1 #103
CC(TB) $\subset$ LC [Kelley and Namioka, 13.3]	b $\subset$ n 12.4 E6
CK 11.3 #20; L7.2.1	cuc.M ⟦D⟧
factor in TBΠ 11.4 #119	
K. θ11.3.7	
Π(TB) 11.4 #122	
ψK 11.3 #22	
$\overline{\text{TB}}$ 11.3 #13	
uc image of TB 11.3 #9	
$\subset$TB	
every uf is Cauchy θ11.3.6	
Cauchy equiv. to TB 11.3 #18	

TABLE 40. TOTALLY DISCONNECTED

(All spaces have ≥2 points)

Implied by†	*Not implied by*‡
$\beta(\aleph_0)$ 5.2 #111	β(TD) Note 26
β (strongly ZD) [Bourbaki(b), IX.6 #1b]	β(ZD) Note 26
β(ZD.Σ.M) [Bourbaki(b), IX.6 #1b, 2b]	ed.T_1 ⟦cof⟧
ed.T_2 5.2 #111	*Spaces*
Π(TD) Note 25	$\beta\omega$ yes 5.2 E9
q by components 6.5 #105	RHO yes 5.2 #111
$\subset$TD	W, W^+, Plank yes 5.2 #111
$(\text{TD.LK.}T_2)^+$ Note 29	
ZD.T_0 5.2 #111	
K.M. unit disc in $C(X)$ is CC of EP	

† See also Table 42.
‡ See also Table 8.

TABLE 41. *US* SPACE

Implied by	*Not implied by*†
KC 5.4 #106	$T_1.\aleph_0$.K.2C ⟦cof⟧
Π(US)	$(US)^+$ 8.1 #207
⊂(US)	
>(US)	*Spaces*
Δ is SF	cof no
seq.(CK ⇒ F) [Franklin]	$\mathbf{Q}^+$ yes 8.1 #203
seq.(SK ⇒ F) [Franklin]	

† See Table 34.

TABLE 42. ZERO DIMENSIONAL

Implied by	*Not implied by*†
$\aleph_0$.R 5.3 #102	$\aleph_0.T_2$ 5.2 E7
$\beta(\aleph_0)$ Note 27	$\aleph_0$.N 4.1 #8 and #103
βX TD [Bourbaki(b), IX.6 #1b]	$\beta(ZD.T_4)$ [Gillman and Jerison, 16P]
D	ed.T_2 14.1 #104
ed.R 14.1 #104	f.N 4.1 #8
ed.SR	TD.LK.T_1 Note 28
f.R 4.1 #104	TD.M.Σ.TG [Gillman and Jerison, 16L]
Π(ZD)	$(ZD.M)^+$ 8.1 #127
TD.LK.T_2 [Hewitt and Ross, p. 12]	>(ZD) 14.1 #104
TD.LpK [Flachsmeyer, p. 152]	ZD ∪ ZD Note 23
$(TD.LK.T_2)^+$ Note 29	
$(ZD.LK)^+$ Note 29	*Spaces*
∃ transitive U 11.1 #203	**Q** yes 5.3 #102
⊂ ZD	**J** yes
K. each point = ⋂ (F.G) 5.4 #207	RHO yes 2.6 E3
$X = \beta Y$, $Y \cup \{t\}$ZD $\forall\, t \in X$ [Gillman and Jerison, 16P]	$\beta\omega$, $\beta\mathbf{Q}$ yes Note 27
K.T_2.$C(X)$ has d set with FD range [Flachsmeyer]	Plank yes Note 38
min prime ideal space [Henriksen and Jerison, p. 200]	W, W^+ yes Note 38

† See also Tables 27 and 40.

Notes

1. An infinite-dimensional Banach space X, when given $\tau(X, X^{\#})$, ([Wilansky(a), p. 247]) is Cat I. ⟦[Kelley and Namioka, p. 95, B].⟧
2. Let B be $(\mathbf{R}, I)$, $X = \mathbf{R} \times B$, $S = \{(1/n, 0): n = 1, 2, 3, \ldots\}$. Then $[(0, 0) \cup S] \cap [(0, 1) \cup S]$ is not compact.
3. The weak* topology on a conjugate Banach space.
4. If X is RK, $C(X)$ is bornological and need not be barreled ⟦Nachbin–Shirota theorem⟧ hence need not be complete. See [Schaefer, 2.8.4].
5. The union of finitely many disjoint open CR subspaces is CR, [Ryeburn].
6. $\{1/n\}$ is a homogeneous subspace of $\mathbf{R}$. Its closure $\{0, 1, \frac{1}{2}, \frac{1}{3}, \ldots\}$ is not.
7. Iseki and Kasahara in Proc. Japan Acad. 1957 relate CK to properties of point and locally finite open covers. See also [Aull, p. 314].
8. Each compact set in an infinite-dimensional TVS is nowhere dense. (Compare Sec. 12.4 E6.) Hence σK implies Cat I.
9. If HK, the space would be LK, hence M, hence not Γ.
10. Map X_α onto itself for each α, carrying x_α to y_α. The resulting map of ΠX_α carries x to y.
11. A k' space is a space X with the property that a function f defined on X must be continuous if $f \mid K$ is continuous for all closed compact subsets K.
12. Let S be a countably infinite subset of $\beta\omega \setminus \omega$. Let $X = \beta\omega \setminus S$. Then X is CK. ⟦Let Y be a countably infinite set in X. If Y has no accumulation point in X, $Y \cup S \supset$ an infinite closed subset of $\beta\omega$. Such a set cannot be countable, by [Gillman and Jerison, 9.12, p. 134].⟧ Also, X is not LK. ⟦If LK, it would be open in $\beta\omega$ by θ5.4.13. But S is not closed, by [Gillman and Jerison, 9.12].⟧
13. In MR*20* #2688; if βX is LΓ, X is also; X is LΓ at x if and only if βX is LΓ at x; X is LΓ if it has a dense subspace d such that βd is LΓ.
14. Let S be either open or closed in a k space X. There exists a LK.T_2 space A and a quotient map $q: A \to X$ ⟦[Dugundji, θ11.9.4]⟧. Now $q^{-1}S$ is open or closed, hence LK, and q is a quotient map from it onto S. ⟦[Dugundji, θ6.2.1].⟧ By [Dugundji, θ11.9.4], S is a k space. (John W. Taylor points out that one can prove this generalization directly: If X is a k space in which every compact set is locally compact, every open subset of X is a k space.)
15. If $C^*(X)$ is separable, the unit disc in its dual is weak* metrizable ⟦12.4 #116⟧. But βX is a subspace of this disc ⟦see [Wilansky(a), pp. 261, 269]⟧, hence X is compact, by Corollary 8.3.1.
16. X is PK if it is a dense subspace of a K.T_2 space Y such that $X \times Y$ is T_4, in particular, if $X \times \beta X$ is T_4. MR*24* #A2365 and MR*25* #5489.
17. $\mathbf{Q}$ is not HK. With the (larger) discrete topology it is HK.
18. Let X be a complete uniform space of nonmeasurable cardinal which is not PK but is such that the set of neighborhoods of the diagonal is a uniformity ⟦MR*21* #5947⟧. This space is normal by 11.1 #107, and RK by [Gillman and Jerison, 15.20].
19. Let E be the Euclidean topology for $\mathbf{R}$, and $T = \{\varnothing, S, \tilde{S}, \mathbf{R}\}$ with $S = (0, 1)$. Then $E \cap T$ is not regular.
20. There are ψK.non-K groups. (See Table 3.)
21. Let G be an open set in $[0, 1]$ which includes $\mathbf{Q} \cap [0, 1]$ and has measure < 1. Consider $[0, 1] \setminus G$.

22. Let B be a reflexive, nonseparable Banach space. Let X be B with the weak topology. The result is contained in 12.4 #102 and [Wilansky(a), 12.2 θ4].
23. $\mathbf{R} = \mathbf{Q} \cup \mathbf{J}$.
24. Let E, C be the Euclidean and cocountable topologies for $\mathbf{R}$ and $T = E \vee C$. For an uncountable set u, $\exists x \in u$ such that $V \cap u$ is uncountable for every E nhd V of x. Then $V \cap W \cap u = (V \cap u) \setminus \tilde{W}$ is uncountable for every C nhd W of x. Thus u is not T-discrete. It is not Σ since C is not and $T \supset C$. The rest is by 6.2 #110.
25. Every projection of a connected set S is connected, hence has only one point. Thus S has only one point.
26. Let X be ZD and βX not ZD. (See Table 42.) Then X is TD. If βX were TD it would be ZD since K.TD.T_2 implies ZD.
27. Let X be $\aleph_0.T_{3\frac{1}{2}}$. Then $X \cup \{t\}$ is $\aleph_0.T_{3\frac{1}{2}}$ $\forall\, t \in \beta X$ hence ZD ⟦5.3 #102⟧. By another entry in ZD table (Table 42), βX is ZD.
28. Let X be $(0, 1, \frac{1}{2}, \frac{1}{3}, \ldots)$ with 0 duplicated as in 3.2 #110. The complement of one of the 0's includes no open closed neighborhood of its 0. Another way to see that X is not ZD is to note that it is T_1 and not T_2, hence not ZD since ZD implies CR.
29. Let G be an open neighborhood of ∞ and $K = \tilde{G}$. Each point of K lies in a compact open set in X since X is ZD and LK. Reducing this cover of K to a finite one and taking its union leads to a compact open set which includes K. Its complement is an open closed neighborhood of ∞ and is a subset of G.
30. Let X be Σ and not L. Let S be a $\aleph_0$.d subset. Then S is L but $\bar{S} = X$ is not.
31. A simple extension of the Euclidean topology is first countable ⟦6.2 #6⟧ hence k, but need not be regular.
32. Let $X = \beta\omega \setminus \{x\}$, $Y = \beta\omega \setminus \omega$, $Z = Y \setminus \{x\}$. If X is normal, Z is C-embedded in X, ⟦θ8.5.3⟧; then Z is C^*-embedded in $\beta X = \beta\omega$ ⟦8.5 E3⟧, hence in Y. But Z is also dense in Y ⟦8.3 #202⟧, thus $\beta Z = Y$ ⟦θ8.5.2⟧ and so by [Fine and Gillman(a), θ4.6], Z is compact. This makes $Z = \beta Z = Y$, which is false. (This result is due to L. Gillman.)
33. If X is σK, βX is TD and $x \in \beta X \setminus X$, then x is not a sequential limit point of its complement Y in βX, by [Snyder, θ4.2]. Thus Y is sequentially closed in βX. For any countable subset S of Y, if S had no accumulation point in Y it would have exactly one in βX. This would make S a sequence converging to x ⟦7.1 #13⟧.
34. On pp. 38 and 39 of [Kister] is shown a non-CK group with a dense CK subset.
35. If X is normal and S is CK, then S is ψK, hence $\bar{S}$ is ψK, and so CK.
36. Let $A_n = \{f\colon |f(x)| \le n \text{ for all } x\}$. Each A_n is nowhere dense, and $C(X) = \bigcup A_n$. Note that T_3 is not sufficient since $C(X)$ could be one-dimensional.
37. The map $(n, a) \to a$ carries both P and P^+ continuously onto W^+. The result follows by 5.3 #7.
38. Let $x \in W$ and let $a < x < b$. Then $[a + 1, x] = (a, x + 1)$ is open and closed, contains x, and is included in (a, b). Thus W, and similarly, W^+ is ZD. Thus $P^+ = \omega^+ \times W^+$ is ZD, and so P is too.
39. There is a local base at each point of CK nhds. These are closed ⟦[Franklin]⟧. Thus the space is regular.
40. Let X be K.KC, not T_2. The diagonal in $X \times X$ is compact ⟦it is homeomorphic with X⟧ but not closed ⟦6.7, E1⟧.

41. A discussion of k spaces is given in [Steenrod] and in [Arhangelskij]. In the latter article it is proved that a T_2 space is hereditarily k if and only if it is sequential.
42. See the abstract of Nancy M. Warren in *Notices Amer. Math. Soc. 16* (1969), 853.
43. See the 1965 Wisconsin Topology Seminar, ed. R. H. Bing and R. J. Bean, p. 105.

Bibliography

ALEXANDROFF, A. D., and P. URYSOHN, "Mémoire sur les espaces topologiques compactes," *Verh. des Konikl Akad. van Wetensch. Te. Amsterdam 14* (1929), no. 1.

ANDERSON, D. R., "On connected irresolvable Hausdorff spaces," *Proc. Amer. Math. Soc. 16* (1965), 463–466.

ARHANGELSKIJ, A., "A characterization of very k spaces," *Czech. Math. Journal 18* (1968), 392–395.

AULL, C. E., (a) "A note on countably paracompact spaces and metrization," *Proc. Amer. Math. Soc. 16* (1965), 1316–1317.

AULL, C. E., (b) "Compactness as a base axiom," *Indag. Math. 29* (1967), 106–108.

AUSLANDER, L., and R. E. MCKENZIE, *Introduction to Differentiable Manifolds* (New York: McGraw-Hill, 1963).

BACON, P., "Extending a complete metric," *Amer. Math. Monthly 75* (1968), 642–643.

BANACH, S., *Théorie des Opérations Linéaires* (Warsaw: PWN, 1932).

BANASCHEWSKI, B., "Über nulldimensionale Räume," *Math. Nachr. 13* (1955), 129–140.

BORGES, C. J. R., "On simple extensions of topologies," *Notices Amer. Math. Soc. 13* (1965), 792.

BOURBAKI, N., (a) *Espaces Vectoriels Topologiques* (Paris: Hermann, 1953).

BOURBAKI, N., (b) *General Topology* (Reading, Mass.: Addison-Wesley, 1966).

BRUNK, H. D., "Problem 4813," *Amer. Math. Monthly 66* (1959), 599.

BUCK, R. C., Editor, *Studies in Modern Analysis* (Englewood Cliffs, N.J.: Prentice-Hall, 1962).

BURGESS, W., "The meaning of mono and epi in some familiar categories," *Canad. Math. Bull. 6* (1965), 759–769.

BUSEMANN, H., "Local metric geometry," *Trans. Amer. Math. Soc. 56* (1944), 200–274.

CECH, E., "On bicompact spaces," *Ann. of Math. 38* (1937), 823–844.

COHEN, H. B., "Injective envelopes of Banach spaces," *Bull. Amer. Math. Soc. 70* (1964), 723–726.

COMFORT, W. W., (a) "Locally compact realcompactifications" in *General Topology and its Relations to Modern Analysis and Algebra*, Edited by J. Novak, Prague Symposium, 1961, pp. 95–100.

COMFORT, W. W., (b) "On the Hewitt realcompactification of a product space," *Trans. Amer. Math. Soc. 131* (1968), 107–118.

COMFORT, W. W., and S. NEGREPONTIS, "Extending continuous functions on $X \times Y$ to subsets of $\beta X \times \beta Y$," *Fund. Math. 59* (1966), 1–12.

COMFORT, W. W., and K. A. ROSS, "Pseudocompactness and uniform continuity in topological groups," *Pacific J. Math. 16* (1966), 483–496.

CORSON, H. H., (a) "The determination of paracompactness by uniformities," *Amer. J. Math. 80* (1958), 185–190.

CORSON, H. H., (b) "The weak topology of a Banach space," *Trans. Amer. Math. Soc. 101* (1961), 1–15.

COURANT, R., *Dirichlet's Principle* (New York: Interscience, 1950).

DUGUNDJI, J., *Topology* (Boston, Mass.: Allyn and Bacon, 1966).

DUNFORD, N., and J. SCHWARTZ, *Linear Operators* (New York: Interscience, 1958).

EVANS, J. W., F. HARARY, and M. S. LYNN, "On the computer enumeration of finite topologies," *Comm. Assoc. Comput. Mach. 10* (1967), 295–298.

FINE, N. J., and L. GILLMAN, (a) "Extension of continuous functions in βN," *Bull. Amer. Math. Soc. 66* (1960), 376–381.

FINE, N. J., and L. GILLMAN, (b) "Remote points in βR," *Proc. Amer. Math. Soc. 13* (1962), 29–36.

FLACHSMEYER, J., "Nulldimensionale Räume" in *General Topology and its Relations to Modern Analysis and Algebra*, Edited by J. Novak, Prague Symposium, 1961, pp. 152–154.

FORT, M. K., "Functions discontinuous on a dense set," *Amer. Math. Monthly 58* (1951), 408–410.

FRANKLIN, S. P., "Spaces in which sequences suffice," *Fund. Math. 57* (1965), 107–115; *61* (1967), 51–56.

FROLIK, Z., (a) "A contribution to the descriptive theory of sets and spaces," in *General Topology and its Relations to Modern Analysis and Algebra*, Edited by J. Novak, Prague Symposium, 1961, pp. 157–173.

FROLIK, Z., (b) "Sums of ultrafilters," *Bull. Amer. Math. Soc. 73* (1967), 87–91.

GAAL, S. A., *Point Set Topology* (New York: Academic Press, 1964).

GILLMAN, L., and M. JERISON, *Rings of Continuous Functions* (Princeton, N.J.: Van Nostrand, 1960).

GLICKSBERG, I., "The representation of functions by integrals," *Duke Math. J. 19* (1952), 253–261.

GOFFMAN, C., and G. PEDRICK, *First Course in Functional Analysis* (Englewood Cliffs, N.J.: Prentice-Hall, 1965).

GROSS, J. L., "A third definition of local compactness," *Amer. Math. Monthly 74* (1967), 1120–1122.

GROTHENDIECK, A., "Critères de compacité," *Amer. J. Math. 74* (1952), 168–186.

HALMOS, P. R., *Naive Set Theory* (Princeton, N.J.: Van Nostrand, 1960).

HANAI, S., "On open mappings," *Proc. Japan Acad. 37* (1961), 233–238.

HENRIKSON, M., and M. JERISON, "The space of minimal prime ideals of a commutative ring" in *General Topology and its Relations to Modern Analysis and Algebra*, Edited by J. Novak, Prague Symposium, 1961, pp. 199–203.

HERRLICH, H., "Wann sind alle stetigen Abbildungen in Y konstant?" *Math. Z. 90* (1965), 152–154.

HEWITT, E., (a) "Rings of real valued continuous functions," *Trans. Amer. Math. Soc. 64* (1948), 54–99.

HEWITT, E., (b) "A class of topological spaces," *Bull. Amer. Math. Soc. 55* (1949), 421–426.

HEWITT, E., (c) "The role of compactness in analysis," *Amer. Math. Monthly 67* (1960), 499–516.

HEWITT, E., and K. A. ROSS, *Abstract Harmonic Analysis. I* (Berlin: Springer, 1963).

HOCKING, J. G., and G. S. YOUNG, *Topology* (Reading: Addison-Wesley, 1961).

HU, S. T., *Theory of Retracts* (Wayne, 1965).

ISBELL, J. R., *Math. Reviews 22* (1961) #7099.

JONES, S. L., "The impossibility of filling E^n with arcs," *Bull. Amer. Math. Soc. 74* (1968), 155–159.

KATETOV, M., "Measures in fully normal spaces," *Fund. Math. 38* (1951), 73–84.

KELLEY, J. L., *General Topology* (Princeton, N.J.: Van Nostrand, 1955).

KELLEY, J. L., and I. NAMIOKA, ET AL., *Linear Topological Spaces* (Princeton, N.J.: Van Nostrand, 1963).

KISTER, J. M., "Uniform continuity and compactness in topological groups," *Proc. Amer. Math. Soc. 13* (1962), 37–40.

KLEE, V. L., "An example in the theory of topological linear spaces," *Arch. Math. 7* (1956).

KLEE, V. L., and M. E. RUDIN, "Certain function spaces," *Arch. Math. 7* (1956), 469–470.

LEVINE, N., "Simple extensions of topologies," *Amer. Math. Monthly 71* (1964), 22–25.

MCCORD, M. C., "A theorem on linear operators and the Tietze extension theorem," *Amer. Math. Monthly 75* (1968), 47–48.

MAY, K. O., *Lectures on Calculus* (San Francisco, Calif.: Holden-Day, 1967).

MICHAEL, E., "Continuous selections," *Ann. of Math. (2) 64* (1956), 562–580.

MISCENKO, A., "Spaces with point-countable base," *Soviet Math. 3* (1962), 855–858.

MROWKA, S. G., "On normal metrics," *Amer. Math. Monthly 72* (1965), 998–1001. (See also *Math. Reviews 33* (1967) #6584.)

MYCIELSKI, J., "α-incompactness of N^α," *Math. Reviews 35* (1968) #2746.

NACHBIN, L., "Topological vector spaces of continuous functions," *Proc. Nat. Acad. Sci., U.S.A. 40* (1954), 471–474.

NAKANO, H., *Topology and Linear Topological Spaces* (Tokyo: Maruzen, 1951).

NIEMYTZKI, V., and A. TYCHONOFF, *Fund. Math. 12* (1928), 118–120.

OXTOBY, J. C., "Cartesian products of Baire spaces," *Fund. Math. 49* (1961), 157–166.

RADEMACHER, H., *Lectures on Elementary Number Theory* (Waltham, Mass.: Blaisdell, 1964).

RAINWATER, J., "Spaces whose finest uniformity is metric," *Pacific J. Math. 9* (1959), 567–570.

RAJAGOPALAN, M., "Fourier transforms in locally compact groups," *Acta Sci. Math. (Szeged) 25* (1964), 86–89.
RANKIN, R. A., "Problem 5137," *Amer. Math. Monthly 71* (1964), 931.
REID, G. A., "On sequential convergence in groups," *Math. Z. 102* (1967), 227–235.
ROSS, K. A., and A. H. STONE, "Products of separable spaces," *Amer. Math. Monthly 71* (1964), 398–403.
ROBERTSON, A. P., and WENDY ROBERTSON, *Topological Vector Spaces* (Cambridge, 1964).
ROY, P., "A countable connected Urysohn space with a dispersion point," *Duke Math. J. 33* (1966), 331–334.
RUDIN, M. E., "A new proof that metric spaces are paracompact," *Proc. Amer. Math. Soc. 20* (1969), 603.
RUDIN, W., *Fourier Analysis on Groups* (New York: Interscience, 1962).
RYEBURN, D., "Finite additivity, etc.," *Amer. Math. Monthly 74* (1967), 148–152.
SANDERSON, D. E., "Solution," *Amer. Math. Monthly 75* (1968), 691.
SCARBOROUGH, C. T., and A. H. STONE, "Products of nearly compact spaces, *Trans. Amer. Math. Soc. 124* (1966), 131–147.
SCHAEFER, H. H., *Topological Vector Spaces* (New York: MacMillan, 1966).
SCHNARE, P. S., "Two definitions of local compactness," *Amer. Math. Monthly* (1965), 764–765.
SEMADENI, Z., "Periods of measurable functions and the Stone–Cech compactification," *Amer. Math. Monthly 71* (1964), 891–893.
SHIROTA, T., "On locally convex vector spaces of continuous functions," *Proc. Japan Acad. 30* (1954), 294–298.
SIERPINSKI, W., (a) "Sur les ensembles connexes et non connexes," *Fund. Math. 2* (1921), 81–95.
SIERPINSKI, W., (b) *General Topology* (Toronto: Toronto Univ. Press, 1952).
SIKORSKI, R., "On the separability of topological spaces," *Colloq. Math. 1* (1947), 279–284.
SIMMONS, G. F., *Introduction to Topology and Modern Analysis* (New York: McGraw-Hill, 1963).
SNYDER, A. K., "The Cech compactification and regular matrix summability," *Duke Math. J. 36* (1969), 245–252.
STEENROD, N., "A convenient category of topological spaces," *Mich. Math. J. 14* (1967), 133–152.
STEPHENSON, R. M., Jr., "Pseudocompact spaces," *Trans. Amer. Math. Soc., 134* (1968), 437–448.
STONE, A. H., (a) "Paracompactness and product spaces," *Bull. Amer. Math. Soc. 54* (1948), 977–982.
STONE, A. H., (b) "Metrizability of decomposition spaces," *Proc. Amer. Math. Soc. 7* (1956), 690–700.
STONE, M. H., "Applications of the theory of Boolean rings to general topology," *Trans. Amer. Math. Soc. 41* (1937), 375–481.
VAROPOULOS, N. TH., "A theorem on the continuity of homomorphisms of locally compact groups," *Proc. Cambridge Philos. Soc. 60* (1964), 449–463.
WARNER, S., "The topology of compact convergence on continuous function spaces," *Duke Math. J. 25* (1958), 265–282.

WATERHOUSE, W. C., "On *UC* spaces," *Amer. Math. Monthly 72* (1965), 634–635.

WHYBURN, G. T., "Open and closed mappings," *Duke Math. J. 17* (1950), 69–74.

WILANSKY, A., (a) *Functional Analysis* (Waltham, Mass.: Blaisdell, 1964).

WILANSKY, A., (b) "Between T_1 and T_2," *Amer. Math. Monthly 74* (1967), 261–266, 1241.

WILANSKY, A., (c) *Topics in Functional Analysis* (Berlin: Springer-Verlag, 1967), Lecture Notes in Math., Vol. 45.

Index

A B C D E F G H I J 5 4 3 2 1 7 0